Surfactants *in* Tribology

VOLUME 2

Edited by
Girma Biresaw
K.L. Mittal

CRC Press
Taylor & Francis Group
Boca Raton London New York

CRC Press is an imprint of the
Taylor & Francis Group, an **informa** business

CRC Press
Taylor & Francis Group
6000 Broken Sound Parkway NW, Suite 300
Boca Raton, FL 33487-2742

First issued in paperback 2019

© 2011 by Taylor and Francis Group, LLC
CRC Press is an imprint of Taylor & Francis Group, an Informa business

No claim to original U.S. Government works

ISBN-13: 978-1-4398-4064-1 (hbk)
ISBN-13: 978-0-367-38289-6 (pbk)

Visit the Taylor & Francis Web site at
http://www.taylorandfrancis.com

and the CRC Press Web site at
http://www.crcpress.com

Contents

Part I Films, Membranes, and Self-Assembled Monolayers: Relevance to Tribological Behavior

Part II Emulsions and Aqueous Systems: Relevance to Tribological Phenomena

Part III Biobased Lubricants

Part IV   General Topics and Applications

Preface

The most widely recognized role of surfactants in tribology is related to their ability to control friction and wear. Surfactants also allow for control of a wide range of lubricant properties such as emulsification/demulsification, bioresistance, oxidation resistance, rust/corrosion prevention, etc. Moreover, surfactants form a variety of organized structures that have interesting tribological properties. These include monolayers, normal/reverse micelles, oil in water (o/w) and water in oil (w/o) microemulsions, and uni- and multi-lamellar vesicles. Another group of organized assemblies that are of great contemporary interest in lubrication are self-assembled monolayers (SAMs). SAMs play a crucial role in the lubrication of a legion of products, including microelectromechanical systems (MEMS) and nanoelectromechanical systems (NEMS).

While there is vast literature separately on the subjects of surfactants and tribology, there is relatively little information on the nexus between surfactants and tribology in spite of the fact that surfactants play many critical roles in tribological phenomena. To fill this major lacuna in the literature, we decided to organize the first symposium on Surfactants in Tribology, which was held as part of the 16th International Symposium on Surfactants in Solution (SIS-2006) in Seoul, South Korea, from June 4 to 9, 2006.

The genesis of the SIS series of biennial events was in 1976, and since then these meetings have been held in many corners of the globe and have been attended by the "who's who" of the surfactant community. These meetings are recognized by the international community as the premier forum for discussing the latest research findings on surfactants in solution. In keeping with the SIS tradition, leading researchers from around the world engaged in unraveling the importance and relevance of surfactants in tribological phenomena were invited to present their latest findings.

The premier symposium on Surfactants in Tribology was a smashing success and it generated a high level of interest in the topic. Thus, we decided to invite leading scientists working in this area, who might or might not have participated in the symposium, to submit written accounts (chapters) of their discoveries on the subject. This culminated in the publication of the first volume of *Surfactants in Tribology* in 2008.

Since the first symposium, interest in the role of surfactants in tribology has continued to grow at an accelerated pace among scientists and engineers working in the areas of both surfactants and tribology. Concomitantly, we decided to organize a follow-up symposium on the subject, which took place during the 17th International Symposium on Surfactants in Solution (SIS-2008) in Berlin, Germany, from August 17 to 22, 2008. Again, because of the tremendous interest and enthusiasm evinced during the second symposium, we decided to bring out a follow-up volume on the subject. We thus sent invitations to leading researchers actively involved in both surfactants and tribology, who might or might not have participated in the second symposium, to submit chapters for *Surfactants in Tribology, Volume 2*. We have been very pleased with the excellent response from the contributors and the quality of chapters submitted for this volume.

This book comprises a total of 20 chapters dealing with variegated aspects of surfactants and tribology, and the connection between these seemingly disparate subjects. Some of the subtopics were not covered earlier in the first volume. The book is divided into four parts as follows. Part I: Films, Membranes, and Self-Assembled Monolayers: Relevance to Tribological Behavior; Part II: Emulsions and Aqueous Systems: Relevance to Tribological Phenomena; Part III: Biobased Lubricants; and Part IV: General Topics and Applications. The topics covered in Part I (Chapters 1 through 6) include tribological, nanotribological, molecular packing, and mechanical properties of silane, thio, phthalocyanine, and phospholipid films, membranes, grafts, and SAMs on gold, silica, and graphite substrates. Part II (Chapters 7 through 11) discusses w/o emulsions used as oil well drilling fluids, wetting and tribological properties of organized surfactant assemblies, surfactants as demulsifiers in enhanced crude oil production from old wells, tribological properties of aqueous solutions of alkyl polyglucosides, and o/w emulsions with biobased surfactants. The topics covered in Part III (Chapters 12 through 16) include the synthesis of novel biobased materials such as estolides, chemically modified fatty acid methyl esters and surface active materials, the characterization of tribological and surface properties of biobased lubricants and surfactants, and the application of modeling and statistical predictive methods in the development of biobased lubricants. Part IV (Chapters 17 through 20) discusses the fundamentals of surface chemistry at tribological interfaces, the role of surface science in magnetic recording tribology, antiwear and friction modifier compounds for automotive applications, and surfactants as antimicrobial agents in lubricants.

Tribological phenomena are of conspicuous importance in many diverse industrial sectors, ranging from seemingly mundane (e.g., behavior of ball bearings) to high-tech (MEMS, NEMS, nanofluidics, and biodevices), and proper control of tribological behavior has serious economic implications. The overall topic of tribology subsumes three important subtopics, namely, friction, lubrication, and wear, and all of these have been accorded due coverage in this volume. Apropos, these days the mantra in all industries is "Think green, go green," and the field of tribology is no exception. Part III in this volume reflects this by focusing on biobased surfactants.

This volume and its predecessor contain a wealth of information and reflect the cumulative wisdom of a contingent of researchers and should be of great value and interest to anyone interested in harnessing surfactants to control tribological phenomena in a host of situations and applications, including the burgeoning fields of MEMS, NEMS, nanofluidics, and biodevices.

Now comes the pleasant task of thanking those who contributed toward materialization of this volume. We are especially thankful to all the contributors for their interest, enthusiasm, and cooperation without which this book would not have seen the light of day. We would also like to thank Barbara Glunn (at Taylor & Francis, publisher) for her interest in this project and for her unwavering support, and the staff at Taylor & Francis for helping in the production of this book.

Girma Biresaw, PhD
Bio-Oils Research Unit
National Center for Agricultural Utilization Research
Agricultural Research Service
United States Department of Agriculture
Peoria, Illinois

K.L. Mittal, PhD
Hopewell Junction, New York

Editors

Girma Biresaw received his PhD in physical organic chemistry from the University of California, Davis, and spent 4 years as a postdoctoral research fellow at the University of California, Santa Barbara, investigating reaction kinetics and products in surfactant-based organized assemblies. He then joined the Aluminum Company of America as a scientist and conducted research in tribology, surface/colloid science, and adhesion for 12 years. Subsequently, he joined the Agricultural Research Service (ARS) of the U.S. Department of Agriculture in Peoria, Illinois, in 1998 as a research chemist, and became a lead scientist in 2002. At ARS, Biresaw conducts research in tribology, adhesion, and surface/colloid science in support of programs aimed at developing biobased products from farm-based raw materials. He is a member of the editorial board of the *Journal of Biobased Materials and Bioenergy*. To date, Biresaw has authored/coauthored more than 200 invited and contributed scientific publications and presentations, including more than 50 peer-reviewed manuscripts, 6 patents, 2 edited books, more than 30 proceedings and book chapters, and more than 100 scientific abstracts.

Kashmiri Lal Mittal received his PhD from the University of Southern California in 1970 and was associated with IBM Corp. from 1972 to 1993. He is currently teaching and consulting worldwide in the areas of adhesion and surface cleaning. He is the editor of 100 published books, as well as others that are in the process of publication, within the realms of surface and colloid science and of adhesion. He has received many awards and honors and is listed in many biographical reference works. Mittal is a founding editor of the *Journal of Adhesion Science and Technology* and has served on the editorial boards of a number of scientific and technical journals. He was recognized for his contributions and accomplishments by the international adhesion community that organized the First International Congress on Adhesion Science and Technology in Amsterdam in 1995 on the occasion of his 50th birthday. In 2002, he was honored by the global surfactant community, which instituted the Kash Mittal Award in the surfactant field in his honor. In 2003, he was honored by the Maria Curie-Sklodowska University, Lublin, Poland, which awarded him the title of doctor *honoris causa*. More recently, he was honored by the international adhesion community on the occasion of the publication of his 100th edited book.

Contributors

Mohamed A. Abdellatif
Plant Polymer Research Unit
National Center for Agriculture
Utilization Research
Agricultural Research Service
United States Department of
Agriculture
Peoria, Illinois

Atanu Adhvaryu
Afton Chemical Corporation
Richmond, Virginia

Ahmed M. Al-Sabagh
Special Application Department
Egyptian Petroleum Research Institute
Nasr City, Egypt

Svajus Joseph Asadauskas
Tribology Group
Institute of Chemistry
Vilnius, Lithuania

David Beck
Department of Chemistry
and
Center for Nanoscience and
Nanotechnology
National Sun Yat-Sen University
Kaohsiung, Taiwan

Girma Biresaw
Bio-Oils Research Unit
National Center for Agricultural
Utilization Research
Agricultural Research Service
United States Department of
Agriculture
Peoria, Illinois

Sophie Bistac
Laboratoire de Chimie Organique,
Bioorganique et Macromoléculaire
Centre National de la Recherche
Scientifique
Université de Haute-Alsace
Mulhouse, France

Steven C. Cermak
Bio-Oils Research Unit
National Center for Agricultural
Utilization Research
Agricultural Research Service
United States Department of
Agriculture
Peoria, Illinois

Henrikas Cesiulis
Department of Physical Chemistry
Vilnius University
Vilnius, Lithuania

Frank J. DeBlase
Petroleum Additives and Fluids
Chemtura Corporation
Middlebury, Connecticut

Kenneth M. Doll
Bio-Oils Research Unit
National Center for Agricultural
Utilization Research
Agricultural Research Service
United States Department of
Agriculture
Peoria, Illinois

Rasha A. El-Ghazawy
Special Application Department
Egyptian Petroleum Research Institute
Nasr City, Egypt

Sevim Z. Erhan
Eastern Regional Research Center
Agricultural Research Service
United States Department of
    Agriculture
Wyndmoor, Pennsylvania

Matthew P. Goertz
Department of Chemistry
University of Minnesota
Minneapolis, Minnesota

J.M. González
Well Construction and Maintenance
    Department
Petróleos de Venezuela S.A.
PDVSA Intevep
Estado Miranda, Venezuela

Ömer Gül
Energy Transfer Technology, LLC
State College, Pennsylvania

Shuchen Hsieh
Department of Chemistry
and
Center for Nanoscience and
    Nanotechnology
National Sun Yat-Sen University
Kaohsiung, Taiwan

Khalil Jradi
Centre Intégré en Pâtes et Papiers
Université du Quebec à Trois-Rivières
Trois-Rivières, Quebec, Canada

Nadia G. Kandile
Department of Chemistry
Ain Shams University
Heliopolis, Egypt

Tom Karis
Hitachi Global Storage Technologies,
    Inc.
San Jose, California

I-Ting Li
Department of Chemistry
and
Center for Nanoscience and
    Nanotechnology
National Sun Yat-Sen University
Kaohsiung, Taiwan

YiLen Liu
Department of Chemistry
and
Center for Nanoscience and
    Nanotechnology
National Sun Yat-Sen University
Kaohsiung, Taiwan

R.L. Márquez
Well Construction and Maintenance
    Department
Petróleos de Venezuela S.A.
PDVSA Intevep
Estado Miranda, Venezuela

Koji Miyake
Advanced Manufacturing Research
    Institute
National Institute of Advanced
    Industrial Science and Technology
Tsukuba, Japan

Miki Nakano
Advanced Manufacturing Research
    Institute
National Institute of Advanced
    Industrial Science and Technology
Tsukuba, Japan

Nabel A. Negm
Petrochemical Department
Egyptian Petroleum Research Institute
Nasr City, Egypt

Mahmoud R. Noor El-Din
Special Application Department
Egyptian Petroleum Research Institute
Nasr City, Egypt

G. Quercia
Well Construction and Maintenance
    Department
Petróleos de Venezuela S.A.
PDVSA Intevep
Estado Miranda, Venezuela

F. Quintero
Well Construction and Maintenance
    Department
Petróleos de Venezuela S.A.
PDVSA Intevep
Estado Miranda, Venezuela

S.D. Rosales
Well Construction and Maintenance
    Department
Petróleos de Venezuela S.A.
PDVSA Intevep
Estado Miranda, Venezuela

Leslie R. Rudnick
Ultrachem Inc.
New Castle, Delaware

Marina Ruths
Department of Chemistry
University of Massachusetts Lowell
Lowell, Massachusetts

Marjorie Schmitt
Laboratoire de Chimie Organique,
    Bioorganique et Macromoléculaire
Centre National de la Recherche
    Scientifique
Université de Haute-Alsace
Mulhouse, France

Brajendra K. Sharma
Bio-Oils Research Unit
National Center for Agricultural
    Utilization Research
Agricultural Research Service
United States Department of
    Agriculture
Peoria, Illinois

Benjamin L. Stottrup
Department of Physics
Augsburg College
Minneapolis, Minnesota

Minda Suchan
Department of Chemistry
and
Center for Nanoscience and
    Nanotechnology
National Sun Yat-Sen University
Kaohsiung, Taiwan

Marian W. Sulek
Department of Chemistry
Technical University of Radom
Radom, Poland

Wilfred T. Tysoe
Department of Chemistry and
    Biochemistry
University of Wisconsin-Milwaukee
Milwaukee, Wisconsin

Tomasz Wasilewski
Department of Chemistry
Technical University of Radom
Radom, Poland

Yutao Yang
Department of Chemistry
University of Massachusetts Lowell
Lowell, Massachusetts

Part I

Films, Membranes, and Self-Assembled Monolayers: Relevance to Tribological Behavior

1 Tribological Properties of Self-Assembled Monolayers

Miki Nakano

CONTENTS

ABSTRACT

The author discusses the effects of the alkyl chain length, chain chemical structure, and number of layers upon the tribological properties of organosulfur self-assembled monolayers (SAMs) on Au surfaces using a conventional pin-on-plate method. The tribological properties of SAMs covalently bonded to Si and alkylsilane SAM were investigated to examine the effect of the head groups upon the tribological properties. For the alkyl chain length effect, the SAMs with longer alkyl chains showed lower and more stable friction coefficients than those with shorter alkyl chains. For the effect of the chain chemical structure, the SAMs with benzene rings and those with alkyl chains had similar tribological behavior. However, the SAMs with benzene rings had higher wear resistance than those with alkyl chains. For the effect of number of layers on tribological properties, the double layer with outer-most layers of methyl groups showed lower friction coefficients than monolayers. For the effect of head group chemistry, the SAMs with Si-C bond and long alkyl chains had low friction coefficients and high wear resistivity.

1.1   INTRODUCTION

The coating of hard surfaces such as metals or semiconductors by organic films has been a well-known method for preparing lubrication films for a long time. At present, the thickness of organic lubrication films used for conventional machines, such as automobile engines, is on the order of micrometers. However, higher accuracy has been required in the field of machinery and thus nanometer-scale lubricants are now needed. For this reason, organic lubrication films with a thickness of a monolayer to a few layers are required. Self-assembled monolayers (SAMs) of organic molecules [1] are expected to be one of the candidates for the new lubrication films. Compared with Langmuir–Blodgett films, the advantages of SAMs are their ease of preparation and strong chemical bonding between molecules and substrate (Figure 1.1). Recently, SAMs have been shown to be effective as lubrication films for microelectromechanical systems (MEMS) [2,3].

It is important to understand the tribological properties of SAMs to apply them as lubrication films. So far, their tribological properties have been investigated using atomic force microscopy (AFM) because the friction force differs among different molecules. Thus, the effect of structure of molecules can be distinguished using AFM [4–7]. In friction measurements using conventional AFM, the desorption of the molecules is not a problem. However, under severe conditions for molecular films such as those at hard disk interface, which is used under noncontact conditions, it is necessary to examine the durability of monolayer films such as SAMs. Therefore, we examined the tribological properties of SAMs using a pin-on-plate method, which presents more severe conditions than conventional AFM [8–11].

It is well known that SAMs can be formed on various substrates such as Au, SiO_2, Si, etc. Typical SAMs that have been reported are summarized in Table 1.1. There are mainly three kinds of SAMs. The first kind of SAM is an organosulfur compound on a noble metal like Au [12]. The second kind is an organic silane compound on an oxide like mica [13]. The third one is an alkene, alcohol, or aldehyde on

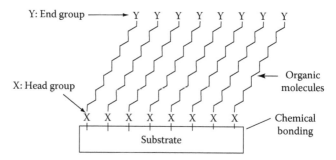

FIGURE 1.1   Schematic diagram of SAM. There is chemical bonding between substrate and film molecules.

TABLE 1.1
Classification of Typical SAMs

Molecules	Substrates	Remarks
Organosulfur: R—SH	Metal/compound semiconductors: Au, Ag, Cu, GaAs, etc.	Ease of preparation
Organosilane: R—SiX₃, X:Cl, OCH₃, OC₂H₅	Oxide/oxide film: glass, mica, SiO₂, etc.	SAM needs to be prepared carefully
Unsaturated hydrocarbon/alcohol/aldehyde: R—CH=CH, R—OH, R—CHO	Hydrogen-terminated Si: H—Si	• Molecules are directly bonded to Si • SAM needs to be prepared carefully

hydrogen-terminated silicon [14]. Very many kinds of SAMs can exist for various combinations of substrates and molecules.

1.2 TRIBOLOGICAL PROPERTIES OF SAMS ON AU

At first we investigated the tribological properties of SAMs formed on Au substrates for the following reasons. The first reason is that there are many kinds of organo-sulfur compounds that can easily form SAMs. Thus, SAMs on Au substrates are suitable for investigating the influence of molecular structure (chain length and functional group) upon the tribological behavior. The second reason is that the SAM formation process on Au substrates is easier than on other substrates such as silicon substrates. In this section, we discuss the effects of the alkyl chain length, chain structure, and number of layers on the tribological properties of organosulfur SAMs on Au surfaces using a conventional pin-on-plate method.

1.2.1 EFFECT OF ALKYL CHAIN LENGTH ON TRIBOLOGICAL PROPERTIES OF ALKANETHIOL SAMS ON AU

It is well known that SAMs with long alkyl chain lengths exhibit low friction coefficients when measured with friction force microscopy (FFM) for the tribological properties of alkanethiol SAMs on Au [15]. We investigated the effects of the alkyl chain length on the macroscopic tribological properties of alkanethiol SAMs on Au surfaces using a conventional pin-on-plate method [8]. Figure 1.2 shows the relationship between the friction coefficient and the sliding time for three sliding runs on dodecanethiol (DDT) SAM (a) and octadecanethiol (ODT) SAM (b) surfaces. For the measurements on the same SAM surface, the sliding position was changed after each run was completed. Consequently, each run was performed at a unique position. The results indicated that the friction coefficient decreased and its scatter decreased with longer alkyl chain. To examine the chemical states of the SAMs, we analyzed the surfaces before and after the friction tests by x-ray photoelectron spectroscopy (XPS) (Figure 1.3). For the DDT SAM with shorter alkyl chains, oxidation of parts of sulfur was observed after the friction tests, as shown in Figure 1.3a. On the other

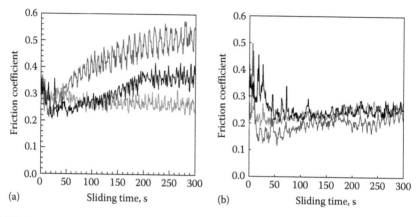

FIGURE 1.2 Friction coefficients of DDT SAMs (a) and ODT SAMs (b) as a function of sliding time. These three friction test runs were performed at different positions on the same sample.

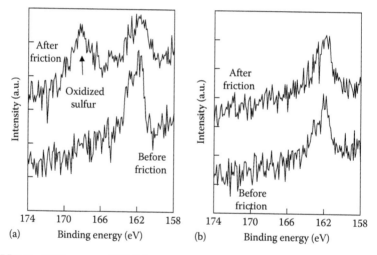

FIGURE 1.3 XPS spectra of DDT SAMs (a) and ODT SAMs (b) in the S (2p) region before and after friction tests.

hand, the chemical states of the ODT SAMs with longer alkyl chains hardly changed after the friction tests (Figure 1.3b).

As mentioned earlier, our results suggest that both chemical structure and the orientation of the molecules of SAMs influence the tribological properties. For the alkanethiol SAMs, the intermolecular interaction of the SAM with the longer alkyl chains was stronger because the intermolecular interaction increased with longer alkyl chains. For the SAMs with shorter alkyl chains, the intermolecular interaction was weaker than that of the alkanethiol with longer alkyl chains [16]. This meant that the pins could more easily contact the substrate and the molecules probably reacted with water in the atmosphere which enabled the tribochemical reaction to occur during friction. Oxidation of a part of the sulfur that directly bonds with the substrate, as shown in Figure 1.3a, is also caused by the tribochemical reaction during friction. Such tribochemical reactions led to the increase of the friction coefficients of the SAMs with shorter alkyl chains. On the other hand, for SAMs with longer alkyl chains, the intermolecular interaction was strong and the pin could scratch only the upper parts of the SAMs. And for this reason, the friction coefficients of the ODT SAMs were low. Therefore, the chemical structure of the molecules of the SAMs affected the tribological properties of the SAMs, and molecules with longer alkyl chains probably formed lubrication films that exhibited stable tribological properties.

1.2.2 EFFECT OF MOLECULAR SPECIES ON TRIBOLOGICAL PROPERTIES OF TERPHENYLMETHANETHIOL SAMs ON AU

Recently, oligophenylene SAMs containing benzene rings have also been investigated. In particular, it has been reported that [1,1′:4′,1″-terphenyl]-4-methanethiol (TP1) SAMs containing benzene rings had higher thermal stability and higher UV irradiation stability than alkanethiol SAMs [17,18]. Therefore, if TP1 SAMs had sufficient durability against friction, these could be expected to form adequate lubrication films. We investigated the tribological behavior of TP1 SAMs on Au surfaces using a conventional pin-on-plate method and FFM and compared them with ODT SAMs [9]. First, we examined the nanotribological properties of TP1 and ODT SAMs using FFM. Figure 1.4 shows the friction force of the SAMs as a function of the normal load and the slope corresponds to the friction coefficients. The friction coefficient of an ODT SAM was only 14% of that of a TP1 SAM. Next, the tribological properties of TP1 and ODT SAMs were measured using the pin-on-plate

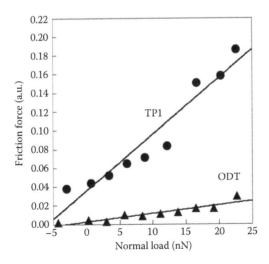

FIGURE 1.4 Variation of friction force with normal load for TP1 SAM and ODT SAM.

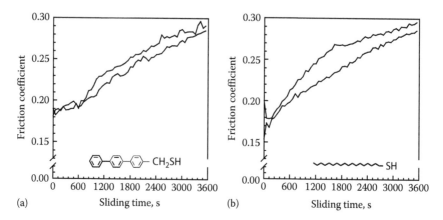

FIGURE 1.5 Friction coefficients from two sliding runs measured using a pin-on-plate tri-bometer on TP1 (a) and ODT SAMs (b) as a function of sliding time. These two friction test runs were performed at different positions on the same sample.

method. Figure 1.5 shows the friction coefficients from two sliding runs as a function of sliding time on the TP1 (a) and ODT SAMs (b). These two friction test runs were performed at different positions on the same sample. The tendencies of the friction behavior were similar for both kinds of SAMs and were different from the results of FFM. From the results of Figure 1.5, the effect of molecular structure in the SAMs was not observed when using the pin-on-plate method. Table 1.2 shows the results from examination of the durability toward friction using XPS. The XPS peak ratio of C(1s)/Au(4f) indicates the amount of adsorbed molecules on the surface. The peak ratio decreased after the friction tests, indicating that the molecules were desorbed by friction. The remaining molecules on the surface of ODT SAM decreased more than those of TP1 SAM. Therefore, this indicated that the TP1 SAM had higher durability than the ODT SAM. Furthermore, the molecular length of TP1 is short at about 1.5 nm, which is close to DDT (1.44 nm) [19]. Therefore, TP1 SAM was considered to be an appropriate coating material for forming a highly durable thin film.

TABLE 1.2

XPS Peak Area Ratio of C(1s)/ Au(4f) after Friction Tests

Sliding Time (h)	TP1	ODT
0	1.63 ± 0.06	2.01 ± 0.08
1	1.54 ± 0.09	1.73 ± 0.09
10	1.22 ± 0.04	1.25 ± 0.03

Note: We measured two to three samples in all conditions.

1.2.3 TRIBOLOGICAL PROPERTIES OF SELF-ASSEMBLED DOUBLE LAYERS ON AU

In Sections 1.2.1 and 1.2.2, we discussed the tribological properties of monolayers. As mentioned earlier, the friction coefficients of the monolayers were 0.2–0.3, which were not low enough. Thus, a decrease in the friction coefficient of the molecular layers is necessary for using SAMs in lubrication. Since it has been reported that the layered structure, which consists of a bonded layer and a mobile layer, contributes to lowering friction and enhancing durability, as with perfluoropolyether [20], the self-assembled multilayer method is an effective means for decreasing the friction coefficient. Thus, we measured the frictional properties of self-assembled double layers [10]. Multilayers reported so far have contained mercaptohexadecanoic acid (MHA) or mercaptoundecanoic acid, which had the carboxyl group and the thiol group at the two ends, and there were copper ions between the upper layers and lower layers [21]. We used MHA, which directly bonds to substrates and ODT or MHA as the molecules in the second layer, as shown in Figure 1.6. Since the outermost layer with low surface energy decreased the friction, we used ODT molecules with the methyl group for the second layer. Figure 1.7 shows the result of the friction tests. The friction coefficients of the monolayer and double layer for both cases that contained MHA for the topmost layer showed similar tendencies. On the other hand, for the topmost layer of ODT, the double layer exhibited lower friction coefficients than the monolayer. To examine the film thickness of the molecular layers, we conducted XPS measurements before and after the friction tests. Table 1.3 shows the film thickness of the self-assembled layers estimated from the XPS C(1s)/Au(4f) ratios before

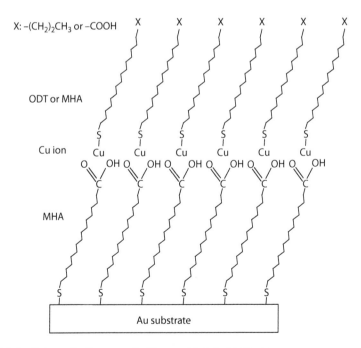

FIGURE 1.6   Schematic diagram of self-assembled double layer.

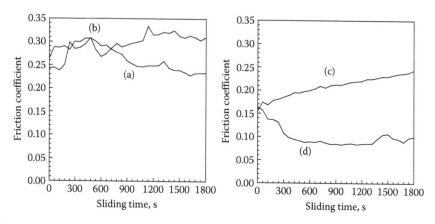

FIGURE 1.7 Friction coefficients of self-assembled layers as a function of sliding time: (a) MHA monolayer, (b) MHA/MHA double layer, (c) ODT monolayer, and (d) ODT/MHA double layer.

TABLE 1.3

Film Thickness (nm) of Self-Assembled Layers Estimated from XPS C(1s)/Au(4f) Ratios

	MHA Monolayer	MHA/MHA Double Layer	ODT Monolayer	ODT/MHA Double Layer
Before friction test	2.62	5.06	2.20	4.63
After friction test (30 mN, 1 h)	2.14	3.73	1.97	2.96

and after the friction tests. The film thicknesses of both double layers after the friction tests were significantly smaller than those before the friction tests, while film thickness changes of both the ODT and the MHA monolayers were smaller before and after friction. The XPS data indicated that the MHA lower layer of double layers might have decomposed only slightly, but the ODT or MHA upper layer did decompose more during friction. Therefore, for the double layer, we think that the molecules of the topmost layer desorbed during friction and that the molecular density of the topmost layer decreased. Moreover, although the frictional coefficients of the ODT/MHA double layers were initially almost the same as those of the ODT monolayer, the friction coefficient of the ODT/MHA double layer gradually decreased. Therefore, this suggested that the desorption of the molecules of the topmost layers was related to lowering the friction coefficient. We thus varied the amount of the adsorbed molecules and found that a certain amount of adsorption led to the lowest friction coefficient. The ODT molecules of the topmost layers bonded weakly to the MHA molecules of the lower layer via the Cu ions but parts of ODT molecules still remained after the friction test. The topmost layer was probably loosely packed and the molecules of topmost layer had enough space to move. Therefore, we believe that these molecules can act as a mobile layer bound with a lower layer and, as such, appropriately loosely packed bound ODT molecules in the ODT/MHA double layers led to the decreased friction.

1.3 TRIBOLOGICAL PROPERTIES OF SELF-ASSEMBLED MONOLAYERS ON SI

Earlier, we discussed the frictional properties of SAMs formed on Au. The frictional properties of SAMs formed on Si or glass substrates have also been investigated because Si has recently become one of the most popular substrates due to the focus on MEMS. The possibility of using SAMs on Si as lubricants for MEMS has been reported. In previous studies on the tribological properties of SAMs on Si [2,3], mainly organosilane SAMs were used. It has been reported that organosilane SAMs reduced the friction coefficient, stiction, and wear. However, the level of wear resistance was not adequate for use as MEMS components, and higher durability was deemed necessary for MEMS lubricants. To improve the durability of the SAMs, we used SAMs covalently bonded to a Si substrate, with the aim of strengthening the bonding between the SAMs and the Si substrate. SAMs covalently bonded to a Si substrate were discovered by Linford and Chidsey in the 1990s [14] and many studies on such SAMs have been performed [14,22–24]. It has been reported that such SAMs have higher chemical resistance than organosilane SAMs and that alkene SAMs exhibited higher resistance than those from alcohol. Therefore, being able to gain an understanding of the tribological properties of SAMs covalently bonded to Si substrates could possibly lead to developing such SAMs for use as a coating material for lubrication. Moreover, the reported results indicated that the head group affected the properties of the SAMs and that the head group possibly affected the tribological properties as well.

1.3.1 TRIBOLOGICAL PROPERTIES OF SELF-ASSEMBLED MONOLAYERS COVALENTLY BONDED TO SI

The tribological properties of SAMs covalently bonded to Si have been examined using AFM. However, the macroscopic tribological properties have not been adequately understood. Using a pin-on-plate method, we examined the tribological properties of the following SAMs covalently bonded to Si: hexadecene $[CH_3(CH_2)_{13}CH_2=CH_2]$ (HD), dodecene $[CH_3(CH_2)_9CH_2=CH_2]$ (DD), and hexadecanol $[CH_3(CH_2)_{14}CH_2OH]$ (HDO) SAMs and alkylsilane SAM, which was octadecyltrichlorosilane $[CH_3(CH_2)_{17}SiCl_3]$ (ODS) SAM. Figure 1.8 shows schematic diagrams of the chemical structures of the HD, HDO, and ODS SAMs. The head groups of the HD and HDO SAMs are Si-C and Si-O-C bond, respectively. HD, HDO, and ODS SAMs have the same end group, similar alkyl chains length, but different head groups.

Figure 1.9 shows typical examples of the friction coefficients on these SAMs and bare Si(111) surfaces as a function of sliding time. The surfaces are represented by the following symbols: HD (squares), HDO (circles), ODS (triangles), and bare

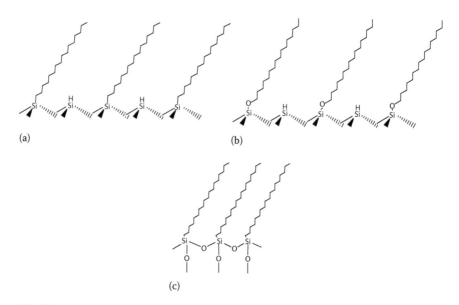

FIGURE 1.8 Schematic diagrams of the chemical structures of HD SAM (a), HDO SAM (b), and ODS SAM (c).

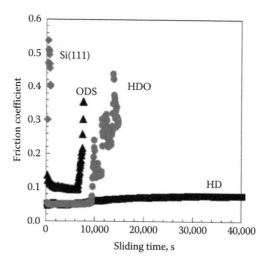

FIGURE 1.9 Friction coefficients of HD, HDO, ODS, and Si(111) surfaces as a function of sliding time. Normal load was 98 mN and average sliding speed was 2 mm/s.

Si(111) (diamonds). The normal load was 98 mN and the values of the friction coefficients were average values for each 1 min of sliding. For the bare Si(111) surface, the friction coefficient increased immediately after the start of the friction test. However, the friction coefficients of the HDO and ODS SAMs were initially 0.04–0.06 and 0.12–0.10, and they increased after 6000 and 9000 s, respectively. Furthermore, the

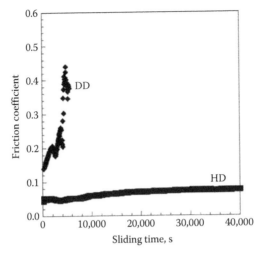

FIGURE 1.10 Friction coefficients of HD and DD SAMs as a function of sliding time. Normal load was 98 mN and average sliding speed was 2 mm/s.

HD SAMs exhibited low friction coefficient of 0.04–0.08 for about 40,000 s, which indicated that the HD SAMs had high durability. Figure 1.10 compares friction coefficients on the HD and DD SAMs as a function of sliding time. The HD and DD SAMs have the same head group but different alkyl chain lengths. Although the DD SAMs had the same head groups as the HD SAMs, the friction coefficient of the DD SAMs increased immediately after the start of the friction test, indicating that both the head group chemistry and the alkyl chain length are key factors when preparing high durable SAMs. Figure 1.11 shows optical microscope images of the HDO (a) and HD SAMs (b) after the friction tests. A wear track was observed on the HDO SAM, and there were wear particles on both sides of the wear track, as shown in Figure 1.11a. For the DD and ODS SAMs, wear and wear particles were also observed similar to the HDO SAMs. On the other hand, wear was not observed on the HD SAM surfaces (Figure 1.11b). One possible reason for the increase of the

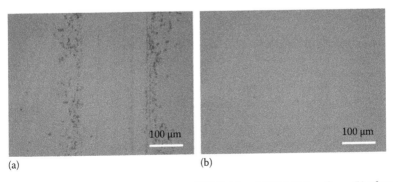

(a)   (b)

FIGURE 1.11 Optical microscope images of HDO (a) and HD SAM surfaces (b) after friction tests for 5 h (a) and 10 h (b) under a load of 98 mN.

friction coefficients for the HDO, ODS, and DD SAMs might be the desorption of molecules during the friction tests. The desorption of molecules could have led to direct contact between the pin and the Si substrate. On the other hand, the HD SAMs exhibited stable friction properties and wear resistance even under severe friction test conditions, and these results clearly indicated that the HD molecules remained on the surface during the entire test period.

1.4 CONCLUSIONS

The possibility of using SAMs as lubrication films is expected in MEMS applications. However, problems with the durability and wear resistance of SAMs still remain. As mentioned in Sections 1.2 and 1.3, the durability and the wear resistance have been improved. Combining chemical, surface, and tribological evaluations of SAMs will likely lead to SAMs with higher durability. In addition, the design of thin films with higher durability might be possible by employing molecules that include fluorine or aromatic groups.

ACKNOWLEDGMENTS

The author gratefully acknowledges Drs. T. Ishida, K. Miyake, and Y. Ando from the National Institute of Advanced Industrial Science and Technology (AIST) of Japan and Prof. S. Sasaki from Tokyo University of Science, Japan, for their significant contributions to this research and for fruitful discussions. The author also acknowledges Prof. H. Sugimura from Kyoto University, Japan, and Dr. H. Sano from AIST for providing SAM samples covalently bonded to Si and for fruitful discussions.

REFERENCES

1. A. Ulman, *An Introduction to Ultrathin Organic Films from Langmuir–Blodgett to Self-Assembly*. Academic Press, Boston, MA (1991).
2. B. Bhushan, A. Kulkarni, and V.N. Koinkar, Microtribological Characterization of Self-Assembled and Langmuir–Blodgett Monolayers by Atomic and Friction Force Microscopy, *Langmuir*, 11, 3189–3198 (1995).
3. V. DePalma and N. Tillman, Friction and Wear of Self-Assembled Trichlorosilane Monolayer Films on Silicon, *Langmuir*, 5, 868–872 (1989).
4. T. Ishida, S.-I. Yamamoto, W. Mizutani, M. Motomatsu, H. Tokumoto, H. Hokari, H. Azehara, and M. Fujihira, Evidence for Cleavage of Disulfides in the Self-Assembled Monolayer on Au(111), *Langmuir*, 13, 3261–3265 (1997).
5. H.I. Kim, T. Koini, T.R. Lee, and S.S. Perry, Systematic Studies of the Frictional Properties of Fluorinated Monolayers with Atomic Force Microscopy: Comparison of CF_3^- and CH_3-Terminated Films, *Langmuir*, 13, 7192–7196 (1997).
6. K. Hayashi, H. Sugimura, and O. Takai, Frictional Properties of Organosilane Self-Assembled Monolayer in Vacuum, *Jpn. J. Appl. Phys.*, 40, 4344–4348 (2001).
7. A. Takahara, K. Kojio, S.-R. Ge, and T. Kajiyama, Scanning Force Microscopic Studies of Surface Structure and Protein Adsorption Behavior of Organosilane Monolayers, *J. Vac. Sci. Tech.*, A14, 1747–1754 (1996).

8. M. Nakano, T. Ishida, T. Numata, Y. Ando, and S. Sasaki, Alkyl Chain Length Effect on Tribological Behavior of Alkanethiol Self-Assembled Monolayers on Au, *Jpn. J. Appl. Phys.*, 42, 4734–4738 (2003).

9. M. Nakano, T. Ishida, T. Numata, Y. Ando, and S. Sasaki, Tribological Behavior of Terphenyl Self-Assembled Monolayer Studied by a Pin-on-Plate Method and Friction Force Microscopy, *Jpn. J. Appl. Phys.*, 43, 4619–4623 (2004).

10. M. Nakano, T. Ishida, T. Numata, Y. Ando, and S. Sasaki, Tribological Behavior of Self-Assembled Double Layer Measured by a Pin-on-Plate Method, *Appl. Surf. Sci.*, 242, 287–294 (2005).

11. M. Nakano, T. Ishida, H. Sano, H. Sugimura, K. Miyake, Y. Ando, and S. Sasaki, Tribological Properties of Self-Assembled Monolayers Covalently Bonded to Si, *Appl. Surf. Sci.*, 255, 3040–3045 (2008).

12. R.G. Nuzzo and D.L. Allara, Adsorption of Bifunctional Organic Disulfides on Gold Surfaces, *J. Am. Chem. Soc.*, 105, 4481–4483 (1983).

13. J. Sagiv, Organized Monolayers by Adsorption. 1. Formation and Structure of Oleophobic Mixed Monolayers on Solid Surfaces, *J. Am. Chem. Soc.*, 102, 92–98 (1983).

14. M.R. Linford and C.E.D. Chidsey, Alkyl Monolayers Covalently Bonded to Silicon Surfaces, *J. Am. Chem. Soc.*, 115, 12631–12632 (1993).

15. A. Lio, D.H. Charych, and M. Salmeron, Comparative Atomic Force Microscopy Study of the Chain Length Dependence of Frictional Properties of Alkanethiols on Gold and Alkylsilanes on Mica, *J. Phys. Chem.*, B101, 3800–3805 (1997).

16. R.W. Carpick and M. Salmeron, Scratching the Surface: Fundamental Investigations of Tribology with Atomic Force Microscopy, *Chem. Rev.*, 97, 1163–1194 (1997).

17. T. Ishida, H. Fukushima, W. Mizutani, S. Miyashita, H. Ogiso, K. Ozaki, and H. Tokumoto, Annealing Effect of Self-Assembled Monolayers Generated from Terphenyl Derivatized Thiols on Au(111), *Langmuir*, 18, 83–92 (2002).

18. T. Ishida, M. Sano, H. Fukushima, M. Ishida, and S. Sasaki, Stability of Terphenyl Self-Assembled Monolayers Exposed under UV Irradiation, *Langmuir*, 18, 10496–10499 (2002).

19. C.D. Bain, E.B. Troughton, Y.-T. Tao, J. Evall, G.M. Whitesides, and R.G. Nuzzo, Formation of Monolayer Films by the Spontaneous Assembly of Organic Thiols from Solution onto Gold, *J. Am. Chem. Soc.*, 111, 321–335 (1989).

20. T. Kato, M. Kawaguchi, M.M. Sajjad, and J. Choi, Friction and Durability Characteristics of Ultrathin Perfluoropolyether Lubricant Film Composed of Bonded and Mobile Molecular Layers on Diamond-like Carbon Surfaces, *Wear*, 257, 909–915 (2004).

21. S.D. Evans, A. Ulman, K.E. Goppert-Berarducci, and L.J. Gerenser, Self-Assembled Multilayers of ω-Mercaptoalkanoic Acids: Selective Ionic Interactions, *J. Am. Chem. Soc.*, 113, 5866–5868 (1991).

22. M.R. Linford, P. Fenter, P.M. Eisenberger, and C.E.D. Chidsey, Alkyl Monolayers on Silicon Prepared from 1-Alkenes and Hydrogen-Terminated Silicon, *J. Am. Chem. Soc.*, 117, 3145–3155 (1995).

23. N. Saito, K. Hayashi, S. Yoda, H. Sugimura, and O. Takai, Chemical Resistivity of Self-Assembled Monolayer Covalently Attached to Silicon Substrate to Hydrofluoric Acid and Ammonium Fluoride, *Surf. Sci.*, 532–535, 970–975 (2003).

24. H. Sugimura, H. Sano, K.-H. Lee, and K. Murase, Organic Monolayers Covalently Bonded to Si as Ultra Thin Photoresist Films in Vacuum UV Lithography, *Jpn. J. Appl. Phys.*, 45, 5456–5460 (2006).

2 Nano-Tribological Properties of Aromatic and Polyaromatic Thiol Self-Assembled Monolayers

Yutao Yang and Marina Ruths

CONTENTS

ABSTRACT

A considerable number of fundamental and applied studies have focused on the nano-tribological properties of confined, molecularly thin films. Despite these efforts, the molecular-level mechanisms of friction and the corresponding functions of actual lubricant additives and surfactants at solid surfaces are only partially understood. To address some of these issues, we have studied aromatic self-assembled monolayers (SAMs) in single-asperity contacts, as model systems for the natural boundary lubrication of solid surfaces by the aromatics content in mineral oil-based fuels and lubricants. Friction force microscopy, a technique based on atomic force microscopy (AFM), was used to investigate the effects of molecular packing, adhesion strength, and sliding speed on the friction of SAMs of the aromatic compounds thiophenol, p-phenylthiophenol, p-terphenyl thiol, 2-naphthalenethiol, and benzyl mercaptan on template-stripped gold surfaces. The adhesion between a monolayer-covered tip and substrate was controlled by immersing the sliding contacts in ethanol or in dry N_2 gas. At low loads L, low adhesion (obtained in ethanol) resulted in a linear dependence of the friction force F on load, $F = \mu L$, whereas in N_2, the higher adhesion between the same monolayers gave a non-linear, apparently area-dependent friction. This nonlinear friction in the adhesive systems was well described by $F = S_c A$, with the contact area, A, calculated for a thin, linearly elastic film confined between rigid substrates using the extended Thin-Coating Contact Mechanics (TCCM) model, and S_c being the critical shear stress, a constant for each monolayer system. With increasing molecular packing, a systematic decrease was found in the friction coefficient, μ, obtained in ethanol and the critical shear stress, S_c, obtained in N_2. To describe these aromatic monolayers with the extended TCCM model, a higher Young's modulus was needed than for fatty acid monolayers of similar packing density. The velocity dependence of F showed that monolayers of benzyl mercaptan gave a more solid-like friction response than the more loosely packed monolayers of thiophenol.

2.1 INTRODUCTION

Self-assembled monolayers (SAMs) are commonly used as model boundary lubricants, i.e., friction-reducing or wear-protecting compounds acting between surfaces in close proximity (in contact or at separations of a few molecular diameters) and at high pressures. Among the properties that are known to affect the friction response of SAMs are their packing density [1–4], molecular chain length [5,6] and rigidity [7], strength of anchoring to the underlying substrate [8], and end-group functionality [6,9], all of which have some effect on the lateral cohesion of the monolayer [3,5,6,10,11], and on the ease with which defects are formed during sliding [1,3–6].

The complex intermolecular and molecule–substrate interactions of aromatic thiols and silanes give rise to a variety of differently ordered SAM structures [10,11]. Aromatic molecules are stiffer than linear or branched alkanes, and their stronger intermolecular interactions make their friction-modifying properties interesting from a fundamental physical–chemical perspective since one current focus of the research on boundary lubrication is on correlating the structure of confined monolayers and thin films with their friction response. Spectroscopic [11–33] and electrochemical [12,16,17,20,23,24,28,34–39] investigations, supported by ellipsometry [11,17,20,28,29,33–35,40–43] and scanning tunneling microscopy [11,16,17,19,22,25, 27,34,36,37,40,44], have provided information on the structure and interesting optical and electronic properties of SAMs of several aromatic compounds. The bonding and the molecular orientation of aromatic thiols to substrates have also been modeled [30,35,45–48]. Because of the stronger intermolecular interactions between polyaromatic molecules, many of them pack more densely in a monolayer than simple aromatic ones. Their rigidity is advantageous for forming monolayers with exposed functional terminal groups, avoiding the gauche defects found in alkanethiol and alkylsilane monolayers [10,11].

Mineral oil-based fuels and lubricants contain small amounts of many different aromatic molecules that are known to protect metal surfaces from oxidation and wear [49–53]. These include nitrogen- and phosphorus-containing molecules, and sulfur-containing ones, such as thiophenes and thiols [54]. Empirically, it is known that the natural lubricity of jet fuel and diesel fuel [54,55] is significantly diminished when their sulfur and aromatics contents are reduced through hydrogenation. To compensate for this loss of lubricity, various aromatic, polyaromatic, and heteroaromatic molecules with polar or aliphatic substituents are incorporated in additive formulations to improve the performance of refined fuels, biodiesel, and oil-based lubricants [49,54–58]. Similar compounds also find use as friction-modifying additives in aluminum-on-steel sliding [50]. Not only the formation of hard protective layers (sulfides [49,59], in the case of sulfur-containing additives) but also the structure of the hydrocarbon part of the molecules plays a role in the wear protection, especially at low loads [49,50,54]. Despite these current applications, and the need for more specific additives to meet increasingly stringent emission standards [60] and address other health concerns [61], only limited information is available on the molecular level lubricating function of simple aromatic model systems [7,8,62–67] and of the aromatics naturally present in oils.

To address some of these topics, we have studied model aromatic and polyaromatic thiol and silane monolayers [8,65,67–69] in single-asperity contacts with atomic force microscopy (AFM) in friction or lateral mode (friction force microscopy). Measurements of boundary friction in single-asperity contacts allow an assessment of the friction-modifying properties of different structures in confined films and of the contributions of adhesion strength, contact area, and pressure. In this chapter, we present results on systems where a flat gold surface and a gold-covered AFM tip carried SAMs of simple aromatic and polyaromatic thiols of different sizes and packing densities: Thiophenol (TP), *p*-phenylthiophenol (PTP), *p*-terphenyl thiol (TPT), 2-naphthalenethiol, and benzyl mercaptan (BM) (cf. Table 2.1). Despite uncertainties in some previous AFM studies on the boundary friction of aromatic molecules (where some experiments were done in ambient air or without calibration of the cantilever), a higher friction than with close-packed methyl-terminated alkanethiols was typically found [7,8,62–70]. These observations correlate with results from macroscopic friction experiments [8,50]. On the other hand, it is known that good

TABLE 2.1
Aromatic Thiol Monolayer Properties. Tip Radius R and Friction Coefficient μ in Experiments in Ethanol

System	c^c (mM)	T (nm)	T_{ave} (nm)	Molecular Area (nm²)	$\theta_{tilt}{}^a$ (°)	R^b (nm)	μ^c
TP	1	0.4–0.7 [18–21,35,41,42]	0.5	0.4–0.7 [13,16,22,34,36,39]	30–65 [14,17,18,21,22,45–48]	52	1.20±0.05
						52	1.44±0.07
PTP	0.01	0.85–1.4 [7,11,12,18,21,27,29,34,41,43,62]	1.0	0.326 [34]	14–32 [7,11,18,21,26,41]	62	0.91±0.03
						62	1.12±0.04
						97	0.89±0.02
TPT	0.01	1.05–1.8 [12,18,19,21,28,30,33,41]	1.5	0.216 [127]	6–36 [12,18,21,30,31,32,33,41]	—	0.59±0.03
						53	0.57±0.04
						53	0.61±0.02
NT	0.01	0.65–0.8 [20,24]	0.7	0.35–0.42 [23,24,44]	30–44 [24,44]	134	1.51±0.07
						186	1.47±0.03
BM	1	0.45–0.6 [25,38]	0.6	0.17–0.22 [25,34,37]	8–19 [25,45,48]	—	0.76±0.03
						106	0.82±0.02

a From the surface normal.
b $\Delta R = 3$ nm ($R < 100$ nm), 5 nm ($R \geq 100$ nm).
c Standard deviations from linear fits to the experimental data in Figures 2.1 and 2.2.

wear protection [8,49,50,71] can be achieved with aromatic additives despite their higher friction. The difference in nanoscopic friction seen between biphenyl thiol and alkanethiol monolayers has been ascribed to the higher rigidity of the aromatic molecules that act as springs under compression [7,62–64]. Here, we explore the effects of change in packing density and monolayer stiffness associated with change in the number of aromatic rings or the addition of a flexible linker, a methylene group, in the molecular structure. Our studies show that increased packing density reduces the friction of aromatic monolayers from the values obtained with ones with intermediate or low packing density, similarly to the trend observed with alkanethiol and alkylsilane SAMs of sufficiently long chain length [5]. The friction is thus lowered as the monolayer structure is made stiffer than that formed by simple aromatic molecules. The initial rationale for this approach came from empirical observations on macroscopic, oil-based multicomponent systems, where stronger intermolecular interactions appeared to lead to more effective lubrication. In particular, triaromatic polar compounds had been found to reduce wear in macroscopic friction tests at concentrations above ca. 2 wt.%, whereas mono- and diaromatic compounds did not have a clear wear-reducing effect [54]. In other studies of macroscopic friction, better packing and stronger intermolecular interactions reduced the friction and wear in simple aromatic systems as well [49,50].

The adhesion strength is commonly found to affect the magnitude and functional form of the friction force, F, vs. load, L [8,67,72–81]. The adhesion in the current systems was modified by immersing the sliding contacts in ethanol, resulting in low adhesion, or in dry N_2 gas, where the adhesion is higher due to stronger van der Waals forces. F was found to be a linear function of L when the adhesion was low and nonlinear when it was higher. No changes of the tip shape or damage to the monolayers were detected. The load dependence of the nonlinear friction force in our adhesive systems (i.e., in dry N_2) was analyzed using the thin-coating contact mechanics (TCCM) model [82,83], a model developed for the macroscopic contact mechanics of a linearly elastic thin film confined between a rigid spherical indenter and a rigid flat substrate. In these adhesive systems, F appeared to be area dependent and was well described by $F = S_c A$, where A was the contact area calculated using the extended TCCM model in a transition regime between its Derjaguin–Muller–Toporov (DMT)- and Johnson–Kendall–Roberts (JKR)-like limits. However, F depended linearly on L in experiments on the same monolayers under conditions of low adhesion (i.e., when immersed in ethanol). The friction, described in terms of friction coefficient, μ, in the nonadhesive systems and the critical shear stress, S_c, in the adhesive systems, decreased with increasing packing density. In the comparison of F measured in the adhesive systems to the calculated $S_c A$ vs. L (with A calculated according to the extended TCCM model), it was found that a larger Young's modulus was needed to describe the functional form of the data from the aromatic systems than the corresponding data from alkane-based (fatty acid) monolayers of similar packing density.

Another key parameter in the understanding of thin film lubrication is the dependence on sliding velocity or shear rate, which is believed to be related to the relaxation time of the confined system. In thin films or adsorbed layers confined between solid substrates, relaxations are several orders of magnitude slower than for the same molecules in the bulk [79,84]. The relaxation time is dependent on the molecular mobility, which, in turn, is reflected in the physical state of the confined substance or mixture, i.e., confinement-induced types of liquid- or fluid-like, amorphous-like, or solid-like states [79,85]. Commonly, a high friction is found if the probing frequency corresponds to some (longest) relaxation time of the film structure, although the detailed nature of the relaxation is not always well known. As a consequence, different friction responses are observed if certain timescales of the experiment differ from the time needed for disentanglement or change in conformation of molecules in the confined film. The relaxation, or physical state, can be altered by either a change in molecular structure (such as double bonds, branching, or the presence of aromatic moieties) or by the addition of a second amphiphile to alter the structure or packing density of an adsorbed monolayer or larger self-assembled structure at the solid–liquid interface. Here, we discuss the effects of sliding velocity on the friction of TP and BM monolayers. The loose-packed TP shows a fluid-like friction response typical of a disordered structure, whereas the more close-packed BM shows a slight plateau in the data on F vs. sliding velocity, indicating a more rigid structure that becomes increasingly solid-like at higher applied loads, L.

2.2 EXPERIMENTAL

2.2.1 SELF-ASSEMBLED MONOLAYERS

We have studied a series of thiols containing an increasing number (one to three) of aromatic rings: TP (Fluka, purity 99.6%), PTP (Oakwood Products, West Columbia, SC, 97.0%), and TPT (Frinton Laboratories, Vineland, NJ). For further comparison with the TP system, we have also studied 2-naphthalenethiol (NT) (Fluka, 99.9%) and BM (Fluka, 99.8%). These two compounds form SAMs of similar thickness to that of TP but with different packing density. Schematic molecular structures of these compounds are shown in Table 2.1. The monolayers were formed by immersing template-stripped gold substrates supported on polystyrene [65,86] and gold-covered AFM tips (CSC38, MikroMasch or CSC12, BioForce Laboratory) in ethanol (Sigma-Aldrich, ≥99.5% or Primalco, Finland, 99.5%) solutions with the concentrations given in Table 2.1. Template-stripped gold was formed by sputtering or evaporating ca. 100 nm of gold (99.99%, Denton Vacuum, or 99.90%, Kultakeskus, Finland) onto freshly cleaved muscovite mica (S & J Trading, Glen Oaks, NJ). Pieces of polystyrene (ca. 2 cm × 2 cm, from a plastic petri dish) were heated on a hot plate until slightly softened. The mica-supported gold was cut in ca. 2 cm × 2 cm pieces and its exposed gold side was pressed against the soft polystyrene [86]. After slowly cooling to room temperature, the mica was removed with tweezers or with adhesive tape to expose the smoother gold surface right before immersing it in thiol solution. The rms roughness of these polystyrene-supported template-stripped gold substrates was 0.2–0.4 nm, measured over 1 μm².

The adsorption kinetics of aromatic thiols is known to be affected by their stiffness and by their conjugation and dipole moment [10,19]. It has been noted that methylbiphenyl thiol and several other 4′-substituted biphenyl thiols formed equilibrium monolayer structures (with a packing density similar to that of PTP) much more slowly than alkanethiols [10,11,87]. Slow equilibration has also been seen for TP [67] and PTP [43]. The thiols were, therefore, allowed to adsorb for 24 h at room temperature, after which the samples were rinsed with ethanol and dried in a stream of N_2 gas. Most of the information available on aromatic SAMs comes from studies of their electronic or optical properties (cf. Section 2.1) and a range of monolayer thicknesses T, molecular areas and tilt angles can be found in the literature (Table 2.1). An average thickness, T_{ave}, was determined from a narrower range of values believed to be most reliable (typically where T was determined against a close-packed alkanethiol standard or calibrated by other means [16,20,21,23,25,27,28,33]), so that the uncertainty ΔT_{ave} was 0.1 nm in all systems except PTP, where it was 0.2 nm. T_{ave} was used in the van der Waals–Lifshitz calculation and contact mechanics model described below. TP is known to form a very poorly packed monolayer where the aromatic ring is strongly tilted from the surface normal (large θ_{tilt}), whereas PTP and TPT form more close-packed, rigid layers with the molecules in a nearly upright orientation. BM is also known to adsorb in an upright orientation, forming more close-packed monolayers than TP because of the rotational freedom around its methylene group [34,45]. NT forms more loose-packed monolayers [20,23,24,44]. For comparison with the molecular areas in Table 2.1, the area occupied by a vertically oriented benzene ring is 0.21 nm² (based on van der Waals and covalent radii) [34,45] and that of a vertically oriented NT molecule is 0.355 nm² [23].

2.2.2 CONTACT ANGLE MEASUREMENTS AND SURFACE ENERGIES

The advancing and receding contact angles of distilled water on the SAMs on template-stripped gold, and the advancing contact angles of methyl iodide, were determined with a Krüss Drop Shape Analysis System 100 (Germany). Typical drop volumes were 5–15 µL. The contact angle data in Table 2.2 are the averages of 3–6 measurements on each of 3–7 different samples of each monolayer, and are given with a typical standard deviation of ±1°. The surface energy was calculated from the advancing contact angles using the Young–Dupré (YD) equation, $W_{SL} = \gamma_{LV}(1 + \cos\theta)$ [88–90], where γ_{LV} is the surface tension of the liquid. Assuming that the work of adhesion, W_{SL}, is composed of the dispersive and polar components of the surface energy of the solid and surface tension of the liquid (the Owens–Wendt approach [90]), $W_{SL} = 2\left(\sqrt{\gamma_S^d \gamma_L^d} + \sqrt{\gamma_S^p \gamma_L^p}\right)$, one obtains

$$2\left(\sqrt{\gamma_S^d \gamma_L^d} + \sqrt{\gamma_S^p \gamma_L^p}\right) = \gamma_{LV}(1 + \cos\theta) \tag{2.1}$$

Using Equation 2.1, the surface energy of the solid surface (i.e., the SAM), $\gamma_{YD} = \gamma_S = \gamma_S^d + \gamma_S^p$, can be obtained from the advancing contact angles in Table 2.2 for liquids with different surface tensions $\gamma_{LV} = \gamma_L^d + \gamma_L^p$ and polarities (water: $\gamma_L^d = 21.8$ mJ/m², $\gamma_L^p = 51$ mJ/m²; methyl iodide: $\gamma_L^d = 48.5$ mJ/m², $\gamma_L^p = 2.3$ mJ/m²) [90]. The values of γ_{YD} are given in Table 2.2, and are around 38 mJ/m², which is only slightly higher than surface energies determined for other aromatic surfaces, such as polystyrene [88,90] and the face (flat side) of the aromatic rings of naphthalene

TABLE 2.2
Contact Angles of Water and Methyl Iodide (MI), and Surface Energies of Aromatic Monolayers. Film Thickness h, Radius R, Critical Shear Stress S_c, Surface Energy γ_{TCCM}, and Transition Parameter ζ in Dry N_2

System	θ_{adv,H_2O} (°)	θ_{rec,H_2O} (°)	$\theta_{adv,MI}$ (°)	γ_{YD} (mJ/m²)	γ_{vdW}^a (mJ/m²)	h^b (nm)	R (nm)	S_c^c (MPa)	γ_{TCCM}^c (mJ/m²)	ζ^c
TP	87	75	40	40	38	1.0	52	770	38	0.010
							52	850	40	0.011
							189	750	38	0.010
PTP	88	73	43	38	39	2.0	62	390	32	0.017
							62	350	40	0.021
TPT	86	72	43	38	40	3.0	53	430	41	0.033
							53	470	40	0.032
							104	360	42	0.034
							300	340	40	0.032
NT	88	71	48	35	42	1.4	300	460 (DMT)	41 (DMT)	0.033
							40	1050	35	0.013
BM	87	72	43	38	40	1.2	97	1000	32	0.012
							135	240	38	0.012

a van der Waals–Lifshitz theory for five-layer system [88] where medium 3 (cf. Equation 2.2) is air.
b $h = 2T_{ave}$.
c From comparisons with the extended TCCM model [82,83] in Figures 2.4 and 2.5, $\gamma_{TCCM} = W/2$. $\Delta\gamma_{TCCM} = 1$ mJ/m², $\Delta S_c = 20\%$ (15% for TPT), $\Delta\zeta = 25\%$ (15% for TPT).

and anthracene molecules (cf. Section 2.4.1) [91]. In all the five systems, the polar contribution to the surface energy of the solid is quite low, $\gamma_s^p \approx 2\,\text{mJ/m}^2$, i.e., the surface energy of the SAMs arises mainly from dispersive intermolecular interactions.

Surface and interfacial energies can also be determined from the interaction force calculated according to the van der Waals–Lifshitz theory [88] for symmetrical five-layer systems consisting of gold/SAM/medium/SAM/gold (materials 1/2/3/2/1 in Equation 2.2), where the medium is dry N_2 gas or ethanol. The values of γ_{vdW} in Table 2.2 were calculated from $\gamma_{vdW} = -F_{vdW}/(4\pi R)$, where F_{vdW} is the nonretarded van der Waals force between a sphere and a flat surface [88],

$$\frac{F_{vdW}}{R} = -\frac{1}{6}\left(\frac{A_{232}}{D^2} - 2\frac{\sqrt{A_{232} \times A_{121}}}{(D+T)^2} + \frac{A_{121}}{(D+2T)^2} \right), \tag{2.2}$$

R is the radius of the sphere, $D=0.165\,\text{nm}$ is the separation at monolayer–monolayer contact [88] and the monolayer thickness is $T = T_{ave}$ from Table 2.1. The Hamaker constants were calculated from bulk refractive indexes and dielectric constants of the aromatic compounds [12,65,67] and N_2 or ethanol. The Hamaker constant for gold interacting across air ($A = 45.5 \times 10^{-20}$ J) [92] was used in the combining relation $A_{121} = \left(\sqrt{A_{11}} - \sqrt{A_{22}} \right)^2$ [88]. The γ_{vdW} values in N_2 are given in Table 2.2, and are not significantly different from those calculated for a three-layer system (SAM/N_2/SAM) (not shown), suggesting that there is not a strong influence from the underlying gold substrate. In ethanol, we obtained $\gamma_{vdW} < 6\,\text{mJ/m}^2$, which corresponds to an adhesion force of $-3\,\text{nN}$ ($R \approx 50\,\text{nm}$) at $D=0.165\,\text{nm}$ (monolayer–monolayer contact) and $-0.6\,\text{nN}$ at $D=0.44\,\text{nm}$ (thickness of a monolayer of ethanol [93]). The measured pull-off force in ethanol was typically -1.0 to $-1.5\,\text{nN}$, which suggests that at the closest separation, there is not a full monolayer of ethanol between the tip and sample. This is consistent with the known reduction or absence of layering of solvents between amorphous or slightly rough monolayers [79].

2.2.3 FRICTION FORCE MICROSCOPY

The lateral or friction force, F, was measured with a Veeco Multimode Nanoscope IIIa atomic force microscope (AFM). Measurements in ethanol were done in a fluid cell. Measurements in N_2 were done with the AFM enclosed in a home-made plastic chamber purged continuously with N_2 gas (Airgas, 99.5%). The humidity inside the chamber was measured with a Vaisala DM70 dewpoint meter, and a relative humidity of $\leq 1.5\%$ was reached after purging for 1 h. The friction force between the monolayer-covered flat substrates and the corresponding functionalized tips was measured at different loads (normal force) L over a scan size of 1 µm at a scan rate of 2 µm/s (cf. Figures 2.1, 2.2, 2.4, and 2.5). In addition, the dependence of F on sliding speed ($v=0.06–50$ µm/s) in the TP and BM systems was investigated over a range of scan sizes (20–750 nm) [65] at loads in the range $L=0.7–114.0$ nN (cf. Figure 2.3).

The normal and lateral spring constants of AFM cantilevers can vary considerably from the manufacturer's specifications, and a calibration is necessary in order to obtain L and F. The normal and lateral spring constants were determined from the resonance frequency and quality factor [94,95] of the cantilevers and from their dimensions [96,97] as measured by a scanning electron microscope (JEOL-7401F). The torsional deflection of the cantilever was recorded in "scope mode" as a voltage vs. sample position (friction loop) and converted to friction force using the calibrated sensitivity of the photodetector as described in Ref. [97] (cf. [67,80]). The height of the sample was recorded simultaneously to ascertain that the measurement was made over a flat area. The gold-covered cantilevers used in this study had normal spring constants in the range 0.05–0.25 N/m and lateral spring constants of 7–40 N/m. Some experiments on the velocity-dependence of F (cf. Figure 2.3) were done with unfunctionalized Si tips carrying a native oxide layer (CSC12 and NSC12, MicroMasch), with normal spring constants of 0.04 and 0.34 N/m and lateral spring constants of 4.6 and 38.0 N/m, respectively. The statistical standard error (standard deviation of the mean) in the F values (from averaging the sliding region of the friction loop) was 0.1–0.2 nN at $F < +20$ nN, and ≤ 0.2–0.5 nN at higher F (error bars are shown only in Figure 2.3, in the other figures the error bars would be of similar height as the symbols or smaller). The geometric mean radius of curvature of the tip, R, was determined by reverse imaging of a calibration sample (TGT01, MikroMasch) at scan angles of $0°$ and $90°$. The values of R in the F vs. L measurements shown in Figures 2.1, 2.2, 2.4, and 2.5 are given in Tables 2.1 and 2.2, and the ones in the experiments on velocity dependence are given in the legend of Figure 2.3. The estimated error [68] in R is $\Delta R = \pm 3$ nm for $R < 100$ and ± 5 nm for larger R. In some cases, the cantilever was accidentally broken during or after an experiment, before R had been determined. This was not of great concern for the analysis of data obtained in ethanol, since similar F vs. L data were obtained with different radii. A few data sets obtained in ethanol with tips of unknown R are shown together with ones with known R (cf. Table 2.1 and legends of Figures 2.1 and 2.2). However, in the adhesive systems (in dry N_2 gas), very different results were obtained with different R values, and it was necessary to know R in order to analyze the data correctly. Occasionally, R was measured both before and after an experiment to ascertain that it did not change appreciably during scanning at the investigated loads.

2.2.4 CONTACT MECHANICS MODEL

In AFM friction experiments on SAMs, the radius of the contact area at low loads can be comparable in size to the thickness of the monolayers, whose elastic modulus is expected to be 10–50 times lower than the bulk modulus of the confining surfaces (for values of the moduli, see Section 2.4.3). The initial deformations in the system are thus expected to occur in the soft, thin film. At increased loads, the effective stiffness of the system is still lower than that of the substrates, and the situation is more complex than that assumed in contact mechanics models for homogeneous bodies [72,98–101]. Several models [82,83,102] have been developed for the contact mechanics of a thin, elastic film confined between stiffer substrates. In nanometer-sized contacts, where the atomic structure of the substrate is expected to affect the pressure distribution [103], it is not yet established how well such models apply, although the influence of the substrate is reduced when a molecularly thin film is introduced [104].

In the TCCM model [82,83], both the spherical indenter and the flat substrate are rigid, i.e., the deformation is allowed to occur only in the confined, thin film. The model describes two limiting cases [82]; a DMT-like response, in systems where the range of adhesion is large compared to the elastic deformations, and a JKR-like case, which describes deformations that are large compared to the range of the adhesion. JKR-like deformations of a thin film have also been previously described for an elastically deforming probe and substrate [102]. For the DMT-like contact, finite element simulations indicated that the rigid indenter and substrate model could be used when the uniaxial strain modulus E_u of the thin film was <5% of the indenter and substrate Young's modulus [82]. A recent extension [83] of the model describes a continuous transition between the DMT- and JKR-like cases. This model has been successfully applied to AFM friction force data [105] and compared with a molecular dynamics simulation [106] of a confined alkylsilane monolayer.

We compare the F vs. L data in our adhesive systems (cf. Figures 2.4 through 2.6) to $F = S_c A$, where S_c is a constant, the critical shear stress (Table 2.2), and A is the contact area at a given load L. The relationship between L, the radius of the contact area (a), and the work of adhesion (W) in the extended TCCM model is given in nondimensional form [107] by [83]

$$\bar{L} = \frac{\pi}{4}\bar{a}^4 - \zeta^{1/2}\pi\bar{a}^2\left(2\bar{W}\right)^{1/2} - 2\pi\bar{W}\left(1 - \zeta\right), \tag{2.3}$$

where $\bar{L} = L/(E_u Rh)$, $\bar{a} = a/\sqrt{(Rh)}$, and $\bar{W} = W/(E_u h)$. $E_u = E(1-v)/[(1+v)(1-2v)]$, where E is Young's modulus (in this case, 7 GPa) and v is Poisson's ratio (0.4). The film thickness, h, is the thickness of two monolayers in contact, $h = 2T_{ave}$ (cf. Tables 2.1 and 2.2). ζ is a transition parameter ($0 \leq \zeta \leq 1$), a measure of the ratio of elastic deformation to the effective range of the surface forces, with $\zeta = 0$ corresponding to the DMT-like and $\zeta = 1$ to the JKR-like limit of the TCCM model. Values of ζ in our systems are given in Table 2.2 and will be discussed together with the choice of Young's modulus in connection with Figure 2.6. In our systems, W in Equation 2.3 represents the work of cohesion for separation of two identical surfaces in dry N_2 gas and can be used to obtain surface energies γ_{TCCM} ($=W/2$), which are also given in Table 2.2. In the current experiments, the propagation of the uncertainties in R, h, W, and E_u to the values of S_c and ζ lead to an uncertainty [68] in S_c of 20% for TP, PTP, NT, and BM, and 15% for TPT. The uncertainty in ζ is 25% (15% for TPT).

2.3 RESULTS

2.3.1 LOAD-DEPENDENT FRICTION AT LOW ADHESION (IN ETHANOL)

The load dependence of the friction force in AFM measurements performed in ethanol is shown in Figures 2.1 and 2.2. The interfacial energy estimated from the pull-off forces was $\gamma \leq 1\,mJ/m^2$ in all five systems, and a linear dependence of F on L was observed, with $F \rightarrow 0$ when $L \rightarrow 0$. Linear fits to the low-load regimes in Figures 2.1 and 2.2 (solid lines) gave the friction coefficients μ listed in Table 2.1. At higher loads, a transition regime in the monolayer was found (marked with arrows in

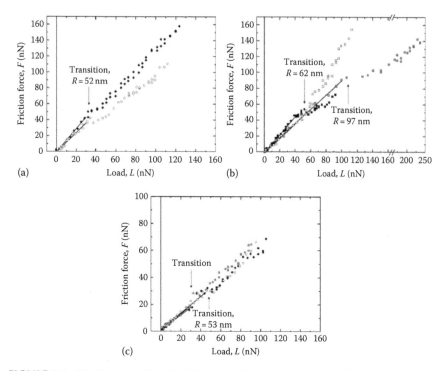

FIGURE 2.1 Friction force F vs. load L measured in ethanol. Both the flat substrate and the tip carried a SAM, and the scan speed was $2\,\mu m/s$. (a) Thiophenol (TP), two separate experiments ($\diamond$, $\blacklozenge$) with tip radius $R=52\,nm$, $\mu=1.2$–1.4 (cf. Table 2.1). (b) p-Phenylthiophenol (PTP), two experiments with $R=62\,nm$ ($\square$, $\blacksquare$) and one with $R=97\,nm$ (gray squares), $\mu=0.9$–1.1. (c) p-Terphenylthiol (TPT), two experiments with $R=53\,nm$ (o, $\bullet$) and one where R is unknown, i.e., the cantilever broke before R could be determined (gray circles), $\mu=0.6$. In all three systems, the low-load regime is reproducible, whereas above the transition regime (marked with an arrow), the scatter and the differences between different experiments are larger.

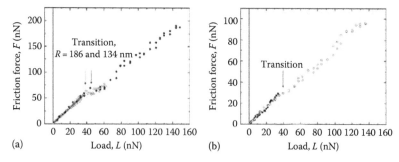

FIGURE 2.2 F vs. L measured in ethanol. Both the substrate and the tip carried a SAM. (a) 2-Naphthalenethiol (NT), tip radius $R=134\,nm$ ($\blacklozenge$) and $186\,nm$ ($\diamond$), $\mu=1.5$. (b) Benzyl mercaptan (BM), one experiment with $R=106\,nm$ ($\bullet$), where the load was not high enough to induce a transition, and one experiment (o) where R is unknown and a transition occurred at $L\approx40\,nN$. In (b), $\mu=0.8$ in both experiments.

the figures), similarly to the response observed for alkanethiol monolayers [108] (cf. Section 2.4.2). The friction was less reproducible at loads above this regime, and in this study we, therefore, focused on only the friction at low loads.

In each system in Figures 2.1 and 2.2, similar friction coefficients were consistently found with different tip radii R—i.e., F did not depend on the probe size. This is consistent with previous observations of friction in contacts with low adhesion [8,65,67,69,73,75,77–80]. In a few cases, R could not be determined due to accidental breakage of the cantilever after the experiment, but the values of μ obtained in those experiments were very similar to the ones where R could be measured. There was a systematic decrease in μ with increasing packing density (Table 2.1)—i.e., μ was the highest for TP and NT, and lower for PTP, TPT, and BM, but μ was still significantly higher than that of a close-packed alkanethiol monolayer under similar conditions ($\mu = 0.02-0.10$, cf. Ref. [5] and discussion in Refs [65 and 67]).

2.3.2 VELOCITY-DEPENDENT FRICTION OF TP AND BM (IN ETHANOL)

The friction force of TP and BM monolayers as a function of sliding velocity [65] is shown in Figure 2.3. These experiments were done in ethanol with bare Si tips (carrying a native oxide layer, Figure 2.3a and c), and with monolayer-covered tips

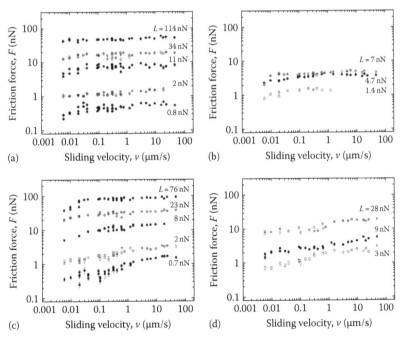

FIGURE 2.3 Velocity dependence of the friction of thiol monolayer systems in ethanol at selected loads: Thiophenol (TP, ◆, ◇) (a) against unfunctionalized Si tips with tip radius $R = 11\,\text{nm}$ (two lowest loads) and $R = 33\,\text{nm}$ (three higher loads) and (b) against a TP-functionalized tip ($R = 13\,\text{nm}$). Benzyl mercaptan (BM, ●, ○) (c) against Si tips with $R = 11$ and 33 nm and (d) against a BM-functionalized tip ($R = 19\,\text{nm}$). (Adapted from Ruths, M., *Langmuir*, 19, 6788, 2003. With permission. Copyright 2003 American Chemical Society.)

(Figure 2.3b and d), at loads within the range $L=0.7–114\,nN$. The data for TP (Figure 2.3a and b) were featureless, as expected for a system with a very fluid-like response of a thin film, with a very small decrease at the very lowest velocities investigated. In the BM system, the data obtained with a bare Si tip showed a plateau that moved toward lower velocity with increasing load, which is an indication that the system (i) shows a more solid-like response than TP and (ii) becomes more solid-like as the load is increased [75,85,96,97]. When the tip also carried a BM monolayer (Figure 2.3d), there was only a slight increase in F similar to that at low velocities in Figure 2.3c, suggesting that the friction response with two monolayers was more fluid-like than with one.

2.3.3 AREA-DEPENDENT FRICTION AT HIGHER ADHESION (IN N₂)

The friction forces of the five monolayer systems in Figures 2.1 and 2.2 were also measured as a function of load in dry N_2 gas (r.h. <1.5%, Figures 2.4 and 2.5), where the adhesion was higher than in ethanol. In some cases, the tips used to obtain the data in Figures 2.1 and 2.2 were used also for these measurements. The values of F

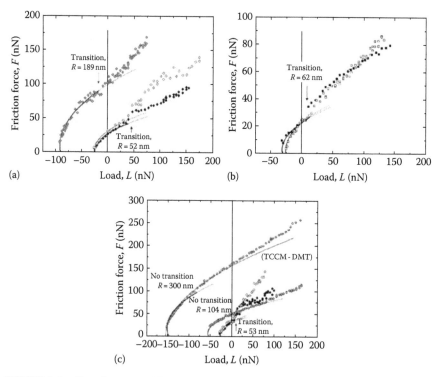

FIGURE 2.4 F vs. L measured in dry N_2 gas. (a) TP, two separate experiments ($\diamond$, $\blacklozenge$) with tip radius $R=52\,nm$ and one with $R=189\,nm$ (gray diamonds). (b) PTP, two experiments with $R=62\,nm$ ($\square$, $\blacksquare$). (c) TPT, two experiments with $R=53\,nm$ (o, ●), one with $R=104\,nm$ (gray circles), and one with $R=300\,nm$ (⊖). The solid curves, intended as comparisons to the low-load regime of the data, are $F=S_cA$, where the contact area A was calculated according to the extended TCCM model with parameters as in Table 2.2 and $E_u=15\,GPa$ ($E=7\,GPa$, $\nu=0.4$). One data set in (c), $R=300\,nm$ (⊖), is better described with the DMT version of the TCCM model.

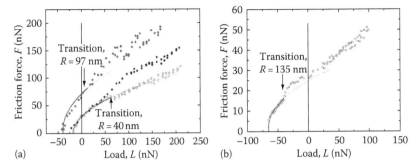

FIGURE 2.5 *F* vs. *L* measured in dry N_2 gas. (a) NT, two separate experiments with tip radius *R*=40 nm (◊, ◆), and one with *R*=97 nm (gray diamonds). (b) BM, *R*=135 nm (o). The low-load regimes are described by the extended TCCM model with the parameters in Table 2.2. In (a), the calculated curve has been extended (dashed curve through ◊) to show an unusual case where it also described the higher load data.

and the shape of the *F* vs. *L* curves in Figures 2.4 and 2.5 were quite different from the data at low adhesion shown in Figures 2.1 and 2.2. We have ascertained that the different results did not arise from tip or monolayer damage, since the different friction responses (linear vs. nonlinear) could be obtained with the same tip when switching from one environment to another (ethanol vs. N_2) and back.

Such a nonlinear dependence of *F* on *L* has been observed previously for adhesive, unfunctionalized surfaces [72,73,75,77,79] and for adhesive SAMs [69,75,78–80]. It is generally believed that the friction under adhesive conditions is area dependent. In such cases, the magnitude of the friction force is expected to depend on the size of the probe, i.e., on the tip radius *R*.

The extended TCCM model was applied to the data in Figures 2.4 and 2.5. The solid curves in Figures 2.4 and 2.5 are intended as comparisons to the low-load regime of the data only (below the transition regimes indicated by arrows) and represent $F = S_cA$, with the contact area *A* calculated from *a* in Equation 2.3 ($A = \pi a^2$) using a uniaxial strain modulus of $E_u = 15$ GPa ($E = 7$ GPa, $v = 0.4$). This model provided a good description of the low-load data in all systems, except for one data set in Figure 2.4c with *R*=300 nm (⊖ symbols), which was better described with the DMT-like limit of the TCCM model. The corresponding critical shear stresses S_c are given in Table 2.2. There was a good agreement between the S_c values obtained with tips of different sizes, supporting the concept that these were cases of area-dependent friction. The values of S_c indicated that the highest friction was obtained in the TP and NT systems, as when measuring in ethanol (μ in Table 2.1), and that the friction in the more closely packed systems was clearly lower.

2.4 DISCUSSION

2.4.1 CONTACT ANGLES AND SURFACE ENERGIES

The advancing and receding contact angles of water were similar on all five SAMs (Table 2.2), and were in good agreement with literature values

[10,12,28,29,33,34,42,65]. Although it is known that contact angles on alkanethiols [109] and on conjugated systems [10] may be affected by a gold substrate, the very similar contact angles obtained on these SAMs arise mainly from the similar chemical structure of their uppermost part. Our advancing contact angles were lower than those found on naphthalene and anthracene crystals cleaved to expose the edges of their aromatic rings (95° and 94°, respectively) [91]. The faces of aromatic rings are more wettable than the edges, and lower packing density thus results in lower contact angles [91]. The critical surface tension (extrapolated value of the surface energy obtained from measurements of contact angles of liquids of different polarity) of the edge-on type of aromatic surface has been reported as ca. 25 mN/m [91], whereas that of the face-on type was suggested to be larger than that of methylene groups ($-CH_2-$), i.e., larger than 31 mN/m [91]. The critical surface tension of polystyrene is 33 mJ/m² [88,90]. Although some of our monolayers (TPT, PTP, and BM) are expected to be more close-packed than others (TP and NT), none is at maximum (crystalline) close-packing for aromatic rings and the faces of the aromatic rings are to some extent accessible to the water, which is reflected in the estimated surface energies we obtained from the contact angle measurements, $\gamma_{YD} \approx 38$ mJ/m². The tilt of the molecules (Table 2.1) is large enough to allow for partial exposure of the faces of aromatic rings at the surface of the monolayer. The very small variation in the contact angle hysteresis ($\theta_{adv} - \theta_{rec} \approx 12°-17°$) is also consistent with a similar chemical structure at the uppermost part of the five monolayers.

Advancing contact angles of 85°, i.e., similar to those on our phenyl-terminated monolayers, have also been found on CH_3- and CF_3-terminated biphenyl thiols [110]. In those systems, the values of the contact angles originated partly from the different dipole moments of the molecules, but also because the terminal methyl group was smaller than the cross-sectional area of the aromatic moiety so that a part of the aromatic ring was exposed to the water [110].

The γ_{YD} values obtained from the contact angle measurements were in good agreement with γ_{vdW} (Table 2.2) calculated for a five-layer system of gold/SAM/dry N_2 gas/ SAM/gold using van der Waals–Lifshitz theory [88]. The interaction of gold across the SAMs was approximated by a Hamaker constant for alkanethiol-on-gold systems (cf. Section 2.2.2), and average monolayer thicknesses T_{ave} (Table 2.1) were used. A further comparison can be made with the values of surface energy obtained from the contact mechanics model. The pull-off force in a thin-layer system is increased above that for homogeneous elastic bodies by a changing distribution of traction and deflection in the contact during the separation of the surfaces [102]. This is accounted for by using the extended TCCM model to obtain γ_{TCCM} ($=W/2$, Table 2.2) from the work of adhesion W needed to reproduce the pull-off (jump-apart) points in the F vs. L data in Figures 2.4 and 2.5. The uncertainty is $\Delta W \approx 0.002$ J/m². A good agreement was found between γ_{TCCM}, γ_{YD}, and γ_{vdW} (cf. Table 2.2).

2.4.2 MONOLAYER TRANSITION REGIME

In each data set in Figures 2.1, 2.2, 2.4, and 2.5, the data at low loads were reproducible from experiment to experiment, whereas this was commonly not the case at higher loads (cf. Figures 2.1, 2.4b and c, and 2.5a). At higher loads, the experimental data were considerably more scattered and also typically deviated from the linear dependence seen at low loads in Figures 2.1 and 2.2, or from the smooth curvature in Figures 2.4 and 2.5. For this reason, the linear fits in Figures 2.1 and 2.2 and the S_cA curves (solid curves) in Figures 2.4 and 2.5 are intended as comparisons to the low-load data only. In Figure 2.5a, one example is shown where the extended TCCM model also describes one data set above the monolayer transition (dashed curve through ◊ symbols), but this was unusual. In another experiment with the same tip (♦ in Figure 2.5a), the high-load data were different.

In most cases, an onset of a different response or a transition region was seen as indicated by arrows in the figures. Measurements done with sharper tips (smaller R) show this onset at lower loads (cf. Figure 2.1b and Refs [65] and [108]), and it is sometimes accompanied by a plateau or dip in the F vs. L data [8,65,67,80]. Work on alkanethiol monolayers [108] and simple aromatics [65] has suggested a reversible displacement of molecules at pressures of 1–2 GPa. In the data shown here, and in previous studies [65,67], the friction response at high loads was not identical to that obtained for a bare substrate, and it is, therefore, unlikely that the monolayers are completely removed from the contact. Above the transition regime, where the compression is >20% [111], it is unlikely that the monolayers are linearly elastic; therefore, we do not expect these regions to be described by the TCCM model. A "plowing" mechanism has been discussed for the friction of octadecyltrichlorosilane monolayers measured with a sharp tip ($R = 15–20$ nm) [1].

2.4.3 MONOLAYER ELASTIC MODULUS

The application of the TCCM contact mechanics model requires an estimate of the monolayer elastic modulus. The spherical indenter (probe) and flat substrate are assumed to be rigid. In our systems, both of these carry a metal layer (20 nm Cr + 20 nm Au on the AFM tip and ca. 100 nm template-stripped Au as the flat surface). The Young's modulus of bulk gold is 78.5 GPa [112]. The modulus of a self-assembled organic film with many defects is likely to be significantly lower than that of the tip and substrate. The value of the uniaxial strain modulus of the monolayer, $E_u = E(1-v)/[(1+v)(1-2v)]$ affects W (and thus γ_{TCCM}) only slightly but has a strong effect on the contact radius a at a chosen L.

The elastic moduli of SAMs are not well known, and a wide range of values is found in the literature. Local compliance measurements with AFM [113,114] suggested Young's moduli of $E = 0.2–0.6$ GPa for close-packed monolayers, and thickness changes on compression of fatty acid monolayers formed by Langmuir–Blodgett deposition gave $E = 1–5$ GPa [115], whereas compression of surfactant monolayers in a surface forces apparatus gave $E = 0.05–0.20$ GPa [116]. Other measurements with AFM- and interfacial force microscopy-based techniques suggest $E = 0.2–15.0$ GPa for close-packed alkanethiol and alkylsilane monolayers [111,117–120], where the higher values were composite moduli that included some substrate deformation (cf. discussion in Ref. [111]). Computer simulations of alkanethiol monolayers indicated moduli around 20 [121] and 36 GPa [112,118]. These higher values are possibly the result of modeling ideal systems with very few defects. In molecular dynamics simulations of a flat plate compressing an alkylsilane monolayer, the monolayer was found to have a uniaxial strain modulus of 3 GPa at 10% nominal strain [122]. Contact areas calculated with the extended TCCM model using this modulus value were successfully compared to molecular dynamics simulations of the contact between an AFM tip and an alkylsilane monolayer [106].

When applying the extended TCCM model to the data in Figures 2.4 and 2.5, it became apparent that neither the JKR-like limit nor very low values of E described the data well. The experimental friction force increased more rapidly at low L than the calculated area in the JKR limit, and the pull-off occurred at a lower F than the position of the apex of JKR-like curves scaled to fit the experimental data between the lowest loads and the monolayer transition regime. This has been seen before for simple aromatic systems [65] and also noted in systems of very loose-packed fatty acid monolayers self-assembled from organic solvent [80], although for the fatty acids a reasonable fit could still be obtained with a JKR-like model [80,102].

This is illustrated in Figure 2.6, where data from an experiment on TP are compared to a self-assembled fatty acid monolayer [80] of similar, low packing density. In Figure 2.6a, curves calculated with the extended TCCM model and different values of the Young's modulus ($E=0.1$, 0.7, and 7 GPa, $\nu=0.4$) were multiplied by different constants (different values of S_c) to bring the curves as close to the experimental data as possible. (The curve for $E=7$ GPa, with $S_c=770$ MPa, is also plotted in Figure 2.4a.) As E was increased, the lower branch of the parabolic curve was reduced and the apex of the curve corresponded better to the pull-off region (lowest data points) in the data. The rise of the calculated curve also better approximated

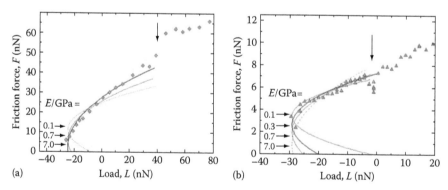

(a)

(b)

FIGURE 2.6 Comparison of $F=S_cA$ curves to F vs. L data for (a) TP (from Figure 2.4a) and (b) a very loose-packed layer of stearic acid physisorbed from n-hexadecane on steel (molecular area 0.4 nm²). (Experimental data in (b) from Ruths, M. et al., *Langmuir*, 24, 1509, 2008.) In (a), $R=52$ nm, $h=1.0$ nm, and $W=0.075$ J/m²; and in (b), $R=58$ nm, $h=1.3$ nm, and $W=0.080$ J/m². In (a), a Young's modulus of $E\geq7$ GPa ($\nu=0.4$, $\zeta\approx0.01$) is needed to reproduce the nonlinear increase in F with L. In (b), the functional form of the data is better reproduced with a lower Young's modulus $E\leq0.7$ GPa ($\nu=0.4$, $\zeta>0.1$). The onsets of monolayer transition regimes are marked with arrows.

the shape of the experimental data. At moduli well over $E = 7$ GPa (not shown), the distinct pull-off at a finite F was lost (i.e., the apex of the curve was at $F = 0$), which corresponded to the DMT-like limit. We found that the data in all our five aromatic systems were best described by a modulus 6–7 GPa $\leq E <$ ca. 15 GPa, except one data set in Figure 2.4c ($R = 300$ nm) that was better reproduced with the DMT-like limit of the model (curve marked TCCM-DMT). The transition parameter ζ was in the range 0.01–0.03, with the higher values obtained for the thicker monolayers (Table 2.2). It has been noted [106] that with a larger tip radius, the monolayer appears to be stiffer (less penetration occurs), which is what is seen for the two larger radii in Figure 2.4c ($R = 104$ and 300 nm), where the pressure was not high enough to induce a detectable monolayer transition.

The opposite trend was seen in the loose-packed fatty acid system in Figure 2.6b. High values of E gave too steep a rise of the calculated curve and did not reproduce the pull-off region well. Reasonable fits were found with 0.1 GPa $< E < 0.7$ GPa in this system and also in other fatty acid systems from ref. 80 (oleic and linoleic acid, not shown) with values of ζ around 0.1. This range of Young's moduli is close to the literature values mentioned above for various SAMs consisting of close-packed alkane chains. The higher values obtained for our aromatic systems are consistent with the concept that aromatic molecules are stiffer than alkanes and might form stiffer monolayers if sufficiently close-packed.

2.4.4 Adhesive vs. Nonadhesive Systems

Our observations of the different functional forms of the F vs. L data at low L in nonadhesive and adhesive systems are consistent with results on other surfaces. A linear increase in F with L has been observed in single-asperity contact between mica sheets experiencing repulsive hydration forces in aqueous electrolyte solution [74,79] and in SAM [8,65,67,69,75,78–80] and polymer [77] systems under conditions where these surfaces adhered very weakly to one another. In adhesive systems, F commonly shows a nonlinear increase with L that is ascribed to an area-dependence [72,73,77–80,105].

Models for the dependence of the friction force on load and contact area have been put forward on the basis of empirical observations. It has been proposed that work has to be done (i) against the external load and (ii) against the adhesion forces (if present) to enable the surfaces to slide past one another (cf. Ref. [79] and references therein). That is, the surfaces must separate slightly (dilate) for sliding to occur. This is often expressed as $F = \mu L + S_c A$, where one of the terms (μL or $S_c A$) might dominate the friction response under certain conditions. In one of the simplest models, only the area-dependent term depends on the adhesion (interfacial or surface energy), which is incorporated into the factor S_c. This implies that the area-dependence of the friction could be largely reduced if the adhesion was decreased and only the μL term would remain. In such cases, the friction would not depend on R, and data obtained with different probe sizes could be directly compared with one another; this appears to be the case in Figures 2.1 and 2.2 and has been demonstrated in work on simpler aromatic systems [65,67] with $\gamma < 4$ mJ/m^2 and with probe sizes differing by 5–6 orders of magnitude [8].

Other models suggest that the friction force always depends on the contact area. The linear dependence of F on L is explained as a nonconstant (pressure-dependent) shear stress. Changes in the shear stress at high pressure are certainly possible, especially in view of the monolayer transition we observe that may be preceded by smaller changes in the orientation of the molecules. However, the linear F vs. L behavior has been observed in a wide variety of nonadhesive systems over wide ranges of loads and pressures, and it is unlikely that all of these systems can be rationalized by a pressure-dependent shear stress, in particular, because investigations on the same monolayer with the same tips but at higher adhesion give a different (nonlinear) F vs. L. Within the accuracy of our experiments, S_c appears to be a constant when determined from the data at low L (Figures 2.4 and 2.5, Table 2.2). Similar observations have been made in fatty acid monolayers systems using a JKR-like model for layered systems [80], and in phenyl-terminated alkanethiols systems using the extended TCCM model [69].

Differences between adhesive and nonadhesive systems are also seen in computer simulations, albeit not as two additive terms. Molecular dynamics simulations of lubricated surfaces (n-hexadecane confined between slightly rough gold surfaces) showed a linear dependence of F on L with a different slope and a shift along the L axis as adhesion was introduced [76]. In a recent molecular dynamics simulation of unlubricated (dry) contacts, sublinear and linear F vs. L were obtained with and without adhesion, respectively [81]. These friction responses were described as F being proportional to the real, atomic-scale area of multi-asperity contacts with roughness at the atomic scale [81]. Since most of our systems have a packing density lower than the maximum close-packing of aromatic rings, there may be slight differences in their molecular-scale roughness even in the dry state (in dry N_2). However, an effect of this was not discernible in our systems because they all showed a nonlinear friction response in adhesive contact. At this point, we do not have information on how the roughness of the monolayers might be altered in contact with a liquid. Previous experiments on BM and close-packed alkanethiols monolayers have suggested (based on measured vs. calculated adhesion strengths and the possibility to obtain identical "atomic resolution" images of close-packed monolayers in dry N_2 gas and in ethanol) that sliding occurs without an intervening full monolayer of solvent [65,67], but some penetration of small solvent molecules into the monolayers is possible.

2.4.5 Dependence of Friction Force on Packing Density

Compared to the considerable amount of data available on the nanoscale friction of alkanethiol and alkylsilane monolayers (see, for example [1,3,5,9,123–126] and references therein) only a few studies have been done of aromatic ones [7,8,62–69]. Although some previous work was done under ambient conditions and using nominal spring constants, relative values clearly indicated that the friction of aromatic monolayers typically was higher than that of alkanethiols and alkylsilanes: The friction coefficient μ of PTP in contact with a Si_3N_4 tip was ca. three times higher than that of hexadecanethiol [7,62], and μ of terphenyl methanethiol was seven to eight times higher than that of octadecanethiol [66]. In ethanol, μ of TP probed with a TP-covered tip was 13–15 times higher than that of octadecanethiol [65,67]. Practical applications of aromatic friction modifiers do not necessarily rely on a particularly low sliding friction but on their wear resistance, which is likely to increase with increasing stability of the SAM. In pin-on-plate experiments [66], it has been found that terphenyl methanethiol monolayers had a higher wear resistance than octadecanethiol. It is also known from macroscopic experiments [49] that dibenzyl sulfide, which forms a BM-like monolayer, protected steel surfaces better from wear than diphenyl disulfide, which forms a TP-like monolayer.

A "spring" model has been proposed [7,62] in which the high friction of PTP was ascribed to the high stiffness of individual aromatic molecules combined with their stronger lateral interactions, which made it difficult to orient and compress the monolayer under the probe tip. In our five systems, the individual aromatic molecules are believed to be stiffer than alkane chains of similar length and thus more difficult to bend or compress. However, as shown in Figure 2.7, we found decreasing friction (decreasing μ and S_c) with increasing lateral interactions (i.e., with improved packing and thus increased stiffness of the monolayer structure). The uncertainties in μ are from the linear fits in Figures 2.1 and 2.2 (Table 2.1), and that in S_c (Table 2.2) is 20% (15% for TPT), as mentioned in Section 2.2.4 describing the contact mechanics model. Possible systematic errors in the measurements of F would affect both μ and S_c in the same direction.

Even in our most close-packed systems (TPT and BM), the friction did not reach the low values seen for close-packed alkanethiols monolayers [5,67]. This overall higher magnitude of the friction might be the result of the inherent stiffness of the constituent molecules. This is consistent with the observed need for a higher Young's modulus to describe the deformations of the confined aromatic monolayers.

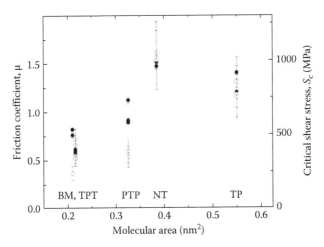

FIGURE 2.7 μ (●) and S_c (○) as a function of average molecular area (cf. Table 2.1). The uncertainties in μ are from the linear fits in Figures 2.1 and 2.2, and the uncertainty in S_c is 20% (15% in the TPT system). Both parameters indicate a lower friction in the more close-packed systems.

2.5 SUMMARY

Atomic force microscopy in friction mode was used to study the effects of adhesion strength and probe size on the friction forces in five aromatic SAM systems with different packing densities. Low adhesion (measurements performed in ethanol, Figures 2.1 and 2.2) resulted in a linear increase in friction force with load, i.e., $F = \mu L$, whereas higher adhesion (in N_2 gas, Figures 2.4 and 2.5) gave an apparent area dependence of the form $F = S_c A$, where S_c is the critical shear stress. By using the extended TCCM model for layered systems to calculate A vs. L, we obtained S_c values at low load in the adhesive systems. μ and S_c were found to decrease with increasing packing density of the monolayers. A larger value of the Young's modulus, $E \geq 7\,\text{GPa}$, was needed to reproduce the F vs. L curves in these systems compared to that in fatty acid systems of similar packing density, where a good agreement was obtained with $E \leq 0.7\,\text{GPa}$. In a comparison of the velocity dependence of the friction of loosely-packed monolayers of TP and the more closely-packed BM, a more fluid-like friction response was seen for TP.

ACKNOWLEDGMENTS

We thank J. Mead for access to the contact angle goniometer and T. Petterson for software for analyzing the AFM data. We also acknowledge the Donors of the American Chemical Society Petroleum Research Fund for support of this research through Grant #45101-G5. This work was also supported through NSF CAREER Award #NSF-CMMI 0645065, by start-up funding through the NSF-funded Center for High-Rate Nanomanufacturing (CHN) (Award #NSF-0425826), and by the Academy of Finland (Grant 48879).

REFERENCES

1. E. E. Flater, W. R. Ashurst, and R. W. Carpick, Nanotribology of Octadecyltrichlorosilane Monolayers and Silicon: Self-Mated versus Unmated Interfaces and Local Packing Density Effects, *Langmuir*, 23, 9242–9252 (2007).
2. P. T. Mikulski and J. A. Harrison, Packing-Density Effects on the Friction of *n*-Alkane Monolayers, *J. Am. Chem. Soc.*, 123, 6873–6881 (2001).
3. M. Salmeron, Generation of Defects in Model Lubricant Monolayers and Their Contribution to Energy Dissipation in Friction, *Tribology Lett.*, 10, 69–79 (2001).
4. S. Lee, Y.-S. Shon, R. Colorado, Jr., R. L. Guenard, T. R. Lee, and S. S. Perry, The Influence of Packing Densities and Surface Order on the Frictional Properties of Alkanethiol Self-Assembled Monolayers (SAMs) on Gold: A Comparison of SAMs Derived from Normal and Spiroalkanedithiols, *Langmuir*, 16, 2220–2224 (2000).
5. A. Lio, D. H. Charych, and M. Salmeron, Comparative Atomic Force Microscopy Study of the Chain Length Dependence of Frictional Properties of Alkanethiols on Gold and Alkylsilanes on Mica, *J. Phys. Chem. B*, 101, 3800–3805 (1997).
6. N. J. Brewer, B. D. Beake, and G. J. Leggett, Friction Force Microscopy of Self-Assembled Monolayers: Influence of Adsorbate Alkyl Chain Length, Terminal Group Chemistry, and Scan Velocity, *Langmuir*, 17, 1970–1974 (2001).
7. B. Bhushan and H. Liu, Nanotribological Properties and Mechanisms of Alkylthiol and Biphenyl Thiol Self-Assembled Monolayers Studied by AFM, *Phys. Rev. B.*, 63, 245412/1–11 (2001).

8. M. Ruths, N. A. Alcantar, and J. N. Israelachvili, Boundary Friction of Aromatic Silane Self-Assembled Monolayers Measured with the Surface Forces Apparatus and Friction Force Microscopy, *J. Phys. Chem. B*, 107, 11149–11157 (2003).
9. H. I. Kim and J. E. Houston, Separating Mechanical and Chemical Contributions to Molecular-Level Friction, *J. Am. Chem. Soc.*, 122, 12045–12046 (2000).
10. A. Ulman, Self-Assembled Monolayers of 4-Mercaptobiphenyls, *Acc. Chem. Res.*, 34, 855–863 (2001).
11. J. F. Kang, A. Ulman, S. Liao, R. Jordan, G. Yang, and G.-y. Liu, Self-Assembled Rigid Monolayers of 4′-Substituted-4-Mercaptobiphenyls on Gold and Silver Surfaces, *Langmuir*, 17, 95–106 (2001).
12. E. Sabatani, J. Cohen-Boulakia, M. Bruening, and I. Rubinstein, Thioaromatic Monolayers on Gold: A New Family of Self-Assembling Monolayers, *Langmuir*, 9, 2974–2981 (1993).
13. D. M. Jaffey and R. J. Madix, Reactivity of Sulfur-Containing Molecules on Noble Metal Surfaces. 4. Benzenethiol on Au(110), *J. Am. Chem. Soc.*, 114, 3020–3027 (1994).
14. C. A. Szafranski, W. Tanner, P. E. Laibinis, and R. L. Garrell, Surface-Enhanced Raman Spectroscopy of Aromatic Thiols and Disulfides on Gold Electrodes, *Langmuir* 14, 3570–3579 (1998).
15. C. M. Whelan, C. J. Barnes, C. G. H. Walker, and N. M. D. Brown, Benzenethiol Adsorption on Au(111) Studied by Synchrotron ARUPS, HREELS and XPS, *Surf. Sci.*, 425, 195–211 (1999).
16. V. Batz, M. A. Schneeweiss, D. Kramer, H. Hagenström, D. M. Kolb, and D. Mandler, Electrochemistry and Structure of the Isomers of Aminothiophenol Adsorbed on Gold, *J. Electroanal. Chem.*, 491, 55–68 (2000).
17. L.-J. Wan, M. Terashima, H. Noda, and M. Osawa, Molecular Orientation and Ordered Structure of Benzenethiol Adsorbed on Gold(111), *J. Phys. Chem. B*, 104, 3563–3569 (2000).
18. S. Frey, V. Stadler, K. Heister, W. Eck, M. Zharnikov, M. Grunze, B. Zeysing, and A. Terfort, Structure of Thioaromatic Self-Assembled Monolayers on Gold and Silver, *Langmuir*, 17, 2408–2415 (2001).
19. T. Ishida, W. Mizutani, H. Azehara, K. Miyake, Y. Aya, S. Sasaki, and H. Tokumoto, Molecular Arrangement and Electrical Conduction of Self-Assembled Monolayers Made from Terphenyl Thiols, *Surf. Sci.*, 514, 187–193 (2002).
20. B. Kim, J. M. Beebe, Y. Jun, X.-Y. Zhu, and C. D. Frisbie, Correlation between HOMO Alignment and Contact Resistance in Molecular Junctions: Aromatic Thiols versus Aromatic Isocyanides, *J. Am. Chem. Soc.*, 128, 4970–4971 (2006).
21. A. Shaporenko, A. Terfort, M. Grunze, and M. Zharnikov, A Detailed Analysis of the Photoemission Spectra of Basic Thioaromatic Monolayers on Noble Metal Substrates, *J. Electr. Spectr. Rel. Phenom.*, 151, 45–51 (2006).
22. D. Käfer, A. Bashir, and G. Witte, Interplay of Anchoring and Ordering in Aromatic Self-Assembled Monolayers, *J. Phys. Chem. C*, 111, 10546–10551 (2007).
23. R. R. Kolega and J. B. Schlenoff, Self-Assembled Monolayers of an Aryl Thiol: Formation, Stability, and Exchange of Adsorbed 2-Naphthalenethiol and Bis (2-Naphthyl) Disulfide on Au, *Langmuir*, 14, 5469–5478 (1998).
24. V. Ganesh and V. Lakshminarayanan, Scanning Tunneling Microscopy, Fourier Transform Infrared Spectroscopy, and Electrochemical Characterization of 2-Naphthalenethiol Self-Assembled Monolayers on the Au Surface: A Study of Bridge-Mediated Electron Transfer in $Ru(NH_3)_6^{2+}|Ru(NH_3)_6^{3+}$ Redox Reactions, *J. Phys. Chem. B*, 109, 16372–16381 (2005).
25. L. Hallmann, A. Bashir, T. Strunskus, R. Adelung, V. Staemmler, Ch. Wöll, and F. Tuczek, Self-Assembled Monolayers of Benzylmercaptan and *p*-Cyanobenzylmercaptan on Au(111) Surfaces: Structural and Spectroscopic Characterization, *Langmuir*, 24, 5726–5733 (2008).

26. W. Geyer, V. Stadler, W. Eck, M. Zharnikov, A. Golzhauser, and M. Grunze, Electron-Induced Crosslinking of Aromatic Self-Assembled Monolayers: Negative Resists for Nanolithography, *Appl. Phys. Lett.*, 75, 2401–2403 (1999).

27. W. Azzam, B. I. Wehner, R. A. Fischer, A. Terfort, and Ch. Wöll, Bonding and Orientation in Self-Assembled Monolayers of Oligophenyldithiols on Au Substrates, *Langmuir*, 18, 7766–7769 (2002).

28. B. de Boer, H. Meng, D. F. Perepichka, J. Zheng, M. M. Frank, Y. J. Chabal, and Z. Bao, Synthesis and Characterization of Conjugated Mono- and Dithiol Oligomers and Characterization of Their Self-Assembled Monolayers, *Langmuir*, 19, 4272–4284 (2003).

29. S. Stoycheva, M. Himmelhaus, J. Fick, A. Korniakov, M. Grunze, and A. Ulman, Spectroscopic Characterization of ω-Substituted Biphenylthiolates on Gold and Their Use as Substrates for "On-Top" Siloxane SAM Formation, *Langmuir*, 22, 4170–4178 (2006). Erratum, *Langmuir*, 24, 2260 (2008).

30. H.-J. Himmel, A. Terfort, and Ch. Wöll, Fabrication of a Carboxyl-Terminated Organic Surface with Self-Assembly of Functionalized Terphenylthiols: The Importance of Hydrogen Bond Formation, *J. Am. Chem. Soc.*, 120, 12069–12074 (1998).

31. C. Fuxen, W. Azzam, R. Arnold, G. Witte, A. Terfort and Ch. Wöll, Structural Characterization of Organothiolate Adlayers on Gold: The Case of Rigid, Aromatic Backbones, *Langmuir*, 17, 3689–3695 (2001).

32. R. Arnold, A. Terfort, and Ch. Wöll, Determination of Molecular Orientation in Self-Assembled Monolayers Using IR Absorption Intensities: The Importance of Grinding Effects, *Langmuir*, 17, 4980–4989 (2001).

33. A. Shaporenko, M. Brunnbauer, A. Terfort, M. Grunze, and M. Zharnikov, Structural Forces in Self-Assembled Monolayers: Terphenyl-Substituted Alkanethiols on Noble Metal Substrates, *J. Phys. Chem. B*, 108, 14462–14469 (2004).

34. Y.-T. Tao, C.-C. Wu, J.-Y. Eu, W.-L. Lin, K.-C. Wu, C.-h. Chen, Structure Evolution of Aromatic-Derivatized Thiol Monolayers on Evaporated Gold, *Langmuir*, 13, 4018–4023 (1997).

35. R. W. Zehner, B. F. Parsons, R. P. Hsung, and L. R. Sita, Tuning the Work Function of Gold with Self-Assembled Monolayers Derived from $X-[C_6H_4-C\equiv C-]_nC_6H_4$-SH ($n=0$, 1, 2; $X=H$, F, CH_3, CF_3, and OCH_3), *Langmuir*, 15, 1121–1127 (1999).

36. T. Sawaguchi, F. Mizutani, S. Yoshimoto, and I. Taniguchi, Voltammetric and In Situ STM Studies on Self-Assembled Monolayers of 4-Mercaptopyridine, 2-Mercaptopyridine and Thiophenol on Au(111) Electrodes, *Electrochim. Acta*, 45, 2861–2867 (2000).

37. T. Baunach and D. M. Kolb, The Electrochemical Characterization of Benzyl Mercaptan-Modified Au(111): Structure and Copper Deposition, *Anal. Bioanal. Chem.*, 373, 743–748 (2002).

38. S. Howell, D. Kuila, B. Kasibhatla, C. P. Kubiak, D. Janes, and R. Reifenberger, Molecular Electrostatics of Conjugated Self-Assembled Monolayers on Au(111) Using Electrostatic Force Microscopy, *Langmuir*, 18, 5120–5125 (2002).

39. K. Bandyopadhyay, V. Patil, M. Sastry, and K. Vijayamohanan, Effects of Geometric Constraints on the Self-Assembled Monolayer Formation of Aromatic Disulfides on Polycrystalline Gold, *Langmuir*, 14, 3803–3814 (1998).

40. A.-A. Dhirani, R. W. Zehner, R. P. Hsung, P. Guyot-Sionnest, and L. R. Sita, Self-Assembly of Conjugated Molecular Rods: A High-Resolution STM Study, *J. Am. Chem. Soc.*, 118, 3319–3320 (1996).

41. H. McNally, D. B. Janes, B. Kasibhatla, and C. P. Kubiak, Electrostatic Investigation into the Bonding of Poly(phenylene) Thiols to Gold, *Superlattices Microst.*, 31, 239–245 (2002).

42. D. Barriet, C. M. Yam, O. E. Shmakova, A. C. Jamison, and T. R. Lee, 4-Mercaptophenylboronic Acid SAMs on Gold: Comparison with SAMs Derived from Thiophenol, 4-Mercaptophenol, and 4-Mercaptobenzoic Acid, *Langmuir*, 23, 8866–8875 (2007).
43. J. M. Tour, L. Jones, II, D. L. Pearson, J. J. S. Lamba, T. P. Burgin, G. M. Whitesides, D. L. Allara, A. N. Parikh, and S. V. Atre, Self-Assembled Monolayers and Multilayers of Conjugated Thiols, α,ω-Dithiols, and Thioacetyl-Containing Adsorbates. Understanding Attachments between Potential Molecular Wires and Gold Surfaces, *J. Am. Chem. Soc.*, 117, 9529–9534 (1995).
44. P. Jiang, A. Nion, A. Marchenko, L. Piot, and D. Fichou, Rotational Polymorphism in 2-Naphthalenethiol SAMs on Au(111), *J. Am. Chem. Soc.*, 128, 12390–12391 (2006).
45. H. H. Jung, Y. D. Won, S. Shin, and K. Kim, Molecular Dynamics Simulation of Benzenethiolate and Benzyl Mercaptide on Au(111), *Langmuir*, 15, 1147–1154 (1999).
46. J. Nara, S. Higai, Y. Morikawa, and T. Ohno, Density Functional Theory Investigation of Benzenethiol Adsorption on Au(111), *J. Chem. Phys.*, 120, 6705–6711 (2004).
47. A. Bilić, J. R. Reimers, and N. S. Hush, The Structure, Energetics, and Nature of the Chemical Bonding of Phenylthiol Adsorbed on the Au(111) Surface: Implications for Density Functional Calculations of Molecular-Electronic Conduction, *J. Chem. Phys.*, 122, 094708/1–15 (2005).
48. T. E. Dirama and J. A. Johnson, Conformation and Dynamics of Arylthiol Self-Assembled Monolayers on Au(111), *Langmuir*, 23, 12208–12216 (2007).
49. E. S. Forbes, The Load-Carrying Action of Organosulfur Compounds—A Review, *Wear*, 15, 87–96 (1970).
50. D. F. Heenan, K. R. Januszkiewicz, and H. H. Sulek, Wear and Friction Characteristics of Aromatic Thiocompounds During Aluminum-on-Steel Sliding Contact, *Wear*, 123, 257–268 (1988).
51. V. K. Verma, R. Singh, A. Bhattacharya, and A. K. Tripathi, Cyclic Alkyl Disulfides as Tribological Additives, *Lubr. Sci.*, 13, 37–44 (2000).
52. C. A. Migdal, Antioxidants, In: *Lubricant Additives: Chemistry and Applications*, L. R. Rudnick, ed., pp. 1–27, Marcel Dekker, New York (2003).
53. E. A. Bardasz and G. D. Lamb, Additives for Crankcase Lubricant Applications, In: *Lubricant Additives: Chemistry and Applications*, L. R. Rudnick, ed., pp. 387–428, Marcel Dekker, New York (2003).
54. D. Wei and H. A. Spikes, The Lubricity of Diesel Fuels, *Wear*, 111, 217–235 (1986), and references therein.
55. E. Kenesey and A. Ecker, How Substituents of Heteroaromatics Influence the Lubricity of Diesel Fuel, In: *Tribology Science and Application*. M. A. Herman, F. Franek, and C. Kajdas, eds., pp. 378–394, Scientific Centre of the Polish Academy of Sciences (PAN), Vienna, Austria (2004).
56. A. M. Kuliev, Synthesis of Lubricant Additives Based on Alkylphenols, Alkylthiophenols, Sulphonates and Other Organic Multifunctional Compounds, In: *Proceedings of the 7th World Petroleum Congress*, Vol. 8, pp. 111–122, Elsevier, Barking, England (1968).
57. D. S. Connor, Fuel for Jet, Gas Turbine, Rocket and Diesel Engines, U.S. Patent No. 7,560,603.
58. R. L. McCormick, J. R. Alvarez, M. S. Graboski, K. S. Tyson, K. Vertin, 1. Fuel Additive and Blending Approaches to Reducing NOx Emissions from Biodiesel, In: *#PT-111: Alternative Diesel Fuels*, D. J. Holt, ed., pp. 51–60, Society of Automotive Engineers, Warrendale, PA (2004).
59. B. A. Baldwin, Relationship between Surface Composition and Wear: An X-Ray Photoelectron Spectroscopic Study of Surfaces Tested with Organosulfur Compounds, *Tribology Trans.*, 19, 335–344 (1976).

60. E. A. Bardasz, F. A. Antoon, E. A. Schieferl, J. C. Wang, and W. Totten, The Impact of Lubricant and Fuel Derived Sulfur Species on Efficiency and Durability of Diesel NO$_x$ Adsorbers, In: *#SP-1894: Oils, Rheology, Tribology, and Driveline*, pp. 151–158, Society of Automotive Engineers, Warrendale, PA (2004).

61. H. B. Krop, *Health and environmental hazards of commonly used additives in lubricants*, IVAM–Chemiewinkel, Consultancy and Research Centre on Chemistry, Work and Environment, University of Amsterdam (2002).

62. H. Liu, B. Bhushan, W. Eck, and V. Stadler, Investigation of the Adhesion, Friction, and Wear Properties of Biphenyl Thiol Self-Assembled Monolayers by Atomic Force Microscopy, *J. Vac. Sci. Technol. A*, 19, 1234–1240 (2001).

63. H. Liu and B. Bhushan, Orientation and Relocation of Biphenyl Thiol Self-Assembled Monolayers under Sliding, *Ultramicroscopy*, 91, 177–183 (2002).

64. H. Liu, B. Bhushan, W. Eck, and A. Kueller, Investigation of Nanotribological Properties of Self-Assembled Monolayers with Alkyl and Biphenyl Spacer Chains (Invited), *Ultramicroscopy*, 91, 185–202 (2002). Erratum, ibid., 96, 125 (2003).

65. M. Ruths, Boundary Friction of Aromatic Self-Assembled Monolayers: Comparison of Systems with One or Both Sliding Surfaces Covered with a Thiol Monolayer, *Langmuir*, 19, 6788–6795 (2003).

66. M. Nakano, T. Ishida, T. Numata, Y. Ando, and S. Sasaki, Tribological Behavior of Terphenyl Self-Assembled Monolayer Studied by a Pin-on-Plate Method and Friction Force Microscopy, *Jpn. J. Appl. Phys.*, 43, 4619–4623 (2004).

67. M. Ruths, Friction of Mixed and Single-Component Aromatic Monolayers in Contacts of Different Adhesive Strength, *J. Phys. Chem. B*, 110, 2209–2218 (2006).

68. Y. Yang and M. Ruths, Friction of Polyaromatic Thiol Monolayers in Adhesive and Nonadhesive Contacts, *Langmuir*, 25, 12151–12159 (2009).

69. Y. Yang, A. C. Jamison, D. Barriet, T. R. Lee, and M. Ruths, Odd-Even Effects in the Friction of Self-Assembled Monolayers of Phenyl-Terminated Alkanethiols in Contacts of Different Adhesive Strength, *J. Adhes. Sci. Tech.*, 24, 2511–2529 (2010).

70. W. A. Hayes, H. Kim, X. Yue, S. S. Perry, and C. Shannon, Nanometer-Scale Patterning of Surfaces Using Self-Assembly Chemistry. 2. Preparation, Characterization, and Electrochemical Behavior of Two-Component Organothiol Monolayers on Gold Surfaces, *Langmuir*, 13, 2511–2518 (1997).

71. S. Nagata, Y. Tomoda, and H. Haga, Lubrication by Aromatic Compounds. III. Wear Determination by a Radioactive Tracer Method, *Techn. Rep. Osaka Univ.*, 6, 31–42 (1956).

72. R. W. Carpick, N. Agraït, D. F. Ogletree, and M. Salmeron, Variation of the Interfacial Shear Strength and Adhesion of a Nanometer-Sized Contact, *Langmuir*, 12, 3334–3340 (1996).

73. R. W. Carpick and M. Salmeron, Scratching the Surface: Fundamental Investigations of Tribology with Atomic Force Microscopy, *Chem. Rev.*, 97, 1163–1194 (1997).

74. A. Berman, C. Drummond, and J. Israelachvili, Amontons' Law at the Molecular Level, *Tribology Lett.*, 4, 95–101 (1998).

75. J. N. Israelachvili and A. D. Berman, Surface Forces and Microrheology in Molecularly Thin Liquid Films, In: *Handbook of Micro/Nanotribology*, 2nd edn., B. Bhushan, ed., pp. 371–432, CRC Press, Boca Raton, FL (1999).

76. J. Gao, W. D. Luedtke, D. Gourdon, M. Ruths, J. N. Israelachvili, and U. Landman, Frictional Forces and Amontons' Law: From the Molecular to the Macroscopic Scale, *J. Phys. Chem. B*, 108, 3410–3425 (2004).

77. C. R. Hurley and G. J. Leggett, Influence of the Solvent Environment on the Contact Mechanics of Tip–Sample Interactions in Friction Force Microscopy of Poly(Ethylene Terephthalate) Films, *Langmuir*, 22, 4179–4183 (2006).

78. T. J. Colburn and G. J. Leggett, Influence of Solvent Environment and Tip Chemistry on the Contact Mechanics of Tip–Sample Interactions in Friction Force Microscopy of Self-Assembled Monolayers of Mercaptoundecanoic Acid and Dodecanethiol, *Langmuir*, 23, 4959–4964 (2007).

79. M. Ruths and J. N. Israelachvili, Surface Forces and Nanorheology of Molecularly Thin Films, In: *Springer Handbook of Nanotechnology*, 3rd edn., B. Bhushan (ed.), Chapter 29, pp. 857–922, Springer-Verlag, Berlin/Heidelberg (2010).

80. M. Ruths, S. Lundgren, K. Danerlöv, and K. Persson, Friction of Fatty Acids in Nanometer-Sized Contacts of Different Adhesive Strength, *Langmuir*, 24, 1509–1516 (2008).

81. Y. Mo, K. T. Turner, and I. Szlufarska, Friction Laws at the Nanoscale, *Nature*, 457, 1116–1119 (2009).

82. E. D. Reedy, Jr., Thin-Coating Contact Mechanics with Adhesion, *J. Mater. Res.*, 21, 2660–2668 (2006).

83. E. D. Reedy, Jr., Contact Mechanics for Coated Spheres That Includes the Transition from Weak to Strong Adhesion, *J. Mater. Res.*, 22, 2617–2622 (2007).

84. S. Granick, Soft Matter in a Tight Spot, *Phys. Today*, 52, 26–31 (1999).

85. H. Yoshizawa, Y.-L. Chen, and J. Israelachvili, Fundamental Mechanisms of Interfacial Friction. 1. Relation between Adhesion and Friction, *J. Phys. Chem.*, 97, 4128–4140 (1993).

86. D. Valtakari, Master's Thesis, Framställning av maleimidterminerade självorganiserade alkantiolatfilmer på guld, Department of Physical Chemistry, Åbo Akademi University, Finland (2002).

87. S. Liao, Y. Shnidman, and A. Ulman, Adsorption Kinetics of Rigid 4-Mercaptobiphenyls on Gold, *J. Am. Chem. Soc.*, 122, 3688–3694 (2000).

88. J. N. Israelachvili, *Intermolecular and Surface Forces*, 2nd edn., Academic Press, London U.K. (1991).

89. A. W. Adamson and A. P. Gast, *Physical Chemistry of Surfaces*, 6th edn., Wiley, New York (1997).

90. H. Y. Erbil, Surface Tension of Polymers, In: *Handbook of Surface and Colloid Chemistry*, K. S. Birdi (ed.), Chapter 9, pp. 265–312, CRC Press, Boca Raton, FL (1997).

91. H. W. Fox, E. F. Hare, and W. A. Zisman, The Spreading of Liquids on Low-Energy Surfaces. VI. Branched-Chain Monolayers, Aromatic Surfaces, and Thin Liquid Films, *J. Colloid Sci.*, 8, 194–203 (1953).

92. J. Visser, Hamaker Constants. Comparison between Hamaker Constants and Lifshitz–van der Waals constants, *Adv. Colloid Interface Sci.*, 3, 331–363 (1972).

93. S. Löning, C. Horst, and U. Hoffmann, Dewatering of Solvent Mixtures—Comparison between Conventional Technologies and a New One, *Chem. Eng. Technol.*, 24, 242–245 (2001).

94. J. E. Sader, J. W. M. Chon, and P. Mulvaney, Calibration of Rectangular Atomic Force Microscope Cantilevers, *Rev. Sci. Instrum.*, 70, 3967–3969 (1999).

95. C. P. Green, H. Lioe, J. P. Cleveland, R. Proksch, P. Mulvaney, and J. E. Sader, Normal and Torsional Spring Constants of Atomic Force Microscope Cantilevers, *Rev. Sci. Instrum.*, 75, 1988–1996 (2004).

96. Y. Liu, T. Wu, and D. F. Evans, Lateral Force Microscopy Study on the Shear Properties of Self-Assembled Monolayers of Dialkylammonium Surfactant on Mica, *Langmuir*, 10, 2241–2245 (1994).

97. Y. Liu, D. F. Evans, Q. Song, and D. W. Grainger, Structure and Frictional Properties of Self-Assembled Surfactant Monolayers, *Langmuir*, 12, 1235–1244 (1996).

98. K. L. Johnson, K. Kendall, and A. D. Roberts, Surface Energy and the Contact of Elastic Solids, *Proc. R. Soc. London A*, 324, 301–313 (1971).

99. B. V. Derjaguin, V. M. Muller, and Yu. P. Toporov, Effect of Contact Deformations on the Adhesion of Particles, *J. Colloid Interface Sci.*, 53, 314–326 (1975).

100. R. W. Carpick, D. F. Ogletree, and M. Salmeron, A General Equation for Fitting Contact Area and Friction vs Load Measurements, *J. Colloid Interface Sci.*, 211, 395–400 (1999).

101. D. S. Grierson, E. E. Flater, and R. W. Carpick, Accounting for the JKR–DMT Transition in Adhesion and Friction Measurements with Atomic Force Microscopy, *J. Adhesion Sci. Technol.*, 19, 291–311 (2005).

102. I. Sridhar, K. L. Johnson, and N. A. Fleck, Adhesion Mechanics of the Surface Forces Apparatus, *J. Phys. D: Appl. Phys.*, 30, 1710–1719 (1997); K. L. Johnson and I. Sridhar, Adhesion between a Spherical Indenter and an Elastic Solid with a Compliant Elastic Coating, *J. Phys. D: Appl. Phys.*, 34, 683–689 (2001); I. Sridhar, Z. W. Zheng, and K. L. Johnson, A Detailed Analysis of Adhesion Mechanics between a Compliant Elastic Coating and a Spherical Probe, *J. Phys. D: Appl. Phys.*, 37, 2886–2895 (2004).

103. B. Luan and M. O. Robbins, The Breakdown of Continuum Models for Mechanical Contacts, *Nature*, 435, 929–932 (2005).

104. S. Cheng, B. Luan, and M. O. Robbins, Contact and Friction of Nanoasperities: Effects of Adsorbed Monolayers, *Phys. Rev. E*, 81, 016102/1–17 (2010).

105. E. D. Reedy, Jr., M. J. Starr, R. E. Jones, E. E. Flater, and R. W. Carpick, Contact Modeling of SAM-Coated Polysilicon Asperities, In: *Proceedings of the 28th Annual Meeting of the Adhesion Society*, pp. 366–368, The Adhesion Society, Blacksburg, VA (2005).

106. M. Chandross, C. D. Lorenz, M. J. Stevens, and G. S. Grest, Simulations of Nanotribology with Realistic Probe Tip Models, *Langmuir*, 24, 1240–1246 (2008).

107. D. Maugis, *Contact, Adhesion and Rupture of Elastic Solids*. Springer, New York (1999).

108. G.-y. Liu and M. B. Salmeron, Reversible Displacement of Chemisorbed *n*-Alkanethiol Molecules on Au(111) Surface: An Atomic Force Microscopy Study, *Langmuir*, 10, 367–370 (1994).

109. W. J. Miller and N. L. Abbott, Influence of van der Waals Forces from Metallic Substrates on Fluids Supported on Self-Assembled Monolayers Formed from Alkanethiols, *Langmuir*, 13, 7106–7114 (1997).

110. J. F. Kang, A. Ulman, R. Jordan, and D. G. Kurth, Optically Induced Band Shifts in Infrared Spectra of Mixed Self-Assembled Monolayers of Biphenyl Thiols, *Langmuir*, 15, 5555–5559 (1999).

111. D. V. Vezenov, A. Noy, and C. M. Lieber, The Effect of Liquid-Induced Adhesion Changes on the Interfacial Shear Strength between Self-Assembled Monolayers, *J. Adhes. Sci. Tech.*, 17, 1385–1401 (2003).

112. A. Lio, C. Morant, D. F. Ogletree, and M. Salmeron, Atomic Force Microscopy Study of the Pressure-Dependent Structural and Frictional Properties of *n*-Alkanethiols on Gold, *J. Phys. Chem. B*, 101, 4767–4773 (1997).

113. R. M. Overney, E. Meyer, J. Frommer, H.-J. Güntherodt, M. Fujihira, H. Takano, and Y. Gotoh, Force Microscopy Study of Friction and Elastic Compliance of Phase-Separated Organic Thin Films, *Langmuir*, 10, 1281–1286 (1994).

114. W. Kiridena, V. Jain, P. K. Kuo, and G.-y. Liu, Nanometer-Scale Elasticity Measurements on Organic Monolayers Using Scanning Force Microscopy, *Surf. Interface Anal.*, 25, 383–389 (1997).

115. V. V. Tsukruk, V. N. Bliznyuk, J. Hazel, D. Visser, and M. P. Everson, Organic Molecular Films under Shear Forces: Fluid and Solid Langmuir Monolayers, *Langmuir*, 12, 4840–4849 (1996).

116. Y. L. Chen, C. A. Helm, and J. N. Israelachvili, Measurements of the Elastic Properties of Surfactant and Lipid Monolayers, *Langmuir*, 7, 2694–2699 (1991).

117. M. Salmeron, G. Neubauer, A. Folch, M. Tomitori, D. F. Ogletree, and P. Sautet, Viscoelastic and Electrical Properties of Self-Assembled Monolayers on Au(111) Films, *Langmuir*, 9, 3600–3611 (1993).

118. R. Henda, M. Grunze, and A. J. Pertsin, Static Energy Calculations of Stress–Strain Behavior of Self-Assembled Monolayers, *Tribology Lett.*, 5, 191–195 (1998).

119. A. R. Burns, J. E. Houston, R. W. Carpick, and T. A. Michalske, Molecular Level Friction As Revealed with a Novel Scanning Probe, *Langmuir*, 15, 2922–2930 (1999).

120. R. C. Major, H. I. Kim, J. E. Houston, and X.-Y. Zhu, Tribological Properties of Alkoxy Monolayers on Oxide Terminated Silicon, *Tribology Lett.*, 14, 237–244 (2003).

121. Y. Leng and S. Jiang, Atomic Indentation and Friction of Self-Assembled Monolayers by Hybrid Molecular Simulations, *J. Chem. Phys.*, 113, 8800–8806 (2000).

122. M. Chandross, personal communication (2008).

123. A. Noy, C. D. Frisbie, L. F. Rozsnyai, M. S. Wrighton, and C. M. Lieber, Chemical Force Microscopy: Exploiting Chemically-Modified Tips to Quantify Adhesion, Friction, and Functional Group Distributions in Molecular Assemblies, *J. Am. Chem. Soc.*, 117, 7943–7951 (1995).

124. J. D. Kiely, J. E. Houston, J. A. Mulder, R. P. Hsung, and X.-Y. Zhu, Adhesion, Deformation and Friction for Self-Assembled Monolayers on Au and Si Surfaces, *Tribology Lett.*, 7, 103–107 (1999).

125. S. C. Clear and P. F. Nealey, Lateral Force Microscopy Study of the Frictional Behavior of Self-Assembled Monolayers of Octadecyltrichlorosilane on Silicon/Silicon Dioxide Immersed in *n*-Alcohols, *Langmuir*, 17, 720–732 (2001).

126. J. A. Harrison, G. T. Gao, R. J. Harrison, G. M. Chateauneuf, and P. T. Mikulski, Friction of Model Self-Assembled Monolayers, In: *Encyclopedia of Nanoscience and Nanotechnology*, Vol. 3, H. S. Nalwa, ed., pp. 511–527, American Scientific Publishers, Stevenson Ranch, CA (2004).

127. W. Azzam, Self-Assembled Monolayers on Gold Made from Organothiols Containing an Oligophenyl-backbone, PhD thesis, p. 108, Ruhr-University Bochum, Germany (2003).

3 Friction of Graphite against Silane-Functionalized Silicon Wafers

Marjorie Schmitt, Sophie Bistac, and Khalil Jradi

CONTENTS

ABSTRACT

The friction and wear properties of graphites are directly linked to the
environment in which they are used: a low friction is observed in the presence of oxygen and mainly in water vapor, whereas under vacuum, the friction coefficient greatly increases. The surface chemistry is then of prime
importance for the friction and wear properties of these materials. One way
to modify surface chemistry is by chemical grafting of surfactant (silane)
molecules.

In order to thoroughly study the role of the surface chemistry, graphite powders of various natures were tested in friction against three counterfaces, each
one with a specific surface chemistry. The first one was an untreated silicon
wafer, which will be the reference for the tests. A chemical modification of
the bulk silicon wafers by grafting silane molecules leads to counterfaces that
are hydrophilic or hydrophobic. These three counterfaces were subjected to
the same experimental conditions on a pin-on-disk tribometer, in the ambient,
with various normal loads.

Observations and analyses realized after the friction indicated that both
the nature of the graphite and the value of the applied load induced modifications in the morphology of the transferred materials. Moreover, the crucial
role of the surface chemistry was clearly shown: an increase of the surface
energy led to a decrease of the friction coefficient of the graphite/silicon
couples. Explanations based on the role of water in graphite friction are
proposed.

3.1 BACKGROUND AND INTRODUCTION

Micro-electromechanical systems (MEMS) were rapidly developed during the last
decade because of their low unit cost and excellent performance [1,2–4], for instance
for applications in areas such as biomedical, communications, mechatronics [1,2],
and magnetic storage devices [5,6]. But the large surface area-to-volume ratio that
induced serious adhesive and frictional problems remained an important drawback
for MEMS [7]. Moreover, the small size of the MEMS components and those of the

gaps between them [8] prohibited the use of both conventional liquid [9] and solid lubricants based on lamellar materials [10].

Recently, self-assembled monolayers (SAMs) were intended to be used as molecular lubricant to solve the previous problems [11]. Their high durability is ensured by both the strong adhesion of the lubricant molecules to the substrate and by the reformation ability of the mobile molecules in the lubricant-depleted zone [12–15].

Langmuir–Blodgett (LB) films were first used to modify the surface [9,16], but they presented a significant disadvantage compared with SAMs: LB films are fixed to the substrate via weak van der Waals and ionic forces whereas SAMs are the result of an adsorption onto the substrate through high-affinity chemisorption [17]. The repeated shearing of LB films induces wear [7] whereas SAMs are more robust; this is why they were most frequently selected as lubricant.

SAMs appeared then to be an ideal candidate for a boundary lubricant [8,9] as their properties in rupture are good thanks to their strong bonding (chemisorption) to the substrate [18,19], which means there will be no transfer from one solid surface to the other, and they are not expected to migrate to the surface [20]. But, the SAMs also present an important drawback: their load-carrying capacity and durability are poor because of their low molecular flexibility and mobility [21–23], and their friction remains low [24,25].

Studies dealing with the general behavior of SAMs have already been carried out and have shown that the tribological properties of SAMs depend on their chemical and physical properties [2,8,9].

Some general trends are given below: it was shown that SAMs with an alkyl chain had a lower coefficient of friction than SAMs with a fluorinated alkyl chain [16]. Moreover, SAMs with longer chains lead to friction lower than the one observed for SAMs with shorter chains [17]. In the case of hydrophilic/hydrophilic contact (among different terminal groups of SAMs), the friction coefficient is high [7]. The introduction of large-size atoms (like fluorine) in mixed monolayers with different chemical functionalities reduces the monolayer order and consequently, increases the friction [26,27]. Concerning the spiroalkanedithiols, when a chain is shortened, the disorder is increased, and consequently the friction too [28,29]. When studying alkanethiols, it was shown that islands made of chains of different lengths induce higher friction (than the one observed for islands constituted of one single length chain), for a given applied normal load [30].

The tribological properties of SAMs on a macroscopic scale will be more particularly detailed first; then works concerning the tribology of these materials at both nano- and macroscopic scales will be addressed. And finally, although this is not the main subject of this chapter, some results dealing with the nano-tribology of SAMs will be presented.

3.1.1 MACRO-TRIBOLOGY

Studies concerning the friction properties of SAMs derived from alkylsilanes were already carried out [15,31,32], but it became essential to find better lubricants for MEMS, which means molecular lubricants that are able to improve the durability and friction properties.

The load-carrying capacity and durability of SAMs on silicon wafers was studied as a function of the multiply alkylated cyclopentane (MAC), a new mobile hydrocarbon lubricant [33]; 1H, 1H, 2H, 2H-perfluorodecyltrichlorosilane SAM (FDTS-SAM) and DTS-SAM were selected to form two double-layer films by spin coating method. They respectively led to MAC-FDTS and MAC-DTS coatings that consist of a nonbonded lubricant (MAC) and of a strongly bonded phase lubricant (SAM). The tribological properties of FDTS-SAM, DTS-SAM, MAC-FDTS, and MAC-DTS deposited on silicon wafers were evaluated on a pin-on-plate tribometer, a steel ball playing the role of the counterface. It was observed that DTS and FDTS SAMs greatly reduced the friction coefficient of the silicon substrate; but their durability and load-carrying capacity remained poor. The deposition of the mobile hydrocarbon lubricant, MAC, on the SAMs to form double-layer films led to a marked reduction of the friction of the DTS-SAM as well as to greatly improved load-carrying capacity and durability of both DTS and FDTS SAMs.

The fact that DTS-SAM has lower values of friction, and higher durability and load-carrying capacity than FDTS-SAM (despite FDTS shows a lower surface energy [34]), in these experimental conditions, is in good agreement with results presented in Ref. [27]: the observed difference in friction may be due to the different sizes of the terminated groups, namely CH_3 and CF_3. CF_3 termination of FDTS induces a more densely packed arrangement than the CH_3 (end group of DTS) since the two SAMs have the same lattice spacing even though CF_3 molecule is larger than CH_3. Another origin for this difference in friction could be attributed to the fact that FDTS coating presents a less uniformity as well as a higher surface roughness compared to the DTS film [35].

In the end, the deposition of MAC on DTS- and FDTS-SAMs greatly decreased the friction of DTS-SAM, and significantly increased the durability and the load-carrying capacity of both films. Better durability and load-carrying capacity can be attributed to the ability of the mobile layer to diffuse and cover the damages and defects of the substrates. The MAC lubricant was indeed trapped into the SAM network when the hydrocarbon molecules were displaced by the motion of the steel ball; the SAM network plays the role of a barrier against displacement of the mobile hydrocarbon molecules. Even though these mobile molecules were mechanically displaced or disrupted during the friction, they could reorganize themselves into the original state or replenish to the lubricant-depleted area due to their high mobility: this is what is called "self-repairing property."

The MAC mobile lubricant was also used on a octadecyltrichlorosilane SAM (OTS-SAM) to prepare a double-layer lubricant film called MAC-OTS. Its tribological properties were also investigated and compared with those of PSPE ($HOCH_2$-CF_2O-[CF_2-CF_2O]$_m$-[CF_2O]$_n$-CF_2-CH_2OH called Zdol-2000) and OTS-SAM [36]. It appeared that under 0.5 N, these three films presented the same friction coefficient, but at 1 N, MAC-OTS and Zdol-2000 showed lower friction coefficient while OTS-SAM presented poor friction properties in the same conditions. The friction-reducing effect in the case of OTS-SAM was only observed at the low load of 0.5 N.

The durability of Zdol-2000 remained always better than the one of OTS-SAM. But, for applied normal loads varying from 0.5 to 3 N, the MAC-OTS double-layer film showed the best durability and load-carrying capacity—this is the reason why this film is intended to be used for lubricating and protecting MEMS at high load.

It was observed that the initial friction coefficient of OTS-SAM was slightly lower than the one of Zdol-2000; this can be linked to the fact that the SAMs are ordered molecular assemblies in which one end of the long chain molecules is strongly bonded to the substrate surface [18]. The low-surface energy could also be at the origin of the difference in friction.

Concerning the poor durability of SAMs under repeated sliding contacts at high load, one must remember that these films behave like solid, and thus have mechanical strength to bear load. But, if the applied load is higher than the threshold, these films could be removed from the surface because of high plastic deformation, which leads to the failure of the films [21,22,37–39].

The role of the length of the alkyl chain in the SAMs, and the effect of their number of expected siloxane bonds on the tribological behavior of these films against a glass ball were evaluated in Ref. [40]; the effect of the number of expected siloxane bonds was first studied through a comparison of the friction properties of OTS, methyloctadecyldichlorosilane (ODS), and dimethyloctadecylmonochlorosilane (OMS) that present, respectively, three, two, and one chlorine atoms in the molecule. OTS and ODS lead to the formation of densely packed and oriented monolayers, which is not the case for OMS. The stable coefficient of friction of OTS and ODS is low, and their durability is quite good compared with those of OMS.

The effect of the length of the alkyl chain was studied by comparing OTS, tetradecyltrichlorosilane (TTS), DTS, and hexadecyltrichlorosilane (HTS): the molecular orientation is improved and the surface density of adsorbed molecules becomes more important when the length of the chain increases. In soft friction conditions, no significant difference was observed in the tribological properties of OTS, TTS, and DTS, but a higher coefficient of friction was noticed in the case of HTS-SAM. Important variations were, however, observed under severe friction.

Although the ability of OTS–SAM to diminish stiction was already used in MEMS to reduce energy loss in micro-motors [41], their tribological properties were not clearly established. This is why the friction (against a steel ball) and stiction of OTS deposited on silicon were studied by Hwan et al. [42]. This work showed that the tribological behavior of these films clearly depended on the OTS coating time: for coating times higher than 1 h, and despite only a partial coverage, the friction coefficient was greatly reduced; moreover, there were no more wear particles along the wear tracks in these conditions. The stiction of the silicon coated with OTS (lower than 0.3) is less sensitive to humidity, applied load, apparent contact area, and soaking time than bare silicon (as high as 1); an origin of this phenomenon could be the hydrophobic nature and low surface energy of OTS films. This means that OTS films deposited in the conditions described in this study can be successfully used as an effective reducer of the stiction and friction in micro-systems applications.

As it was important to better know the failure mode of MEMS, Ding et al. have studied the tribological properties of OTS-SAM sliding against a sapphire ball [43]; they also proved that these films showed coefficient of friction lower than that of uncoated silicon, as well as good antistiction properties. The stable friction of OTS-SAM makes it an ideal candidate to prevent stick-slip motion in MEMS.

Moreover, it was shown that the use of OTS-SAM induced an increase of the average fracture strength of the polysilicon films, up to 32%; the fracture strength could be related here again to the hydrophobic properties of the substrates coated with OTS.

Antiwear tests were also carried out and it appeared that bare hydrophilic polysilicon substrates adsorbed an inorganic contamination layer that would change the surface into a hydrophobic one and thus decreased the friction. This contaminant layer is physically adsorbed, so it can be easily worn. On the contrary, OTS-SAM is chemically combined with the substrate, which leads to a more durable surface.

3-Mercaptopropyltrimethoxysilane (MPTS) was deposited on silicon, and its tribological properties were studied [44]; the $-SH$ terminated group can be completely oxidized into sulfonate group $(-SO_3H)$ in 30% nitric acid solution. For deposition time shorter than 1 h, the friction coefficient was as high as 0.7, but for longer coating times, the friction coefficient was in the range of 0.1–0.2. It was also noticed that in the case of SAMs with $-SO_3H$ functional group, the coefficient of friction is higher than the one observed for the mixed SAM: SAMs having $-SH$ group present the lowest friction coefficient of all samples. This was attributed to the low surface energy and hydrophobic properties of MPTS films.

An increase of the applied normal load as well as a rise of the sliding speed lead to a decrease of the coefficient of friction. The self-assembled molecule that is chemically adsorbed on the substrate can be compared with an assembled molecular spring anchored to the substrate. In that case, when a tip slides on the surface of a SAM, it is as if the tip moved on the top of "molecular springs or brushes" [45]. The orientation of these "molecular springs or brushes" under the application of a normal load consequently reduces the shearing force at the interface, which finally leads to a decrease of the friction.

The better wear resistance and ability to reduce the friction of MPTS-SAMs can be linked to their viscoelastic properties and also to the orientation of the molecules adsorbed on the substrate.

Compared with an OTS-SAM, the friction coefficient of MPTS-SAM is higher, but remains lower than that of a bare silicon substrate [19,46]. The lower coefficient of friction of the MPTS can be explained by the fact that $-CH_3$ functional group has a lower friction than other functional groups [22,47]. Moreover, another origin of the difference in friction of MPTS films compared with OTS ones may be the difference in surface roughness and uniformity [19].

3-Aminopropyltriethoxysilane (APTES) films were deposited on hydroxylated silicon substrate by self-assembling process [48]; the sliding of these APTES-SAM films against a steel ball leads to a very low friction (about 0.2) and to a relatively long wear life. When the coating time is longer than 3 h, the friction coefficient is in the range of 0.177–0.2, whereas for shorter duration, the friction coefficient is as high as 0.8. Moreover, the coefficient of friction increases with both the normal load and the sliding speed. The wear of APTES-SAM is characterized by low abrasion and microcracks (whereas it is rather severe abrasion and brittle fracture for bare silicon). It was thought that the better friction and wear properties of the APTES-SAM were due to the orientation of the molecules adsorbed on the substrate and also to the good adhesion of these molecules to it.

A new kind of lubrication layer was developed by Luzinov et al. [49], by functionalizing triblock copolymer, poly[styrene-*b*-(ethylene-*co*-butylene)-*b*-styrene] (SEBS). This polymer is chemically anchored to the silicon oxide surface thanks to the interfacial epoxy-terminated monolayers. Its tribological properties were found to be better than those of other molecular coatings and SAMs: its friction coefficient is low, as well as its stiction; its wear stability is quite good compared with that of other materials.

3.1.2 TRIBOLOGY AT BOTH NANO- AND MACROSCOPIC SCALES

The tribological properties of one-component hexyltrichlorosilane (HTS), DTS, OTS SAMs, and of their two-component mixtures were studied at micro- and nano-scales [50]. At both scales, the pure SAMs and their mixtures were shown to outperform the bare silicon wafer; moreover, it was shown that they were effective lubricants. Concerning the monolayer systems, the nano- and micro-tribological properties were shown to greatly depend on the chain length; the smaller number of chains found far from the mixtures seems to have an essential role in the improvement of their nano-tribological properties.

As the number of atoms is more important in the OTS chains in the outermost region, the (OTS+DTS) mixture has lower nano-adhesion force and nanofriction than those of (OTS+HTS) mixture. The microfriction properties of the mixed monolayers are in between those of their individual components; this can be due to the different role of the long- and short-chain components that are in contact with the counterface. In the case of uncoated silicon wafer, the higher interfacial energy induces a higher friction force (because of the intrinsic adhesion force and contact area) [51]. At nanoscale, friction occurs in conditions where the contribution of the intrinsic adhesion outweighs the one of asperities deformation [52]. Various attractive forces (capillary, electrostatic, van der Waals, etc.) can increase the contribution of this intrinsic adhesion force; amongst all the previously mentioned forces, the strongest is the capillary one, due to the condensation of the water vapor from the environment [8]. This capillary force dominates the inherent adhesion in the case of the bare silicon wafer (because of its hydrophilic character), which leads to high friction [53]. This is also the reason why bare silicon wafers show high adhesion [52]. On the contrary, if SAMs have a hydrophobic character, the effect of the capillary force is not important anymore. In these conditions, as the capillary force does not dominate the adhesion force, the van der Waals forces would play an important role [54].

Various models can be found in the literature to explain the tribological behavior of SAMs. One possibility is to consider a pure SAM as "molecular spring assembly" [55]; but this model cannot be applied to mixed SAMs because of the absence of the atoms from the neighboring species in the outer layer. Another model based on the number of atomic contacts within an area of contact was also proposed [28,29]: here, the loosely packed SAMs can strongly interact with an AFM tip, as the van der Waals interactions increased, compared with the well-packed SAMs. In these conditions, there is an increase of the shear force per unit area, that can lead to a rise of the friction force. But here again, this is only available for pure SAMs and not in the case of mixed ones.

The low friction values of the mixed SAMs can be attributed to the smaller number of chains in the outer region; there is consequently less resistance to the motion of the AFM tip (lower shear strength) [56]. As long as the AFM tip moves on the outer layer of the mixtures, a contact between this tip and the atoms along the backbone of the molecular chains can be envisaged; but even in this case, the number of atomic contacts within a unit area could be less important in mixed SAMs than in pure monolayers. The lesser number of chains in the outer region of the mixed layers plays then an important part in the friction (compared with the nonmixed layers) [52].

The decrease of the friction of the monolayers is primarily due to the molecular chains of the SAMs: they have an important freedom of swing and can consequently rearrange along the sliding direction (under shear stress); this could lead to lower resistance during sliding and thus decrease the friction [21,57].

In a similar way to what happened at the nanoscopic scale, the friction of a one-component monolayer at the microscopic scale is less important as long as the length of the chain increases, which was also found in Ref. [58]. The applied normal load is higher at the microscale than at the nanoscale; the counterface is then also in contact with the short chain length. In the case of mixed monolayers, at this scale, the friction could therefore be influenced by the long- and short-chain components; this can explain why the friction values are in between those of the pure monolayers.

The development of mechano-chemical micro-fabrication technique, based on mechanical abrasion and chemical etching, needs the knowledge of the tribological behavior of the SAMs. To obtain patterns with nano-scale dimensions, it is essential that the mechanical abrasion process is extremely controlled to remove the resist material where chemical etching is expected to remove the workpiece material. From this perspective, the tribological properties of FDTS, OTS, and single-chain alkanethiol SAMs with various chain lengths were studied, in order to find the best mechanical scribing conditions for micromachining applications [22]. It appeared then that even if the surface energy of FDTS was lower than that of OTS, the surface of FDTS was damaged about 20% quicker than the OTS surface. It was also shown that a normal load of only a few nN was necessary to remove thiol from a copper surface; moreover, the native oxide layer on the metals greatly influenced the nano-tribological properties of alkanethiol SAMs. Finally, FDTS was proved to be used as resistant for silicon.

Singh et al. have studied the nano- and micro-tribological properties of a silicon wafer, dimethyldichlorosilane (DMDC), diphenyldichlorosilane (DPDC), perfluorooctyltrichlorosilane (PFOTS), and perfluorodecanoic acid (PFDA) coated on (100) silicon wafer [54]; they have highlighted the following points:

- As long as the glass ball size increases, at the microscale, the coefficient of friction increases too (larger contact area), irrespective of the materials. With the exception of PFDA, the silicon wafer has a higher friction than the SAMs, because of its higher surface energy.
- Only bare silicon wafer and PFDA show wear; the presence of PFDA film has a great influence on the friction. The PFDA wear was linked to the effect of the moisture, whereas that of the silicon wafer was attributed to solid–solid adhesion.

- The physical/chemical properties of the SAMs have a great influence on their nano-friction, at both nano and microscopic-scales.
- At the two studied scales, it appeared that only DMDC monolayer showed superior friction properties.
- At nanoscale, an increase of the normal load induces a rise of the friction, as the contact area increases. The lower interfacial energy of the SAMs influences their inherent adhesion and contact area, leading to lower friction properties than those of the bare silicon wafer.
- The phenyl-terminated SAMs show high friction force because of the high stiffness due to the benzene rings; the fluorinated SAMs also have high nanofriction, but due to the effect of the larger van der Waals radius of the fluorine atom.

Nano- and micro-tribological properties of a novel self-assembled dual-layer were studied in Ref. [59]; PFDA molecules on 3-APTES SAM with terminal amino group films were prepared on single-crystal silicon wafer by chemisorption. After being coated with PFDA film, the surface appeared relatively smooth and homogeneous. The PFDA-APS film greatly reduces the friction force at both the nano and microscales. Moreover, the film shows a better antiwear durability than the dodecanoic acid—APS self-assembled dual-layer film that has the same length chain and similar structure.

3.1.3 Nano-Tribology

Alkylsilane monolayers (C6, C8, C10) and mobile perfluoropolyether (PFPE) lubricants were deposited on hydrogenated amorphous carbon surface; the nano-tribological properties of the mixed lubricants on the carbon surface were investigated and compared with a 2 nm thick PFPE lubricant-coated carbon surface [39]. It appears that both the friction and durability of the mixed lubricants on carbon surface greatly depend on the silane monolayer—the friction force is lower and the critical load becomes more important when the alkylsilane chain length increases (although the amount of mobile lubricant decreases). However, it offers a higher durability due to both the higher surface coverage by the chemisorbed molecules and the rigid nature of the silane monolayers. It was also shown that an optimum thickness of the mobile lubricant could be determined to achieve a low friction; this thickness must be comparable to the radius of gyration of the molecules (if the thickness of the mobile layer exceeds the radius of gyration, the capillary force increases, and the friction becomes higher).

SAMs of heptadecafluoro-1,1,2,2-tetradecyltriethoxysilane (FTE) were formed on diamond-like carbon (DLC) by immersion process; the friction of unlubricated, SAM-coated and 2 nm thick PFPE-coated DLC surfaces was measured by lateral force microscopy (LFM) and compared [25]. The SAM-coated surfaces present the worst frictional characteristics, compared with the PFPE film, which is attributed to the mobile characteristic of the liquid lubricants (indicating that here the hydrophobicity of the surface is not the only factor that influences the friction). But compared with the unlubricated DLC surface, the SAM films offer lower frictional characteristics, probably because of the hydrophobic nature of the FTE SAM surface.

Brushes of polystyrene (PS) were grafted on single-crystal silicon substrate, but prior to the radical chain-transfer reaction, MPTS (chain-transfer agent) was self-assembled on the silicon substrate. The graft-polymerization of the styrene was then carried out from the mercapto-functionalized silicon surface [60]. It appeared that after an exposure to toluene for 12 h, the presence of the PS brushes greatly reduces the nano-friction force. It was also shown that the chemically adsorbed PS films have a better scratch resistance than that of spin-coated films.

The nano-tribological properties of trimethylsilyl (TMS) and pentamethyl-disilyl (PMDS)-terminated surfaces were studied and compared with those of Si(111)-(1 × 1):H surface and of a SAM on a silicon surface with long hydrocarbon chains (prepared with OTS) [61]. AFM measurements were carried out with an OTS-terminated silicon tip. The lateral forces of TMS and PMDS surfaces appeared to be similar, and about three times higher than that of OTS-SAM and H-terminated silicon surface. This higher value for TMS and PMDS films can be attributed to the relatively low surface density of the TMS and PMDS groups that modifies the interaction forces: this means that not only methyl species but also small hydrophilic $Si(OH)_x$ and Si-O-Si end groups are found on the surface. The comparable friction behavior of OTS monolayer and Si(111)-(1 × 1):H surfaces can probably be attributed to similar hydrophobic interaction and to a smooth morphology at the scale of the tip.

The chemical grafting method can also be applied to modify nano-alumina, silicon carbide, and silicon nitride by introducing polyacrylamide (PAAM) onto the particles [62]; the use of untreated particles leads to nanoparticles/epoxy composites with lower friction coefficient and lower specific wear rate (than those of unfilled epoxy). But it appeared that due to the strengthening of the agglomerated nanoparticles and also to the enhancement of the filler/matrix interface, the grafted nanoparticles reinforced composites had the lowest frictional properties and the highest wear resistance (than unfilled epoxy and untreated nanoparticles/epoxy composites). The grafting treatment of nanoparticles seems thus to be more efficient in improving the tribological properties of the composites than the conventionally used silane treatment.

Studies dealing with the frictional behavior of graphite/steel couples revealed that there existed a very strong relationship between the tribological properties of these materials and the interfacial interactions that were created between the graphite pins and the steel [63–65]. It appears, therefore, essential to understand these mechanisms of interaction, which necessitates knowledge, as completely as possible, of the physicochemical state of the surfaces in contact.

However, it is not easy to study the influence of the chemistry of the surface of the steel, mainly for two reasons:

1. It is not possible to ignore the nanometric roughness of this surface (R_a of about 10 nm, even after polishing).
2. The chemical modification of the surface (to obtain different surface energies) is difficult.

To overcome these problems, silicon wafers were selected as counterfaces.

3.2 EXPERIMENTAL DETAILS AND CHARACTERIZATION TECHNIQUES

3.2.1 MATERIALS

3.2.1.1 Graphite Pins

Three types of graphite were tested in this work; the following is the specific designation of these powders.

Graphite A: This powder was made of synthetic graphite and was available in the form of particles of different sizes, i.e., 15 μm (A15) and 75 μm (A75).

Graphite C: It was made of natural graphite powder, with an average particle size of 10 μm (C10).

All these powders were compacted into the shape of small graphite pins according to the following procedure: 120 mg of graphite powder, without binder, were introduced into a cylindrical mould of 5 mm diameter. All was then placed in a hydraulic press where the powder was first degassed for 3 min (primary vacuum) before compacting. An uniaxial load of one ton was then gradually applied for 5 min (during which the primary pump was still working). The resulting pins were 5 mm in diameter and 3 mm in height (Vickers hardness $H_V = 1.4$ kg.mm^2; $R_a = 1.4$ μm).

3.2.1.2 Silicon Wafers

The mechanical and chemical properties of these wafers made them interesting substrates to replace the steels disks whose drawbacks have been previously mentioned. These substrates showed very smooth surfaces (R_a around 0.2 nm) and were coated with a silicon oxide film that can, after activation, create covalent bonds with grafted molecules.

These wafers (from Mat Technology, Essonne, France) were (100) oriented, and one of their face is polished.

3.2.2 MODIFICATION OF SURFACE CHEMISTRY

In order to study the influence of surface chemistry during friction, two methods were used to modify the surface chemistry of silicon wafers, without any modification of the topography (the substrates were infinitely stiff but had a controlled chemistry):

- *Hydroxylation*: The surface of the silicon wafer was activated in order to create hydroxyl (OH) sites; the so-obtained surfaces were hydrophilic.
- *Grafting*: An organosilane with a controlled end was grafted on the silicon wafer which was hydrolyzed beforehand; a homogeneous molecular film was thus formed, leading to a hydrophobic surface.

These two processes are detailed in the following.

3.2.2.1   Hydroxylation (Piranha Solution)

The silicon wafers were first cleaned with acetone, in an ultrasonic bath, for 4 min; they were then dried with nitrogen and afterwards rinsed with bi-distilled and deionized water. They were finally dried again with nitrogen.

The silicon wafers were immersed in a Piranha solution formed of 70% sulfuric acid (H_2SO_4, 96% purity) and 30% hydrogen peroxide (H_2O_2—30%) and heated at 50°C for 30 min. These substrates were then cleaned with bi-distilled and deionized water in an ultrasonic bath; they were finally dried with nitrogen.

The so-obtained surfaces showed a considerable density of silanol groups (5 Si-OH/nm^2) [66], and were hydrophilic. These modified surfaces were immediately used in order to avoid any contamination by the environment, as the Si-OH functional group is very reactive.

3.2.2.2   Grafting

SAMs can be achieved using the following two different ways: in the vapor phase or by immersing the substrate in the solution containing molecules (silanes) that will interact with its surface; the latter approach was adopted here. The silanes were hydrolyzed by the water that was either adsorbed on the surface or present in small amount in the solvent used to dissolve the silanes. This procedure was simple but very long—the immersion of the substrate can take 12 h in the case of organosilanes, or about 18 h when organothiols were used.

The organosilane used in this work was a monofunctional one as this kind of molecules can avoid the reactions of polycondensation, that can lead to the formation of a monomolecular layer; it was HTS (purchased from ABCR, 99% purity), with a methyl end group.

This process can only be realized on surfaces which were already hydrolyzed otherwise strong bonds with the hydrolyzable parts of the silanes cannot be formed. The high density of silanol groups obtained at the end of the previous step will allow adsorption of the organosilanes by hydrolysis [67].

After being hydrolyzed with the Piranha solution, the silicon wafers were immersed in a 3:1 solution of carbon tetrachloride and HTS for 12 h; this step was carried out in an ultrasonic bath in order to remove the silanes that were not grafted or physisorbed.

The formation of SAMs in liquid phase occurred in different stages [68]: as a first step, there was hydrolysis of the Si–X bonds (X=Cl or O–CH_3) of the silane molecules by water. Second, strong interactions, like hydrogen bonds, were formed between the silanols and the surface of the substrate [69]. Then siloxane (covalent) bonds were created through a reaction of condensation [70], and water was eliminated (chemisorption). Finally, a two-dimensional network was formed near the substrate surface, due to lateral chemical bonds between the silanols. The so-modified surface presented a hydrophilic or hydrophobic character, depending on the nature of the terminal group of the organosilane used. Here, a methyl end group was utilized: consequently, the surface was a hydrophobic one.

3.2.2.3   Chemically Controlled Surfaces

The first studies dealing with the grafting of a long carboxylic acid chain onto a rigid substrate, in order to modify the surface chemistry, was carried out by Blodgett

[71,72]. Zisman [73] also studied SAMs, and a more detailed work devoted to the history of these SAMs was written by Gaines [74] and Ulman [20].

The molecules involved in the formation of SAMs consist of three parts:

1. *The head group of the molecule*: It gives rise to a strong interaction between the molecule and the substrate; the nature of the link can vary: covalent bond (for instance Si–O for organosilanes), strongly covalent but partially ionic bond (Au–S for organothiols), or very strong ionic bond (carboxylic acid–Ag).
2. *The alkyl chain*: Its length can vary from one molecule to another. In order to obtain self-assembled chains, the number of carbon atoms of this chain must be between 11 and 18. The interactions between these chains are van der Waals ones; they are at the origin of the organization of these molecules between them.
3. The head group and the alkyl chain are essential for the creation of SAMs. The terminal end, however, allows the modification of the chemical properties: if a hydroxyl end is used instead of a methyl one, the surface will be hydrophilic, and not hydrophobic.

In our study, the application of the previously mentioned processes led to two substrates (silicon wafers) of different surfaces chemistries:

- *Hydrophilic ones*: Obtained through the hydroxylation of the surface to create hydroxyl groups
- *Hydrophobic ones*: An organosilane with a methyl end was grafted onto the silicon wafer

3.2.3   CHARACTERIZATION OF SAMPLES BEFORE AND AFTER FRICTION

3.2.3.1   Wettability

This is an easy and efficient technique to determine the surface energy of a solid, and, consequently, to confirm the hydrophilic or hydrophobic nature of a surface.

Two modes of measurements can be used:

- *The dynamic mode*: The hysteresis measurement of the contact angle (difference between the advancing angle and the receding one) gives information on both the heterogeneity and the roughness of the studied surface.
- *The static mode (which was selected in this study)*: A drop of a liquid is placed on the studied surface, and the contact angle (θ), which is the angle between the tangent of the drop profile and the surface of the substrate, is measured.

Various informations are obtained through the measurement of the contact angles [75]:

- If the liquid used is water, the hydrophobic (high angle, low surface energy) or hydrophilic (low angle, high surface energy) character of the substrate can de deduced.
- With other liquids, more particularly with apolar ones, the surface energy of the substrate can be calculated.

When the contact angles are known, the Fowkes theory [76] allows the calculation of both the dispersion and nondispersion (or polar) components of the surface energy, leading thus to the calculation of the surface energy of the studied substrates.

3.2.3.2   Tribometer

The graphite pins fabricated from the above-described powders were subjected to sliding tests against silicon wafers on a commercial CSM pin-on-disk tribometer.

The tests were run at an ambient temperature of 25°C, at a relative humidity of 30%–40% for a duration of 40 min. The applied loads were 2, 5, 10, and 15 N; the sliding speed was kept constant at a value of 10 rpm (which corresponded to a linear speed of 1.73 cm s^{-1}). Each test was carried out three times to check the reproducibility of the results.

3.2.3.3   Chemical Analyses and Topographical Observations

The chemical composition of the silicon wafers surface was analyzed by x-ray photoelectron spectroscopy (XPS). This technique was used to check the efficiency of grafting because unlike energy dispersive spectroscopy (EDS) whose analysis depth is on the order of a micrometer, the measurements here are carried out to a depth of only about 10 nm. On the other hand, EDS analysis can only determine the nature of the chemical elements (C, Si, etc.), while XPS provides these data supplemented by the way these elements are linked together (C=O, C–O–C, R–OH, etc.).

The wafer surface was examined by scanning electron microscopy (SEM) after the friction tests in order to characterize the morphology of the transferred material.

3.3   RESULTS AND DISCUSSION

3.3.1   Characterization of Grafting

The chemical composition of the subsurface (first 10 nm under the surface) of both hydrophilic and hydrophobic silicon wafers was analyzed by XPS (Figure 3.1).

The photopeak observed at 285 eV in the spectrum of the hydrophobic substrate is typical of carbon C1s of the alkyl chain ($-CH_3$): it confirms that the organosilane molecules (HTS) were chemically adsorbed on the silicon wafer.

Moreover, wettability measurements can be used to determine the surface energy of the various substrates: the hydrophobic wafer shows surface energy of 21 mJ m^{-2}, whereas for the hydrophilic substrate it reaches 73 mJ m^{-2}. The surface energy of the hydrophilic silicon wafer is clearly higher than that of the bare silicon (47 mJ m^{-2}), while that of the hydrophobic wafer remains well below the reference substrate; this tends to prove that both processes of hydroxylation and grafting were carried out with success.

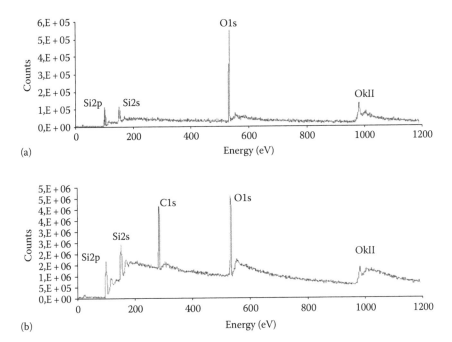

(a)

(b)

FIGURE 3.1 XPS spectra of (a) hydrophilic and (b) hydrophobic silicon wafers.

The three counterfaces used in friction experiments have a very low surface roughness but have different surface energies; this helps to test the influence of the surface chemistry on the tribological behavior of graphite/silicon wafer couples.

3.3.2 TRIBOLOGICAL TESTS: ROLE OF NATURE OF GRAPHITE

Two general trends emerge from the study of the influence of the nature of the graphite on its tribological behavior (Figure 3.2):

1. *Bare and hydrophilic wafers*: The friction coefficient of natural graphite is clearly higher than those of synthetic graphites; graphite A75 shows the smallest friction coefficient, and that of graphite A15 shows intermediate values. These observations are valid for all the applied normal loads, except for 2 N for which the highest friction is observed for graphite A75:

   μ (C10) > μ (A15) > μ (A75), irrespective of the applied normal load,
   except for 2 N: μ (A75) > μ (C10) > μ (A15)

2. *Hydrophobic wafer*: The behavior observed here is the opposite of what was previously noticed: μ (A75) > μ (A15) > μ (C10)

Two specific cases were particularly studied as they highlighted particular friction mechanisms.

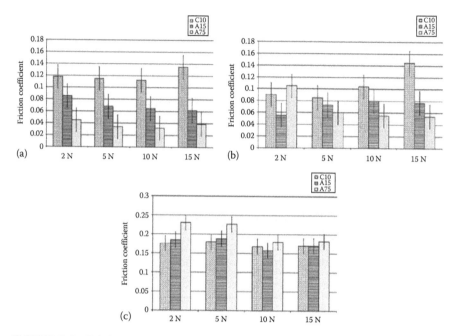

FIGURE 3.2 Friction coefficient as a function of the applied normal load, for graphites C10, A15, and A75, in the case of (a) hydrophilic, (b) nontreated, and (c) hydrophobic wafers.

3.3.2.1 Hydrophilic Wafer: 10 N

In these experimental conditions, it was found that the highest value of the friction coefficient was observed for natural graphite; when synthetic graphites are tested, the material with the largest particles leads to the lowest friction (regardless of the applied normal load).

SEM images of the transferred particles on the silicon wafers are in agreement with the above observations: the material transferred from natural graphite is made of large plates (Figure 3.3), which clearly do not facilitate the movement of the pin on the substrate (hence the high friction); but in the case of graphite A15, even if the transferred material is always made up of plates, these are much more spread out, and the surface irregularities that were observed in the transferred material in the case of natural graphite were practically missing here (Figure 3.4). The transferred material seems to be smooth, with no significant roughness—this can explain the lower friction in this case. The transferred material in the case of graphite A75 is made of rounded elements, with little bumps scattered on the track (Figure 3.5); such a morphology can be favorable to the sliding of the pin on the silicon, which can explain why the friction coefficient is lower here.

The variations due to the different particles sizes could be related to the fact that the more the grain size increases, the larger the distance between the nodes in each grain is [77]; this leads to a decrease of the mechanical properties of these materials, as confirmed by the data in Table 3.1.

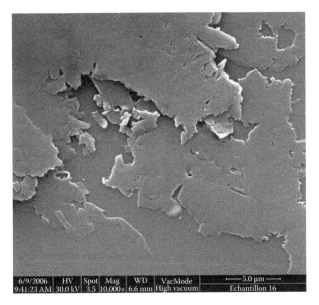

FIGURE 3.3   SEM image of the transfer of natural graphite C10 to hydrophilic wafer under 10 N.

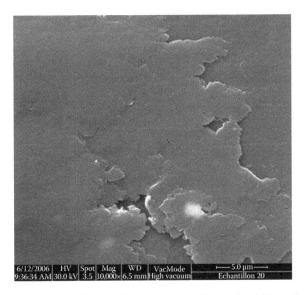

FIGURE 3.4   SEM image of the transfer of graphite A15 to hydrophilic wafer under 10 N.

As a result, the intragranular cohesion of graphite A75 is lower than that of graphite A15, which implies an easier reorientation of the basal planes. The presence of many of these planes, which are known to act as water reservoir [78,79] will accentuate the lubricant character of graphite in the presence of the water vapor contained in ambient air, and thereby reduce friction.

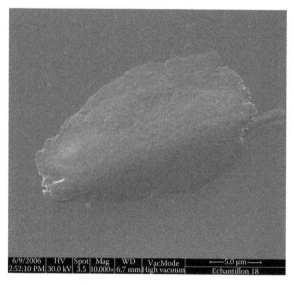

FIGURE 3.5 SEM image of the transfer of graphite A75 to hydrophilic wafer under 10 N.

TABLE 3.1
Vickers Hardness and Compression
Modulus of A15 and A75 Graphites

	A15	A75
Vickers hardness (kg mm^{-2} ± 0.2)	3.50	2.30
Compression modulus (GPa ± 0.03)	3.48	2.12

Another hypothesis justifying the lowest coefficient of friction of graphite A75 (compared to that of graphite A15) could be the low adhesion between the large crystallites, which would make the shearing easier [80].

The difference between natural and synthetic graphites may have a dual origin:

- Natural graphite has better mechanical properties (intergranular cohesion, mobility of crystallites, etc.) than synthetic graphites, which could explain its higher values of friction.
- The chemical analysis before friction experiments indicates the presence of impurities (Al, Si, Ca, Fe, etc.) in small quantities in natural graphite. These impurities could play the role of binder and thus increase the intergranular cohesion, and on the other hand, could interact with the silicon wafer modifying hence the nature of interfacial interactions (and therefore the adhesion), thereby increasing friction.

3.3.2.2   Hydrophobic Wafer: 10 N

In the case of synthetic graphites, the material with the smallest particle size leads to the lowest coefficient of friction; for graphites having the same particle size, the natural graphite shows the highest coefficient of friction:

$$\mu\ (C10) = 0.168;\ \mu\ (A15) = 0.158;\ \mu\ (A75) = 0.1870$$

3.3.2.2.1   Influence of Particle Size: Comparison between Synthetic Graphites A15 and A75

SEM images of the friction track of graphite A75 against the hydrophobic silicon wafer show that the clusters forming the transferred material are made of graphitic planes that are bonded together, the torn particles are easily visible and their outlines are clearly sharp. These transferred clusters are bulky rather than spread as the elements pulled out from the pin tend to pile up (instead of being crushed) (Figure 3.6a). This may be related to the large size of the constituent particles of the graphite that are more difficult to spread and be crushed during the friction experiment. This specific shape of these clusters may cause a more difficult motion of the pin (low load-bearing capacity), and therefore, a high coefficient of friction.

On the contrary, the particles forming A15 graphite are much smaller than those of graphite A75. Once pulled out from the pin, they spread out and gather relatively easily to form clusters, even if the boards of basal planes are still visible at some points, as if the process was not completely finished (Figure 3.6b). It is possible that the small size of the particles forming the torn pieces facilitates their "fragmentation," and induces the formation of spread agglomerates that oppose less the moving of the pin (leading thus to a stabilized friction coefficient lower than that of graphite A75).

The size of the graphite particles thus has a great influence on the morphology of the transferred material and consequently on the friction behavior—particles of modest size tend to amalgamate to form one single and long agglomerate, and the graphitic planes are still visible after the friction of a graphite made of large particles.

3.3.2.2.2   Influence of the Graphite Origin: Comparison between Graphites A15 and C10

The transferred material after the friction tests of the natural graphite has a morphology that could be the result of a spreading process that is unfinished (Figure 3.7): graphitic planes are visible, especially along the edges, while their outlines are more difficult to distinguish at the center of the agglomerates, where spreading resulting from the friction phenomenon has embedded them into the cluster.

This finding, already established for graphite A15, is, however, less marked in the case of natural graphite; spreading is obviously more advanced in the case of graphite A15, whereas it seems unfinished in the case of natural graphite.

It is possible that the hardness of the graphite can be the origin of this phenomenon of unfinished spreading: natural graphite ($H_v = 5.3\ kg\,mm^{-2}$) has a greater hardness than that of graphite A15 ($H_v = 3.5\ kg\,mm^{-2}$), which could explain that its

(a)

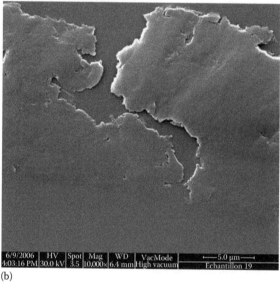

(b)

FIGURE 3.6 SEM images of the transfer of (a) A75 and (b) A15 graphites to hydrophobic wafer under 10 N.

particles are more difficult to spread, resulting in a higher friction coefficient, despite a smaller particle size.

3.3.3 Tribological Tests: Role of Applied Normal Load

3.3.3.1 Bare Silicon

Variations of the normal load do not induce significant modifications of the coefficient of friction for synthetic graphites A15 and A75; only the values observed at

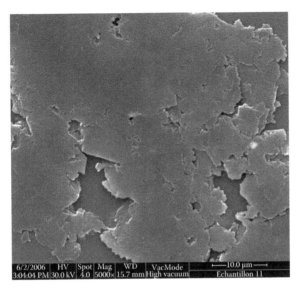

FIGURE 3.7   SEM image of the transfer of graphite C10 to hydrophobic wafer under 10 N.

2 N are different (Figure 3.2b). However, the coefficient of friction of graphite C10 increases, in a moderate way, when the normal load increases. Under the application of higher loads the difference between the graphites is more glaring—the values reported in the case of synthetic graphites are lower than those observed for natural graphite.

Two phenomena could explain the different behavior observed at 2 N. First, it is possible that in the experimental device, the coplanarity between the pin and the silicon wafer is not sufficient, which could alter the geometry of the contact. Moreover, 2 N may not be sufficient enough to result in reorientation of the large crystallites of graphite A75—adhesion phenomena would be promoted by the higher number of prismatic sites on the surface, which leads to higher values of the coefficient of friction.

3.3.3.2   Hydrophilic Silicon

The histogram in Figure 3.2a indicates that irrespective of the graphite considered, the coefficient of friction is only slightly dependent on the applied normal load. The tribological behavior observed in the case of a hydrophilic wafer tends to be similar to the one observed during the tests against the bare silicon, except for experiments performed at 2 N.

This could be due to the fact that the bare silicon is naturally slightly hydrophilic. The constant friction behavior observed in the case of hydrophilic wafer is a consequence of a better control of the surface chemistry, which means that the nature of the surface species (mixed of Si-OH, Si-O-S for the bare wafer, whereas only OH termination for the hydrophilic silicon), and the density of sites (5 sites Si-OH per nm^2 on the hydrophilic wafer) are fully determined.

3.3.3.3 Hydrophobic Silicon

3.3.3.3.1 General Trends

The behavior observed in the case of hydrophobic wafers differs from what was noticed for bare and hydrophilic wafers (Figure 3.2c):

- While the coefficients of friction varied between 0.05 and 0.15 for tests carried out against the bare and hydrophilic substrates, the values reported here are higher (0.15–0.25).
- The friction of natural graphite does not appear to depend on the variations of the applied normal load, while synthetic graphites show values that are relatively close at 2 and 5 N, then decrease (but remain related) when applying higher loads. This phenomenon is particularly marked in the case of graphite A75, which presents the highest coefficient of friction over the entire range of applied load values studied.

The especial tribological behavior of this couple will be more particularly studied below.

3.3.3.3.2 Specific Case of Graphite A75/Hydrophobic Wafer Couple

Under the application of 10 and 15 N, the stabilized coefficient of friction is in the order of 0.180, whereas it reaches 0.230 and 0.226 under the application of 2 and 5 N, respectively.

The SEM observations of the transferred material in both cases indicate that this is not the morphology of the transferred particles to the silicon wafer that has an influence, but their size. Actually, whatever the applied load, the transferred material consists of rounded and rather large agglomerates, made of graphitic plates that are still visible. However, at 10 and 15 N, these agglomerates are three times smaller than those observed at 2 and 5 N, and they are much less numerous (Figure 3.8a and b). These few small and rounded particles could be a factor favorable to the motion of the pin on the substrate, like balls in a bearing.

This could explain the lower coefficient of friction observed at higher loads. The effect of the applied normal load remains relatively modest in these experimental conditions, for both the transferred particles and the values of the coefficient of friction.

3.3.4 TRIBOLOGICAL TESTS: ROLE OF SURFACE CHEMISTRY

The histograms in Figure 3.9 summarize all the stabilized coefficients of friction observed at the end of the friction tests on the three graphites A15, A75, C10 against the bare, hydrophobic, and hydrophilic silicon wafers. It appears that as long as the surface energy increases, the friction coefficient decreases, regardless of the applied load.

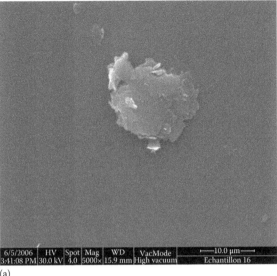

(a)

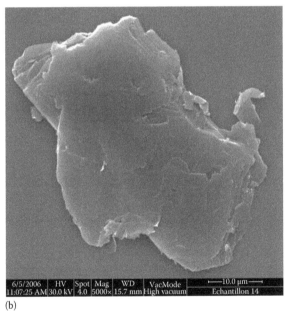

(b)

FIGURE 3.8 SEM images of the transfer of graphite A75 to hydrophobic wafer under (a) 15 N and (b) 5 N.

It is necessary to point out that according to the uncertainty in the tribological measurements (±0.02), the reduction of the friction coefficient is assumed to be still valid, especially in the case of the natural graphite, or for the results obtained on bare and hydrophobic wafers. Four cases illustrating the most representative behaviors in all these tests are particularly detailed.

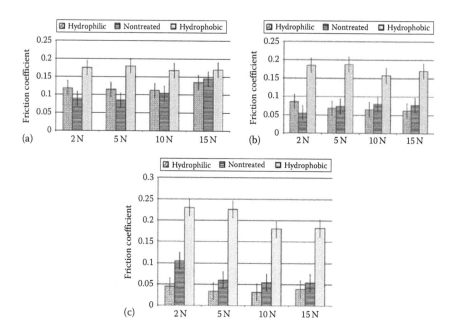

FIGURE 3.9 Friction coefficient as a function of the applied normal load, for hydrophilic, nontreated and hydrophobic wafers, for graphites (a) C10, (b) A15, and (c) A75.

3.3.4.1 Graphite C10: 10 N

The coefficients of friction identified during sliding against bare and hydrophilic substrates are very close and remain lower than the characteristic value of the friction coefficient against the hydrophobic wafer:

$$\mu \text{ (bare Si)} = 0.104; \mu \text{ (hydrophilic Si)} = 0.112; \mu \text{ (hydrophobic Si)} = 0.168$$

The SEM images can help to envisage an explanation for this difference. The graphite transferred against the hydrophilic substrate has a smooth appearance, the motion of the pin has clearly spread out the material that has fallen from it (Figure 3.10); the surface of the bare wafer (not shown here) showed the same characteristics.

As for the hydrophobic wafer, its friction track is composed of chaotic elements, disorganized assembly of particles torn from the pin (some of these particles have still the form of plates); their size is important, and they seem bulky (Figure 3.11).

In presence of such clusters, the motion of the pin is much more difficult than against a hydrophilic wafer, which explains the higher values of the coefficient of friction in the case of the hydrophobic wafer.

3.3.4.2 Graphite A15: 10 N

In the same way as before, the surface whose energy is higher (hydrophilic) has lower coefficient of friction (even if here the values for bare and hydrophilic wafers

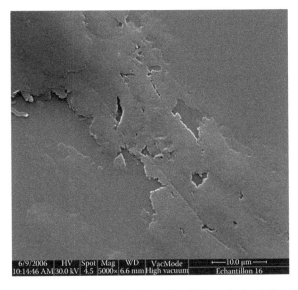

FIGURE 3.10   SEM image of the transfer of graphite C10 to a hydrophilic wafer under 10 N.

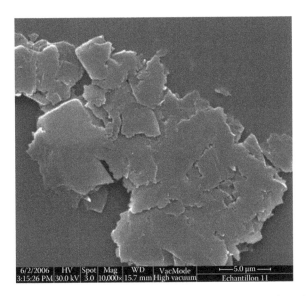

FIGURE 3.11   SEM image of the transfer of graphite C10 to hydrophobic wafer under 10 N.

are relatively close): μ (bare Si) = 0.080; μ (hydrophilic Si) = 0.065; μ (hydrophobic Si) = 0.158.

This behavior can be attributed to the role of moisture. SEM images show that the transferred material on the hydrophilic wafer is composed of parallel tracks, which are made up of clusters spread in length, as plastically deformed, and whose edges are more rounded (Figure 3.12).

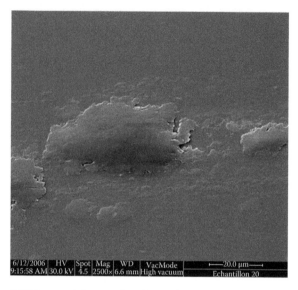

FIGURE 3.12   SEM image of the transfer of graphite A15 to hydrophilic wafer under 10 N.

On the other hand, the clusters that constitute the transferred material on the bare silicon reveal graphitic planes on some places, and their extremities look like little "strips" (Figure 3.13). This kind of morphology may suggest that the process of spreading of the material was in progress, but that the "strip" had not been sufficiently crushed to be integrated in the cluster and be agglomerated with it. Such scattered clusters are not favorable to the movement of the pin and could

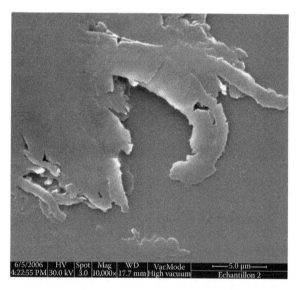

FIGURE 3.13   SEM image of the transfer of graphite A15 to nontreated wafer under 10 N.

explain the higher coefficient of friction in the case of the bare silicon (compared to hydrophilic wafer).

Hydrophobic silicon, after rubbing, shows transferred material made on the one hand of plates with irregular shapes and relatively large size—*type a* (Figure 3.14a), and, on the other hand, of clusters relatively spread out and rounded—*type b* (Figure 3.14b). These two types of transferred materials coexist along the all friction track. The particles of *type b*, because of their rounded morphology and their smoothness would tend to induce a low friction, but these particles are in minority. In contrast,

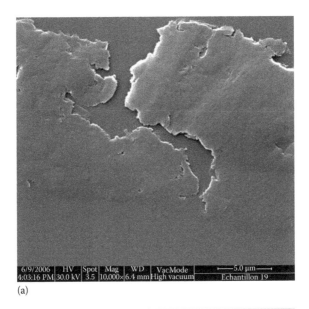

(a)

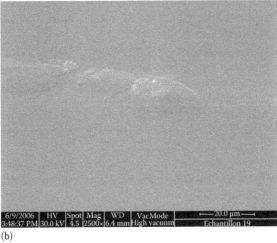

(b)

FIGURE 3.14   SEM images of the transfer of graphite A15 to hydrophobic wafer under 10 N: (a) *type a* and (b) *type b*.

the agglomerates of *type a* are predominant, and their irregular appearance could be a possible explanation for the high friction.

It is possible that initially the transferred clusters have the appearance of particles of *type a*; then, as the friction test goes on, they undergo changes (mechanical and thermal) leading to the appearance of particles of *type b*. It is clear that this change occurs only slowly, probably because of the hydrophobic character of the substrate (low surface energy), which explains that the particles of *type a* are the majority and control the friction, leading to a high steady-state value of the coefficient of friction.

3.3.4.3   Graphite A75: 5 N

In these experimental conditions, it appears that the coefficient of friction of A75 graphite/hydrophilic Si couple is two times lower than that achieved against a bare Si substrate, whereas in the case of hydrophobic silicon, the coefficient is almost four times higher than the one corresponding to the test against the untreated wafer:

$$\mu \text{ (bare Si)} = 0.060; \mu \text{ (hydrophilic Si)} = 0.034; \mu \text{ (hydrophobic Si)} = 0.226$$

The extremely low coefficient of friction noticed in the case of hydrophilic silicon might seem surprising, as sliding on hydrophilic surfaces, i.e., highly polar, is generally more dissipative than that observed on hydrophobic nonpolar surfaces. This is the characteristic behavior of graphite in the presence of water, notably the lubricating action of water, which may explain these results and confirm the significant role of the adsorbed moisture.

After Piranha treatment, the surface of the silicon wafer is indeed covered with OH groups; these polar groups are favorable to the adsorption of water vapor on the wafer. The amount of water vapor in the contact is then, in the case of the hydrophilic substrate, greater than during the friction on the bare reference surface. This was confirmed by wettability measurements: the surface energy of hydrophilic wafers was comparable to that of water.

With water acting as a lubricant for graphite, the coefficient of friction in the presence of a hydrophilic surface is lower than in the case of a bare surface, since for the latter, the number of OH groups naturally present on the surface of the material is not sufficient to cause adsorption of water and thus to induce the process of lubrication. The friction coefficient of the (graphite/bare Si) couple thus is higher than that of the (graphite/hydrophilic Si) one but remains, however, lower than that obtained after sliding against the hydrophobic wafer.

It appears, therefore, that the higher the surface energy of the silicon wafers, the lower the coefficient of friction of the tribological couple, which underlines the unusual frictional behavior of graphite. The friction on an apolar surface will be more difficult than on a polar one inclined to promote the self-lubrication of graphite through the presence of water at the interface.

SEM images recorded on the bare and hydrophilic wafers (Figure 3.15a and b) show friction tracks formed by parallel and concentric strips (which is not true for the hydrophobic wafer (Figure 3.15c)).

The tracks observed on the bare silicon wafer are made of elements in which the outline of graphitic planes is visible, which can be interpreted as the result of spreading of the pieces that were torn from the graphite (Figure 3.15a).

In the case of hydrophilic silicon (Figure 3.15b), it is rather accumulation of small rounded, indeed oval, elements, making the transferred film relatively compact and

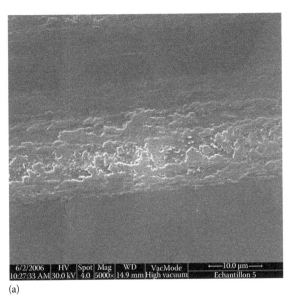

(a)

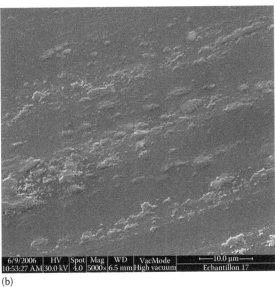

(b)

FIGURE 3.15 SEM images of the transfer of graphite A75 to (a) nontreated, (b) hydrophilic, and (c) hydrophobic wafers under 5 N.

(*continued*)

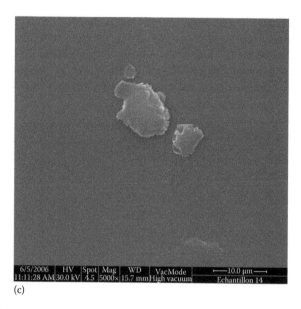

(c)

FIGURE 3.15 (continued)

continuous, while on bare silicon, the tracks in the friction zone are clearly separated from each other. Moreover, these small rounded elements could be likened to "load-carrier" agglomerates, supporting the pressure of the pin, and ensuring its motion: friction occurs, therefore, in a relatively fluid, almost rolling, way, hence such a coefficient of friction.

The transferred material observed on the hydrophobic silicon (Figure 3.15c) is different from that obtained on the bare and hydrophilic wafers—the parallel tracks, which were clearly visible for the other friction couples, do not appear here, and the transferred elements have the appearance of small clusters formed from graphite plates with clean outlines. The morphology of the particles that constituted the transfer in this case could be the origin for the friction increase, as these torn elements could oppose the motion of the pin, and consequently make sliding more difficult.

The transfer on hydrophobic surface appears then to be more difficult, as if there was no (or little) affinity between graphite and the CH_3 groups covering the hydrophobic wafer.

3.3.4.4 Graphite A75: 10 N

The coefficient of friction against the hydrophobic surface is six times greater than that obtained during friction test against the hydrophilic wafer, and more than three times higher than that resulting from the friction test against a bare silicon:

$$\mu \text{ (hydrophilic Si)} = 0.032; \mu \text{ (bare Si)} = 0.055; \mu \text{ (hydrophobic Si)} = 0.180$$

The differences that were observed in the morphology of the transferred material are again the key to explain such variations in the friction values: the clusters which formed the transferred material when sliding against a hydrophobic silicon are made of stacking of graphitic plates with irregular contours (Figure 3.16a), creating thus large and rough elements, inclined to slow down the motion of the pin on the substrate, resulting in high friction.

The agglomerates constituting the transferred material observed on the bare substrate have a size, which is equivalent to the one observed on the hydrophobic

(a)

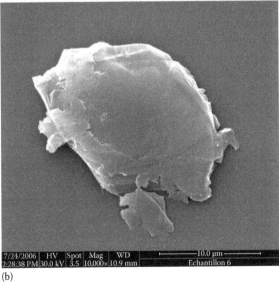

(b)

FIGURE 3.16   SEM images of the transfer of graphite A75 to (a) hydrophobic, (b) non-treated, and (c) hydrophilic wafers under 10 N.

(*continued*)

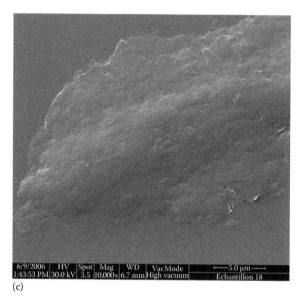

6/9/2006 | HV | Spot | Mag | WD | VacMode | ⊢——5.0 μm——⊣
1:43:53 PM | 30.0 kV | 3.5 | 10,000× | 6.7 mm | High vacuum | Echantillon 18

(c)

FIGURE 3.16 (continued)

surface, but their appearance is much smoother: the contours are more rounded and they seem more flat (Figure 3.16b). The movement of the pin is thus less difficult in these conditions, and, therefore, the friction becomes lower. Finally, the transferred material on hydrophilic substrate is formed of relatively long particles, with no asperity (Figure 3.16c): the pin can practically slide on this kind of surface, which leads to a very low friction.

There is, therefore, an obvious link between the surface energy of the substrate, the changes in the morphology of the transferred particles and the value of the stabilized coefficient of friction. The lower the surface energy, the more irregular and rough the transferred particles, and the higher the friction. On the contrary, when the surface energy is high, the transferred material can be done only in so-called soft conditions (smooth and relatively flat clusters), the motion of the pin has no obstacle, and thus the coefficient of friction is low.

3.3.5 Mechanisms

3.3.5.1 Graphite/Hydrophilic Silicon Couples

The structure of graphite is lamellar, the surface of a crystal is made of cleaved and edge faces. The cleaved faces are characterized by a low surface energy (they are mainly constituted of basal planes) [81]; on the other hand, the dangling bonds of the edge planes are very active and will react with the water and oxygen of the atmosphere to create a surface whose terminated bonds are various oxygenated groups [82,83]. These groups have a polar character and confer thus on graphite a tendency toward hydrophilicity [84].

In these experimental conditions, it is possible to simplify the considered couple as the contact between a rather hydrophilic surface, and another that is highly hydrophilic. The nature of this contact will promote a significant adsorption of oxygen molecules, but above all of the water vapor of the atmosphere. And yet, one of the specific properties of the graphite is the lubricating effect that water has (or other condensed vapor) on it [85,86]. As a result, the surfaces of both the graphite pin and the graphitic transferred film in the contact are enriched in water, which facilitated the shear strength of graphite planes, resulting in a lower energy dissipation, and therefore a reduced friction.

Furthermore, during the contact between two surfaces which tend to be hydrophilic, the adhesive properties are modified. In the experimental conditions of this study (30% humidity and 25°C), the water in the air can condense between the graphite pin and the silicon wafer (specifically between the asperities of these two members) to form a capillary meniscus. Such a meniscus should theoretically increase the adhesion strength between these members, the friction should consequently be higher. But friction tests against hydrophilic wafers lead to low coefficient of friction. This contradiction could be explained by the porous appearance of the graphitic materials: water molecules could come in between the graphitic planes and thus reduce the interlamellar strength [87], which on the one hand would facilitate the detachment of particles and, on the other hand, would lead to a low friction.

In some respects, the previous observations show analogies with a fundamental phenomenon of physicochemistry: the Rehbinder effect. This theory studies the influence of the surface active medium on the mechanical properties of the materials in contact [88–90]. This phenomenon includes the decrease of the solids strength (due to adsorption), an easier deformation of the solids (indeed their destruction) linked to the physicochemical action of the environmental medium. Water could be considered here as the active medium that modified the physicochemical nature of the contact, inducing thus the observed modifications of the friction behavior.

In a hydrophilic/hydrophilic contact, the presence of water at the interface has two effects: first, it facilitates the shearing of graphitic sheets, and second, its adsorption on crystallites will help to reduce the adhesion between them (and also between the graphene layers of each crystallite). All of these mechanisms contribute to low friction.

3.3.5.2 Graphite/Hydrophobic Silicon Couples

In the same way as before, the surface of graphite can be considered as rather hydrophilic: a hydrophilic/hydrophobic contact can thus represent the studied couple. The mechanisms involved in this case are more complex. The hydrophobicity of the wafer reduces the capacity for adsorption of water molecules. There are thus fewer water molecules that may come in between the graphitic planes, which consequently promote shearing, leading thus to a greater friction.

Unlike OH sites (present on the surface of hydrophilic wafer in the form of a very thin film of 1 Å thickness), which are able to adsorb water molecules and thereby induce lubrication of the contact, the hydrocarbon molecular film covering the surface of the hydrophobic wafer could act as an antilubricant film during the friction test.

During a hydrophilic/hydrophobic contact, the CH_3 molecular film (16 Å thickness) plays a fundamental role, preventing direct contact between silicon and graphite, which can lead to an increase of the mechanical energy which is dissipated because of the deformation of Si–CH$_3$ alkanes chains present at the interface. The shear and cleavage of the basal graphitic planes are then more difficult and thus result in an increase of the friction.

3.4 CONCLUSIONS

In order to understand the role of the surface chemistry in the tribological behavior of graphite, tests were carried out with three different (in terms of their origin and size of particles) graphite powders sliding against bare, hydrophilic, and hydrophobic silicon wafers, under various applied normal loads.

The results of this study allowed to highlight the following points:

- The size of the graphite particles as well as the normal load (but to a more modest degree) have an influence on the morphology of the transferred particles.
- By modifying the surface chemistry, and therefore the surface energy, the coefficient of friction of the considered couples can be varied; the lowest friction coefficient is obtained after sliding against the substrate with the highest surface energy.
- When using a hydrophilic wafer, the self-lubricant ability of graphite is intensified.

The results of these experiments indicate that it is essential to continue the investigation of chemically treated substrates, more particularly to elucidate the specific role of the molecular hydrocarbon film (–CH$_3$) during the friction against hydrophobic wafer.

A nanoscopic approach (AFM measurements) would also be interesting since a study at the level of the graphitic planes could help to better understand the influence of the surface chemistry and to refine the mechanisms involved during the friction test.

ACKNOWLEDGMENTS

The authors are grateful to the French Ministry of Education and Research for the financial support of this work.

REFERENCES

1. S. Spearing, Materials issues in microelectromechanical systems (MEMS), *Acta Mater.*, 48, 179–196 (2000).
2. K. Komvopoulos, Surface engineering and microtribology for microelectromechanical systems, *Wear*, 200, 305–327 (1996).
3. S. Patton, W. Cowan, K. Eapen, J. Zabinsky, Effect of surface chemistry on the tribological performance of a MEM electrostatic lateral output motor, *Tribol. Lett.*, 9, 199–209 (2000).
4. R. Maboudian, W. Ashurst, C. Carrato, Tribological challenges in micromechanical systems, *Tribol. Lett.*, 12, 95–100 (2002).

5. A. Homola, Lubrication issues in magnetic disk storage devices, *IEEE Trans. Magn.*, 32, 1812–1818 (1996).
6. A. Khurshudov, R. Waltman, Tribology challenges of modern magnetic hard disk drives, *Wear*, 251, 1124–1132 (2001).
7. B. Bhushan, A. Kulkarni, M. Boehm, V. Koinkar, L. Odoni, C. Martelet, M. Belin, Microtribological characterization of self-assembled and Langmuir–Blodgett monolayers by atomic and friction force microscopy, *Langmuir*, 11, 3189–3198 (1995).
8. R. Maboudian, R. Howe, Critical review: Adhesion in surface micromechanical structure, *J. Vac. Sci. Technol. B*, 15, 1–20 (1997).
9. B. Bhushan (ed.), *Modern Tribology Handbook*, Vol. 2, CRC Press, Boca Raton, FL (2001).
10. S. Henck, Lubrication of digital micromirror devices, *Tribol. Lett.*, 3, 239–247 (1997).
11. H. Liu, I. Ahmed, M. Scherge, Microtribological properties of silicon and silicon coated with diamond like carbon, octadecyltrichlorosilane and stearic acid cadmium salt films: A comparative study, *Thin Solid Films*, 381, 135–142 (2001).
12. J. Ruhe, V. Novotny, T. Clarke, G. Street, Ultra-thin perfluoropolyether films—Influence of anchoring and mobility of polymers on the tribological properties, *ASME Trans. J. Tribol.*, 118, 663–668 (1996).
13. R. Wang, R. White, S. Meeks, B. Min, A. Kellock, A. Homola, D. Yoon, The interaction of perfluoro-polyether lubricant with hydrogenated carbon, *IEEE Trans. Magn.*, 32, 3777–3779 (1996).
14. B. Bhushan, Z. Zhao, Friction/stiction and wear studies of magnetic thin film disks with two polar perfluoroether lubricant, *IEEE Trans. Magn.*, 33, 918–925 (1997).
15. K. Eapen, S. Patton, J. Zabinski, Lubrication of microelectromechanical systems (MEMS) using bound and mobile phase of Fomblin Zdol, *Tribol. Lett.*, 12, 35–41 (2002).
16. E. Yoon, R. Singh, H. Kong, Enhancement of nano/micro-tribological behaviour by surface modification, in: *Proceedings of the 1st International Conference on Manufacturing, Machine Design and Tribology (ICMDT)*, Seoul, Korea, pp. 1–4 (2005).
17. A. Ulman, Formation and structure of self-assembled monolayers, *Chem. Rev.*, 96, 1533–1554 (1996).
18. M. Wang, K. Liechti, Q. Wang, J. White, Self-assembled silane monolayers: Fabrication with nanoscale uniformity, *Langmuir*, 21, 1848–1857 (2005).
19. V. Depalma, N. Tillman, Friction and wear of self-assembled trichlorosilane monolayer films on silicon, *Langmuir*, 5, 868–872 (1989).
20. A. Ulman, *An Introduction to Ultrathin Organic Films: From Langmuir–Blodgett to Self-Assembly*, Academic Press, Amsterdam, the Netherlands (1991).
21. S. Ren, S. Yang, Y. Zhao, J. Zhou, T. Xu, W. Liu, Friction and wear studies of octadecyltrichlorosilane SAM on silicon, *Tribol. Lett.*, 13, 233–239 (2002).
22. H. Sung, J. Yang, D. Kim, S. Shin, Micro/nano-tribological characteristics of self-assembled monolayer and its application in nano-structure fabrication, *Wear*, 255, 808–818 (2003).
23. D. Janssen, R. Palma, S. Verlaak, P. Heremans, W. Dehaen, Static solvent contact angle measurements, surface free energy and wettability determination of various self-assembled monolayers on silicon dioxide, *Thin Solid Films*, 515, 1433–1438 (2006).
24. J. Ruhe, V. Novotny, K. Kanazawa, T. Clarke, G. Street, Structure and tribological properties of ultrathin alkylsilane films chemisorbed to solid surfaces, *Langmuir*, 9, 2383–2388 (1993).
25. J. Choi, T. Ishida, T. Kato, S. Fujisawa, Self-assembled monolayer on diamond-like carbon surface: Formation and friction measurements, *Tribol. Int.*, 36, 285–290 (2003).
26. H. Kim, T. Koini, T. Randall Lee, P. Scott, Systematic studies of the frictional properties of fluorinated monolayers with atomic force microscopy: Comparison of CF_3- and CH_3-terminated films, *Langmuir*, 13, 7192–7196 (1997).

27. H. Kim, M. Graupe, O. Oloba, T. Koini, S. Imaduddin, T. Lee, S. Perry, Molecularly specific studies of the frictional properties of monolayer films: A systematic comparison of CF$_3$-, (CH$_3$)$_2$CH-, and CH$_3$-terminated films, *Langmuir*, 15, 3179–3185 (1999).

28. S. Lee, Y. Shon, R. Colorado, R. Guenard, T. Lee, S. Perry, The influence of packing densities and surface order on the frictional properties of alkanethiol self-assembled monolayers (SAMs) on gold: A comparison of SAMs derived from normal and spiroalkanedithiols, *Langmuir*, 16, 2220–2224 (2000).

29. Y. Shon, S. Lee, R. Colorado, S. Perry, T. Lee, Spiroalkanedithiol-based SAMs reveal unique insight into the wettabilities and frictional properties of organic thin films, *J. Am. Chem. Soc.*, 122, 7556–7563 (2000).

30. E. Barrena, C. Ocal, M. Salmeron, A comparative AFM study of the structural and frictional properties of mixed and single component films of alkanethiols on Au(111), *Surf. Sci.*, 482–485, 1216–1221 (2001).

31. S. Ren, S. Yang, Y. Zhao, Micro- and macro-tribological study on a self-assembled dual-layer film, *Langmuir*, 19, 2763–2767 (2003).

32. M. Choudross, G. Grest, M. Stevens, Friction between alkylsilane monolayers: Molecular simulation of ordered monolayers, *Langmuir*, 18, 8392–8399 (2002).

33. J. Ma, J. Liu, Y. Mo, M. Bai, Effect of multiply-alkylated cyclopentane (MAC) on durability and load-carrying capacity of self-assembled monolayers on silicon wafer, *Colloids Surf. A*, 301, 481–489 (2007).

34. J. Brzoska, I. Azouz, F. Rondel, Silanization of solid substrates: A step toward reproducibility, *Langmuir*, 10, 4367–4372 (1994).

35. J. Kushmerick, M. Hankins, M. De Boer, P. Clews, R. Carpick, B. Bunnker, The influence of coating structure on micromachine stiction, *Tribol. Lett.*, 10, 103–108 (2001).

36. J. Ma, C. Pang, Y. Mo, M. Bai, Preparation and tribological properties of multiply-alkylated cyclopentane (MAC)-octadecyltrichlorosilane (OTS) double-layer film on silicon, *Wear*, 263, 1000–1007 (2007).

37. B. Yu, F. Zhou, Z. Mu, Y. Liang, W. Liu, Tribological properties of ultra-thin ionic liquid films on single-crystal silicon wafers with functionalized surfaces, *Tribol. Int.*, 39, 879–887 (2006).

38. J. Choi, M. Kanwaguchi, T. Kato, Nanoscale lubricant with strongly bonded phase and mobile phase, *Tribol. Lett.*, 15, 353–358 (2003).

39. J. Choi, H. Morishita, T. Kato, Frictional properties of bilayered mixed lubricant films on an amorphous carbon surface: Effect of alkyl chain length and SAM/PFPE portion, *Appl. Surf. Sci.*, 228, 190–200 (2004).

40. M. Masuko, H. Miyamoto, A. Suzuki, Tribological characteristics of self-assembled monolayer with siloxane bonding to Si surface, *Tribol. Int.*, 40, 1587–1596 (2007).

41. K. Deng, R. Collins, M. Mehregany, C. Sukenik, Performance impact of monolayer coating of polysilicon micromotors, *J. Electrochem. Soc.*, 142, 1278–1285 (1995).

42. K. Hwan, D. Kim, Investigation of the tribological behavior of octadecyltrichlorosilane deposited on silicon, *Wear*, 251, 1169–1176 (2001).

43. J. Ding, P. Wong, J. Yang, Friction and fracture properties of polysilicon coated with self-assembled monolayers, *Wear*, 260, 209–214 (2006).

44. B. Tao, C. Xian-Hua, Investigation of the tribological behavior of 3-mercaptopropyl trimethoxysilane deposited on silicon, *Wear*, 261, 730–737 (2006).

45. H. Liu, B. Bhushan, Investigation of nanotribological properties of self-assembled monolayers with alkyl and biphenyl spacer chains, *Ultramicroscopy*, 91, 185–202 (2002).

46. E. Yoon, S. Yang, H. Han, H. Kong, An experimental study on the adhesion at a nanocontact, *Wear*, 254, 974–980 (2003).

47. H. Ahn, P. Cuong, S. Park, Y. Kim, J. Lim, Effect of molecular structure of self-assembled monolayers on their tribological behaviors in nano- and microscales, *Wear*, 255, 819–825 (2003).

48. G. Qinlin, C. Xianhua, Tribological behaviors of self-assembled 3-aminopropyl-triethoxysilane films on silicon, *Current Appl. Phys.*, 8, 583–588 (2008).
49. I. Luzinov, D. Julthongpiput, V. Gorbunov, V. Tsukruk, Nanotribological behavior of tethered reinforced polymer nanolayer coatings, *Tribol. Int.*, 34, 327–333 (2001).
50. R. Arvind Singh, J. Kim, S. Yang, J.-E. Oh, E.-S. Yoon, Tribological properties of tri-chlorosilane-based one- and two-component self-assembled monolayers, *Wear*, 265, 42–48 (2008).
51. E. Yoon, R. Singh, H. Oh, H. Kong, The effect of contact area on nano/micro-scale friction, *Wear*, 259, 1424–1431 (2005).
52. C. Mastrangelo, Adhesion-related failure mechanisms in micromechanical devices, *Tribol. Lett.*, 3, 223–238 (1997).
53. R. Singh, E. Yoon, H. Oh, H. Kong, Nano-scale adhesion and friction in Si-wafer with the tip size using AFM, *KSTLE Int. J.*, 5, 1–6 (2004).
54. R. Singh, E. Yoon, H. Han, H. Kong, Friction behavior of chemical vapor deposited self-assembled monolayers on silicon wafer, *Wear*, 262, 130–137 (2007).
55. B. Bhushan, H. Liu, Nanotribological properties and mechanisms of alkylthiol and biphenyl thiol self-assembled monolayers studied by AFM, *Phys. Rev. B*, 63, (245412), 1–11 (2001).
56. F. Bowden, D. Tabor, *The Friction and Lubrication of Solids*. Clarendon Press, Oxford, U.K. (1950).
57. K. Tupper, D. Brenner, Molecular dynamics simulations of friction in self-assembled monolayers, *Thin Solid Films*, 253, 185–189 (1994).
58. W. Zisman, Friction, durability and wettability properties of monomolecular films on solids, in: *Proceedings of Symposium of Friction and Wear*, R. Davies (ed.), pp. 110–148, Elsevier, Amsterdam, the Netherlands (1959).
59. Y. Mo, M. Zhu, M. Bai, Preparation and nano/microtribological properties of perfluo-rododecanoic acid (PFDA)-3-aminopropyltriethoxysilane (APS) self-assembled dual-layer film deposited on silicon, *Colloids Surf. A*, 322, 170–176 (2008).
60. J. Zhao, M. Chen, Y. An, J. Liu, F. Yan, Preparation of polystyrene brush film by radical chain-transfer polymerization and micromechanical properties, *Appl. Surf. Sci.*, 255, 2295–2302 (2008).
61. A. Schmohl, A. Khan, P. Hess, Functionalization of oxidized silicon surfaces with methyl groups and their characterization, *Superlatt. Microstruct.*, 36, 113–121 (2004).
62. M. Rong, M. Zhang, G. Shi, Q. Ji, B. Wetzel, K. Friedrich, Graft polymerization onto inorganic nanoparticles and its effect on tribological performance improvement of poly-mer composites, *Tribol. Int.*, 36, 697–707 (2003).
63. K. Jradi, Study of the tribological behaviour of graphites powders: Transfer at nano and macroscopique scales, Doctoral thesis. Université de Haute–Alsace, France (1997).
64. M. Schmitt, S. Bistac, K. Jradi, Tribological behaviour of graphite powders at nano- and macroscopic scales, *J. Phys. Conf. Ser.*, 61, 1032–1036 (2007).
65. K. Jradi, M. Schmitt, S. Bistac, Surface modifications induced by the friction of graph-ites against steel, *Appl. Surf. Sci.*, 255, 4219–4224 (2009).
66. D. Allara, A. Parikh, F. Rondelez, Evidence for a unique chain organization in long chain silane monolayers deposited on two widely different solid substrates, *Langmuir*, 11, 2360–2387 (1995).
67. M. Chabinyc, X. Chen, R. Holmlin, H. Jacobs, H. Skulason, C. Frisbie, V. Mujica, M. Ratner, M. Rampi, G. Whitesides, Molecular rectification in a metal-insulator-metal junction based on self-assembled monolayers, *J. Am. Chem. Soc.*, 124, 11730–11736 (2002).
68. M. Azzopardi, Reactions of organosilanes at the silicon—Solution interface. In situ study by infrared spectroscopy, Doctoral thesis. Université Paris VI, France (1994).
69. C. Tripp, M. Hair, An infrared study of the reaction of octadecyltrichlorosilane with silica, *Langmuir*, 8, 1120–1126 (1992).

70. M. Azzopardi, H. Arribart, In situ FTIR study of the formation of an organosilane layer at the silica/solution interface, *J. Adhesion*, 46, 103–115 (1994).
71. K. Blodgett, Films built by depositing successive monomolecular layers on a solid surface, *J. Am. Chem. Soc.*, 57, 1007–1022 (1935).
72. K. Blodgett, Built-up films of barium stearate and their optical properties, *Phys. Rev.*, 51, 964–982 (1937).
73. W. Bigelow, D. Pickett, W. Zisman, Oleophobic monolayers. I. Films adsorbed from solution in non-polar liquid, *J. Colloid Interface Sci.*, 1, 513–538 (1946).
74. G. Gaines, *Insoluble Monolayers at Liquid-Gas Interfaces*. Interscience, New York (1966).
75. F. Etzler, Characterization of surface free energies and surface chemistry of solids, in: *Contact Angle, Wettability and Adhesion*, Vol. 3, K. Mittal (ed.), pp. 219–264, VSP, Utrecht, the Netherlands (2003).
76. F. Fowkes, Attractive forces at interfaces, *Ind. Eng. Chem.*, 56, 40–52 (1964).
77. P. Diss, Study of the physical and chemico-physical mechanisms depending on the sliding speed in a dry graphite/steel tribo-contact. Consequences for the friction force and the topography of the third body, Doctoral thesis. Université de Haute–Alsace, France (1997).
78. J. Lancaster, J. Pritchard, On the dusting wear regime of graphite sliding against carbon, *J. Phys. D*, 13, 1551–1564 (1980).
79. J. Lancaster, J. Pritchard, The influence of environment and pressure on the transition to dusting wear of graphite, *J. Phys. D*, 14, 747–762 (1981).
80. J. Midgley, D. Teer, Surface orientation and friction of graphite, graphitic carbon and non-graphitic carbon, *Nature*, 4766, 735–736 (1961).
81. J. Skinner, N. Gane, D. Tabor, Micro-friction of graphite, *Nat. Phys. Sci.*, 232, 195–196 (1971).
82. B. Marchon, J. Carrazza, H. Heinemann, G. Somorjai, TPD and XPS studies of O_2, CO_2, and H_2O adsorption on clean polycrystalline graphite, *Carbon*, 26, 507–514 (1988).
83. H. Harker, J. Horsley, R. Dobson, Active centres produced in graphite by powdering, *Carbon*, 9, 1–9 (1971).
84. P. Delahès, *Graphite and Precursors (World of Carbon)*, Vol. 1. Lavoisier, Paris, France (2000).
85. R. Savage, Graphite lubrication, *J. Appl. Phys.*, 19, 1–10 (1948).
86. R. Savage, D. Shaffer, Vapor lubrication of graphite sliding contact, *J. Appl. Phys.*, 27, 136–138 (1956).
87. G. Rowe, Some observations on the frictional behaviour of boron nitride and of graphite, *Wear*, 3, 274–285 (1960).
88. P. Rehbinder, E. Shchukin, Surface phenomena in solids during deformation and fracture processes, *Prog. Surf. Sci.*, 3, 97–188 (1972).
89. V. Savenko, E. Shchukin, New applications of the Rehbinder effect in tribology. A review, *Wear*, 194, 86–94 (1996).
90. E. Shchukin, Physical–chemical mechanics in the studies of Peter A. Rehbinder and his school, *Colloids Surf. A*, 149, 529–537 (1999).

4 Frictional Properties of Physisorbed Layers of Self-Organized Molecules at Solid–Liquid Interface

Koji Miyake

CONTENTS

ABSTRACT

In this chapter, nanoscale to macroscale investigation of the frictional properties of physisorbed layers of self-organized phthalocyanine derivatives is discussed. First, frictional properties of self-assembled monolayers and Langmuir–Blodgett films are briefly summarized. Then, frictional properties of physisorbed layers of self-organized molecules observed at solid-liquid interface are discussed. The discussion is in terms of comparing the frictional properties of the physisorbed layers of self-organized molecules on nanoscale with the properties on macroscale.

4.1   INTRODUCTION

Lubricants and additives are used for maintaining friction and wear at optimal levels. They are physisorbed on the surfaces of sliding materials and prevent direct contact between these materials. As a result, they reduce friction and improve wear resistance. The molecular structures of these lubricants and/or additives are determinants in producing a desired effect. However, at the molecular level, the relationships between the molecular structure, surface adsorbed structure, and the effects obtained are not yet clear. To understand the behavior of physisorbed molecules, the selection of molecules is important. For example, controllable adsorbed molecular structure is needed to understand the relationship between the surface adsorbed structure and frictional property.

Recent progress in measurement technology involving the surface force apparatus (SFA) [1] and scanning probe microscopy (SPM) [2] makes it possible to analyze the origin of friction at the atomic and/or molecular level [3–6], and intriguing observations such as superlubricity [7,8] and molecular bearing [9] have been made. In addition, recent progress in modeling also contributes to the understanding of the frictional properties at the atomic and/or molecular level [10–13]. Here, nanoscale to macroscale investigation of the frictional properties of physisorbed layers of self-organized phthalocyanine derivatives is discussed. First, frictional properties of self-assembled monolayers (SAMs) and Langmuir–Blodgett (LB) films are briefly summarized. Then, we focus on the frictional properties of alkyl-substituted phthalocyanine derivatives physisorbed on highly oriented pyrolytic graphite (HOPG).

4.2   FRICTIONAL PROPERTIES OF SELF-ASSEMBLED MONOLAYERS

According to the frictional properties of SAMs on solid surfaces analyzed by friction force microscopy (FFM), the friction force decreases with reduced adhesion [14–17]. In addition, the frictional properties of SAMs correlate to molecular packing at the surface [12,18–22]. The results indicate that friction force decreases as the molecular packing density increases. A detailed discussion of frictional properties of SAMs can be found in Chapters 1 and 2.

4.3 FRICTIONAL PROPERTIES OF LANGMUIR–BLODGETT FILMS

AFM and FFM measurements on LB films have shown that a molecular resolution can be achieved [23]. FFM can also be used to image and identify compositional domains, i.e., additional information, such as chemical composition, provided by the friction measurement. Furthermore, LB films were used as a model to study lubrication on a microscopic scale and to examine the initial stage of wear [24]. One- and two-bilayer films of cadmium arachidate are used as a model lubricant. A reduction in friction was observed for LB film–covered surfaces, compared to the bare substrates. Wearless friction was observed for low load. With higher loads, the onset of wear was observed as a displacement of clusters of molecules. The observed ability of the molecules to remain in the ordered state illustrates one of the fundamental origins of boundary lubrication. Furthermore, it is known that a longer life can be obtained with cadmium arachidate films that have a longer alkyl chain [25].

Generally, the effect of load on the friction behavior of a LB film is much complicated. It depends on the film-forming material and thickness and the magnitude of load [26]. About the effect of the number of layers on the friction behavior of a LB film, there is a threshold in the number of layers of LB film for effective lubrication for a given environment and substrate. Below this threshold, the greater the number of layers is, the better the lubrication performance of LB film. Otherwise, above this threshold, the number of layers in the LB film has no effect on its frictional properties [27]. The tribological properties of the LB films, SAMs, and the molecular deposition films were reviewed in Ref. [28].

4.4 FRICTIONAL PROPERTIES OF PHYSISORBED LAYERS OF SELF-ORGANIZED MOLECULES OBSERVED AT SOLID–LIQUID INTERFACE

The frictional properties of SAMs and LB films were measured under ambient conditions in many cases. In other words, test conditions were very different from conditions in practical use. Analysis of the frictional behavior of lubricants/additives and the relationship between the nanoscopic and macroscopic frictional behaviors needs to measure friction of physisorbed molecules at the solid–liquid interface. Despite the fact that the tribological properties of LB films and SAMs have been studied extensively, the relationship between the nanoscale and macroscale frictional properties is still unclear. To understand accurately the lubrication behavior in a range from nanoscale to macroscale, it is important what kinds of molecules and what kinds of apparatuses we use.

4.4.1 EXPERIMENTAL APPARATUS FOR COMPARING NANOSCOPIC TO MACROSCOPIC FRICTIONAL PROPERTIES

The geometry of a pin-on-plate tribometer [29] (Figure 4.1) is analogous to that of AFM, i.e., a spherical-shaped surface slides against a flat surface. Therefore, the relationship

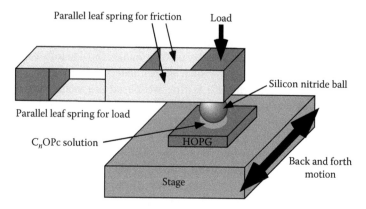

FIGURE 4.1 (See color insert.) Schematic model of pin-on-plate tribometer. (From Miyake, K. et al., *Tribol. Lett.*, 31, 9, 2008. With permission.)

between the nanoscopic and macroscopic frictional behaviors will be clarified by comparing the results of pin-on-plate tribometer with those of AFM. Consequently, AFM and pin-on-plate tribometer should be useful for measuring the nanoscopic and macroscopic frictional properties of physisorbed molecules, respectively.

4.4.2 MOLECULES

It is well known that alkyl-substituted porphyrins and phthalocyanines are arranged in lines or discrete states with regular spacing on a HOPG surface under ambient conditions [30,31]. These molecules physisorb on each substrate with their molecular planes parallel to the substrate surface. In addition, phthalocyanines are known as solid lubricants [32]. Therefore, these molecules will be ideal models of monolayer lubricants. Here, two phthalocyanine derivatives, dodecyloxy-substituted ($C_{12}OPc$) and octadecyloxy-substituted ($C_{18}OPc$) phthalocyanines, were used as models of lubricant molecules [29,33]. The molecular structures are shown in Figure 4.2.

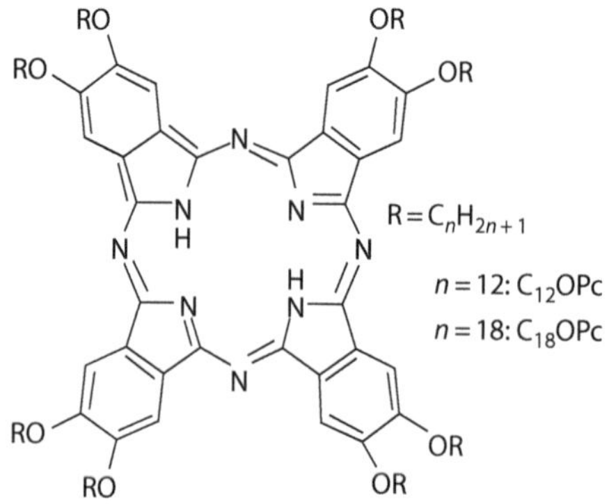

$$R = C_nH_{2n+1}$$

$$n = 12: C_{12}OPc$$
$$n = 18: C_{18}OPc$$

FIGURE 4.2 Molecular structure of alkyl-substituted phthalocyanine derivative. *R* is alkyl substituent. *n* is the number of carbon atoms of alkyl substituent.

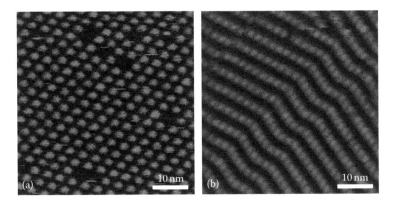

FIGURE 4.3 STM images of self-organized structures of (a) $C_{12}OPc$ and (b) $C_{18}OPc$.

Liquid samples were prepared by dissolving the molecules in phenyloctane at concentrations of less than 1 mM. These two molecules were physisorbed on the substrate with their molecular planes parallel to the substrate surface. The STM images of $C_{12}OPc$ and $C_{18}OPc$ are shown in Figure 4.3. The relationships between the alkyl chain length and the 2D self-organized structures are discussed in Ref. [34].

4.4.3 NANOSCOPIC FRICTIONAL PROPERTIES OF PHYSISORBED PHTHALOCYANINE DERIVATIVES MEASURED BY AFM

The normal load dependence of the friction force of phthalocyanine derivatives consists of two regions when measuring nanoscopic frictional behavior using AFM [29,33]. Figure 4.4 shows the friction force versus normal load curves for (a) $C_{12}OPC$ and (b) $C_{18}OPc$. In one region, the friction force varies as a power law with an exponent of approximately 0.6–0.7 (Region I), and in the other, the friction force is approximately constant (Region II). In order to clarify the origin of the dependence of the friction force on normal load, we simultaneously observed the topographic and frictional images.

Figure 4.5 shows (a) the topographic and (b) the frictional images of $C_{18}OPc$ at approximately 7 nN, which is the normal load below the transition point (~21 nN). Bright and dark regions were observed in both images. Figure 4.5c and d shows the topographic and the frictional images of $C_{18}OPc$ at approximately 27 nN, which is the normal load beyond the transition point (~21 nN). A slow scan was performed from bottom to top. The normal load was changed from 27 to 7 nN at the dashed line drawn in Figure 4.5c and d. When the normal load exceeded the transition point, the patterns of height difference observed at low load disappeared, and both topographic and frictional images became flat. Furthermore, the patterns of height difference were observed again after the normal load fell below the transition point.

Figure 4.5e shows the average height and friction force variation in the area surrounded by rectangles in Figure 4.5a and b. The height difference between two triangles in Figure 4.5e is about 0.25 nm, which is different from the single-step height of the graphite substrate. Furthermore, the topographically high region shows

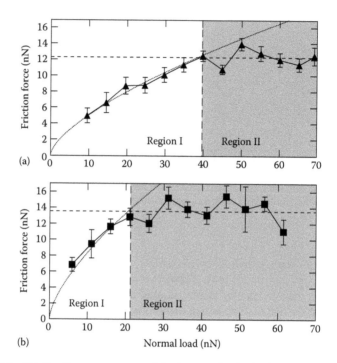

FIGURE 4.4 Friction force versus normal load curves for (a) $C_{12}OPc$ and (b) $C_{18}OPc$ measured using AFM. The dashed line in the figure indicates the average value of friction force in the high-load region (Region II). In the low-load region (Region I), the dotted line indicates the fitted curve of friction force to the 2/3 power of the applied normal load predicted by the Hertz model. (From Miyake, K. et al., *Jpn. J. Appl. Phys.*, 44, 5403, 2005. With permission.)

low friction and the topographically low region shows high friction. In the case of the clean HOPG surface, approximately 0.3 nm height step structures, which coincided with the single-step height of graphite, were observed. In addition, the friction force observed at the upper and lower terraces had the same value. If the height difference observed on the phthalocyanine-adsorbed surface corresponds to the graphite steps, the friction force of the upper and lower terraces must have the same value. These results indicate that physisorbed phthalocyanines may exist on the topographically high regions observed at the normal load below the transition point.

STM results show that disordered structures coexist with ordered structures, as shown in Figure 4.6a. The bare HOPG surface was also observed on the same sample. The height difference between ordered structures and the bare HOPG surface is about 0.20–0.25 nm, as shown in Figure 4.6b. Even though STM contrast does not reflect the real height of adsorbed molecules, the height difference observed by STM coincides with that observed by AFM. These results led us to conclude that the higher regions in the topographic images observed at the normal load below the transition point correspond to a monolayer of physisorbed phthalocyanines. At high load, substantially constant friction force was observed and the forces were independent of the length of the substituted alkane moieties. This indicates that the

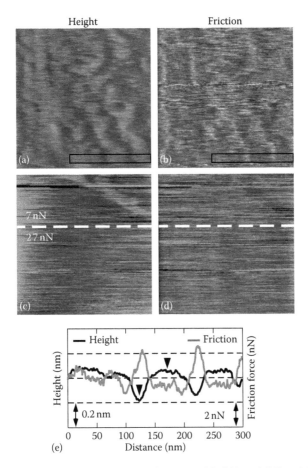

FIGURE 4.5 Simultaneous observation of topographic [(a) and (c)] and frictional [(b) and (d)] images of $C_{18}OPc$. The normal load during observation was set at 7 nN for (a) and (b) and at 27 nN for (c) and (d). A slow scan was performed from bottom to top for all images. The normal load was changed from 27 to 7 nN at the dashed line drawn in (c) and (d). (e) Average height and friction force variation in the area surrounded by rectangles in (a) and (b). (From Miyake, K. et al., *Jpn. J. Appl. Phys.*, 44, 5403, 2005. With permission.)

monolayer of phthalocyanine derivatives was completely desorbed from the surface at high load.

Figure 4.7 shows the schematic illustrations of the dynamic layer structures of C_nOPc films. In the low-load region, shear may be accomplished by the sliding of the monolayer of C_nOPc. Therefore, the force required to shear the contact area may relate to the intermolecular interaction of phthalocyanine molecules and the interaction between molecules and substrate. The monolayer of C_nOPc was completely desorbed from the surface at high load. Therefore, the tip may slide on the graphite surface. However, additional energy dissipation for order–disorder transition and/or desorption of molecules may be needed at high load.

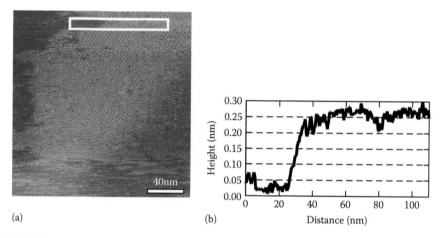

FIGURE 4.6   (a) STM image of $C_{18}OPc$. (b) Average height variation in the area surrounded by the rectangle in (a).

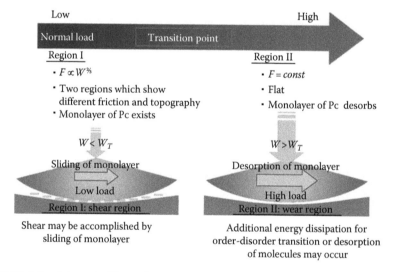

FIGURE 4.7   Schematic illustrations of the dynamic layer structures of phthalocyanine films.

4.4.4   Macroscopic Frictional Properties of Physisorbed Phthalocyanine Derivatives Measured by Pin-on-Plate Tribometer

Contrary to the nanoscopic results, the friction force was proportional to the normal load and the coefficient of friction was less than 0.1. Furthermore, the frictional property of $C_{12}OPc$ was nearly identical to that of $C_{18}OPc$. These results strongly suggest that the origins of the macroscopic and nanoscopic frictional behaviors were different. Figure 4.8 shows the friction force versus the normal load curves for $C_{12}OPc$ and $C_{18}OPc$ measured using pin-on-plate tribometer.

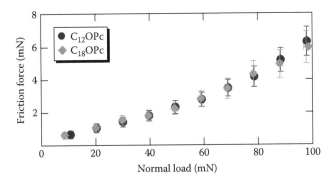

FIGURE 4.8 Friction force versus normal load curves for $C_{12}OPc$ and $C_{18}OPc$ measured using pin-on-plate tribometer.

4.4.5 COMPARISON OF NANOSCOPIC BEHAVIOR TO MACROSCOPIC BEHAVIOR

To compare the nanoscopic behavior with the macroscopic behavior universally, a Stribeck curve, which is generally used to discuss lubrication regimes that are dependent on contact pressure, the sliding velocity, and the viscosity of lubricant, is useful. The Stribeck curve represents the relationship between the coefficient of friction μ and the Stribeck parameter $\eta V/W$, where η is the viscosity of the liquid, V is the sliding velocity, and W is the normal load. Because the coefficient of friction is a dimensionless quantity, the Stribeck parameter must also be a dimensionless quantity. Therefore, the normal load per arbitrary unit length (W/L) is generally used instead of W.

Assuming the lubricants are a Newtonian liquid, the shear stress, F/A, is expressed as

$$\frac{F}{A} = \eta \frac{V}{h},$$ (4.1)

where
  F is the friction force
  A is the contact area
  η is the viscosity of the liquid
  h is the separation between the two sliding materials

By dividing Equation 4.1 by the normal load W, we obtain the coefficient of friction μ:

$$\mu = \frac{F}{W} = \frac{1}{h}\frac{\eta V}{W/A} = \frac{1}{h}\frac{\eta V}{P},$$ (4.2)

where $P = W/A$ is the contact pressure. Assuming the separation between the two sliding materials h is approximated by the elastic deformation in the Z-direction (perpendicular to the sample surface) δ, Equation 4.2 becomes

$$\delta = \frac{a^2}{R}, A = \pi a^2 \qquad\qquad (4.3)$$

$$\mu = \pi \eta \frac{V}{W/R}. \qquad\qquad (4.4)$$

Here, a is the contact radius and R is the radius of the curvature of the tip. Equation 4.4 clearly indicates that we can universally compare the macroscopic frictional behavior with the nanoscopic frictional behavior using a Stribeck curve with a parameter of $\eta\ V/(W/R)$. This also indicates that the radius of the curvature of the tip was chosen as an arbitrary unit length. Generally, the contact radius is used as the unit length. When the contact radius is chosen as the unit length, the unit length is affected by the elastic modulus of materials and normal load. When the radius of the tip curvature was chosen as arbitrary unit length, the normal load, arbitrary unit length, and velocity are directly measurable and independent variables. Therefore, the choice of the radius of the tip curvature as the arbitrary unit length is suitable for comparing frictional properties over a wide range [29].

4.4.6 BEHAVIOR OF CONFINED LIQUID

It is well known that the effective dynamic viscosity of liquids confined between parallel plates with separations of less than 1 nm are many orders of magnitude larger than the bulk value [35–41]. Therefore, it is likely that the effective viscosity of lubricants containing additives increases when the separation between the two sliding objects is reduced by increasing the contact pressure or shear rate. While hydrodynamic lubrication has been observed for tip-substrate separation distances as small as a few nanometers using SPM [42], a solid-like response is typically observed for polymer lubricants when using SFA [43–45]. Furthermore, the results of shear measurements using SFA have shown that the effective dynamic viscosity of liquids confined between parallel plates with separations less than 1 nm was many orders of magnitude larger than the bulk value [35]. The difference in the behaviors of SPM and SFA is most likely related to the vastly different geometries of the SFA and SPM experiments. For SPM, the geometry is one of spherical-shaped surfaces sliding against flat surfaces, but for SFA, liquids are sheared between two parallel and smooth surfaces. Anyway, the change in the viscosity of the liquid depending on the experimental setup must be considered when comparing the nanoscopic behavior with macroscopic behavior of frictional properties.

4.4.7 STRIBECK CURVE

Figure 4.9 shows the Stribeck curves measured by pin-on-plate tribometer and AFM. Here, it is assumed that the viscosity of the liquid was constant, i.e., $V/(W/R)$ was used as the Stribeck parameter for descriptive purposes. An AFM tip does destroy or push away the molecules on the graphite at a higher load as mentioned above.

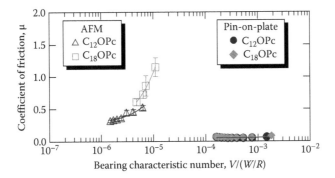

FIGURE 4.9 (See color insert.) Stribeck curves with a parameter of $V/(W/R)$, where V is the sliding velocity, W is the normal load, and R is the radius of the curvature of the tip. Open triangles and squares correspond to the data obtained using AFM for $C_{12}OPc$ and $C_{18}OPc$, respectively. Filled circles and diamonds correspond to the data obtained using pin-on-plate tribometer for $C_{12}OPc$ and $C_{18}OPc$, respectively. The thick lines indicate the fitted curve by linear correlation.

The critical normal load was about $40\,nN$ for $C_{12}OPc$ and about $20\,nN$ for $C_{18}OPc$. Therefore, only the AFM data below the critical normal load were plotted in Figure 4.9. The AFM results are clearly different from those of the pin-on-plate results. This difference can be explained by the change in the intermolecular interactions of alkanes and the interaction between the alkanes and graphite substrate. The effect of the viscosity of the liquid on the frictional properties was not considered in Figure 4.9. To explain the difference between nanoscopic and macroscopic behaviors, the effect of the change in the viscosity of the liquid depending on the experimental setup must be considered. Here, the contact pressure in the AFM measurement was much higher than that in the pin-on-plate tribometer. The effective viscosity under the AFM conditions is likely to be much larger than that under the tribometer conditions.

To estimate the effective viscosity of $C_{12}OPc$ and $C_{18}OPc$ under both the AFM and tribometer conditions, Stribeck curve was fitted by a linear correlation between μ and $V/(W/R)$. The thick lines in Figure 4.9 indicate the Stribeck curve fitted by linear correlation. In the case of pin-on-plate tribometer measurements, the effective viscosity of $C_{12}OPc$ ($0.69\,Pa \cdot s$) was nearly identical to that of $C_{18}OPc$ ($0.67\,Pa \cdot s$). On the other hand, $C_{18}OPc$ ($2.6 \times 10^4\,Pa \cdot s$) showed a higher effective viscosity than $C_{12}OPc$ ($1.4 \times 10^4\,Pa \cdot s$) under the AFM conditions. Furthermore, the effective viscosities of C_nOPc under AFM conditions are four orders of magnitude larger than those under tribometer conditions. The observed effective viscosity is related to the molecular behavior of additives under AFM conditions.

To consider the effect of the effective viscosity on the lubrication behavior, the Stribeck curve is redrawn with the parameter $\eta_{eff}\,V/(W/R)$. Figure 4.10 shows a Stribeck curve with the parameter $\eta_{eff}\,V/(W/R)$. It appears that the curves for $C_{12}OPc$ and $C_{18}OPc$ vary continuously. Thus the macroscopic frictional behavior can be universally compared with nanoscopic frictional behavior using a Stribeck curve

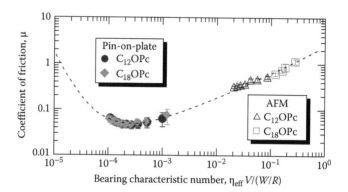

FIGURE 4.10 (See color insert.) Stribeck curve with a parameter of $\eta_{eff} V/(W/R)$, where η_{eff} is the effective viscosity, V is the sliding velocity, W is the normal load, and R is the radius of the curvature of the tip. Open triangles and squares correspond to the data obtained using AFM for $C_{12}OPc$ and $C_{18}OPc$, respectively. Filled circles and diamonds correspond to the data obtained using pin-on-plate tribometer for $C_{12}OPc$ and $C_{18}OPc$, respectively. The dashed line traces the data on the basis of the shape of the Stribeck curve.

with a parameter $\eta_{eff} V/(W/R)$. These results indicate that hydrodynamic lubrication occurred at a contact pressure of about 1 GPa under the AFM tip.

The selection of the parameter $\eta_{eff} V/(W/R)$, which accounts for changes in the effective viscosity depending on experimental conditions, is effective for comparing the macroscale and nanoscale frictional properties of the physisorbed layers of self-organized molecules. In addition, a Stribeck curve may be useful as a master curve for analyzing the viscoelastic properties of lubricants and/or additives under various sliding conditions.

4.5 SPECTROSCOPIC ANALYSIS

Complementary information on the local chemical composition, rheology, or molecular conformation has proved difficult to obtain. One solution is to use molecular vibrational spectroscopy, either infrared (IR) or Raman, to examine behavior in and around the contact region. These techniques can also give information on molecular conformation [46], alignment [47], and the effects of pressure [48].

IR spectroscopy was applied to in-contact lubrication studies of a simple hydrocarbon fluid [49]. IR spectra were taken from the Hertzian region of a sliding contact lubricated by poly-α-olefin fluids (Figure 4.11). The results indicate that in the high-pressure region there is an increase in gauche defects in the alkyl chain, implying a more globular molecule with a lower volume but higher energy conformation. In thin lubricant films, a structured liquid phase is formed in the near-surface region where the alkyl chains are aligned parallel to the substrate in the extended conformation. In their experiments, it was possible to obtain good quality in-contact spectra from the CH stretch region for hydrocarbon films of less than 50 nm.

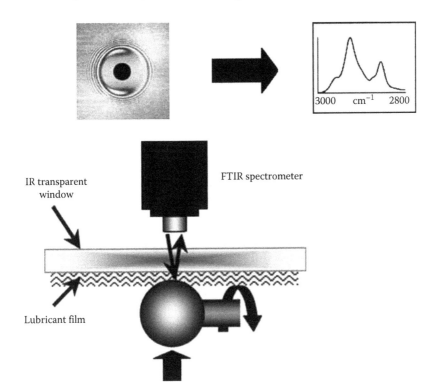

FIGURE 4.11 Sampling configuration for in-contact spectroscopy. (From Cann, P.M. and Spikes, H.A., *Tribol. Lett.*, 19, 289, 2005. With permission.)

IR spectroscopy cannot be used to monitor physisorbed molecular monolayer films, in situ, with an applied pressure or shear due to their limited sensitivity. Sum frequency generation (SFG) spectroscopy [50] is an ideal tool to study buried interfaces because of its high surface specificity. This method has been used to study the structure of organic molecules that are confined and compressed between a lens and a flat surface (Figure 4.12) [51,52]. For LB films and SAMs, relatively low pressure is needed to alter the end configuration of the chains and this change is completely reversible [51]. When various alkanethiols on gold are observed, well-ordered layers that differ only in the length of their alkane tails behave similarly [52]. However, when they are poorly packed or defective, the strong intermolecular stabilization of thicker layers does not protect them. The results suggest that the less dense packing of these physically formed layers allows them to relax significantly when compressed. An increased population of terminal gauche conformers is identified as the disordering mechanism under pressure.

SFG can be used to provide orientational information for molecular monolayers under high static pressure. Therefore, it is hoped that SFG can be used as a complementary approach for the analysis of physisorbed layers of self-organized molecules at the solid–liquid interface.

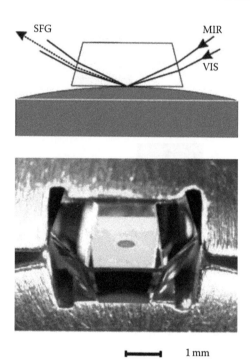

|⊢——————⊣| 1 mm

FIGURE 4.12   Schematic cross-sectional diagram of the pressure cell (upper figure), and photograph of the contact along the optic axis. The spherical surface is a gold-coated bronze ball with a 25 mm radius of curvature. The flat surface is a synthetic sapphire prism. These are forced together with a pneumatic piston, forming a round area of contact visible against total internal reflection from the prism face. An applied force of 95 N, as here, produces a contact 550 μm in diameter, with a central pressure of 660 MPa. Also shown are the paths of the visible (VIS), mid-infrared (MIR), and sum frequency generation (SFG) beams. (From Berg, O. and Klenerman, D., *J. Am. Chem. Soc.*, 125, 5493, 2003. With permission.)

4.6   SUMMARY

Nanoscale to macroscale investigation of the frictional properties of physisorbed layers of self-organized phthalocyanine derivatives has been discussed. Despite the fact that the tribological properties of LB films and SAMs have been studied extensively, the relationship between the nanoscale and macroscale frictional properties is still unclear. In addition, the frictional properties of SAMs and LB films were measured under ambient conditions in many cases. Analysis of the frictional behavior of lubricants/additives and the relationship between the nanoscopic and macroscopic frictional behaviors needs to measure the friction of physisorbed molecules at the solid–liquid interface. The present studies have shown that the physisorbed layers of alkyl-substituted phthalocyanine derivatives have the potential for the nanoscale to macroscale investigation of the frictional properties of physisorbed molecules. We hope that these molecular films lead to an understanding of the universality of friction from nanoscale to macroscale levels and will be widely used for tribological applications and nanolubrication in the near future.

REFERENCES

1. B. Bhushan, J. N. Israelachivili, and U. Landman, Nanotribology: Friction, Wear and Lubrication at the Atomic Scale, *Nature*, 374, 607–616 (1995).
2. R. W. Carpick and M. Salmeron, Scratching the Surface: Fundamental Investigations of Tribology with Atomic Force Microscopy, *Chem. Rev.*, 97, 1163–1194 (1997).
3. C. M. Mate, Force Microscopy Studies of the Molecular Origins of Friction and Lubrication, *IBM J. Res. Develop.*, 39, 617–627 (1995).
4. R. Luthi, E. Meyer, H. Haefke, L. Howald, W. Gutmannsbauer, and H. J. Guntherodt, Sled-Type Motion on the Nanometer Scale: Determination of Dissipation and Cohesive Energies of C_{60}, *Science*, 266, 1979–1981 (1994).
5. S. Fujisawa, K. Yokoyama, Y. Sugawara, and S. Morita, Analysis of Experimental Load Dependence of Two-Dimensional Atomic-Scale Friction, *Phys. Rev. B*, 58, 4909–4916 (1998).
6. A. Socoliuc, R. Bennewitz, E. Gnecco, and E. Meyer, Transition from Stick-Slip to Continuous Sliding in Atomic Friction: Entering a New Regime of Ultralow Friction, *Phys. Rev. Lett.*, 92, 134301 (2004).
7. M. Hirano, K. Shinjo, R. Kaneko, and Y. Murata, Observation of Superlubricity by Scanning Tunneling Microscopy, *Phys. Rev. Lett.*, 78, 1448–1451 (1997).
8. M. Dienwiebel, G. S. Verhoeven, N. Pradeep, J. W. M. Frenken, J. A. Heimberg, and H. W. Zandbergen, Superlubricity of Graphite, *Phys. Rev. Lett.*, 92, 126101 (2004).
9. K. Miura, S. Kamiya, and N. Sasaki, C_{60} Molecular Bearings, *Phys. Rev. Lett.*, 90, 055509 (2003).
10. N. Sasaki, M. Tsukada, S. Fujisawa, Y. Sugawara, S. Morita, and K. Kobayashi, Load Dependence of the Frictional-Force Microscopy Image Pattern of the Graphite Surface, *Phys. Rev. B*, 57, 3785–3786 (1998).
11. H. Tamura, M. Yoshida, K. Kusakabe, C. Young-Mo, R. Miura, M. Kubo, K. Teraishi, A. Chatterjee, and A. Miyamoto, Molecular Dynamics Simulation of Friction of Hydrocarbon Thin Films, *Langmuir*, 15, 7816–7821 (1999).
12. P. T. Mikulski and J. A. Harrison, Packing-Density Effects on the Friction of n-Alkane Monolayers, *J. Am. Chem. Soc.*, 123, 6873–6881 (2001).
13. B. Park, C. D. Lorenz, M. Chandross, M. J. Stevens, G. S. Grest, and O. A. Borodin, Frictional Dynamics of Fluorine-Terminated Alkanethiol Self-Assembled Monolayers, *Langmuir*, 20, 10007–10014 (2004).
14. D. C. Frisbie, L. F. Rozsnyai, A. Noy, M. S. Wrighton, and C. M. Lieber, Functional-Group Imaging by Chemical Force Microscopy, *Science*, 265, 2071–2074 (1994).
15. A. Noy, D. C. Frisbie, L. F. Rozsnyai, M. S. Wrighton, and C. M. Lieber, Chemical Force Microscopy—Exploiting Chemically-Modified Tips to Quantify Adhesion, Friction, and Functional-Group Distributions in Molecular Assemblies, *J. Am. Chem. Soc.*, 117, 7943–7951 (1995).
16. R. W. Carpick, N. Agrait, D. F. Ogletree, and M. Salmeron, Variation of the Interfacial Shear Strength and Adhesion of a Nanometer-Sized Contact, *Langmuir*, 12, 3334–3340 (1996).
17. V. V. Tsukruk, M. P. Everson, L. M. Lander, and W. J. Brittain, Nanotribological Properties of Composite Molecular Films: C_{60} Anchored to a Self-Assembled Monolayer, *Langmuir*, 12, 3905–3911 (1996).
18. X. D. Xiao, J. Hu, D. H. Charych, and M. Salmeron, Chain Length Dependence of the Frictional Properties of Alkylsilane Molecules Self-Assembled on Mica Studied by Atomic Force Microscopy, *Langmuir*, 12, 235–237 (1996).
19. A. Lio, D. H. Charych, and M. Salmeron, Comparative Atomic Force Microscopy Study of the Chain Length Dependence of Frictional Properties of Alkanethiols on Gold and Alkylsilanes on Mica, *J. Phys. Chem. B*, 101, 3800–3805 (1997).

20. K. Boschkova, A. Feiler, B. Kronberg, and J. J. R. Stalgren, Adsorption and Frictional Properties of Gemini Surfactants at Solid Surfaces, *Langmuir*, 18, 7930–7935 (2002).

21. X. J. Yang and S. S. Perry, Friction and Molecular Order of Alkanethiol Self-Assembled Monolayers on Au(111) at Elevated Temperatures Measured by Atomic Force Microscopy, *Langmuir*, 19, 6135–6139 (2003).

22. S. L. Ren, S. R. Yang, J. Q. Wang, W. M. Liu, and Y. P. Zhao, Preparation and Tribological Studies of Stearic Acid Self-Assembled Monolayers on Polymer-Coated Silicon Surface, *Chem. Mater.*, 16, 428–434 (2004).

23. R. M. Overney, E. Meyer, J. Frommer, D. Brodbeck, R. Luthi, L. Howald, H. J. Guntherodt, M. Fujihira, H. Takano, and Y. Gotoh, Friction Measurements on Phase-Separated Thin-Films with a Modified Atomic Force Microscope, *Nature*, 359, 133–135 (1992).

24. E. Meyer, R. Overney, D. Brodbeck, L. Howald, R. Luthi, J. Frommer, and H. J. Guntherodt, Friction and Wear of Langmuir–Blodgett Films Observed by Friction Force Microscopy, *Phys. Rev. Lett.*, 69, 1777–1780 (1992).

25. V. Novotny, J. D. Swalen, and J. P. Rabe, Tribology of Langmuir–Blodgett Layers, *Langmuir*, 8, 485–489 (1989).

26. Y. Liu, D. F. Evans, Q. Song, and D. W. Grainger, Structure and Frictional Properties of Self-Assembled Surfactant Monolayers, *Langmuir*, 12, 1235–1244 (1996).

27. B. Bhushan, A. V. Kulkarni, and V. N. Koinkar, Microtribological Characterization of Self-Assembled and Langmuir–Blodgett Monolayers by Atomic and Friction Force Microscopy, *Langmuir*, 11, 3189–3198 (1995).

28. S. Zhang and H. Lan, Developments in Tribological Research on Ultrathin Films, *Tribol. Int.*, 35, 321–327 (2002).

29. K. Miyake, M. Nakano, A. Korenaga, Y. Hori, T. Ikeda, M. Asakawa, T. Shimizu, S. Sasaki, and Y. Ando, Nanoscale to Macroscale Investigation of the Frictional Properties of Physisorbed Layers of Self-Organized Phthalocyanine Derivatives, *Tribol. Lett.*, 31, 9–15 (2008).

30. X. H. Qiu, C. Wang, Q. D. Zeng, B. Xu, S. X. Yin, H. N. Wang, S. D. Xu, and C. L. Bai, Alkane-Assisted Adsorption and Assembly of Phthalocyanines and Porphyrins, *J. Am. Chem. Soc.*, 122, 5550–5556 (2000).

31. T. Ikeda, M. Asakawa, M. Goto, K. Miyake, T. Ishida, and T. Shimizu, STM Observation of Alkyl Chain-Assisted Self-Assembled Monolayers of Pyridine-Coordinated Porphyrin Rhodium Chlorides, *Langmuir*, 20, 5454–5459 (2004).

32. D. G. Flom, A. J. Haltner, and C. A. Gaulin, Friction and Cleavage of Lamellar Solid in Ultrahigh Vacuum, *ASLE Trans.*, 8, 133–145 (1965).

33. K. Miyake, Y. Hori, T. Ikeda, M. Asakawa, T. Shimizu, T. Ishida, and S. Sasaki, Alkyl Chain Length Dependence of Frictional Properties of Alkyl-Substituted Phthalocyanines Physisorbed on Graphite Surface, *Jpn. J. Appl. Phys.*, 44, 5403–5408 (2005).

34. K. Miyake, Y. Hori, T. Ikeda, M. Asakawa, T. Shimizu, and S. Sasaki, Alkyl Chain Length Dependence of the Self-Organized Structure of Alkyl-Substituted Phthalocyanines, *Langmuir*, 24, 4708–4714 (2008).

35. J. Van Alsten and S. Granick, Molecular Tribometry of Ultrathin Liquid Films, *Phys. Rev. Lett.*, 61, 2570–2573 (1988).

36. S. Granick, Motions and Relaxations of Confined Liquids, *Science*, 253, 1374–1379 (1991).

37. A. L. Demirel and S. Granick, Glasslike Transition of a Confined Simple Fluid, *Phys. Rev. Lett.*, 77, 2261–2264 (1996).

38. Y. Zhu and S. Granick, Viscosity of Interfacial Water, *Phys. Rev. Lett.*, 87, 096104 (2001).

39. Y. Zhu and S. Granick, Reassessment of Solidification in Fluids Confined between Mica Sheets, *Langmuir*, 19, 8148–8151 (2003).

40. W. H. Briscoe, S. Titmuss, F. Tiberg, R. K. Thomas, D. J. McGillivray, and J. Klein, Boundary Lubrication under Water, *Nature*, 444, 191–194 (2006).

41. H. Sakuma, K. Otsuki, and K. Kurihara, Viscosity and Lubricity of Aqueous NaCl Solution Confined between Mica Surfaces Studied by Shear Resonance Measurement, *Phys. Rev. Lett.*, 96, 046104 (2006).

42. C. M. Mate, Atomic-Force-Microscope Study of Polymer Lubricants on Silicon Surface, *Phys. Rev. Lett.*, 68, 3323–3326 (1992).

43. M. L. Gee, P. M. McGuiggan, J. N. Israelachivili, and A. M. Homola, Liquid to Solidlike Transitions of Molecularly Thin Films under Shear, *J. Chem, Phys.*, 93, 1895–1906 (1990).

44. E. Manias, I. Bitsanis, G. Hadziioannou, and G. T. Brinke, On the Nature of Shear Thinning in Nanoscopically Confined Films, *Europhys. Lett.*, 33, 371–376 (1996).

45. F. Mugele, B. Persson, S. Zilberman, A. Nitzan, and M. Salmeron, Frictional Properties of Straight-Chain Alcohols and the Dynamics of Layering Transition, *Tribol. Lett.*, 12, 123–129 (2002).

46. R. Begum and H. Matsuura, Conformational Properties of Short Poly(Oxyethylene) Chains in Water Studied by IR Spectroscopy, *J. Chem. Soc. Faraday Trans.*, 93, 3839–3848 (1997).

47. L. Bokobza, T. Buffeteau, and B. Desbat, Mid- and Near-Infrared Investigation of Molecular Orientation in Elastomeric Networks, *Appl. Spec.*, 54, 360–365 (2000).

48. K. K. Zhuravlev and M. D. McCluskey, Infrared Spectroscopy of Biphenyl under Hydrostatic Pressure, *J. Chem. Phys.*, 117, 3748–3752 (2002).

49. P. M. Cann and H. A. Spikes, In-Contact IR Spectroscopy of Hydrocarbon Lubricants, *Tribol. Lett.*, 19, 289–297 (2005).

50. Y. R. Shen, *The Principles of Nonlinear Optics*. John Wiley & Sons, New York (1984).

51. Q. Du, X. D. Xiao, D. Charych, F. Wolf, P. Frantz, Y. R. Shen, and M. Salmeron, Nonlinear-Optical Studies of Monomolecular Films under Pressure, *Phys. Rev. B*, 51, 7456–7463 (1995).

52. O. Berg and D. Klenerman, Vibrational Spectroscopy of Mechanically Compressed Monolayers, *J. Am. Chem. Soc.*, 125, 5493–5500 (2003).

5 Temperature Dependence of Molecular Packing in Self-Assembled Monolayer Films

*Shuchen Hsieh, YiLen Liu, David Beck,
Minda Suchan, and I-Ting Li*

CONTENTS

ABSTRACT

Self-assembled monolayers (SAMs) are promising candidates for lubrication films in micro- and nanoelectromechanical systems (MEMS and NEMS) devices. Characterizing changes in the structure and stability of these films due to thermal cycling is important for optimizing their performance. In this report, we have used hexyltrichlorosilane (C6), dodecyltrichlorosilane (C12), and octadecyltrichlorosilane (C18) to form SAMs on silicon, to study the molecular packing and tribological behavior of each type of film as a function of temperature. Fourier transform infrared spectroscopy (FTIR), contact angle, and atomic force microscopy (AFM) were used to characterize films molecular packing, surface energy, and roughness/friction, respectively. FTIR results show the peak position changes in discrete steps for C12 and C18 indicating structural transitions that result in a change in the molecular packing of the films. On the other hand, the C6 film peak position was constant to 500 K, indicating that no change in the molecular packing state occurred over this temperature range. Friction studies have shown that the molecular packing and tribological behavior of SAM films at elevated temperatures is an important consideration for MEMS and NEMS devices, in order to optimize performance and increase lifetimes.

5.1   INTRODUCTION

Self-assembled monolayers (SAMs) [1] and Langmuir–Blodgett films [2,3] are used in a wide variety of device fabrication processes. In recent years, extensive efforts have been made to better understand the properties, behavior, and performance of these monolayer films [4,5]. Alkanethiol molecules self-assembled on gold surfaces, for example, have been researched extensively [6–10], and studies show that the physical and chemical properties of the surface can be tuned by the appropriate selection of the attaching molecule [11]. For example, saturated hydrocarbon molecules can be used to create a hydrophobic surface [12] or conjugated molecules to improve charge transfer [13]. Further, functional monolayers can be developed by proper design and careful control over the surface reactions that create the film. This type of surface can then be used in micro- and nanoelectromechanical systems (MEMS/NEMS), photovoltaic devices [13–15], interface lubricants [16–20], liquid crystalline alignment [21,22], surface patterning [23,24], electrochemistry [25], and in biological chemistry [26].

In nanotribology, alkylsilane molecules self-assembled on silicon have been used to reduce friction in MEMS and NEMS devices. Ambient temperature AFM studies have shown that the friction coefficient for such films decrease with increasing molecular chain length. However, temperature-dependent friction studies are still lacking. In the past, several researchers have studied temperature effects on SAMs using different techniques [27–29] and have reported that higher temperatures alter the structure and/or chemical properties of the SAM films. In this chapter, we show that the molecular packing, surface roughness, and friction properties of SAM films on silicon are highly temperature dependent and that implementation of SAM films in practical devices (MEMS, NEMS) must account for these tribological changes in order to optimize performance and increase lifetimes.

5.2 MATERIALS AND METHODS

5.2.1 MATERIALS

Hexyltrichlorosilane (C6), dodecyltrichlorosilane (C12), and octadecyltrichlorosilane (C18) were purchased from Gelest, Inc. (Morrisville, PA) and used without further purification. Hexadecane was obtained from ACROS (Morris Plains, NJ), chloroform was obtained from TEDIA (Fairfield, OH), and anhydrous ethanol (water $<0.1\%$) was obtained from Merck (Germany). Pure water was prepared using a Millipore Milli-Q purification system (resistivity $\geq 18.2\,M\Omega$). Double-sided polished silicon (100) wafer substrates (B-doped) were obtained from MEMC Electronic Materials (Pearl Drive, MO) with resistance of $1-10\,\Omega$ cm and were cut into 40×20 mm pieces using a diamond-tipped stylus. All silicon samples were sequentially cleaned with a neutral detergent solution (Merck, Germany) and ethanol and then plasma cleaned (Harrick Scientific Corp., Ossining, NY, Model PDC-32G) at medium power for 2 min to increase the hydroxyl density on the silicon surface. All reagents used in this experiment were anhydrous.

5.2.2 SAM FORMATION

SAMs were prepared from 1 mM silane solutions of C6, C12, and C18 in hexadecane. The glassware used in this process was successively cleaned in a detergent solution and ethanol and then baked for 24 h to remove residual water from the containers. A final plasma treatment was performed on the glassware immediately before preparing the alkylsilane solutions. Clean silicon wafer samples were placed into the C6, C12, and C18 silane solutions for 24 h, followed by ultrasonic cleaning in chloroform, isopropanol, and water, sequentially. The samples were then dried under a stream of nitrogen and annealed at 110°C for 10 min. Samples used in temperature studies were annealed as follows: the samples were heated to each target temperature over 30 min using a hot plate with an accuracy of ±5°C and then cooled down to room temperature over 20 min and immediately used in the experiments.

5.2.3 TRANSMISSION FTIR

Transmission FTIR spectra were acquired (8 cm^{-1} resolution, 1000 scans) at ambient temperature using a nitrogen-purged Perkin-Elmer FTIR spectrometer (Spectrum 100) equipped with a liquid-nitrogen–cooled MCT (mercury–cadmium–telluride) detector. Unpolarized light was used at an incident angle of 74° with respect to the surface normal. This was accomplished experimentally by rotating the sample so that the substrate normal was 74° away from the incident IR beam [30]. The spectrum from a freshly plasma-cleaned silicon wafer sample was collected before each measurement to obtain the background spectrum.

5.2.4 Contact Angle Measurement

Water contact angle measurements were made on the SAM samples under ambient conditions using a PG-X Measuring Head from Rycobel (Deerlijk, Belgium) and MilliQ water (R $\geq$ 18.2 MΩ). An autopipetting system was used to deliver a fixed volume of Millipore water (1 μL) onto the sample surface. A total of 20 contact angle measurements were made for each sample and annealing temperature and averaged to obtain the mean contact angle value.

5.2.5 Atomic Force Microscopy

All topographical images and lateral force microscopy (LFM) measurements were performed with an atomic force microscope (Asylum Research, MFP-3D™) under ambient conditions. A silicon probe (Olympus, AC240-TS) with a measured spring constant of 1.71 N/m was used for all experiments and was calibrated using the thermal method [31,32]. Friction force measurements were performed by recording "friction loops," where the trace and retrace data from a single scan line are used to calculate a scalar value for friction along that line according to the formula: [friction force = (average trace – average retrace)/2]. Tip velocities of 9 μm/s were used for all friction loop experiments. No attempt was made to calibrate the lateral spring constant.

5.3 RESULTS

5.3.1 Temperature Effects on Molecular Packing in SAM Films

FTIR spectroscopy was used to characterize the degree of ordering in the SAM films on silicon. The C–H vibrational modes for the alkyl chains in the films are very sensitive to the chemical environment, with peaks occurring at higher wavenumber for disordered films and at lower wavenumber for ordered films. Figure 5.1 shows FTIR spectra acquired at 300 K for the CH_2 symmetric, ν_s (CH_2), and asymmetric stretch, ν_{as} (CH_2), peaks for C6, C12, and C18 films. The asymmetric stretch peak was centered at 2928, 2924, and 2918 cm^{-1} for these films, respectively. The peak position for the C6 film at 2928 cm^{-1} was similar to the position observed for the same peak in a liquid phase sample of C18, suggesting a disordered, or loosely packed, film structure [33]. For C18, the peak at 2918 cm^{-1} was consistent with other literature reports for solid-like C18 monolayer films [18,34,35]. The C12 film peak at 2924 cm^{-1} was intermediate between the two. This trend demonstrates a shift to lower wavenumber with increasing alkyl chain length, which can be explained by an increase in intermolecular stabilization energy due to the longer chains.

SAM film samples were annealed at a series of increasing temperatures to characterize the thermal effects on molecular packing. The resulting FTIR spectra for these annealed samples are shown in Figure 5.2. At room temperature, the C6 film peak was centered at 2928 cm^{-1} and remained constant to 500 K and then shifted higher to 2932 cm^{-1} at 515 K. The peak intensity began to decrease at 425 K and was completely attenuated at 525 K. The C12 film peak centered at 2924 cm^{-1} was constant to 460 K, then shifted higher 2928 cm^{-1} at 500 K, and further to 2930 at

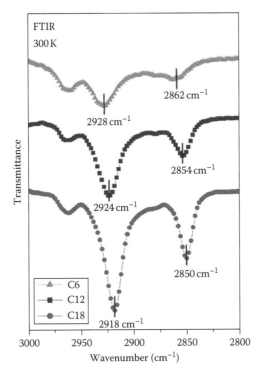

FIGURE 5.1 Transmission FTIR spectra of the CH stretch region showing the symmetric, v_s (CH$_2$), and asymmetric, v_{as} (CH$_2$), stretching modes for C6, C12, and C18 SAM films at room temperature. The v_{as} (CH$_2$) peak is at 2928, 2924, and 2918 cm^{-1} for C6, C12 and C18, respectively.

525 K. The peak intensity began to decrease at 425 K and was completely attenuated at 550 K. The C18 film peak centered at 2918 cm^{-1} at room temperature, shifted to 2920 cm^{-1} at 425 K, to 2924 cm^{-1} at 535 K, and reached a maximum of 2934 cm^{-1} at 565 K. The peak intensity began to decrease at 425 K and was completely attenuated at 575 K.

The asymmetric CH$_2$ stretching mode peak position relates to the molecular packing of each film. The peak positions for all films as a function of annealing temperature are shown graphically in Figure 5.3. The C6 film peak position was constant to 500 K, indicating that no change in the molecular packing state occurred over this annealing temperature range. This is likely due to the weaker intermolecular stabilization energy for the short alkyl chains, resulting in more degrees of freedom by which the energy can dissipate as heat. The peak position for the C12 monolayer changed in discrete steps, with no changes observed in the annealing temperature range 300–460 and 500–515 K. The C18 monolayer exhibited similar behavior, but the stable annealing temperature intervals for this film were 300–415, 425–485, and 515–550 K. The discrete peak position steps observed for C12 and C18 indicate structural transitions that result in a change to the molecular packing of the films.

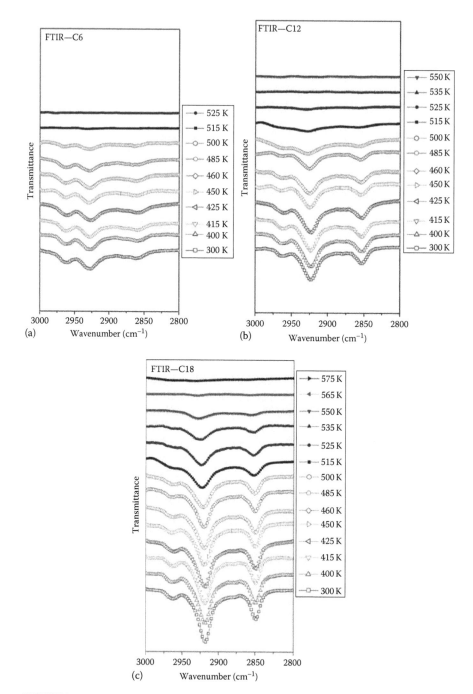

FIGURE 5.2 FTIR spectra as a function of annealing temperature for (a) C6, (b) C12, and (c) C18 SAM films. The ν_{as} (CH$_2$) and ν_s (CH$_2$) peak intensity decreased for all the films as the temperature was increased and were completely attenuated at temperatures of 525, 550, and 575 K, respectively.

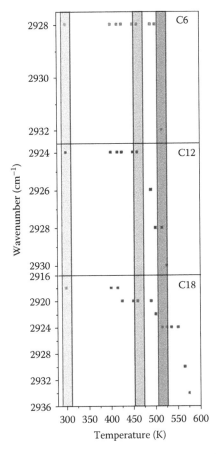

FIGURE 5.3 FTIR asymmetric CH_2 peak position as a function of annealing temperature for C6, C12, and C18 monolayer films. The vertical bands in the figure correspond to the approximate temperature at which friction data were acquired. The change in peak position was not gradual but occurred in a stepwise fashion with the peak position constant over a range of temperatures for each film. C6 (top) 300–500 K; C12 (middle) 300–460 and 500–515 K; and C18 (bottom) 300–415, 425–485, and 515–550 K.

5.3.2 SURFACE ENERGY OF SAM-MODIFIED SILICON AS A FUNCTION OF TEMPERATURE

The water contact angle for each of the alkylsilane-modified silicon samples annealed at a series of increasing temperatures was measured at room temperature to monitor changes in the surface energy as a function of temperature. Figure 5.4 shows that the water contact angles for the C6, C12, and C18 films were $101.4 \pm 0.2°$, $103.2 \pm 0.5°$, and $101.7 \pm 1.0°$ at room temperature, respectively. The corresponding surface energies for the three films, and for bare silicon, were 28.3, 27.7, 28.4, and $56.0 \, mJ/m^2$. The surface energy was calculated using the ASTM D5946 method [36]. This change from a hydrophilic (bare silicon) to a hydrophobic surface (SAM) was

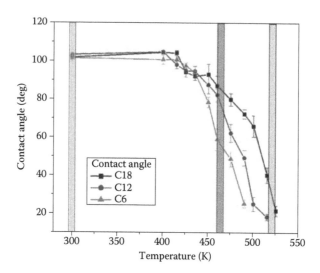

FIGURE 5.4 Water contact angles for C6, C12, and C18 monolayers on silicon as a function of anneal temperature. At room temperature, the contact angles were 101.7°, 103.2°, and 101.4°, respectively. The contact angle was undetectable (<10°) for all films annealed at 500, 525, and 535 K, respectively.

expected, due to the methyl termination of the precursors and complete monolayer formation of the film. The samples were subsequently annealed to a series of temperatures. No changes in contact angle were observed for any of the films annealed up to 400 K. As the annealing temperature was increased further, the contact angle for all three film samples decreased with a concomitant increase in surface energy. The C6, C12, and C18 samples reached a minimum measurable value of approximately 20° contact angle after 490, 500, and 525 K annealing temperatures, respectively. The FTIR spectra in Figure 5.2 showed that the asymmetric CH_2 peak intensity began to decrease at 425 K for all three films. The peak intensities decreased further with increasing anneal temperature and were completely attenuated at 525, 550, and 575 K for the C6, C12, and C18 films, respectively. No new peaks were observed in the FTIR spectra in the range of 400–4000 cm^{-1} during the experiments, showing that no new molecules were formed due to decomposition or reaction.

5.3.3 TEMPERATURE EFFECT ON SURFACE TOPOGRAPHY

Figure 5.5 shows topographical AFM images for the C6, C12, and C18 films that were acquired at 300 K and after annealing to 460 and 515 K. These temperatures were selected based on FTIR results (Figure 5.3), which showed them to be regions of thermally stable molecular packing. All images were acquired using contact mode with a small applied force of 22.5 nN to minimize damage to the surface. The scan size was $5 \times 5 \, \mu m^2$ and the height scale was fixed at 2 nm for all images.

Images in the first column of Figure 5.5a through c show the surface topography for C6, C12, and C18 films at 300 K, respectively. The C6 and C12 film surfaces exhibited variations in topography, while the C18 film was nearly featureless. The

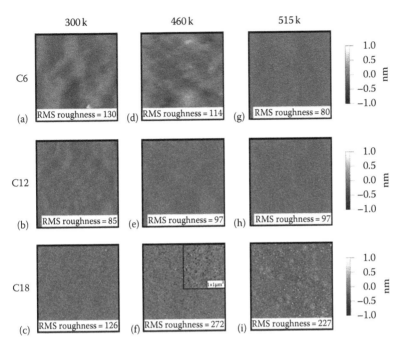

FIGURE 5.5   AFM topographical images ($5 \times 5\,\mu m^2$) for C6, C12, and C18 films on silicon (rows) after annealing to 300, 460, and 515 K (columns). The inset in (f) shows a zoomed-in image ($1 \times 1\,\mu m^2$) of the C18 monolayer at 460 K to more clearly show that defects have formed in the film. RMS roughness values shown at the bottom of each image are in units of pm.

RMS roughness values for the three films were 130, 85, and 126 pm, respectively, indicating that all three films were complete monolayers.

Images in the second column of Figure 5.5d through f were acquired after annealing the samples at 460 K. The morphology of the C6 and C12 films was very similar to the 300 K samples, and the roughness changed only slightly to 114 and 97 pm for these films, respectively. However, the C18 sample developed pits in the film (Figure 5.5f), which led to doubling of the surface roughness from the value at 300 K. The inset in Figure 5.5f shows a zoomed-in view of the surface ($1 \times 1\,\mu m^2$). The average diameter of the pits in the inset was about 18.5 nm, and the average depth was about 1.5 nm. Images in the third column of Figure 5.5g through i were acquired on samples annealed at 515 K. The C6 film morphology is very flat (Figure 5.5g) and the roughness (80 pm) is reduced compared to the C6 films at 300 and 460 K. The C12 film image (Figure 5.5h) was also very flat with a measured surface roughness of 97 pm, very similar to the 460 K C12 film. The C18 film image (Figure 5.5i) shows small protrusions and large morphology variations quite different from the C18 film at 460 K. The roughness was 227 pm, which was lower than the 460 K film but still much higher than the bare silicon substrate (<100 pm).

The observed changes in surface morphology and roughness as a function of annealing temperature are consistent with our FTIR and contact angle results.

From FTIR data in Figure 5.2, the C6, C12, and C18 peak areas at 515 K decreased to 7.5%, 11.0%, and 73.0% of their values at 300 K. For the C6 and C12 films, this suggests significant degradation of these films during annealing at 515 K, leading to morphology and surface roughness that are characteristic of the silicon substrate. The C18 film peak area at 515 K was 73% of its value at 300 K, which showed that the alkyl chains were still present on the surface. The morphological and surface roughness changes at this temperature, however, indicated that the molecular packing of the film had degraded. The contact angles of the C18 and C12 films at 515 K decreased to 40.3° and 18.5°, and the contact angle of the C6 monolayer was below the detection limit (~10°) of our instrument. A gradual change from hydrophobic to hydrophilic surfaces is consistent with degradation of the film.

5.3.4 TEMPERATURE EFFECT ON SURFACE FRICTION

Using AFM in lateral force mode, we measured the friction force as a function of applied load on the C6, C12, and C18 films annealed at 300, 460, and 515 K. The results of these measurements are shown in Figure 5.6 for all films and for a bare silicon sample. All data points shown in Figure 5.6a–c are the result of "friction loop" line scans that were acquired as described in the Experimental section above.

All three films exhibited very low friction force at 300 K relative to that on a bare silicon sample. In addition, the friction force on the SAM film samples was constant over the entire range of applied load despite the bare silicon sample exhibiting an increase. After annealing each sample to 460 K, however, the relative friction force on the C6 and C18 films increased and also exhibited some dependence on applied load. The friction force for the C12 film, however, was nearly unchanged at this temperature, and only a very small load dependence was observed. After annealing the samples to 515 K, the friction force on the C6 and C12 films was very close to the bare silicon sample. The C18 film, however, was lower, both in relative value and in load dependence. These results are discussed later in terms of their respective friction coefficients.

The linear relationship between friction force and applied load allows us to use the generalized form of Amontons' law [37,38], where the slope of the friction force versus load plot yields the friction coefficient. As a benchmark for these measurements, a silicon control sample was analyzed, which yielded a friction coefficient of 145×10^{-6}.

For the C6, C12, and C18 films annealed at 300 K, the friction coefficients were 5.7×10^{-6}, 1.9×10^{-6}, and 2.5×10^{-6}, respectively. After annealing the samples to 460 K, the friction coefficients increased to 42×10^{-6}, 4.5×10^{-6}, and 53×10^{-6}, representing an increase of 7.4, 2.4, and 21.2 times the 300 K values. The large increase in friction coefficient for the C18 film is consistent with the formation of pits in the film (Figure 5.5), which also leads to an increase in surface roughness.

After annealing the samples at 515 K, the friction coefficients for C6, C12, and C18 films increased further to 151×10^{-6}, 124×10^{-6}, and 62×10^{-6}, or 3.6, 27.6, and 1.2 times their values at 460 K. The 515 K annealing temperature induced a significant increase in the C6 and C12 friction coefficients, which produced values

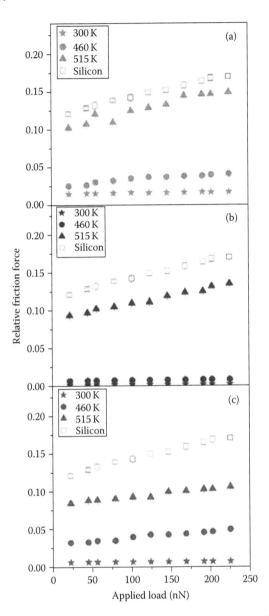

FIGURE 5.6 Friction force versus applied load for the C6 (a), C12 (b), and C18 (c) monolayers annealed to 300, 460, and 515 K. The results from a silicon control sample are included in all graphs for comparison.

that were similar to that of bare silicon. FTIR results at 515 K showed that the C6 and C12 asym-CH_2 peak intensities decreased to 7.5% and 11% from their values at 300 K, respectively. In addition, water contact angle measurements for the C6 film were below the detection limit of our instrument and the C12 film contact angle was reduced to 18.5°. The friction coefficient, FTIR, and water contact angle changes at

515 K annealing temperature indicate that the C6 film was destroyed and the C12 film was significantly degraded.

The C18 monolayer had the smallest friction coefficient at 515 K annealing temperature of the three films. From FTIR and water contact angle results, the C18 monolayer asym-CH_2 peak intensity decreased to 73% of its value at 300 K, and the water contact angle was reduced to 40.3°. This suggests a change in the molecular packing state of the C18 molecules due to thermal degradation of the film. At 515 K, only the C18 monolayer reduced surface friction significantly.

5.4   DISCUSSION

5.4.1   TEMPERATURE EFFECTS ON MOLECULAR PACKING STATE AND NANOTRIBOLOGICAL PROPERTIES OF SAMs

In this work, the thermal stability of C6, C12, and C18 monolayers on the oxidized surface of silicon (100) was studied in order to characterize temperature effects on the molecular packing and nanotribological properties of the SAM films. Self-assembled alkylsilane monolayers are stabilized by intermolecular attractive forces from van der Waals interactions between the alkyl chains. Therefore, longer alkyl chains have stronger attractive forces providing the monolayer with higher stabilization energy [39]. This results in more densely packed and, hence, ordered monolayer film structure [8].

FTIR spectra in Figure 5.1 show that molecular packing at room temperature depends on the alkyl chain length. The asym-CH_2 peak for the C6 film occurred at higher wavenumber than C12 and C18 due to higher degrees of freedom of the shorter alkyl chains. Also, Table 5.1 shows that the friction coefficient of the C6 monolayer was higher than the other two films, indicating that molecular packing of the C6 film was more disordered than the C12 and C18 monolayers at 300 K.

From Figure 5.3, FTIR shows an upward shift of the asym-CH_2 vibrational peak with increasing annealing temperature for all the films. The thermal stability of the SAM films is directly related to molecular packing. An upward shift

TABLE 5.1
Friction Coefficients and FTIR Peak Centers for C6, C12, and C18 Monolayer Films as Function of Annealing Temperature

Temperature	300 K		460 K		515 K	
	$\mu \times 10^6$	IR (cm⁻¹)	$\mu \times 10^6$	IR (cm⁻¹)	$\mu \times 10^6$	IR (cm⁻¹)
C6	5.7	2928	42	2928	151	2932
C12	1.9	2924	4.5	2926	124	2928
C18	2.5	2918	53	2920	62	2924
Silicon	145					

Note: The friction coefficient for silicon at 300 K is included for reference.

of the asym-CH_2 peak for the C6 monolayer did not occur until the sample was annealed to 515 K, indicating that the molecular packing state was stable up to this temperature. Also from Figure 5.3, the C12 monolayer exhibited two stable packing state temperature intervals and the C18 monolayer exhibited three such intervals. The abrupt transition between each of these intervals indicates thermally induced changes in molecular packing. With increased annealing temperature, the molecular packing of all the films changed to a more liquid-like structure, thereby becoming more disordered.

For the C18 monolayer, the pits observed in AFM images (Figure 5.5f inset) result from molecular reorganization, and LFM shows a 20.9-fold increase in the friction coefficient as the annealing temperature increased from 300 to 460 K. Both the AFM and LFM results demonstrate that the C18 monolayer becomes more disordered with increased annealing temperature. This is consistent with the FTIR results where the asym-CH_2 peak shifted upward from 2918 to 2920 cm^{-1}. For the C6 and C12 monolayers, AFM images showed only slight changes in the topography, and LFM showed 7.4 and 2.3 times increase in friction coefficient as temperature increased to 460 K, respectively. The FTIR asym-CH_2 peak, and thus molecular packing, was unchanged for C6 and C12 films over this temperature interval.

Combining the results from AFM, LFM, and FTIR shows that the C6 and C12 SAMs were more thermally stable than the longer chain C18 SAMs. The C6 mono-layer has higher degrees of freedom, allowing the heat to dissipate more easily [39,40]. The C6 monolayer also has a more disordered structure than C12 and C18 films, yielding a higher friction coefficient. Thus, a more disordered SAM, such as the C6 monolayer, may be thermally more stable [37,39,41,42], but it had higher friction coefficients at 300 K and at elevated temperatures, compared to ordered SAM films (C12 and C18).

Although the C6 and C12 monolayers were thermally more stable than the C18 monolayer, they both disappeared at lower annealing temperatures than the C18 monolayer. Figure 5.3 shows that the asym-CH_2 peak of the C18 monolayer disappeared at >575 K and the asym-CH_2 peak of the C6 and C12 monolayers disappeared at >515 and >525 K, respectively. Figure 5.4 shows that the water contact angle on the C6, C12, and C18 monolayers was measurable on samples annealed up to temperatures of 500, 525, and 535 K, respectively. These results (FTIR and contact angle) demonstrate that the C18 monolayer survived higher annealing temperatures than the C6 and C12 monolayers despite its denser molecular packing and less efficient heat dissipation capability.

5.5   CONCLUSION

This study provides new insight into the temperature effects on the molecular packing state of C6, C12, and C18 alkylsilane SAMs on silicon (100). At room temperature, FTIR revealed that the short chain-length C6 molecules exhibited nearly "liquid-like" behavior, while the C18 film behaved like a solid. The C12 films were intermediate between these two, suggesting both solid- and liquid-like properties. With increasing temperature, the C6 film was stable up to 500 K. However, the C12 and C18 monolayers exhibited abrupt FTIR peak shifts at temperatures between

300 and 575 K, with stable packing observed over several temperature ranges. Water contact angle measurements showed that the C6, C12, and C18 films changed from hydrophobic to hydrophilic with increasing anneal temperature. Atomic force microscopy revealed that the surface roughness and friction coefficients for all three films increased with increasing anneal temperature. However, the C18 film exhibited an interesting anomaly characterized by pits forming in the film, as observed in topographical images, and by an abrupt change in the friction coefficient at 460 K. This was likely due to the molecular reorganization of C18 molecules on the surface. These results show that molecular packing and tribological behavior are highly temperature dependent and must be considered in the design phase of MEMS and NEMS devices in order to optimize performance and increase lifetimes.

ACKNOWLEDGMENTS

The authors would like to thank the National Science Council (NSC 95-2113-M-110-022 and NSC 97-2113-M-110-007) of Taiwan and the National Sun Yat-sen University Center for Nanoscience and Nanotechnology for financial support of this work.

REFERENCES

1. W. C. Bigelow, D. L. Pickett, and W. A. Zisman, Oleophobic monolayers, *J. Colloid Interface Sci.* 1, 513 (1946).
2. K. B. Blodgett, Films built by depositing successive monomolecular layers on a solid surface, *J. Am. Chem. Soc.* 57, 1007 (1935).
3. I. Langmuir, The constitution and fundamental properties of solids and liquids. 11. Liquids., *J. Am. Chem. Soc.* 39 (9), 1848–1906 (1917).
4. J. C. Love, L. A. Estroff, J. K. Kriebel, R. G. Nuzzo, and G. M. Whitesides, Self-assembled monolayers of thiolates on metals as a form of nanotechnology, *Chem. Rev.* 105 (4), 1103–1169 (2005).
5. A. Ulman, Formation and structure of self-assembled monolayers, *Chem. Rev.* 96 (4), 1533–1554 (1996).
6. R. G. Nuzzo and D. L. Allara, Adsorption of bifunctional organic disulfides on gold surfaces, *J. Am. Chem. Soc.* 105 (13), 4481–4483 (1983).
7. G. E. Poirier and E. D. Pylant, The self-assembly mechanism of alkanethiols on Au(111), *Science* 272 (5265), 1145–1148 (1996).
8. M. D. Porter, T. B. Bright, D. L. Allara, and C. E. D. Chidsey, Spontaneously organized molecular assemblies. 4. Structural characterization of normal-alkyl thiol monolayers on gold by optical ellipsometry, infrared-spectroscopy, and electrochemistry, *J. Am. Chem. Soc.* 109 (12), 3559–3568 (1987).
9. L. H. Dubois and R. G. Nuzzo, Synthesis, structure, and properties of model organic-surfaces, *Annu. Rev. Phys. Chem.* 43, 437–463 (1992).
10. H. A. Biebuyck, C. D. Bian, and G. M. Whitesides, Comparison of organic monolayers on polycrystalline gold spontaneously assembled from solutions containing dialkyl disulfides or alkenethiols, *Langmuir* 10 (6), 1825–1831 (1994).
11. C. D. Bain, J. Evall, and G. M. Whitesides, Formation of monolayers by the coadsorption of thiols on gold—Variation in the head group, tail group, and solvent, *J. Am. Chem. Soc.* 111 (18), 7155–7164 (1989).

12. S. A. Kulkarni, S. A. Mirji, A. B. Mandale, R. P. Gupta, and K. P. Vijayamohanan, Growth kinetics and thermodynamic stability of octadecyltrichlorosilane self-assembled monolayer on Si (100) substrate, *Mater. Lett.* 59 (29–30), 3890–3895 (2005).
13. T. B. Cao, L. H. Wei, S. M. Yang, M. F. Zhang, C. H. Huang, and W. X. Cao, Self-assembly and photovoltaic property of covalent-attached multilayer film based on highly sulfonated polyaniline and diazoresin, *Langmuir* 18 (3), 750–753 (2002).
14. B. Oregan and M. Gratzel, A low-cost, high-efficiency solar-cell based on dye-sensitized colloidal TiO_2 films, *Nature* 353 (6346), 737–740 (1991).
15. S. Takenaka, Y. Maehara, H. Imai, M. Yoshikawa, and S. Shiratori, Layer-by-layer self-assembly replication technique: Application to photoelectrode of dye-sensitized solar cell, *Thin Solid Films* 438, 346–351 (2003).
16. Y. H. Liu, D. F. Evans, Q. Song, and D. W. Grainger, Structure and frictional properties of self-assembled surfactant monolayers, *Langmuir* 12 (5), 1235–1244 (1996).
17. B. D. Beake and G. J. Leggett, Friction and adhesion of mixed self-assembled monolayers studied by chemical force microscopy, *Phys. Chem. Chem. Phys.* 1 (14), 3345–3350 (1999).
18. O. P. Khatri and S. K. Biswas, Friction of octadecyltrichlorosilane monolayer self-assembled on silicon wafer in 0% relative humidity, *J. Phys. Chem. C* 111 (6), 2696–2701 (2007).
19. V. V. Tsukruk, V. N. Bliznyuk, J. Hazel, D. Visser, and M. P. Everson, Organic molecular films under shear forces: Fluid and solid Langmuir monolayers, *Langmuir* 12 (20), 4840–4849 (1996).
20. Q. Zhang and L. A. Archer, Boundary lubrication and surface mobility of mixed alkylsilane self-assembled monolayers, *J. Phys. Chem. B* 107 (47), 13123–13132 (2003).
21. H. Schonherr, F. J. B. Kremer, S. Kumar, J. A. Rego, H. Wolf, H. Ringsdorf, M. Jaschke, H. J. Butt, and E. Bamberg, Self-assembled monolayers of discotic liquid crystalline thioethers, discoid disulfides, and thiols on gold: Molecular engineering of ordered surfaces, *J. Am. Chem. Soc.* 118 (51), 13051–13057 (1996).
22. R. Viswanathan, L. L. Madsen, J. A. Zasadzinski, and D. K. Schwartz, Liquid to hexatic to crystalline order in Langmuir–Blodgett-films, *Science* 269 (5220), 51–54 (1995).
23. R. Maoz, S. R. Cohen, and J. Sagiv, Nanoelectrochemical patterning of monolayer surfaces: Toward spatially defined self-assembly of nanostructures, *Adv. Mater.* 11 (1), 55–61 (1999).
24. S. Hoeppener, R. Maoz, and J. Sagiv, Constructive microlithography: Electrochemical printing of monolayer template patterns extends constructive nanolithography to the micrometer-millimeter dimension range, *Nano Lett.* 3 (6), 761–767 (2003).
25. J. T. Sullivan, K. E. Harrison, J. P. Mizzell, and S. M. Kilbey, Contact angle and electrochemical characterization of multicomponent thiophene-capped monolayers, *Langmuir* 16 (25), 9797–9803 (2000).
26. E. Ostuni, B. A. Grzybowski, M. Mrksich, C. S. Roberts, and G. M. Whitesides, Adsorption of proteins to hydrophobic sites on mixed self-assembled monolayers, *Langmuir* 19 (5), 1861–1872 (2003).
27. A. Faucheux, A. C. Gouget-Laemmel, P. Allongue, C. H. de Villeneuve, F. Ozanam, and J. N. Chazalviel, Mechanisms of thermal decomposition of organic monolayers grafted on (111) silicon, *Langmuir* 23 (3), 1326–1332 (2007).
28. J. Frechette, R. Maboudian, and C. Carraro, Thermal behavior of perfluoroalkylsiloxane monolayers on the oxidized Si(100) surface, *Langmuir* 22 (6), 2726–2730 (2006).
29. H. K. Kim, J. P. Lee, C. R. Park, H. T. Kwak, and M. M. Sung, Thermal decomposition of alkylsiloxane self-assembled monolayers in air, *J. Phys. Chem. B* 107 (18), 4348–4351 (2003).

30. S. Sambasivan, S. Hsieh, D. A. Fischer, and S. M. Hsu, Effect of self-assembled monolayer film order on nanofriction, *J. Vac. Sci. Technol. A* 24 (4), 1484–1488 (2006).
31. H. J. Butt and M. Jaschke, Calculation of thermal noise in atomic-force microscopy, *Nanotechnology* 6 (1), 1–7 (1995).
32. J. L. Hutter and J. Bechhoefer, Calibration of atomic-force microscope tips, *Rev. Sci. Instrum.* 64 (7), 1868–1873 (1993).
33. H. O. Finklea, L. R. Robinson, A. Blackburn, B. Richter, D. Allara, and T. Bright, Formation of an organized monolayer by solution adsorption of octadecyltrichlorosilane on gold—Electrochemical properties and structural characterization, *Langmuir* 2 (2), 239–244 (1986).
34. O. P. Khatri, C. D. Bain, and S. K. Biswas, Effects of chain length and heat treatment on the nanotribology of alkylsilane monolayers self-assembled on a rough aluminum surface, *J. Phys. Chem. B* 109 (49), 23405–23414 (2005).
35. T. Ishizaki, N. Saito, S. Lee, and O. Takai, Comparative study of the molecular aggregation state of alkyl organic monolayers prepared on Si and hydrogen-terminated Si substrates, *Nanotechnology* 19 (5) (2008).
36. V. Mangipudi, M. Tirrell, and A. V. Pocius, Direct measurement of the surface-energy of corona-treated polyethylene using the surface forces apparatus, *Langmuir* 11 (1), 19–23 (1995).
37. N. J. Brewer, B. D. Beake, and G. J. Leggett, Friction force microscopy of self-assembled monolayers: Influence of adsorbate alkyl chain length, terminal group chemistry, and scan velocity, *Langmuir* 17 (6), 1970–1974 (2001).
38. T. T. Foster, M. R. Alexander, G. J. Leggett, and E. McAlpine, Friction force microscopy of alkylphosphonic acid and carboxylic acids adsorbed on the native oxide of aluminum, *Langmuir* 22 (22), 9254–9259 (2006).
39. X. D. Xiao, J. Hu, D. H. Charych, and M. Salmeron, Chain length dependence of the frictional properties of alkylsilane molecules self-assembled on mica studied by atomic force microscopy, *Langmuir* 12 (2), 235–237 (1996).
40. P. T. Mikulski and J. A. Harrison, Packing-density effects on the friction of n-alkane monolayers, *J. Am. Chem. Soc.* 123 (28), 6873–6881 (2001).
41. M. Salmeron, Generation of defects in model lubricant monolayers and their contribution to energy dissipation in friction, *Tribol. Lett.* 10 (1–2), 69–79 (2001).
42. A. Lio, D. H. Charych, and M. Salmeron, Comparative atomic force microscopy study of the chain length dependence of frictional properties of alkanethiols on gold and alkylsilanes on mica, *J. Phys. Chem. B* 101 (19), 3800–3805 (1997).

6 Mechanical Properties of Phospholipid-Based Biolubricant Films and Membranes

Matthew P. Goertz and Benjamin L. Stottrup

CONTENTS

ABSTRACT

Phospholipid assemblies such as liposomes and supported bilayers have received considerable attention as models of the plasma cell membrane and as biolubricants. Amphiphilic molecules, such as lipids, are naturally occurring surfactants that self-assemble to create films and act as the structural component of the plasma cell membrane. However, there is still a gap in the understanding of how mechanical properties correlate with specific functions that occur within the membrane. Using interfacial force microscopy to measure the

mechanical properties of phospholipid monolayers and bilayers, we can correlate the 2D structure of these systems with their nanoscale mechanical behavior. By measuring the elastic, plastic, and friction properties, we observe that
the close packed gel phase systems behave similar to a 2D solid, while disordered fluid phase systems resemble a 2D liquid. Relevant to their applications
in nanotechnology, these results provide a foundation for how supported lipid
films interact with objects in the ~500 nm–1 μm size range.

6.1 INTRODUCTION

The plasma cell membrane is a complex composite fluid that provides structure to
localize proteins at the interface of the cell and its surroundings, an encapsulation
of biosynthesis pathways, as well as a protective barrier for the cell [1]. The diversity of membrane functions is the result of the collective behavior of hundreds of
different lipid types, sterols, and proteins working together in a 2D system. Lateral
mobility within the membrane is crucial to its function. There are two "liquid-like"
fluid states of interest in biophysical studies: the liquid-disordered phase in which
the hydrophobic lipid chains are melted and fairly disordered and the liquid-ordered
phase where sterol molecules or proteins introduce an increased ordering to saturated phospholipid chains [2]. The fluid nature of the membrane allows it to remain
flexible and heal in response to perturbations.

In the past half century, our understanding of lateral organization within this
fluid membrane has advanced significantly. Two contributions stand out: the fluid
mosaic model [3] and the lipid raft model [4]. Both models regard the cell membrane
to be composed of a sea of phospholipids acting as a host to other biomolecules such
as sterols and proteins. These models also rely on the phase behavior of lipids to
explain membrane properties. The major difference between these models is how
they regard the organization of lipids, sterols, and proteins within the phospholipid
bilayer. In the fluid mosaic model, molecules are randomly distributed and laterally
diffuse within the bilayer [4]. While this model may accurately describe the time-
average compositional structure of the cell membrane, the lipid raft model imparts
further organization to all the components of the cell membrane [4]. According to
the lipid raft model, biomolecules laterally organize to form heterogeneities that perform a specific function such as cellular signaling or endocytosis [4]. These organized rafts are theorized to be only a few nanometers in size and exist only for a
short time period. As a result, rafts may be too small to image with fluorescence
microscopy and too short-lived to be imaged with scanning probe techniques. The
challenge of observing lipid rafts in vivo has created significant debate about the
validity of the lipid raft model [5]. However, the observation of miscibility phase
separation observed in model lipid membranes offers a physical mechanism for the
selected lateral organization [6]. In this work, we discuss the differences between
two phases commonly observed in phospholipid bilayers.

The structure of the plasma cell membrane is the result of the self-assembly of
amphiphilic proteins, sterols, and sphingolipids and phospholipids into a bilayer in
the presence of water. In addition to bilayers, phospholipids form other supramolecular structures including monolayers and micelles. Membrane biophysicists utilize

these structures to study the physical properties of phospholipids and fabricate biomimetic materials. Capitalizing on their biomimetic properties, supported lipid films can be used to create useful devices such as biomolecule microarrays, biocatalysts, biolubricants, and ultrathin electrical insulators. Understanding the mechanical properties of lipid films is essential for improving their usefulness and reliability in these devices. For example, a recent success has utilized the area condensation observed in phospholipid/cholesterol monolayers to create lipid bilayers that resist delaminating upon transfer through the air–water interface [7]. It was found that by tethering the bottom leaflet of a supported lipid bilayer (SLB) to a polymer cushion using a cholesteryl anchor, an SLB remains intact and fluidic even after several passes through the air–water interface [7]. This is attributed to cholesterol's ability to impose local conformational ordering, which increases the packing density of neighboring lipid molecules [8]. In terms of mechanical properties, the addition of cholesterol increases the compressive modulus and bending energy of a lipid film [9]. This finding allows SLBs to be used in a variety of tools that rely on a robotic spotting machine. For example, a protein microarray has been shown to offer high-throughput analysis during protein–protein interaction studies [10].

Lipids are also found to play a key role in the tribology of many naturally occurring systems. In many of these systems, lipids are utilized in a similar manner as traditional surfactants used in the field of tribology. One example of this is the lubrication of joints. Synovial fluid present in joints is known to contain lipids that help to bind glycoproteins, such as Lubricin. Acting together, lipids and proteins create a self-healing boundary lubricant that prevents wear and lowers friction of joints [11,12]. Lipids have also been linked to several other physiological sites where lubrication is needed, for example, the pleura, the pericardium, and the ocular surface [13]. The ability to be used as boundary lubricants and their inherent biocompatibility may allow lipid films to be used for lubrication of implantable devices such as lab-on-a-chip and bio-microelectromechanical systems (bioMEMS) [14].

While these self-assembled structures exist in natural systems and are being used in devices, there exists a gap in the correlation of structure and properties in nanoscale materials. Specifically, how does the phase behavior of 2D materials such as supported lipid films influence their mechanical properties? Understanding and quantifying the mechanical properties of lipid films and membranes is essential for creating robust and reliable devices that incorporate these materials. In this chapter, we will highlight the findings from two studies that probed the mechanical properties of both supported lipid monolayers [15] and bilayers [16] of phospholipids in both the fluid and the gel phases.

6.2 PROPERTIES OF PHOSPHOLIPIDS

6.2.1 Phospholipid Structure

We will begin with a brief examination of phospholipids. Phospholipids are one classification in a broad category of molecules called lipids. Lipids are biomolecules and surfactants that are insoluble (or only sparingly soluble) in water but dissolve in organic solvents such as alcohol or chloroform. In addition to phospholipids, other broad lipid classifications include fatty acids, wax esters, sterols, and sphingolipids. These molecules perform a range of biological functions. For example, sterols such as cholesterol and ergosterol are important to many biosynthesis pathways in mammals and yeast, respectively [17]. Waxes are naturally found as coatings on plants and animals alike and fatty acids provide an effective way to store energy and serve as insulation at low temperatures. Most relevant to our discussion, phospholipids and sphingolipids form lipid bilayers which are the structural backbone of plant and animal membranes [2].

The amphiphilic structure of phospholipids allows for self-assembly in an aqueous environment. All phospholipids have both hydrophobic and hydrophilic domains. This property enables them to create a variety of structures: most importantly the lipid bilayer that localizes proteins at the interface between the cell and its surroundings. Self-assembly is driven by both energetic and entropic forces [18,19]. For this reason, our research in membrane biophysics has used phospholipids as a key building block for in vitro systems. Lipid monolayers at the air–water interface allow researchers to probe molecular packing and lateral organization of a single membrane leaflet. In vivo monolayers serve to reduce the surface tension at the air–tissue interface within the alveoli in the lungs [20].

The specific structure of a phospholipid is composed of three components: a diglyceride, a simple polar organic moiety, and a phosphate group linking the two. The structures of these components determine the material properties of lipid bilayers. One important characteristic of lipid membranes is their phase. Specifically, are the lipids locked tightly against each other in the bilayer with little freedom to move, or does thermal energy provide enough movement to the lipids that the bilayer can be considered a 2D liquid? A gel or solid-like phase, S_o, is observed at low temperatures for lipids with long saturated chains. In this phase, the lipids are arranged in a hexagonal lattice and possess long-range order that has been observed by scanning probe methods [21]. As the temperature is increased past the main chain transition (T_m), the lipid bilayer will adopt a fluid or liquid crystalline (L_α) state in which the hydrocarbon chains are free to move. A schematic representation of a lipid bilayer illustrates these two phases in Figure 6.1. A liquid phase is consistent with the behavior of the cell membrane as discussed above, whereas the gel state is usually not considered biologically relevant. Beyond lateral order, fluorescence microscopy

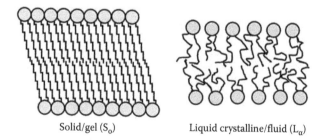

Solid/gel (S_o)   Liquid crystalline/fluid (L_α)

FIGURE 6.1 Schematics depicting the difference in hydrocarbon chain ordering between the solid or gel phase and the liquid crystalline phase. The liquid crystalline phase bilayer is thinner and possesses a larger area per molecule.

has shown that the lateral diffusion of lipids in the fluid phase is approximately 100 times faster than in the gel phase [22]. Other important distinguishing features are the average bilayer thickness, which is ~38 Å for the liquid crystalline phase of 1,2-dipalmitoyl-sn-glycero-3-phosphocholine (DPPC) and 44 Å for the gel phase [23]. The change in bilayer thickness is due to both an increase in the order of the lipid's alkyl chains and a concerted change in the molecular tilt angle.

There are several ways that molecular structure and properties influence phase behavior. First let us consider the influence of the hydrocarbon chain. Two simple modifications of an alkyl chain are to change the length or introduce double bonds between carbon molecules. These modifications will influence how regularly the chains pack and to what extent. Increasing the length of the hydrocarbon tail increases the magnitude of van der Waals interactions and, therefore, increases the main chain transition temperature. The relationship between main chain transition temperatures for several phospholipids with saturated alkyl chain tails is shown in Figure 6.2. Here, we see that transition temperature can change by ~80°C as the number of carbons in the alkyl chains increases from 12 to 24 [24]. A second example of how the structure of lipid chains determines the physical properties of lipid membranes is the addition of an unsaturated bond. An unsaturated bond within the hydrocarbon tails will increase the chain disorder and hence reduce the main chain melting temperature.

Charge, head-group size, and number of hydrogen bond acceptors and donors may also determine the self-assembly properties of lipids. For example, recent studies illustrate the importance of head-group charge on the supramolecular assembly of collapsed phases [25]. Head-group size, charge, and structure also play important

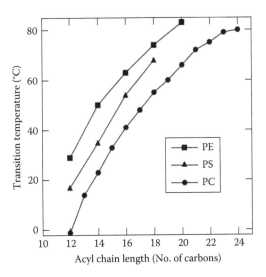

FIGURE 6.2 Plot of transition temperature as a function of the number of carbon repeat units in the alkyl chains of phosphatidylethanolamine (PE), phosphatidylserine (PS), and phosphatidylcholine (PC). (Reproduced from NIST Standard Reference Database 34, *Lipid Thermotropic Phase Transition Database*.)

roles in phase behavior as illustrated in Figure 6.2. The trend between head-group interactions is seen by comparing the transition temperatures of lipids with varying head groups (keeping the alkyl chain length constant). We see that the smaller head group of phosphatidylethanolamine produces a higher transition temperature than phosphatidylcholine, both of which are zwitterionic and have a neutral charge (in most biological buffers). Figure 6.2 also shows that as head-group interactions become stronger because of electrostatic interactions for the negatively charged phosphatidylserine, the transition temperature concomitantly increases relative to neutral phosphatidylcholine [24].

Most relevant to our discussion of the mechanical properties of supported lipid membranes here is the ability of the lipid membrane to respond and re-form when perturbed. We would expect monolayers and bilayers in the fluid phase to more easily re-form and adjust in the presence of an interfacial force microscope (IFM) tip. Therefore, we will simplify our experimental investigation by studying two phospholipids identical in every way but hydrocarbon chain length.

6.2.2 FORMATION OF SUPPORTED LIPID FILMS

The two most common techniques to coat solid substrates with phospholipid membranes (monolayers and bilayers) are Langmuir–Blodgett (LB) deposition and vesicle fusion. Each of these methods has potential applications where they may be advantageous. LB deposition was originally developed at General Electric by Irving Langmuir and Katherine Blodgett in the 1930s. In the late 1970s and early 1980s, McConnell et al. pioneered the use of LB deposition to create SLBs [26]. There are several comprehensive references that describe LB deposition in detail [27]. To describe the technique briefly (Figure 6.3), a solid substrate is drawn through a surfactant bearing interface. During this process, a single molecular layer of the surfactant will be deposited on the substrate. There are several putative advantages of using LB deposition. For example, LB deposition provides the ability to form SLBs with distinct compositions in the upper and lower leaflets. LB deposition also allows the experimenter to control the packing density of the deposited film. This technique has been used to investigate several lipid membrane–related phenomena.

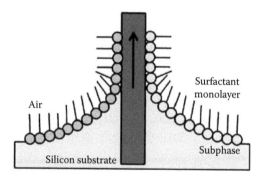

FIGURE 6.3 Illustration of the LB deposition process used in our experiments. The surfactant monolayer is deposited on a solid silicon substrate at the air–water interface.

For example, phase-separated monolayers have been used to model the behavior of lipid rafts [28]. Supported lipid membranes formed with LB deposition have also been used to investigate protein–protein interactions [29] and the asymmetric nature of the cell membranes [30].

A more recent technique to fabricate SLBs is the vesicle fusion method [31]. In this method, lipid vesicles are prepared and introduced above a substrate (glass, mica, and silicon). The spontaneous rupture of the vesicles and subsequent healing of surface defects have been shown to form high-quality lipid bilayers. This technique also provides an opportunity to introduce membrane proteins into the supported system. A potential limitation of this technique is that it is difficult to control the molecular area occupied by a lipid within the SLB. Another limitation is that due to the stochastic nature of bilayer formation, it is not typically feasible to form asymmetric bilayers.

The work reviewed here examines the mechanical properties of supported lipid films as models of the cell membrane. We use interfacial force microscopy [32,33] to probe the elastic, plastic, and friction properties of both monolayers [15] and bilayers [16] of phospholipids. This technique allows for quantitative determination of these mechanical properties and forms an underlying foundation of knowledge about how a simple model of the cell membranes responds to mechanical stress. This chapter highlights the properties of two fairly similar phospholipids: 1,2-dilauroyl-sn-glycero-3-phosphocholine (DLPC) and 1,2-distearoyl-sn-glycero-3-phosphocholine (DSPC). The basic structure of these molecules is the same. A neutral charge zwitterionic phosphatidylcholine head group is connected to twin saturated alkyl tails through a glycerol linker, see Figure 6.4. Only the length of the alkyl tails differs between the two molecules with DLPC and DSPC having 12 and 18 repeat carbon units, respectively. While similar in molecular structure, these two molecules assemble into films that have distinctly different mechanical properties. As discussed above, the longer alkyl chains of DSPC tend to form closely packed gel phase films, while the shorter DLPC lipid tends to form disordered fluid phase film. Note that of the two fluid phases of lipid bilayer mentioned above, DLPC forms liquid-disordered phase films. However, we refer to these films simply as fluid phase. In the following, both a quantification of the mechanical properties of the films formed

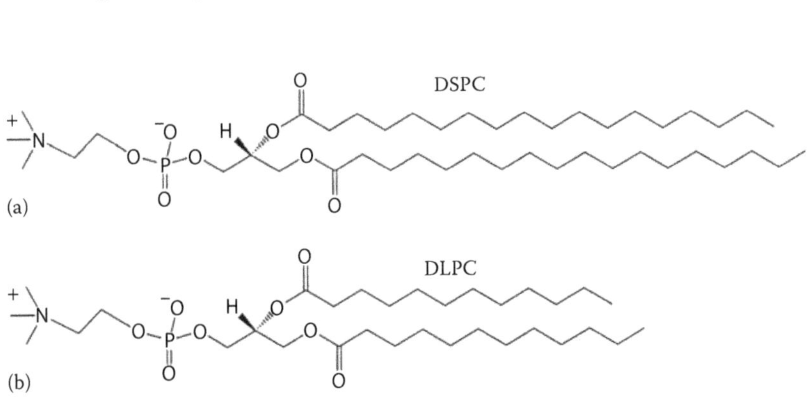

FIGURE 6.4 Schematic representation of the molecular structure of DSPC (a) and DLPC (b).

by these two lipids and the underlying physical properties that contribute to the mechanics are discussed.

6.3   INTERFACIAL FORCE MICROSCOPY

The IFM was developed at Sandia Labs to investigate the fundamental nature of interfacial adhesion [34]. At the heart of the IFM is the differential capacitor sensor shown in Figure 6.5. Currently, this sensor is manufactured by placing an etched Si common plate above two Au pads that are deposited on a glass substrate, creating a differential capacitor. The capacitance of each side of the sensor is defined primarily by the gap between the common Si plate and the underlying Au pad. A probe tip is attached on one side of the sensor's top plate. Operating without feedback, when a force is applied to the probe, the top plate pivots around a set of torsion bars. This teeter-totter motion creates a differential between the capacitors on each side of the sensor. This imbalance is proportional to the applied force and is detected using an AC-bridge (operating at ~1 MHz). Demodulating the bridge imbalance and feeding the resulting DC signal to a feedback controller allows for proportional-integral-derivative (PID) control of the electrostatic restoring force that is placed on one side of the sensor and is again proportional to the force applied to the probe tip. This sensor design, coupled with the force feedback controller, creates effectively a compliance-free sensor that can measure both adhesive and repulsive forces from tens of nanonewton up to several micronewton with Ångstrom level spatial sensitivity.

   There are several advantages to using a compliance-free instrument for measuring mechanical properties. First and foremost is the ability to measure adhesive

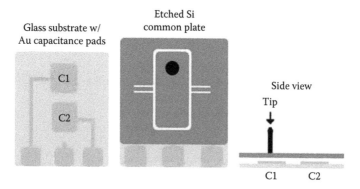

FIGURE 6.5 (See color insert.)   Schematic of the IFM sensor. Gold capacitance pads (C1 and C2) are bonded to a glass substrate (a), to which an etched Si common plate is bonded (b). Also shown are three Au leads on the bottom of the glass substrate to which electrical connections are made. The left and right leads connect to capacitance pads, while the center is bonded to the common plate. Etched W tips are bonded to the common plate above one of the capacitance pads. (c) Side view of the common plate with tip suspended above the substrate.

forces without the mechanical instabilities that accompany spring cantilever-based instruments such as the atomic force microscope (AFM) [35] and surface force apparatus (SFA) [18]. These instabilities result when an adhesive force overcomes the cantilever's spring restoring force and the probe uncontrollably contacts the surface, this event is commonly referred to as "jump-in." The reverse process occurs during retraction and is referred to as "pull-off." This artifact of the experimental approach creates difficulties. For example, it is a common practice to use the magnitude of the "pull-off" force to quantify the work of adhesion between the probe and sample. However, this is problematic as a contact mechanics model must be employed to relate the "pull-off" force to the work of adhesion. This is not necessary when using a compliance-free sensor as employed in the IFM. Instead, the true work of adhesion can be calculated by integration of the force profile.

A compliance-free instrument is also advantageous when measuring the plastic deformation (a.k.a. failure, collapse, yield, and breakthrough) of materials. During operation, a spring-based instrument will store elastic energy in the cantilever. Unfortunately, some of this energy is transferred to the sample making it difficult to analyze the fundamental process behind a relaxation event, such as plastic deformation. The unique sensor design of the IFM does not store elastic energy while performing a force measurement, allowing for a more direct measurement of the relaxation process.

In addition, the differential capacitor sensor of the IFM balances the total applied torque. Therefore, using a technique similar to shear modulation force microscopy [36] performed with the AFMs, it is possible to simultaneously measure both normal and lateral (friction) forces. This is accomplished by moving the probe with a sinusoidal lateral modulation while recording a force profile. Extracting the friction using a lock-in amplifier allows for both the magnitude and the phase angle (relative to the modulation) of the lateral force to be collected. Analyzing the phase angle of the lateral force has proven to be invaluable in determining the origins of lateral interactions such as sliding friction, slip-stick, and viscous forces.

A typical IFM force profile experiment (force vs. distance) begins with the tip held in contact with the sample at a very light load. The tip is then retracted a fixed distance from the surface. The force profile begins with an approach of the sample at a constant speed until a force set point is reached. This is followed by retraction of the tip from the surface at the same speed and distance as used during the approach. The force on the tip is recorded during this entire process. Unless noted otherwise, all force profiles were performed at a constant approach/retract rate of 2 nm/s.

6.4 EXPERIMENTAL DETAILS

IFM experiments of measuring forces normal to supported lipid films can be used to answer simple questions about the deposition, such as was the deposition successful? If so, how thick is the lipid film? It can also be used for extracting mechanical properties such as elastic modulus, adhesion, electrical properties, and yield force. The remainder of this chapter constitutes discussion of mechanical properties of both supported lipid monolayers and bilayers.

Experiments on monolayers were conducted with an electrochemically etched W (tungsten) tip (radius = 1.3 µm) on samples formed from the LB deposition at several surface pressures using a commercial LB trough (Nima Model 612D, U.K.). To test the influence of packing density, films were prepared by incrementally raising the surface pressure of the deposition for both DLPC and DSPC from 10 to 50 mN/m (increments of 10 mN/m). All lipids were used as received (Avanti Polar Lipids, Inc., Alabaster, AL) and stored in a freezer at approximately −20°C. Supported bilayers were deposited from a vesicle solution and kept hydrated during the entire experiment. SLBs were also characterized using a W tip (radius = 0.5 µm). Lipid vesicle solutions were made by mixing stock solutions of dissolved lipids in chloroform. The solution was dried under N_2 for ~1 h and suspended in an aqueous buffer (100 mM NaCl, 50 mM Tris, 5 mM CaCl2, pH ~7.5) and sonicated for ~1 min to create a 100 mM vesicle suspension. We prepared solutions of small unilamellar vesicles (SUVs) by extruding the lipid solution 11 times through a 100 nm polycarbonate filter. A sufficient amount of the SUV solution was then incubated with a clean silicon surface for 3 h at room temperature to form an SLB. Excess vesicles were removed from the surface by flushing with copious amounts of water.

All films (monolayers and bilayers) were deposited on Piranha (3:1 conc. H_2SO_4/30% H_2O_2; Caution! Piranha solution is a strong oxidant and reacts violently with organic substances) cleaned Si wafers. This process cleans the native oxide of Si of all organic contaminants and produces an extremely hydrophilic surface. Cleaned Si wafers with no deposited lipids are used as control surfaces and characterized in the same matter as the supported lipid films.

Friction measurements were performed by adding a ~2 nm (peak-to-peak) sinusoidal lateral modulation to the tip. The dissipative component of the lateral force signal in quadrature (90° of a phase) with the drive is used as the friction signal. Force profiles shown for lipid monolayers are averages of three profiles in three separate locations. Averaging was not performed on lipid bilayers force profiles as there are several fine features that would be smoothed out by subtle changes in the location and magnitude of these features present on all indentations. Note that for all force profiles presented here, the displacement axis is always relative and the "0" point is only for reference.

6.5 MECHANICAL PROPERTIES

6.5.1 NORMAL FORCE PROPERTIES

6.5.1.1 Supported Lipid Monolayers

Figure 6.6 shows typical force profiles for LB-deposited monolayers of DLPC and DSPC deposited at a specific surface pressure (40 mN/m). Also shown are fits to the Johnson–Kendall–Roberts (JKR) model of contact mechanics [37]. Contact mechanics models are used to extract mechanical properties such as the elastic modulus from force profiles. It is helpful to use the method outlined by Carpick et al. to determine which model of contact mechanics, if any, is appropriate for a specific system [38]. The two models considered are those of JKR and

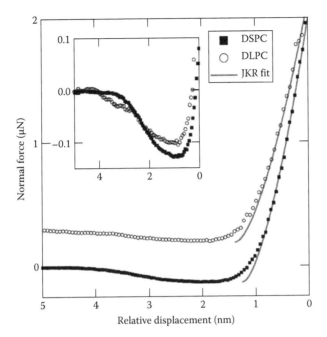

FIGURE 6.6 Normal force profiles during approach on LB-deposited monolayers of DSPC (solid squares) and DLPC (open circles). DLPC data are offset vertically by 0.3 μN for clarity. Solid lines are fits to the JKR model of contact mechanics. The inset shows an expanded view of the adhesion of both force profiles.

Derjaguin–Müller–Toporov (DMT) [39]. Both are based on the Hertz model [40] with additional contributions to contact area due to adhesion. The JKR model considers only in-contact forces that cause deformations of the tip and substrate. In contrast, DMT theory considers long-range forces while maintaining Hertzian contact geometry. As a rule of thumb, the JKR model is appropriate for large soft contacts and the DMT model works best for small hard contacts [41]. Note that both models assume a linearly elastic response over the entire contact range, a condition not necessarily met by viscoelastic thin films. These models were also developed for monolithic systems and may not be appropriate for layered samples. Despite limitations, these models are useful in comparing similar systems and their associated properties.

According to Carpick et al.'s method, an α parameter is first calculated by fitting the contact area vs. load. Since a direct measurement of contact area is experimentally difficult, it is estimated through other properties such as friction or contact resistance. One method is to assume that friction (F) is linearly proportional to contact area (A) through the friction shear strength (τ) by the Bowden and Tabor (BT) model [42]

$$F = \tau A. \tag{6.1}$$

Since the IFM can measure both friction and load in a single experiment, we sub-
stitute friction for contact area into Carpick et al.'s method. The parameter α can be
determined by fitting the friction vs. load to the following equation:

$$\sqrt{F} = \sqrt{F_0} \left(\frac{\alpha + \sqrt{1 - L/L_c}}{1 + \alpha} \right)^{2/3}, \qquad (6.2)$$

where
    F is the friction force
    F_0 is the friction force at zero applied load
    L is the load
    L_c is the load at maximum adhesion

Accordingly, as α goes to 0, the DMT model describes the contact, and as α goes
to 1, the JKR model is appropriate. It is important to note that substituting friction for contact area in the method implies that τ is independent of load. This
assumption is still debated in an attempt to understand friction laws for nanoscale
contacts [43].

Applying Equation 6.1 to the force profiles in Figure 6.6 yields an α of 0.9–1.1
for both DLPC and DSPC for all the depositions studied. This indicates that the
JKR model is appropriate for these films. Indeed, the JKR model fits the data well
after a small initial compression of the monolayers, see Figure 6.6. By breaking up
the force profile into sections, we can understand the dominant processes behind
each region of the force profile. Note these general trends hold true for all the monolayer depositions studied.

As the tip approaches the monolayer, adhesion dominates the force profile, this is
shown in the inset of Figure 6.6. Since the lipid monolayers are standing with their
hydrophobic tails pointing toward the tip, van der Waals interactions likely dominate the adhesion. A quantitative analysis proves difficult as many other adhesive
forces potentially exist. For example, capillary forces from the hydrophilic W tip and
the possibility that weakly adsorbed lipid molecules may respond dynamically and
reorient under attractive forces further complicate the analysis.

As the tip compresses the monolayer, the force begins to rise repulsively.
Establishing the exact point of contact is difficult, though a reasonable estimation is
somewhere near the point of maximum adhesion, or critical force. For all force profiles, the location of the critical force is slightly (~5Å) before the location of contact
predicted by the JKR model and (<1 nm) before the JKR model accurately models
the data. As noted above, JKR theory models all adhesive interactions occurring
only during contact and, of course, is not an accurate description of van der Waals
forces. This causes the discrepancy between the critical force and the JKR estimation of contact.

For all force profiles, we observe that the normal force vs. indentation rises relatively slowly for the first nanometer before rising sharply to meet the JKR model.
This behavior is attributed to two main causes. First, lipid monolayers may have a

dynamic response and reorient under stress into a more ordered film, causing the first nanometer of indentation to appear relatively soft (low modulus) compared to further indentation, which induces high compressive strain [44]. Second, as previously noted, most contact mechanics models fail to treat layered systems such as supported thin films. In general, indentations of only 10% of a film's thickness are an accurate representation of the film's properties. Upon further indentation, the mechanical properties of the underlying substrate and tip dominate the force profile. Unfortunately, applying this rule to a ~2 nm lipid monolayer allows meaningful data for only 2 Å of the monolayer itself.

However, by systematically varying the properties of a thin film, it is still useful to compare mechanical properties derived through contact mechanics. Applying the JKR model yields composite modulus values for the compressed lipid monolayers of 33 ± 4 GPa for DLPC and 49 ± 6 GPa for DSPC. The values of the composite modulus were found to be independent of the surface pressure and/or associated packing density of the LB depositions. Note that these values are much higher than previously reported for LB monolayers [45]. Fitting only the first nanometer of indentation provides a reasonable estimate for the compliance of the LB monolayer. Applying the JKR model for the first nanometer of compression for all depositions of DLPC and DSPC yields a composite modulus of ~1 GPa. This value is close to the previous values measured for a supported lipid monolayer by Brillouin spectroscopy [45].

Several other insights can be gained from the normal profiles of supported lipid monolayers. First, normal force profiles of a control surface (bare Si) yield a composite modulus of 110 GPa (not shown). This indicates that the LB film is not squeezed out between the tip and the substrate during compression. Second, the force profiles at a given location are reproducible, indicating that the lipid monolayer is able to completely recover within the experimental timescale (~1 min). Third, forces measured during retraction from the surface indicate little to no hysteresis (data not shown). This suggests that the tip only elastically stresses the films during a force profile and leads to our final observation. We do not observe any sharp features in the force profiles corresponding to plastic deformation/failure of the monolayers as previously observed with AFM [46]. This discrepancy is not unexpected as the tips used here were nearly two orders of magnitude larger and correspondingly produced much less stress during indentation.

6.5.1.2 Supported Lipid Bilayers

Increasing in both complexity and relevance to biomaterials, the normal force measurements on SLBs indicate how these films respond to applied stress. Figure 6.7 shows force profiles for SLBs of DSPC and DLPC. These force profiles exhibit behavior not observed for monolayers, specifically the onset of plastic deformation and significant approach/retract hysteresis. There are also distinct differences between DSPC and DPLC, which are not observed in monolayers. Because of this, we must analyze DSPC separately from DLPC.

Beginning with a force profile approach of a gel phase DSPC SLB, we observe a rise in the repulsive force until ~2 μN, top panel Figure 6.7. This response suggests a near elastic stressing of the SLB. Upon further indentation, the normal

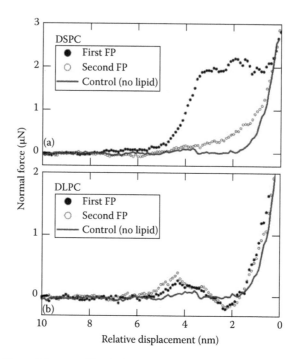

FIGURE 6.7 Normal force profiles during an approach (filled circles)/retract (open circles) cycle for supported lipid bilayers (SLBs) of DSPC (a) and DLPC (b). Approach on silicon is shown as a control (solid line). (Reproduced from Goertz, M.P. et al., *J. Phys. Chem. B*, 113, 9335, 2009. With permission from ACS.)

force remains constant as the SLB undergoes plastic deformation from the applied stress. Plastic deformation continues as the SLB is squeezed out between the tip and underlying substrate for the next ~3 nm of indentation. After the bilayer is squeezed out/removed, the tip contacts the underlying Si substrate and the force profile closely resembles that of the bare Si control surface. Upon reaching the force set point and reversing direction significant hysteresis is observed. The hysteresis indicates that the DSPC SLB is not able to heal and recover to its original structure within the timescale of the experiment (each force profile takes ~1 min to complete).

Several key differences are observed by performing the same experiment on a fluid DLPC SLB, bottom panel of Figure 6.7. As with DSPC, as the tip first approaches the DLPC SLB, there is a region of elastic compression of the film. However, this film collapses at only ~0.3 μN and further indentation does not produce a constant normal force. Instead, we observe the normal force to decrease and become adhesive as lipids begin to create a bridging structure between the tip-substrate gap. Finally, once the lipids are completely squeezed out from under the tip, the force profile again resembles that of the bare Si control. Figure 6.7 shows that retraction produces significant hysteresis, though in a different manner than for DSPC. Here, all of the features found during approach are reproduced during retraction with a broadened length scale. This suggests there is partial recovery of the fluid phase SLB between approach/retract profiles.

The repetition of multiple force profiles in a single location allows for the recovery of SLBs to be probed on a longer timescale than approach/retract experiments. Figure 6.8 shows how both DSPC and DLPC bilayers respond to repeated stress. The time between repeated approaches is ~10 min compared to the <1 min between an approach/retract cycle. As discussed in Section 6.2, we do not expect the gel phase DSPC SLB to recover on this experimental timescale, and this is indeed what is observed in Figure 6.8a. The solid-like behavior of DSPC reduces the mobility of the bilayer, so that no recovery is observed after the initial plastic deformation of the SLB. The same experiment on an SLB of DLPC shows a distinct contrast. Figure 6.8b demonstrates a complete recovery of the fluid DLPC bilayer after plastic deformation. Sequential force profiles in the same location for DLPC are nearly identical, indicating the bilayer has completely healed.

Similar to supported monolayers, it is helpful to break the force profiles of SLBs into several regions, Figure 6.9. This allows us to separate and quantify the dominant material response in each of the four regions of contact. In region "1," the tip

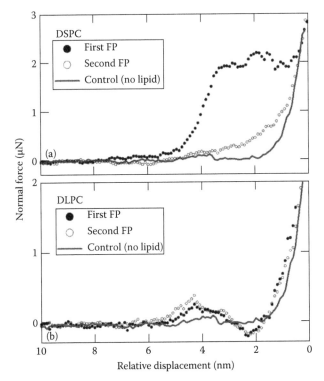

FIGURE 6.8 Successive approach force profiles in a single location of supported lipid bilayers (SLBs) of DSPC (a) and DLPC (b). Silicon is shown as a contrast in both figures (solid line). Significant hysteresis is observed only on DSPC. (Reproduced from Goertz, M.P. et al., *J. Phys. Chem. B*, 113, 9335, 2009. With permission from ACS.)

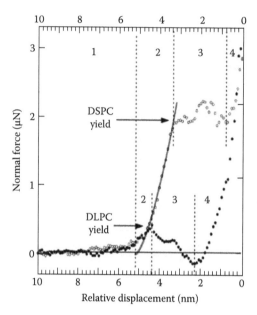

FIGURE 6.9 Normal force profiles for supported lipid bilayers (SLBs) of DSPC (open circles) and DLPC (filled circles). The solid line shows a fit to the Hertz theory of contact mechanics. Interaction regions as defined in the text are shown for both lipids. (Reproduced from Goertz, M.P. et al., *J. Phys. Chem. B*, 113, 9335, 2009. With permission from ACS.)

is not yet in direct contact with the bilayer. Further indentation initiates an elastic compression of the bilayer, region "2," and transitions into plastic deformation and failure observed in region "3." Finally, direct contact occurs between the tip and the underlying Si substrate in region "4." Note that the force profiles in Figure 6.9 are shifted along the horizontal axis in order to normalize contact with the SLB, i.e., at the same relative displacement. This is done by shifting the force profiles laterally, so there is overlap between the leading edge of the rise in normal force in region 2. Separating force profiles into regions also allows estimation of the SLB thickness. A rough estimate of bilayer thickness can be found by adding the width of regions 2 and 3. This yields bilayer thicknesses of 4.4 and 2.9 nm for DSPC and DLPC, respectively. These thicknesses are similar to those previous observed using AFM [47].

For both lipid bilayers, we observe a small amount (>100 nN) of force in region 1. To elucidate the nature of these long-range weak forces, the approach speed was varied from 2 to 12 nm/s during which no change was observed (see Figure 6.10). As a result, the origin of this interaction remains unclear and is most likely the result of many contributions including a dynamic roughness of the SLB due to thermal fluctuations [18], contaminants adsorbed on the tip, and solvation forces associated with disrupting and/or removing water near the head groups of the SLB [21].

The transition between regions 1 and 2 is characterized by an elastic stress in the SLB and a rapid rise in the normal force. As with indentation of monolayers, the elastic compression of SLBs can be described by contact mechanics. The lack

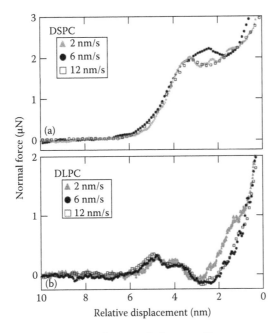

FIGURE 6.10 Speed dependence of approach force profiles on supported lipid bilayers (SLBs) of DSPC (a) and DLPC (b).

of adhesion during approach simplifies the analysis and the Hertz model can be directly applied [40]. The fit to this model for DSPC yields a composite modulus of 35 GPa, similar to the 49 GPa found for monolayers of the same lipid, Figure 6.6. Unfortunately, the limited data points on elastic deformation for DLPC do not allow for an accurate fitting with the Hertz model. However, the repulsive rise in normal force in region 2 overlays that of DSPC until collapse occurs. This suggests that the two SLBs have similar elastic properties but differ significantly in the amount of stress required for plastic deformation.

The transition between regions 2 and 3 occurs as the bilayers yield under the applied stress of the IFM tip. We calculate the yield stress for both SLBs by normalizing the load by the contact area at the onset of plastic deformation. According to the Hertz model [40], the contact area (A) is related to the tip radius (r) and the indentation depth (h) by

$$A = \pi rh. \tag{6.3}$$

Applying this method to DSPC, which at an indentation depth of 1.8 nm plastically yields at force of 2 μN, produces a yield stress of 710 MPa. A similar treatment of DLPC, which yields at 0.3 μN and 0.8 nm, shows a lower yield stress of 240 MPa. While there has been significant work using the AFM to probe the failure of SLBs, we are unaware of any previous evaluation of the yield stress of a gel phase bilayer. Our measurement of the yield stress of DLPC is remarkably close to the yield stress of another fluid phase SLB, dioleoyloxypropyl-trimethylammonium chloride

(DOTAP). Previous work by Franz et al. has shown that an SLB of DOTAP on mica has a yield stress of 259 MPa as probed with AFM [48]. This result is of considerable interest because of the significantly different tip sizes used in the two experiments. Our IFM tip was ~10× larger than that in the AFM study, which at the onset of plastic deformation has ~100× larger contact area. Therefore, although there are ~100× more lipid molecules undergoing plastic deformation in our experiment, the yield stress is almost identical to the previous AFM study. This result shows that material properties such as yield stress are highly dependent on nearest neighbor molecular interactions.

While the behaviors of DSPC and DLPC are similar until collapse occurs, they show drastically different responses as they are squeezed out of the tip/substrate gap. In region 3, we observe that the normal force remains repulsive during the collapse of DSPC while DLPC becomes adhesive. While an analysis of force profiles does not provide insight at the single molecule level of the SLB, this behavior can be explained through the mechanism shown in Figure 6.11. During collapse of the DSPC SLB, the bilayer structure fails but the majority of molecules remain in the tip/substrate gap creating a repulsive force. The opposite behavior is seen in DLPC, where the same fluidic nature of the SLB that led to faster recovery times. This allows the lipid bilayer to re-form to create a bridge between the tip and substrate. The negative curvature of this lipid bridge is similar to the creation of liquid meniscus and produces the observed adhesive force.

To examine the theory that DLPC forms a "meniscus" lipid bridge to the tip, we will consider the bending modulus (k_b) of a DLPC bilayer and its related adhesion [18]. The energy per unit area (E_b) of a curved lipid bilayer is defined by two principal radii: the contact radius (R_c) and the radius of curvature defined by the lipid bridge (R_l)

$$E_b = \frac{1}{2} k_b \left(\frac{1}{R_l} + \frac{1}{R_c} \right)^2. \tag{6.4}$$

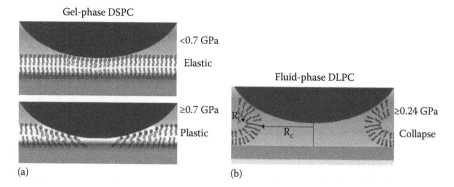

FIGURE 6.11 Schematic representation of the collapse of supported lipid bilayers (SLBs) of DSPC (a) and DLPC (b). (Reproduced from Goertz, M.P. et al., *J. Phys. Chem. B*, 113, 9335, 2009. With permission from ACS.)

The 500 nm probe used in our studies creates large contact area, where the contribution to the bending energy from the contact radius is negligible, as $R_c \gg R_l$, and Equation 6.4 reduces to

$$E_b = \frac{1}{2} k_b \left(\frac{1}{R_l} \right)^2.$$ (6.5)

While the bending modulus for several lipids has been measured macroscopically using micropipette techniques [49], much less is known about the mechanics of bending a membrane on the nanoscale or near a substrate. However, we can estimate R_l by considering the maximum adhesive force for DLPC ($-0.19~\mu N$). The relationship between the force (F) and energy per unit area (E_a) acting between a spherical probe (radius = R_t) and a flat surface is given by the Derjaguin approximation [39]

$$E_a = \frac{F}{2\pi R_t}.$$ (6.6)

Using the previously measured k_b for DLPC [49] of 6.9×10^{-20} J, we can estimate R_l by setting Equations 6.5 and 6.6 equal. This yields a lipid bridge curvature of $R_l \sim -1$ nm, a reasonable value considering the geometry of the experiment. Note that the negative value signifies that the bridge curvature is in the opposite direction of the contact radius as shown in Figure 6.11 and is responsible for the bridge being an attractive interaction.

It is of interest to compare our results for the plastic deformation of SLBs to previous nanoindentation studies (for a comprehensive review, see [50]). In AFM studies, it is common to use the threshold force and "jump-in" distance to characterize the failure of SLBs. However, as previously mentioned, performing these measurements with an open-loop cantilever-based instrument is problematic. During plastic deformation, as the tip ruptures the SLB, the tip becomes unstable and suddenly "jumps" into contact with the underlying substrate. During this process, some of the elastic spring energy stored in the cantilever may be transferred to the SLB. As a result, these experiments provide only little information about how the bilayer behaves during collapse. The analysis approach taken by the majority of bilayer failure studies using AFM is also significantly different. In these studies, the rupture event is taken to be a statistical process with the tip having a certain probability to break through the SLB at a certain applied load [50]. According to this model, the tip needs to overcome an activation energy to initiate hole formation in the SLB. After this occurs, it is energetically favorable for the tip to penetrate the SLB. Calculating the activation energy is accomplished by performing force profiles with increasing loading rates. The larger tips used in our studies make it unlikely that the collapse of the SLB proceeds by the mechanism of hole formation under the tip. This is justified considering two observations. First, according to the theory presented above, the tip should actually lose contact with the SLB during the process of hole formation. If contact with the SLB was lost, we expected to observe the normal force to suddenly decrease, which we did not. Second, according to the theory of hole formation, the yield force should increase with loading rate [50]. Figure 6.10 shows a constant yield force for both SLBs as the loading rate is varied from 2 to 12 nm/s.

6.5.2 FRICTION FORCE PROPERTIES

6.5.2.1 Supported Lipid Monolayers

Further mechanical properties of supported lipid films can be extracted from friction force measurements. Figure 6.12 shows the friction vs. load behavior of supported monolayers of DSPC and DLPC (both deposited at 40 mN/m). The fits are to the BT model (Equation 6.1), with the contact area as defined by the JKR model. The excellent agreement between fit and experiment suggests the contact is a true single asperity and at constant friction shear strength. In other words, the friction force per unit area does not depend on the applied load. In order to produce the fits shown in Figure 6.12, we found that the addition of a small vertical offset was necessary. The sign of the offset was positive for DSPC and negative for DLPC monolayers and was always <4 nN. These offsets suggest that using Equation 6.1 (BT model) simultaneously with the JKR model underestimates the friction and/or contact area for DSPC while overestimates for DLPC. Using the above fitting method, we extract a friction shear strength coefficient at a variety of packing densities for LB-deposited lipid monolayers. Figure 6.13 shows the relationship between friction shear strength and

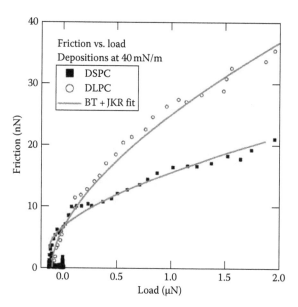

FIGURE 6.12 Friction vs. load relationship for LB-deposited monolayers (at 40 mN/m) of DSPC (filled squares) and DLPC (open circles). Solid lines are fit to the BT model of friction with contact area described by the JKR model. (Reproduced from Goertz, M.P. et al., *J. Phys. Chem. A*, 111, 12423, 2007. With permission from ACS.)

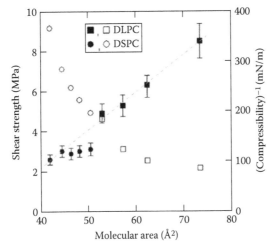

FIGURE 6.13 Friction shear strength (filled symbols) and compressibility^{-1} (open symbols) vs. molecular area for DLPC (squares) and DSPC (circles) monolayers on oxide terminated silicon. The dashed line is a guide to the eye for the trend in friction shear strength. (Reproduced from Goertz, M.P. et al., *J. Phys. Chem. A*, 111, 12423, 2007. With permission from ACS.)

molecular area (inverse of packing density). The molecular area is estimated from the pressure-area isotherms taken before LB deposition. We find a linear relationship between friction and molecular area for both lipids. This trend can be explained by analyzing the origin of friction at the molecular level.

Previous work addressing the molecular origins of friction includes scanning probe methods, spectroscopic measurements, and simulations. These studies have shown that the major energy dissipation mechanisms that create friction from mechanical contributions include intramolecular vibrations, concerted changes in molecular tilt angle, and conformational changes [51–54], as well as chemical effects such as interfacial bonding [55]. Note that these mechanisms are dominant below the wear threshold of these films. At high applied shear stress, irreversible damage (wear) is the major cause of friction [42]. However, as noted above, the reproducibility of force profiles and lack of plastic deformation in supported monolayers show that the stresses used did not produce irreversible damage.

The three major mechanical contributions established above correlate strongly with both crystalline order and the conformational freedom provided by the molecular area of a monolayer. Increasing the molecular area of a lipid decreases the crystalline order and increases the conformational freedom. These two effects lower the steric hindrance of the mechanical contributions to friction. This creates more channels for energy dissipation and, in turn higher, friction. The relationship between packing density and friction was observed previously but has been limited to covalently bonded self-assembled monolayers (SAMs) [53,56–59]. Varying the packing density of these films is usually accomplished by varying the length of their alkyl chains and hence the interchain van der Waals attraction. This technique only allows for a very limited range of packing densities to be probed and necessitates the use of different molecules, which may not have the same intrinsic friction properties. Our use of LB deposition allows for the study of friction over a much broader range of packing densities using the same molecules.

The increase of friction with molecular area is more obvious for the fluid phase DLPC monolayer relative to DSPC. This is attributed to both the larger variation in molecular area of DLPC deposition and the change in phase between the two films. Previous surface sensitive spectroscopy experiments have shown that there is a "strong reorientation" of the alkyl tails of a lipid monolayer during the 2D fluid to gel phase transition [60,61]. This reorientation creates additional crystalline order in the gel vs. fluid phase lipid film, see Figure 6.1.

We can also investigate the correlation between the compressibility of a monolayer and its frictional properties. The compressibility (C) for a 2D film measures how easy it is to energetically change the surface pressure of the film [62,63] and is defined as

$$C = -(1/A)*(dA/d\pi),\qquad(6.7)$$

where
   A is the molecular area
   π is the surface pressure defined by the π vs. A isotherm

The inverse compressibility (C^{-1}) is a measure of the 2D compressive modulus (sometimes called the area expansion modulus) that relates the amount of 2D stress per unit 2D strain required to compress a Langmuir film. Figure 6.13 also shows the relationship between inverse compressibility and friction. We find that these two measurements are inversely correlated. This indicates that the same mechanisms that give rise to friction also allow for easier (lower energy) compressibility of an LB monolayer.

6.5.2.2 Supported Lipid Bilayers

The friction properties of SLBs in Figure 6.14 show dramatic differences when compared to monolayer films of the same composition. Unlike monolayers, where DLPC always produces higher friction than DSPC, we find that DLPC shows almost no measurable friction while it is stressed elastically or during SLB collapse. Because of differences in experimental conditions, this contrast is not completely unexpected. As established above, when probing lipid monolayers, the tip primarily contacts the hydrophobic alkyl tails of the lipids. The friction originates at this interface and is associated with the conformational freedom of the tails. This is not the case when probing SLBs, the tip now primarily contacts the hydrophilic head groups of the lipids. Further, by performing friction measurements in water to maintain the SLB hydration, it is likely that the slip plane changes [64]. The slip plane is the actual location of any movement that occurs during sliding friction. It has been previously theorized that hydration of a surfactant film changes the slip plane from the tip/alkyl interface (dry) to the head-group/tip or head-group/substrate interface (wet) [64]. Note that there is a well-established thin film (<1 nm) of

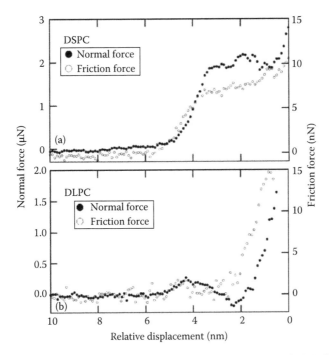

FIGURE 6.14 Overlay of normal (filled circles) and friction (open circles) force profiles for SLBs of DSPC (a) and DLPC (b) during approach. (Reproduced from Goertz, M.P. et al., *J. Phys. Chem. B*, 113, 9335, 2009. With permission from ACS.)

interfacial water that acts as a cushion between the head group of the bottom leaflet and the substrate. This water layer may play a significant role in allowing for SLB movement (slip) under shear.

Another key difference between monolayer and bilayer experiments is the onset of plastic deformation in SLBs. As noted in the discussion above, the major mechanical contributions to friction are significant only below the onset of plastic deformation. While there is a small amount of friction observed for DSPC SLBs in the elastic regime, most of the friction occurs during collapse of the bilayer. As the bilayer fails, the tip begins to slide through the film instead of on it. Unfortunately, a lack of models for friction which incorporate plastic deformation makes it difficult to quantify the friction properties of an SLB in a manner similar to our treatment of monolayers. However, the same properties that determine the yield strength can be used for qualitative analysis of friction.

The longer alkyl chains of DSPC, which produce a 2D gel state not only produce a relatively high yield stress but also provide significant resistance to the IFM tip sliding through them. This effect produces the friction observed during the failure of the DSPC SLB. On the other hand, the nanoscale fluid lipid bridge formed during the collapse of DLPC creates only very little resistance to sliding, at least on the length scale probed with the IFM.

6.6 SUMMARY AND CONCLUSIONS

These results show how two very similar lipids can create films with drastically different mechanical properties. By simply adding six more carbon units to the alkyl chains of phospholipids, we observe the change from a disordered fluid phase film (DLPC) to a well-ordered gel phase film (DSPC). The increased van der Waals interactions between the alkyl chains of DSPC induce crystalline order that is not present in fluid phase DLPC. This change in structure between these two phases results in the varying mechanical properties of supported monolayer and bilayer films.

The major mechanical differences highlighted in this work are as follows:

1. The crystalline order of the gel phase SLBs creates a 2D solid with a relatively high yield stress. The gel phase DSPC SLB also has a slow lateral mobility that does not allow for healing of the film after plastic deformation has occurred (within experimental timescale).
2. Fluid phase SLBs show drastically different properties than gel phase films due to the significant disorder of their alkyl chains above the main chain transition temperature. These films are similar to 2D fluid and show a ~3× smaller yield stress than gel phase films. However, fluid films still show an elastic response at low forces. The fluidic nature of DLPC and relatively fast lateral mobility allow for complete healing within minutes of plastic deformation and even creates an adhesive fluid lipid bridge to the IFM tip.
3. Sliding friction on the alkyl chains of supported lipid monolayers correlates strongly with the packing density. The molecular mechanisms that create friction are hindered in highly ordered and densely packed films. Therefore, the fluid phase monolayers always exhibit higher friction than gel phase films.
4. Sliding friction for the head groups of hydrated SLBs shows the opposite trend of monolayers of the same lipids. The relatively low mobility and high yield strength of gel phase SLBs create significant friction for the tip sliding on or through the film. The nanoscale lipid bridge formed while in contact with fluid phase DLPC offers only very little resistance to the shear imposed by the IFM tip.

Utilizing the unique mechanical properties of these two phases of lipid films may help in the creation of stable and reliable devices that incorporate thin lipid films.

ACKNOWLEDGMENTS

The majority of this work was performed in the laboratory of Professor Xiaoyang Zhu at the University of Minnesota; the authors would like to express their profound gratitude to Professor Zhu for his guidance and many stimulating discussions. The authors would also like to thank Jack Houston for his continued support of the IFM program at the University of Minnesota. BLS is supported by a grant from The Research Corporation. Portions of this work were performed under support of Sandia National Laboratories. Sandia is a multiprogram laboratory operated by

Sandia Corp., a Lockheed Martin Co., for the United States Department of Energy's National Nuclear Security Administration under Contract DE-AC04-94AL85000.

REFERENCES

1. P. R. Bergethon, *The Physical Basis of Biochemistry: The Foundations of Molecular Biophysics*. Springer, New York (1998).
2. D. H. Boal, *Mechanics of the Cell*. Cambridge University Press, Cambridge, U.K. (2002).
3. S. J. Singer and G. L. Nicolson, Fluid mosaic model of structure of cell-membranes, *Science*, *175*(4023), 720–731 (1972).
4. K. Simons and E. Ikonen, Functional rafts in cell membranes, *Nature*, *387*(6633), 569–572 (1997).
5. S. Munro, Lipid rafts: Elusive or illusive? *Cell*, *115*(4), 377–388 (2003).
6. S. L. Veatch and S. L. Keller, Seeing spots: Complex phase behavior in simple membranes, *Biochimica Et Biophysica Acta—Molecular Cell Research*, *1746*(3), 172–185 (2005).
7. Y. Deng, Y. Wang, B. Holtz, J. Y. Li, N. Traaseth, G. Veglia, and B. J. Stottrup et al., Fluidic and air-stable supported lipid bilayer and cell-mimicking microarrays, *Journal of the American Chemical Society*, *130*(19), 6267–6271 (2008).
8. K. Simons and W. L. C. Vaz, Model systems, lipid rafts, and cell membranes, *Annual Review of Biophysics and Biomolecular Structure*, *33*, 269–295 (2004).
9. J. Henriksen, A. C. Rowat, E. Brief, Y. W. Hsueh, J. L. Thewalt, M. J. Zuckermann, and J. H. Ipsen, Universal behavior of membranes with sterols, *Biophysical Journal*, *90*(5), 1639–1649 (2006).
10. O. Stoevesandt, M. J. Taussig, and M. Y. He, Protein microarrays: High-throughput tools for proteomics, *Expert Review of Proteomics*, *6*(2), 145–157 (2009).
11. B. A. Hills, Oligolamellar lubrication of joints by surface-active phospholipid, *Journal of Rheumatology*, *16*(1), 82–91 (1989).
12. B. A. Hills, Remarkable antiwear properties of joint surfactant, *Annals of Biomedical Engineering*, *23*(2), 112–115 (1995).
13. K. Larrson, *Lipid: Molecular Organization, Physical Functionals, and Technical Application*. The Oily Press, Dundee, U.K. (1994).
14. W. Wang and S. A. Soper, *Bio-MEMS: Technologies and Application*. CRC Press, Boca Raton, FL (2007).
15. M. P. Goertz, B. L. Stottrup, J. E. Houston, and X. Y. Zhu, Density dependent friction of lipid monolayers, *Journal of Physical Chemistry. A*, *111*(49), 12423–12426 (2007).
16. M. P. Goertz, B. L. Stottrup, J. E. Houston, and X. Y. Zhu, Nanomechanical contrasts of gel and fluid phase supported lipid bilayers, *Journal of Physical Chemistry. B*, *113*(27), 9335–9339 (2009).
17. B. Alberts, *Molecular Biology of the Cell*. Garland Science, New York (2008).
18. J. N. Israelachvili, *Intermolecular and Surface Forces*. Academic Press, London, U.K.; San Diego, CA (1991).
19. D. Chandler, Interfaces and the driving force of hydrophobic assembly, *Nature*, *437*(7059), 640–647 (2005).
20. R. Veldhuizen, K. Nag, S. Orgeig, and F. Possmayer, The role of lipids in pulmonary surfactant, *Biochimica Et Biophysica Acta-Molecular Basis of Disease*, *1408*(2–3), 90–108 (1998).
21. M. J. Higgins, M. Polcik, T. Fukuma, J. E. Sader, Y. Nakayama, and S. P. Jarvis, Structured water layers adjacent to biological membranes, *Biophysical Journal*, *91*(7), 2532–2542 (2006).

22. M. R. Alecio, D. E. Golan, W. R. Veatch, and R. R. Rando, Use of a fluorescent choles-terol derivative to measure lateral mobility of cholesterol in membranes, *Proceedings of the National Academy of Sciences of the United States of America—Biological Sciences*, *79*(17), 5171–5174 (1982).

23. J. F. Nagle and S. Tristram-Nagle, Structure of lipid bilayers, *Biochimica Et Biophysica Acta—Reviews on Biomembranes*, *1469*(3), 159–195 (2000).

24. J. R. Solvius, *Lipid–Protein Interaction*. John Wiley, New York (1982); also found in NIST Stand Reference Database 34—*Lipid Thermotropic Phase Transition Database*.

25. A. Gopal, The collapse of phospholipids Langmuir monolayers: Implications for biolog-ical surfactants, PhD thesis, Department of Chemistry, University of Chicago, Chicago, IL (2004).

26. H. M. Mcconnell, L. K. Tamm, and R. M. Weis, Periodic structures in lipid monolayer phase-transitions, *Proceedings of the National Academy of Sciences of the United States of America—Physical Sciences*, 81(10), 3249–3253 (1984).

27. M. C. Petty, *Langmuir–Blodgett Films: An Introduction*. Cambridge University Press, New York (1996).

28. B. L. Stottrup, D. S. Stevens, and S. L. Keller, Miscibility of ternary mixtures of phos-pholipids and cholesterol in monolayers, and application to bilayer systems, *Biophysical Journal*, *88*(1), 269–276 (2005).

29. A. Mardilovich, J. A. Craig, M. Q. McCammon, A. Garg, and E. Kokkoli, Design of a novel fibronectin-mimetic peptide-amphiphile for functionalized biomaterials, *Langmuir*, *22*(7), 3259–3264 (2006).

30. V. Kiessling, J. M. Crane, and L. K. Tamm, Transbilayer effects of raft-like lipid domains in asymmetric planar bilayers measured by single molecule tracking, *Biophysical Journal*, *91*(9), 3313–3326 (2006).

31. T. M. Bayerl and M. Bloom, Physical-properties of single phospholipid-bilayers adsorbed to micro glass-beads—A new vesicular model system studied by H-2-nuclear magnetic-resonance, *Biophysical Journal*, *58*(2), 357–362 (1990).

32. S. A. Joyce and J. E. Houston, A new force sensor incorporating force-feedback control for interfacial force microscopy, *The Review of Scientific Instruments*, *62*(3), 710–715 (1991).

33. J. E. Houston and T. A. Michalske, The interfacial-force microscope, *Nature*, *356*(6366), 266–267 (1992).

34. J. E. Houston, Interfacial force microscopy: Selected applications. In: *Applied Scanning Probe Methods*, B. Bushan, H. Fuchs, and S. Hosaka (eds.), pp. 41–74, Springer-Verlag, Berlin, Germany (2004).

35. G. Binnig, C. F. Quate, and C. Gerber, Atomic force microscope, *Physical Review Letters*, *56*(9), 930–933 (1986).

36. R. W. Carpick, D. F. Ogletree, and M. Salmeron, Lateral stiffness: A new nanomechani-cal measurement for the determination of shear strengths with friction force microscopy, *Applied Physics Letters*, *70*(12), 1548–1550 (1997).

37. K. L. Johnson, K. Kendall, and A. D. Roberts, Surface energy and contact of elastic sol-ids, *Proceedings of the Royal Society of London Series A—Mathematical and Physical Sciences*, *324*(1558), 301 (1971).

38. R. W. Carpick, D. F. Ogletree, and M. Salmeron, A general equation for fitting con-tact area and friction vs. load measurements, *Journal of Colloid and Interface Science*, *211*(2), 395–400 (1999).

39. B. V. Derjaguin, V. M. Muller, and Y. P. Toporov, Effect of contact deformations on the adhesion of particles, *Journal of Colloid and Interface Science*, *53*(2), 314–326 (1975).

40. H. Hertz, Ueber die Berührung fester elastischer Körper, *Journal für die Reine und Angewandte Mathematik*, *92*, 156 (1882).

41. D. S. Grierson, E. E. Flater, and R. W. Carpick, Accounting for the JKR-DMT transition in adhesion and friction measurements with atomic force microscopy, *Journal of Adhesion Science and Technology*, *19*(3–5), 291–311 (2005).

42. F. P. Bowden and D. Tabor, *The Friction and Lubrication of Solids*. Clarendon Press, Oxford, U.K. (1986).

43. B. N. J. Persson, I. M. Sivebaek, V. N. Samoilov, K. Zhao, A. I. Volokitin, and Z. Zhang, On the origin of Amonton's friction law, *Journal of Physics—Condensed Matter*, *20*(39), 1–11 (2008).

44. M. J. Wang, K. M. Liechti, V. Srinivasan, J. M. White, and P. J. Rossky, Nano mechanical analysis of IFM force profiles on self-assembled monolayers. In: *Mechanics of the 21st Century*, W. Gutkowski and T. A. Koalweski (eds.), pp. 217–228, Springer, Berlin, Germany (2005).

45. R. Zanoni, C. Naselli, J. Bell, G. I. Stegeman, and C. T. Seaton, Elastic properties of Langmuir–Blodgett-films, *Physical Review Letters*, *57*(22), 2838–2840 (1986).

46. G. Oncins, J. Torrent-Burgues, and F. Sanz, Lateral force microscopy study of Langmuir–Blodgett films of a macrocyclic compound, *Tribology Letters*, *21*(3), 175–184 (2006).

47. W. C. Lin, C. D. Blanchette, T. V. Ratto, and M. L. Longo, Lipid asymmetry in DLPC/DSPC-supported lipid bilayers: A combined AFM and fluorescence microscopy study, *Biophysical Journal*, *90*(1), 228–237 (2006).

48. V. Franz, S. Loi, H. Muller, E. Bamberg, and H. H. Butt, Tip penetration through lipid bilayers in atomic force microscopy, *Colloids and Surfaces B—Biointerfaces*, *23*(2–3), 191–200 (2002).

49. N. Kucerka, Y. F. Liu, N. J. Chu, H. I. Petrache, S. T. Tristram-Nagle, and J. F. Nagle, Structure of fully hydrated fluid phase DMPC and DLPC lipid bilayers using X-ray scattering from oriented multilamellar arrays and from unilamellar vesicles, *Biophysical Journal*, *88*(4), 2626–2637 (2005).

50. H. J. Butt, B. Cappella, and M. Kappl, Force measurements with the atomic force microscope: Technique, interpretation and applications, *Surface Science Reports*, *59*(1–6), 1–152 (2005).

51. M. Salmeron, Generation of defects in model lubricant monolayers and their contribution to energy dissipation in friction, *Tribology Letters*, *10*(1–2), 69–79 (2001).

52. Q. Du, X. D. Xiao, D. Charych, F. Wolf, P. Frantz, Y. R. Shen, and M. Salmeron, Nonlinear-optical studies of monomolecular films under pressure, *Physical Review B*, *51*(12), 7456–7463 (1995).

53. P. T. Mikulski and J. A. Harrison, Packing-density effects on the friction of n-alkane monolayers, *Journal of the American Chemical Society*, *123*(28), 6873–6881 (2001).

54. J. I. Siepmann and I. R. Mcdonald, Monte-Carlo simulation of the mechanical relaxation of a self-assembled monolayer, *Physical Review Letters*, *70*(4), 453–456 (1993).

55. H. I. Kim and J. E. Houston, Separating mechanical and chemical contributions to molecular-level friction, *Journal of the American Chemical Society*, *122*(48), 12045–12046 (2000).

56. N. J. Brewer, B. D. Beake, and G. J. Leggett, Friction force microscopy of self-assembled monolayers: Influence of adsorbate alkyl chain length, terminal group chemistry, and scan velocity, *Langmuir*, *17*(6), 1970–1974 (2001).

57. A. Lio, D. H. Charych, and M. Salmeron, Comparative atomic force microscopy study of the chain length dependence of frictional properties of alkanethiols on gold and alkylsilanes on mica, *Journal of Physical Chemistry. B*, *101*(19), 3800–3805 (1997).

58. N. J. Brewer, T. T. Foster, G. J. Leggett, M. R. Alexander, and E. McAlpine, Comparative investigations of the packing and ambient stability of self-assembled monolayers of alkanethiols on gold and silver by friction force microscopy, *Journal of Physical Chemistry. B*, *108*(15), 4723–4728 (2004).

59. M. Chandross, E. B. Webb, M. J. Stevens, G. S. Grest, and S. H. Garofalini, Systematic study of the effect of disorder on nanotribology of self-assembled monolayers, *Physical Review Letters*, *93*(16), 1–4 (2004).

60. P. Guyotsionnest, J. H. Hunt, and Y. R. Shen, Sum-frequency vibrational spectroscopy of a Langmuir film—Study of molecular-orientation of a two-dimensional system, *Physical Review Letters*, *59*(14), 1597–1600 (1987).

61. G. Ma and H. C. Allen, Real-time investigation of lung surfactant respreading with surface vibrational spectroscopy, *Langmuir*, *22*(26), 11267–11274 (2006).

62. F. Behroozi, Theory of elasticity in two dimensions and its application to Langmuir–Blodgett films, *Langmuir*, *12*(9), 2289–2291 (1996).

63. N. W. Tschoegl, Elastic moduli in monolayers, *Journal of Colloid Science*, *13*(5), 500–507 (1958).

64. W. H. Briscoe, S. Titmuss, F. Tiberg, R. K. Thomas, D. J. McGillivray, and J. Klein, Boundary lubrication under water, *Nature*, *444*(7116), 191–194 (2006).

Part II

Emulsions and Aqueous Systems: Relevance to Tribological Phenomena

7 Influence of Surfactants on Wetting and Colloidal Processes of Lubricant Emulsions on Metal Surfaces

Svajus Joseph Asadauskas and Henrikas Cesiulis

CONTENTS

ABSTRACT

This review focuses on several theoretical and technological aspects of water-based lubricants, primarily their appearance, aging, lubricity, and wetting on metal surfaces. Colloidal properties of major categories of lubricant emulsions are compared, describing the importance of micelles, lamellae, vesicles, droplets, and other colloidal structures. Molecular mechanisms are suggested to explain the aging processes of lubricant concentrates and emulsions as well as lubricant film formation both on the open surface, as in wetting, and between the moving surfaces. Attention is also given to the chemical structures of surfactants, used to control wetting of metal surfaces by water-based lubricants.

7.1 INTRODUCTION: COLLOIDAL CATEGORIES OF WATER-BASED LUBRICANTS

The main purpose of a lubricant, which separates sliding surfaces pressed against each other in order to reduce frictional resistance and wear, is served quite well by straight oils. Nevertheless, oil-in-water (O/W) emulsions were also found suitable for this purpose and became common as lubricants in the 1940s. Metal processing industries have benefited the most from diluting the lubricants with water. This not only reduced costs but also took advantage of the cooling capability of water. Due to its higher heat capacity and enthalpy of evaporation, water is about three times more effective than oil in taking the heat away from surfaces. Environmental benefits [1] also promote the utilization of water-based lubricants. In addition, one more advantage is a compact storage of a lubricant concentrate, some of which can be diluted as many as 50 times with water to make the ready-to-use lubricant.

However, from the technical standpoint, water-based lubricants have numerous disadvantages compared to neat oils. Their range of operating temperatures is below 100°C. Volatility is very high, although water vapors pose essentially no hazard as compared to those of oils. It is much harder to control viscosity, which is the key property for lubricants. Many additives are not water soluble and need to be emulsified, thus lacking uniformity and stability. Those additives that dissolve or emulsify easily also often foam severely. Biological contamination is a particularly troublesome issue. Water does not have any inherent lubricity; on the contrary, it promotes such detrimental mechanisms as tribocorrosion and hydrogen wear [2]. The latter is caused by protons, diffusing from the lubricant into the metal substrate, acquiring electrons and producing hydrogen gas, which evolves and acts very destructively on the surface, resulting in pronounced wear [3]. Water-based liquids release protons much easier than straight oils. In comparison with oils, aqueous emulsions have many more problems when considered for lubricant application, and the above list is far from complete.

Despite numerous technical issues, water-based lubricants are gaining market share at the expense of traditional "straight oils." This is especially obvious in metalworking industry, where "straight oils" are rapidly being replaced by water-based lubricants, broadly termed "soluble oils," "synthetic" and "semisynthetic" fluids [4]. Such definitions of water-based lubricants are accepted in other industries as well, e.g., mining oils or fire-resistant hydraulic fluids. Although derived mostly from the visual appearance of the ready-to-use lubricant emulsions, these definitions follow the rules of colloidal systems quite well:

- "Synthetic fluids" represent aqueous solutions of salts, soaps, polymers, etc.
- "Soluble oils" cover O/W emulsions.
- "Semisynthetic fluids" represent microemulsions.

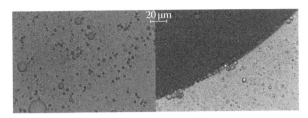

FIGURE 7.1   Photomicrographs of two "soluble oils," diluted in water to 10% "solids." *Left*: a "milky, tight" lubricant emulsion, designed to remain stable for several weeks. *Right*: a "milky, coarse" lubricant emulsion, designed to form a cream phase in about 1 h.

When relating the water-based lubricants to traditional colloids, some caution should be exercised, since the actual lubricants are usually more complex. Synthetic fluids, in addition to being true solutions of, e.g., triethanol amine and its borates, often contain micelles, thus meeting the definition of microemulsions. Semisynthetic fluids might contain relatively large, less-stable colloidal particles, e.g., vesicles, in addition to thermodynamically stable micelles. Figure 7.1 shows photomicrographs of two typical soluble oils, whose dilutions are used as lubricants for metal forming [5]. Many types of emulsions are used as metalworking fluids. Some are "tight" in appearance, i.e., spread uniformly on a clean glass container walls without any disruption of homogeneous liquid film or appearance of any visible particles, such as floc, cream, droplets. The thin layers of "tight" emulsions seem relatively translucent in appearance, although the bulk emulsions are usually white, i.e., "milky."

The opposite of "tight" emulsions is "coarse" emulsions, whose thin films appear much less translucent and often show some visible, albeit very small particles. The white color of thin layers of coarse emulsions is more solid, although the bulk emulsion might appear similarly milky at the same dilution level as the "tight" emulsion. Colloquially, in lubricant industry the dilution level is measured in "percent solids," which indicates the weight of low-volatility components in the diluent of higher volatility (such as water, mineral solvents, alcohol, etc.). Although the pure low-volatility components quite often are liquids, suggesting that the term "solids" is technically not adequate, such terminology is retained in this report.

The images in Figure 7.1 show that many dispersed particles are much larger than classical O/W micelles. This is not in line with the expectation that water-based lubricants represent just micelles in water. Clearly, the actual lubricant emulsions are much more complex and a good understanding of colloids is necessary to interpret their composition properly.

7.2   WATER–LUBRICANT OIL–SURFACTANT (EMULSIFIER) SYSTEM

Development, formulation, in-field application, and maintenance of water-based lubricants strongly relate to the knowledge of colloidal structures that can be formed in the mixtures of water, surfactant (emulsifier), and lubricant. Water-based lubricants usually represent an emulsion, which is a mixture (dispersion) of oil and water, i.e., two immiscible liquids. By definition, the classical emulsion is a thermodynamically unstable system with one liquid dispersed in the other as tiny droplets. In order

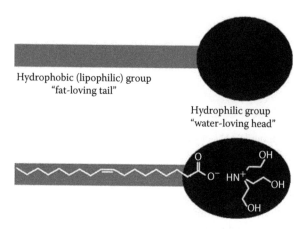

Hydrophobic (lipophilic) group
"fat-loving tail"

Hydrophilic group
"water-loving head"

FIGURE 7.2 *Top*: schematic representation of lipophilic/hydrophilic portions in the molecule. *Bottom*: embedded actual molecular structure of the one of the most prevalent [6,19] emulsifiers in water-based lubricants—TEA oleate (anionic surfactant).

to render the system stable, surfactants (a.k.a. emulsifiers) are added. The emulsifiers change properties and structure of particles and may transform conventional emulsions into various colloidal systems, e.g., microemulsions. Many lubricant components, being amphiphilic, possess both hydrophilic and lipophilic (fat-loving)/hydrophobic fragments and can act as emulsifiers (surfactants). A schematic representation of an amphiphilic molecule and a classical surfactant molecule is shown in Figure 7.2.

In surfactants used for lubricant emulsions, the lipophilic group typically is a large hydrocarbon moiety, such as a long chain of the form $CH_3(CH_2)_n$, usually with $4 < n < 20$. The hydrophilic group is polar (either ionic or uncharged) and falls into one of the following categories:

1. Charged groups
   a. *Anionic*. This is by far the most popular type of surfactants used in lubricant emulsions, with the hydrophilic group represented by alkaline metal (e.g., K^+ or Na^+) or various alkanol ammonium salts (soaps) of
      i. Carboxylates: $RCOO^-$ (animal, vegetable [6] or tall oil [7] fatty acids, in particular)
      ii. Sulfonates: RSO_3^-
      iii. Phosphates: RPO_4^{2-} or $R_2PO_4^-$
      iv. Other anions (sulfates, borates, ionized heterocycles, etc.)
   b. *Cationic*. These surfactants are used less frequently in lubricants, but they still are very important in control of colloidal properties. As the core of their hydrophilic groups, various ammonium chemistries are utilized, starting with RNH_3^+ up to quaternary NR_4^+. These cations are usually neutralized with carboxylates, sometimes by sulfonates, hydrocarbonates, etc. Alkyl chains of both amine and counter-ion moieties serve as lipophilic portions of the surfactant.

2. Polar, uncharged groups; in lubricants, this category is mostly represented by
   a. Alcohols, such as oleyl alcohol, glycerol mono-oleate
   b. Polyethers, such as poly ethylene oxide or Poly Alkylene Glycols (PAGs)
   c. PAG ethers or esters with alkyl chains, colloquially called "nonionic surfactants" in lubricant industry
   d. Amides, mostly derived from alkanol amines
   e. Less frequent chemistries (sulfonamides, succinates, etc.)
3. Amphoteric/zwiterionic compounds; these may contain two charged groups of different signs, e.g., ammonium and carboxylate (as in alkyl betaine) or ammonium and phosphate (as in lecithin). Just as cationic surfactants, these are used less often in lubricants, but can impart very important performance characteristics

In accordance with Bancroft rule, emulsion type depends more on the nature of emulsifying agent than on the relative proportions of oil and water present or the method of emulsion preparation. The phase in which an emulsifier is more soluble usually constitutes the continuous phase. In O/W emulsions, most surfactants are more soluble in water than in oil, whereas in water-in-oil (W/O) emulsions (a.k.a. "inverse emulsions") the majority of emulsifying agents are more soluble in oil than in water. Although there is a variety of mechanisms, which result in the colloid formation, it can be generalized that surfactants function in two different ways, i.e., for emulsification and solubilization.

7.2.1 EMULSIFICATION MECHANISM IN COLLOID FORMATION

According to classical colloid theory, emulsifiers adsorb at the oil/water interface due to their hydrophilic/hydrophobic character. This helps stabilize the oil droplets in the bulk water phase by electrostatic and/or steric repulsion mechanisms (Figure 7.3). Electrostatic stability of the emulsion is provided by the adsorption of anionic surfactants around an oil droplet resulting in an electrically negative charged layer, which is balanced by positive counterions, i.e., the double layer. When one droplet comes close to another, the electric double layers cause them to repel each other, preventing the two droplets from coalescing.

Emulsifiers used in the lubricant industry are often anionic surfactants, so the emulsion is stabilized by electrostatic repulsion (double layers). Therefore, water hardness, pH, and ion concentration (electric conductivity) will greatly affect the emulsion stability. However, no correlation has been firmly established between droplet charge and emulsion stability in inverse (W/O) emulsions. If an emulsion contains significant levels of nonionic or polymeric emulsifiers, the system can also be stabilized by steric effects. In the case of nonionic emulsifiers, the charges may be induced due to adsorption of ions from the aqueous phase or by contact charging (phase with higher dielectric constant is charged positively). Steric effects generate van der Waals repulsion forces when two particles are very close. The size and type of emulsifiers (or polymers), pH, and temperature are important parameters in the stability of these emulsions [8]. In many O/W systems, both electrostatic and steric

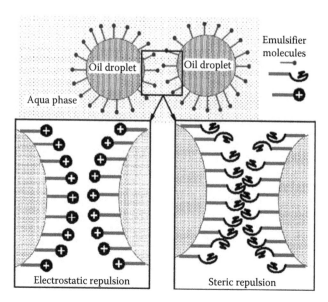

FIGURE 7.3 Prevention of oil droplet coalescence in O/W emulsion. *Top*: illustration of droplet separation by repulsion forces between the shells of adsorbed emulsifiers. *Bottom*: illustration of electrostatic and steric repulsion mechanisms at molecular level.

mechanisms play roles in the stability. However, inverse emulsions (W/O) are mainly stabilized by the steric mechanism because of the low dielectric medium of the continuous oil phase. It must be noted that in addition to electrostatic and steric repulsion mechanisms, water-based lubricants sometimes use more exotic mechanisms of stabilization, such as formation of gel-like network (sometimes based on hard water soaps), solid phase insertion (often silica, waxes, or fats), etc.

The nature of emulsifiers, their concentration and ratios are key parameters governing many properties of emulsions, such as appearance, stability, droplet size distribution, surface and interfacial tension, contact angle, spreading coefficient, or work of adhesion. The composition of emulsions is clearly related to their lubricating performance, being a topic of continuous development. Some characteristic results have been presented in [9]. It has been demonstrated that emulsifier content exerts a strong influence on the emulsion properties, such as stability, droplet size distribution, surface and interfacial tension, wetting ability, etc. Lubricating behavior is shown to relate to emulsion properties: emulsions with droplet size ~10 μm and strong adsorption at the steel/oil/aqueous solution interface exhibit better frictional performance in rolling operations.

From an industrial lubricant application point of view, viscosity and flow behavior are among the most important properties of emulsions. The trend of steady increase in emulsion viscosity as the average drop size decreases can be applied for rheology control by altering the droplet size distribution. In a typical batch stirring process, the average droplet size decreases with time, concurrently with a rise in viscosity. Droplet size is important not only to emulsion processing (handling, mixing, pumping, or draining) but also to its physical appearance and feel, such as consistency and creaminess [10].

Emulsification can be improved by other methods in addition to electrostatic and steric repulsion mechanisms. It is known that solid colloidal particles, such as chalk, waxes, pyrite, coal, soot, mica, finely dispersed clays, and others, can aid emulsification [11–14]. The thermodynamic analysis of the adsorption–desorption of solid particles onto liquid droplets shows that high energy require to move particle from interface into one of the bulk phase, and the values of energy exceed 10^3 kT, and for small solid particles (100 nm) is on the order of 10^6 kT. Such values of energy (kT is a product of Boltzmann constant k and absolute temperature T) mean that thermal power is not enough to remove adsorbed solid particles from the surface of droplet. As the adsorption energy is high, the formation of closely packed two-dimensional structure is an energetically favorable process. Therefore, colloidal powder collects at the interface and forms a structural–mechanical barrier. This barrier prevents coalescing of oil or water droplets, depending on the type of emulsion (O/W or W/O). In a concentrated system, the type of emulsion is governed by the Hydrophilic–Lipophilic Balance (HLB) of the colloidal powder. If HLB is shifted toward hydrophilicity, the O/W emulsion is formed (Figure 7.4), and vice versa, hydrophobic powder stabilizes the inverse (W/O) emulsion.

In general, the effectiveness of the solid emulsion stabilizers depends on particle size, shape, concentration, surface energy, and the interactions between particles. Also, particle surface energy may be modified by adsorption of suitable surfactants, in some cases leading to emulsion phase inversion [15].

A number of lubricating formulations containing oil, solid particles as stabilizers and water are known. Garvey et al. [16] describe lubricant formulation comprising oil droplets as a lubricating material and solid silica particles as a stabilizer. According to the patent, this lubricant may be used as a metalworking fluid. The silica, used in this formulation, was modified to be partially hydrophobic by treatment with a silanating agent, e.g., trimethylchlorosilane or an organic cationic compound cetyl trimethyl ammonium bromide. The resulting size of the modified silica powder particles was <1 μm, and the size of stabilized oil droplets was <50 μm.

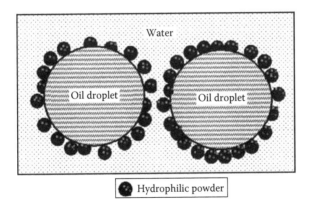

FIGURE 7.4 Schematic picture explaining O/W emulsion stabilization by the hydrophilic colloidal powder. If hydrophobic powder were used, it would completely be dispersed in the oil phase without preventing droplets from coalescing.

These particles form a coating on the oil droplets and produce an emulsion with good stability. It must be noted that depending on HLB, the particle condensation at the oil–water interface might be observed, eventually forming a permanent, but brittle two-dimensional interfacial shell around the oil droplets. It was desirable that the coated droplets of the emulsion when subjected to friction zone conditions undergo an irreversible rupturing and thereby form a tribofilm of droplet material on the lubricated surfaces. Under certain circumstances, it was preferable that the aqueous phase also included hydrophobic agent to render the particle sufficiently lipophilic to adsorb at the oil–water interface.

Some experimental and analytical results suggest that, under high loads, W/O emulsions perform similarly to straight oils [17]. Even more complex emulsion systems are proposed for lubricant applications, such as a system of O/W emulsion dispersed in a continuous oil phase, forming so-called oil-in-water-in-oil (O/W/O) emulsion [18]. However, the fundamentals of these observations need further investigation. Clearly, colloid formation through the emulsification mechanism can be very complex and bear significant implications on technical properties of water-based lubricants.

7.2.2 SELF-AGGREGATION AND SOLUBILIZATION

Self-aggregation of amphiphilic molecules results in the formation of entities called micelles in the liquid. Micelle is a multimolecular aggregate, primarily formed of self-assembled relatively short surfactant (emulsifier) molecules, dissolved in a liquid. Micelles only form when the concentration of the dissolved surfactant is greater than the critical micelle concentration (cmc), and the temperature of the system is higher than the critical micelle temperature, or Krafft temperature. Below the Krafft temperature, the solubility of amphiphilic molecules is lower than cmc, and micelles do not form. The micelle formation in water is illustrated in Figure 7.5. Triethanol ammonium (TEA) oleate is taken as an example due to its widespread usage and excellent emulsification properties [19]. Micelles are always in equilibrium with "free" amphiphilic molecules present in the solution at concentration of cmc. Thermodynamic analysis shows that micelles are thermodynamically stable, and the

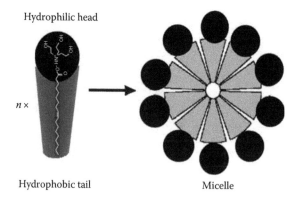

FIGURE 7.5 Self-ordering of TEA oleate surfactant molecules into a spherical micelle in continuous aqueous phase above cmc concentration and Krafft point temperature.

lower the cmc, the more stable the micelles [20]. Microemulsions are also thermodynamically stable, as opposed to conventional O/W emulsions. In oil, the exposure of the hydrophilic head groups to the surrounding liquid is energetically unfavorable, thus giving rise to a W/O system. Consequently, hydrophilic groups are aligned in the micelle core and the lipophilic chains extend away from the center. Such micelles are called inverse micelles. They are less likely to form when head groups are highly charged since collection of the hydrophilic heads in a confined core of the micelle would be resisted due to unfavorable electrostatic repulsion. For example, it is harder to produce an inverse emulsion with sodium oleate, compared to TEA oleate (Figure 7.2), which is less polar and has a relatively bulky counterion moiety.

In water, self-aggregation of molecules is controlled by two opposing forces. The hydrophobic force causes repulsion of the lipophilic chain from the aqueous medium with formation of an organic droplet as a micelle core. The polar head groups extend into the water phase from this core, and the force of repulsion of the head groups limits self-association to relatively small aggregates [21].

The size of the micelle formed as such mainly depends on the chain size of the surfactant molecules but is typically in the range of colloidal domain. Microemulsions, used in lubricant industry, most often employ surfactants, which have 10–40 carbon atoms, based on various glycols, ether, ester, amide, and other derivatives. Hence, the diameter of such micelles would be on the order of 10 nm. The shape of individual micelles varies with concentration. Close to the cmc, they are spherical and, at higher concentration, they have been shown [22] to form lamellar or rod-like structures, Figure 7.6.

Often surfactants assemble into aggregates, larger than micelles and form globular bilayers. Such colloidal structures are usually called vesicles. A variety of other terms are used for specialized types of similar colloidal structures: liposomes, niosomes, coacervates, etc. In 1989, Kaler et al. [23] revealed the vesicle formation from mixed cationic and anionic surfactants using cetyltrimethylammonium tosylate/sodium dodecylbenzene sulfonate with single alkyl chains. Vesicles are often spherical and can be mono-lamellar (a.k.a. uni-lamellar) or multi-lamellar, see Section 7.3.2.

Unlike micelles, vesicles may not be thermodynamically stable. The formation of vesicles from, e.g., lecithin, usually requires input of some form of external energy,

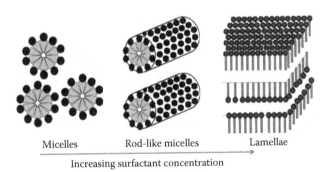

Micelles   Rod-like micelles   Lamellae

→ Increasing surfactant concentration

FIGURE 7.6 Structural changes in colloids as the surfactant concentration is increased and formation of larger aggregates: rod-shaped micelles and lamellae.

like extensive stirring or ultrasonic agitation. The vesicles thus formed are believed to be metastable and easy to fuse, i.e., one combining with another. On the other hand, vesicles formed in cationic/anionic surfactant mixtures are believed to be thermodynamically stable, and their size, charge, or permeability can be adjusted by varying the relative amounts and/or chain lengths of the two surfactants [24]. Another important difference between vesicles and micelles is that vesicles can have an inside core that encloses some of the continuous phase and outside bilayers. The existence of a critical vesiculation concentration, above which the micelles of some surfactants would start transforming into vesicles, is sometimes mentioned. This is still highly questionable, however. At low concentrations, surfactants first start forming micelles that may turn into rod-shaped micelles, lamellae and vesicles at higher concentrations [25]. Schematic representation of micelles and lamellae is given in Figure 7.6, while vesicles are discussed in more detail in Section 7.3.2.

Mixtures of surfactants in aqueous solutions exhibit different levels of synergism toward formation of colloidal structures. The structures formed depend on the charge and molecular architecture of the individual surfactants. In particular, mixtures of cationic and anionic surfactants provide a rich source of phase behavior, such as spherical and rod-like micelles, vesicles, lamellae, etc.

Water-based lubricants employ a large variety of colloids. The improvement of the tribological performance for the lamellar solution in comparison with vesicles for WC/Cu system is discussed in [26]. The presence of lamellar crystallites leads to the formation of a multilayer highly resistant protective film. It has been shown that the presence of condensed lamellar structures in microemulsions improves the performance of an aqueous extreme-pressure lubricant. Particularly pronounced tribological effectiveness of the lamellar structure is reported for the cases when one of the two friction surfaces was made of brass. The coupling between the lamellar structure and brass led to a multilayer film, which was particularly resistant to shear at high loads.

Another important phenomenon which takes place in micellar or vesicular systems is oil solubilization. Surfactant micelles are able to increase the solubility of most organic molecules in water. The mechanism, by which the oil solubilization occurs, is the incorporation of the oil molecules into the micelle core. The solubilization mechanism is schematically shown in Figure 7.7. According to thermodynamic considerations of oil solubilization, it appears that the lower cmc surfactants will result in more stable solubilized systems [20].

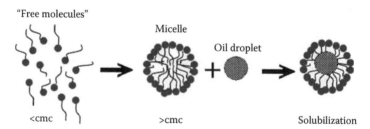

FIGURE 7.7 Schematics of micelle formation at concentrations above the cmc and oil-droplet solubilization by micelle in aqua phase.

Consequently, significant amounts of oil can be incorporated into the emulsion not only in a form of micellized droplets but also as integral portions of the micelle/lamellae/vesicle layers. Oil molecules can also serve secondary functions of lipophilic linkers [27,28]. This phenomenon sometimes produces unexpected improvements in emulsion stability and other colloidal effects.

7.2.3 THERMODYNAMIC CONSIDERATIONS OF WATER–LUBRICANT OIL–SURFACTANT SYSTEMS

The correlation between actual performance of industrial lubricant emulsions and macroscopic thermodynamic data is rarely directly evident. Nevertheless, thermodynamic processes can represent an equilibrium state of emulsion systems, including water—lubricant oil—surfactants. Such equilibrium may or may not be achieved in actual applications. However, it describes the target state of the emulsion system toward which the colloidal processes are driven thermodynamically. A brief review of thermodynamic aspects is presented below.

As a model system for soluble oils, a lubricant O/W emulsion can be assumed to contain three main components: water, oil, and surfactant. The characteristics in the friction zone and lubrication properties of emulsions are determined by the inner structure of formed droplets, micelles and vesicles, the oil volume percentage and especially the additives, capable of withstanding the extreme pressure developed in the area of contact. However, even the mechanisms of straight oil lubrication are complex and not yet well understood. Thus, tribological mechanisms in emulsions, whose colloidal structures are poorly characterized, have been interpreted only with a great degree of speculation.

Still, some generic concepts, derived from a long-term experience with water-based lubricants, have already been postulated to explain the effectiveness of O/W emulsions as lubricants. First of all, the ability of emulsions to form a deposited or "plated out" film of oil has been stated as an important requirement to give a satisfactory performance as a lubricant for repeatedly loaded contacts [29]. It has been recognized for more than 30 years that an emulsion containing a relatively small fraction of oil can form a lubricant film nearly as thick as that of a straight oil [30]. This plate-out concept assumes that when the oil droplets are attached to the surfaces, they tend to spread to an angle of θ, i.e., wetting angle. In this case, the emulsifier concentration is an important parameter in producing an effective emulsion: too much of the surfactant may produce an emulsion which is too stable to deposit a film. This is particularly important for water-based rolling oils. Films that attach to rollers can be partially carried back to the work zones for bearings or work rolls, which are not directly immersed in the water-based rolling oil.

Another widespread concept relates to improvement of the O/W emulsion lubricity by increasing the ability of dispersed oil droplets to wet polar metal surfaces and thus promote elastohydrodynamic lubrication (EHL), without starvation up to high speeds [31]. It has been proposed that a monomolecular layer separates the contacting surfaces. This monomolecular layer is expected to be formed by the adsorption of the lubricant or, more precisely, the lubricant additives on the worn surfaces.

Several explanations can be found to describe more precisely the mechanisms that are the bases for both concepts. Kimura and Okada [32] relate the film-forming capability of emulsions to the interactions between the dispersed droplets and the surfaces. The changes in the minimum EHL film thickness have been determined by varying oil contents, emulsifying agent nature and concentration, and blending of different surfactants. These results can be quantitatively explained by a theory that assumes trapping of oil droplets between steel surfaces with a certain trapping probability. It is then shown that this trapping probability is correlated with the "displacement energy"—a thermodynamic measure of the tendency of oil to displace water from the surfaces at the oil–water–metal interface. Thus, trapping probability and the entire lubricating process is governed by wetting ability of metallic surface (substrate), which can be expressed either by the spreading coefficient or displacement energy, DE, defined [33] as

$$DE = \gamma_{OS} - \gamma_{WS}, \tag{7.1}$$

where

DE is the displacement energy

γ_{OS} and γ_{WS} are the oil/solid and water/solid surface tensions, respectively

Thus, in theory the term DE expresses the tendency of oil droplets to wet a surface. The spreading wetting is the phenomenon when a liquid in contact with a substrate spreads over the substrate and displaces another fluid from the surface. For oil to displace water from solid surfaces, the spreading coefficient, S, is used as a measure of the driving force. If the spreading coefficient is positive, the system decreases in the surface-free energy during the spreading process, and the displacement of water by oil process can then occur spontaneously. Consequently, if the spreading coefficient becomes negative, the less the absolute value of S is, the more readily will such a displacement occur. When the substrate is a solid, the spreading coefficient for a liquid in contact with a solid substrate is denoted as S_{SL}, the expression for this coefficient is given using conveniently determined values and employing Young equation as

$$S_{E/S} = \gamma_{SA} - (\gamma_{SL} + \gamma_{LA}) = \gamma_{LA}\left[\cos\theta_{LS} - 1\right], \tag{7.2}$$

where

γ_{SA} is the solid/air surface tension

γ_{SL} is the solid/liquid interfacial tension

γ_{LA} is the liquid/air surface tension

θ_{LS} is the liquid/solid contact angle

The liquid/solid contact angle θ_{LS} in Equation 7.2 is certainly related to the extent of coverage of the solid surface by the emulsifier. This angle varies with the emulsifier concentration in the emulsion. The spreading coefficient is positive, if spreading is accompanied by a decrease in free energy; that is, the spreading is spontaneous.

As described by Rieffe [34], the key to producing an effective O/W emulsion in terms of film formation is to maximize the tendency of the oil droplets to wet the solid surfaces, which is quantified by the displacement energy Equation 7.1. For oil to displace water from solid surfaces DE must be negative, and the more negative the value of DE is, the more readily will such displacement take place. An application of the Young equations for oil/solid and water/solid surfaces is given by Chen [33]

$$DE = \gamma_{WA} \cos\theta_{WS} - \gamma_{OA} \cos\theta_{OS} \qquad (7.3)$$

This equation indicates that the displacement energy is simply the difference between the work of the adhesion of oil and water, respectively, on the solid surface. The O/W emulsion with an optimal tendency for oil to wet the surface should have a minimum value of this expression [34]. Experimental results [33] relate an increase in spreading coefficient or reduction in displacement energy of emulsion to better lubricity in terms of lower wear losses.

7.3 IMPORTANCE OF COLLOIDAL PROCESSES IN KEY PROPERTIES OF LUBRICANT EMULSIONS

Since quite often water-based lubricants are viewed as classical O/W emulsions, their performance tends to be interpreted with excessive oversimplification and incorrect usage of thermodynamic principles. Emulsion lubricity is explained as its ability to produce an oil layer. Wetting on a metal surface is viewed mostly as a process dictated solely by surface tension forces. Observations during the aging of lubricant concentrates are often generically attributed to constituent oxidation or crystallization. Importance of complex colloidal transformations is often underestimated. Below, some colloidal processes are evaluated with more attention to the key processes in water-based lubricants: aging, wetting, and tribofilm formation.

7.3.1 EFFECTS OF COLLOIDS ON LUBRICATION AT MOLECULAR LEVEL

At the molecular level, some suggestions can be proposed to explain the initial stages of the lubricating effect or friction reduction of the emulsions. Once the surfaces are exposed to the emulsion, formation of a multilayer film begins. Polar organic substances greatly accelerate formation of such films on metallic surfaces. The polarity of compounds from the outer side of the surfactant shell around the micelle is sufficient to facilitate the film formation. This is further enhanced by the presence of "free" amphiphilic molecules in the solution. Such film formation is the basis for the lubrication mechanism since adsorption of organic molecules having polar groups produce low friction on the surface [33]. The polar group of a molecule, e.g., carboxyl ion of a fatty acid "$-COO^-$," is strongly attracted to the polar metallic surface. The nonpolar tail, which often is an alkyl group "$-C_nH_{2n+1}$," is repelled by hydrophilic substances and forms the core of a micelle in the shape of a solubilized oil droplet. Strong adsorption ensures that almost every available surface site is occupied by the fatty acid to first produce cores for micelles and eventually lead to a dense and robust film.

The thermodynamic theory of wetting and spreading does not incorporate these colloidal aspects of the film formation process on the surface. However, micellar and lamellar mechanisms may affect the buildup of the lubricant film very significantly. The spreading/wetting performance is not the only affected parameter. The repulsion or weak bonding between the contacting alkyl groups ensures relatively low shear strength of the interface. So, it has a pronounced tribological influence on the lubricant applications, where the stationary surfaces are initially exposed to the emulsion, and their movement begins only later. Such applications include fire-resistant hydraulic fluids, water-based metal forming lubricants, rolling oils, water-based rust prevention fluids, etc.

This colloidal mechanism also accounts for the EHL behavior of lubricating emulsions and its dependence on oil concentration in rolling oils, which are reported to display two critical speeds [30,35] at point contact. At low rolling speed, the emulsion behaves similarly to straight oil because the adsorption plays an important role. The first critical speed for the emulsion seems to depend on the initial oil concentration. In the low-speed region, an emulsion with high oil concentration forms lubricant films in a similar manner as straight oil because enough oil is plated out on solid surfaces. In contrast, if the oil concentration is low, oil starvation occurs even at low speeds, so that the film thickness is smaller than that of the straight oil. Below the first critical speed, the traction behavior of emulsion is essentially the same as that of the neat oil. Above the first critical speed value, the emulsion film thickness decreases with increasing speed. At the second critical rolling speed value, the emulsion film thickness begins to increase again with increasing speed. The oil adsorption on the solid surface is restricted by the high flow rate of the fluid. When the oil concentration is extremely low, the reason for the film formation is that only a tiny volume of oil is needed to form a relatively thick film. Above the second critical speed, it was observed that the more dilute emulsions resulted in higher traction, but this was attributed to asperity contacts, not to lubricant behavior. Emulsion particle size had a noticeable influence on film thickness and the traction characteristics. As a result, an increase in particle size led to improved lubrication.

Clearly, the molecular aspects of colloidal phenomena in water-based lubricants are not less important than thermodynamic parameters. In industrial lubricants, understanding of former might provide more insight for deciding which surfactant system might serve a particular application more successfully.

7.3.2 Influence of Colloidal Transformations on Lubricant Appearance and Storage

In order to emulsify oil droplets by adding a surfactant to the stirred dispersion of water and oil, a relatively little amount of surfactant might be needed, compared to the amount of the emulsified oil. The amount of oil in the droplet is much larger than the amount of surfactants in the shell surrounding it. Consequently, if the surfactant amount is increased, while oil content kept the same, the micelle size should be smaller because in larger micelles, the ratio of surfactants to oil is lower. Under appropriate HLB and with intensive stirring of O/W emulsion, addition of less than 10% surfactants (with remaining 90% comprising oil) to the system should be sufficient to produce a relatively stable O/W emulsion. Such approach is sometimes used in the manufacture of preformulated water-based lubricant emulsions.

However, the ease of emulsification is an important performance characteristic of lubricants supplied as oil-based concentrates. In order to form an emulsion, the oil concentrate is diluted with water by subjecting it to only limited mechanical mixing. If the surfactant percentage is low in the concentrate, very extensive agitation might be required. This could lead to substantial technological issues, such as extended mixing duration, high foaming or air entrainment, excessive energy consumption, varying quality of produced emulsion, etc. During storage of the oil concentrate, the surfactant usually arranges itself into W/O micelles, forming a clear "inverse microemulsion," where oil is the continuous phase. For so-called self-emulsification to occur, it is necessary to have surfactant molecules readily available. So when surfactant concentration is low, self-emulsification is difficult and this presents a major obstacle when diluting the lubricant concentrate with water. Consequently, manufacturers employ low-cost surfactants, e.g., TEA oleate (Figure 7.2), at treat rates which are often much higher than 10% ratio of surfactant to oil [4]. Thus, the excess surfactants are likely to assemble into bilayer lamellae, multi-lamellar vesicles and other complex colloidal structures.

On occasion, the major structural elements of water-based lubricants are vesicles [26,36–38]. Several mechanisms are available to explain formation of the basic structure of vesicles, which also serve as the basis for liposomes, lisosomes, coacervates, and other related colloidal entities. Their sizes and properties are very diverse. The diameter of mono-lamellar vesicles may still be in the same order of magnitude as that of a classical micelle [39], but multi-lamellar vesicles can be much larger. In its core, a vesicle might contain a droplet of either water or oil, whose size can vary widely. Also, in the lamellar sheaths both oil and aqueous phases can be entrapped. It has been described [25,39] that layer curvature of a multi-lamellar vesicle is different for inner and outer shells. Thus, several surfactants with different or variable packing patterns are needed to attain the needed curvature. So the diversity of vesicle types, sizes, structures, and compositions is very broad. An illustration of formation of mono-lamellar and multi-lamellar vesicles with oil cores is shown in Figure 7.8.

In general, colloidal transformations are much slower than chemical reactions, especially, if they occur in the oil phase. It is not unusual for a lubricant concentrate to remain clear and uniform for months until a sudden, seemingly "mysterious" phase separation. The mechanism of such "mystery" often involves colloidal phase transformation and formation of thermodynamically more stable vesicles. In such cases, after long storage (e.g., several weeks) as a clear liquid, the oil-based concentrate produces a rather sudden (within several days) separation, e.g., floc, cream, sediment, and a hazy layer. The new phase can in fact be clear, if the vesicle sizes are sufficiently low. Unsuspecting users might blame contamination, wrong shipment, oxidation, poor quality, etc. Unfortunately, a very long and resource-intensive

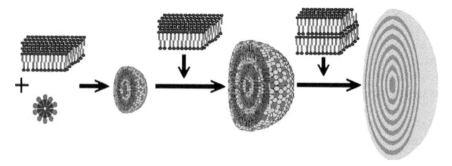

FIGURE 7.8 Formation of mono-lamellar and multi-lamellar vesicles, containing several species of surfactants.

analysis would be needed to determine that vesicle formation was responsible for the phase separation.

Colloidal transformations in water phase are generally faster but might also take several days or weeks. Since water molecules are much smaller than those of oil, the lamellar structure formation in the water phase is more rapid, thus less perplexing and sometimes predictable. Rules of thumb have been devised to counteract the formation of lamellar phases, often without actually realizing the underlying mechanism. One widely known practice is related to producing a water-rich concentrate of a synthetic or semisynthetic lubricant. As a rule of thumb, it is recommended to add water into the oil phase very slowly and with good mixing. Slow phase inversion produces smaller micelles and does not result in lamellar structure formation, which might compromise the appearance of the water-rich concentrate and its stability.

Soluble oil dilutions in water contain micelles, vesicles; also they are likely to contain emulsified oil droplets. Vesicles might be present in semisynthetic lubricant dilutions as well, just their diameters are less than 1 μm on average. Micelles are very important, but the vesicles and emulsified droplets are responsible for the most properties of the lubricant dilutions in water, appearance, and stability in particular. In soluble oils, suspended oil particles can be especially large (right frame of Figure 7.1) and often might even be visible on the glass container walls. Water-based lubricants used in rolling of metal sheets (rolling oils) are known to perform better, if they are "coarse," i.e., the emulsion film on the glass container walls is not very transparent or very uniform, with suspended particles possibly visible to the naked eye. In this unstable emulsion, see Figure 7.9, right, some cream or oil layer separates

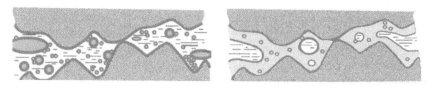

FIGURE 7.9 Two different lubrication mechanisms for water-based lubricants. *Left*: a "tight" metalworking fluid for high speed operation with good cooling capability. *Right*: a "coarse" rolling oil for oil film-assisted lubrication.

rapidly, and an oil film (more likely, a W/O emulsion film) forms between the surfaces. On the contrary, emulsions for high-speed metalworking operations have to be "tight," i.e., their films must have uniform appearance; bulk emulsions must remain stable, but not completely clear, because their milky texture seems to benefit the lubricity. Slight addition of surfactant to stabilize the emulsion on occasion [4] may result in much poorer lubricity.

Metalworking fluids may need lubricity additives delivered to the nascent surfaces, which are freshly formed during the operation and are likely to experience some lubricant starvation. Mobility of water is much higher than that of oil phase or vesicles. Thus, small micelles help lubricity because they provide the water-insoluble additives, which build tribofilms on the surfaces. In the rolling oil application, the tribofilm formation is not very necessary because lubricity is primarily dictated by the oil film presence. The oil, captured within large vesicles and droplets, adhering to the lubricated surfaces, is released easier compared to micellized oil because the vesicles and droplets are not stable thermodynamically and are much less resistant to mechanical shear. The high pressure rheology of the oil film assures good lubricity. Consequently, loose, unstable emulsions are much more beneficial in this respect. Some oil film lubricity is often necessary in metalworking fluids as well. Thus, soluble oils have an advantage over the synthetic fluids, which do not have any oil droplets with a capability to produce an oil film in the friction zone. It must be noted, however, that fresh surface formation releases a large quantity of heat, which needs to be carried away. As mentioned in Section 7.1, water is almost three times better in removing heat than oil. Thus, for such application a tight emulsion might be more suitable than a coarse emulsion.

7.3.3 COLLOIDAL PROCESSES DURING SURFACE WETTING

Clearly, the nature of an emulsion system is highly important for the tribological performance of a water-based lubricant. It affects not only the lubricity but also the entrainment of the additive film into the friction zone. As discussed above in molecular terms, in order to assure proper spreading and wetting, the surfactants must adhere to the surface asperities and start building a lipophilic core, which later propagates to film-like structures, rich in oil phase additives. This process undoubtedly affects the subsequent lubricant entrainment into the friction zone. Also, wetting is very important for those water-based lubricant applications, where the lubricant film has to stay uniform for prolonged time, such as metal-forming lubricants, aqueous rust preventives, or low-friction surface coatings. An example of the importance of spreading and wetting for the water-based low-friction coating on stainless steel (AISI 316) surface is shown in Figure 7.10.

A recent report [5] described the spreading and wetting of the industrial water-based lubricant ("synthetic fluid" type), whose concentrate was mostly made up of alkanol amine and potassium soaps (>50%), including TEA oleate (Figure 7.2). The remainder consisted of an oil-free proprietary additive package with a sharp red color. This concentrate was diluted in deionized water to make emulsions at 10% and 2% wt. "solids" (i.e., low volatility components). The produced liquids appeared completely clear and stable at room temperature, thus constituting microemulsions.

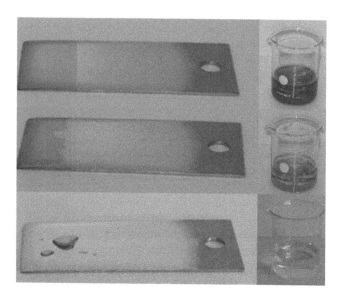

FIGURE 7.10 (See color insert.) Wetting of a steel plate by water-based lubricant emulsions with 10%, 2%, and 0% solids (top to bottom). *Left*: coated plates (actual size). *Right*: beakers with the "synthetic fluid" microemulsions (reduced image size).

A freshly polished, cleaned, and degreased plate (stainless steel AISI 316) was immersed into the beaker with the microemulsion for 5–10 s and smoothly pulled out within 0.5–1 s, while held vertically. The few droplets of the microemulsion were drained by touching the bottom edge of the vertical plate with a paper cloth and laid down horizontally within less than 5 s. Initially, both microemulsions maintained uniform films for about 30 s. Afterward, the 2% microemulsion began forming pools and dry areas, as shown in the center photograph, Figure 7.10. The film dried out in about 3 min at 23°C and 50% relative humidity. The empty areas did not have any visual oily film, while the coated areas had a clearly visible oily coating. On the other hand, the 10% microemulsion dried out in somewhat shorter time and produced a film, which appeared to be completely uniform and continuous. For comparison, deionized water was "coated" in the same manner. Its film stayed on the surface for only a few seconds before pooling into several large droplets.

Higher solids percentage in the microemulsion clearly favored wetting and spreading. It can be expected that some surfactants might be able to escape micelles and adsorb on the metal surface as a monolayer. The deposited film appears much thicker than a monolayer, so it is obvious that it must be composed of many layers, most likely being lamellar in structure. Thus, there is a high probability that transition from micelles to lamellae is favored at a certain concentration. Most likely, such concentration is below 10% solids for the microemulsion in Figure 7.10. Above that concentration, equilibrium is attained when some dispersed phase is deposited on the metal surface and the other part remains as the micelles in the bulk emulsion. With water evaporating, some micelles are adsorbed on the multilayer film and are incorporated into the lamellae. Eventually, the water concentration becomes low enough to favor the microemulsion inversion. The film turns into W/O-type colloidal system with much slower water evaporation. The "film drying rate" becomes nearly zero and the steel plate surface attains "oily" appearance.

At lower concentrations, e.g., 2% solids, the film buildup is already much slower, while the surface tension forces are higher. Transition from micelles to lamellae or adsorbed films is less favored thermodynamically. Consequently, the liquid may start retracting at a faster rate than that for the formation of a viable multilayer film. In most areas, the film is still thick enough to achieve the same equilibrium as in the case of 10% microemulsion. However, in some spots the surface tension forces overtake the film buildup process and produce some "empty" areas before a viable film can be produced. On the panel edges, where the surface tension forces are the highest, some pooling can be clearly observed. In the case of DI water, no film formation occurs and within few seconds surface forces lead to the pooling and droplet formation.

In water-based lubricants, the solids concentration, which favors micelle transition into lamellae, has to be as low as possible. Often this concentration determines how far the lubricant can be diluted with water in the field. In many cases, less than 1 μm thick "oily film" is sufficient for lubrication. Therefore, the ability of the surfactant system to undergo micelle-to-lamella transformation can be crucial in determining the dilution ratio of industrial water-based lubricants.

Significant amounts of industrial efforts are devoted to developing functional additives, which are required to improve the spreading and wetting of water-based lubricants, paints, coatings, rust preventives, and other emulsions. Many of these additives are selected based on accidental observations and then their structure is optimized by the trial-and-error approach. Several hypotheses to explain their spreading and wetting behaviors could be found in literature. A "triple line" concept is sometimes used when describing the wetting mechanisms [40,41]. Rosen and Dahanayake [42] propose that two polar heads are important for wetting and spreading. Rosen [43–45] also suggests that alternating anionic and cationic surfactants attach to metal surface and enable "superspreading." It is too early to decide, however, which of these hypotheses can evolve into a substantially founded theory.

7.3.4 KEY SURFACTANTS FOR SURFACE WETTING CONTROL

Nevertheless, the question of "how to improve wetting of water-based emulsions" already has several viable answers, albeit without a clear explanation on "why these additives improve wetting." Some surfactants are marketed specifically for the purpose of better wetting. Fluorinated surfactants seem to have the highest effectiveness and are sometimes even considered "superspreaders." But it must be noted that not all fluorinated surfactants have such ability in every colloidal system. For example, a fluorinated fatty soap might be effective in O/W emulsions, but not as effective in microemulsions. Fluoroalkyl phosphate (Figure 7.11, top) might behave completely opposite to perfluoro fatty acids. Nevertheless, in industry it was observed that the fluorosurfactants were much more effective than other surfactants in achieving satisfactory wetting and spreading at lower concentrations. Their high costs (at least

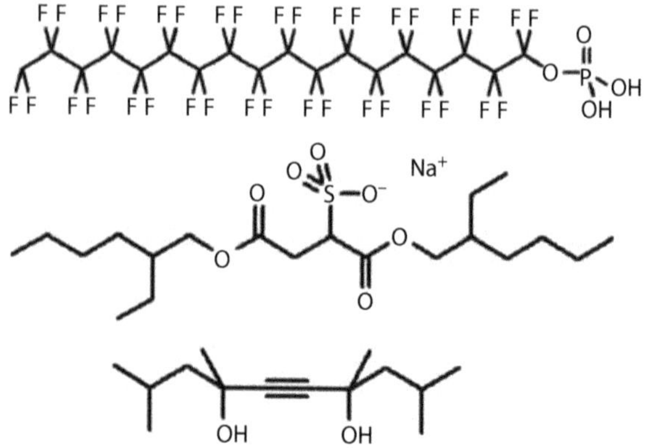

FIGURE 7.11 Classes of surfactants, representing the most effective additives for improving the wetting and spreading of industrial water-based lubricants. *Top*: fluorosurfactants (e.g., acidic fluoroalkyl phosphate). *Middle*: sulfosuccinates (e.g., sodium di-2-ethyl hexyl sulfosuccinate). *Bottom*: alkyne diols (e.g., dimethyl isobutyl 1,4-butynediol).

10 times more expensive than regular surfactants) and regulatory restrictions, however, limit their industrial applications.

Sodium di-isooctyl sulfosuccinate is also known for improving wetting and spreading. Its other alkyl derivatives might be even more effective but exhibit prohibitively excessive foaming. This makes them hard to handle because of foamy lubricant and jeopardizes incompressibility and film rheology. Foam interferes with the uniform emulsion film formation because the bubbles in the film produce dry spots (a.k.a. "fish eyes") with water evaporation. Nevertheless, due to its relatively low cost, sodium di-isooctyl sulfosuccinate is one of the most widely used additives for improved wetting and spreading. Another series of additives designed for this purpose are alkyne diols. It is not clear how the triple bond in these molecules contributes to their effectiveness. However, relatively low concentrations of these surfactants seem to deliver a pronounced improvement in wetting and spreading. These surfactants command relatively high price but appear to gain recognition among water-based lubricant formulators.

The influence of these three groups of additives on transition from micelles to lamellae has not been yet systematically investigated. However, these surfactants might be able to facilitate lamella formation on the metal surfaces. Both alkyne diols and sulfosuccinates have four C atoms between their hydrophilic groups. It is likely that upon incorporation into the surface film, one lipophilic group of this surfactant remains in one lamellar layer, while the other lipophilic chain extends outward and initiates formation of the next layer. Fluorosurfactants do not follow such trend; however, they might facilitate micelle-to-lamella transition via other mechanisms.

Several observations suggest [5] that some otherwise effective surfactants might in fact inhibit emulsion wetting and spreading. Among these are lecithin (a phospholipid) and dimer acid (e.g., oleic dimer) soaps. Both of these low cost surfactants are well-established additives in industrial emulsions and clearly demonstrate good capabilities to emulsify oils and improve emulsion stability. However, these good emulsification properties do not correlate with the detrimental effects on wetting and spreading. One explanation for this perplexing finding might be that these surfactants prevent the lamella formation on metal surfaces. They could be excellent components of the micelle or vesicle shells, but due to some steric or other factors, the lamella formation might be restricted.

Uncertainties concerning the factors affecting wetting and spreading are serious barriers for wider industrial application of water-based lubricants. It appears that these properties are related to the types and formation tendencies of colloidal particles in emulsions, just as is their appearance. Once again, limited understanding of colloidal processes in water-based lubricants and the lack of studies in this field do not allow for proposing a more comprehensive overview of the roles of surfactants in emulsions. A better understanding of emulsion types, compositions and transformations in this field would have a pronounced impact not only on lubricants, but also on paints, coatings, rust preventives, food ingredients, and other consumer and industrial products.

7.4 CONCLUSIONS

Water-based lubricants are often considered as classical emulsions, suggesting that their colloidal properties follow the established rules of surface chemistry and thermodynamics. However, several important aspects have to be considered for more realistic interpretation of technical performance of water-based lubricants:

- Presence of vesicles and other lamellar structures in water-based lubricants dictates their appearance and aging characteristics.
- Lubricity of such emulsions primarily depends on (1) their ability to deliver antiwear additives into the friction zone in order to form a tribofilm and (2) formation of an oil phase layer between moving surfaces.
- Wetting of lubricant emulsions on metal surfaces might be closely related to colloidal transitions, such as from micelles to lamellae, which must compete with liquid film rearrangement processes dictated by surface tension forces.

Many other important properties of water-based lubricants can also be explained using principles of the colloid theory. Therefore, application of surface chemistry concepts in the research on water-based lubricants promises a number of answers to unresolved technical issues, mentioned in numerous publications.

ACKNOWLEDGMENTS

The grant of the Research Council of Lithuania B-34/2009 "Nanotribosuspensijos" was very instrumental in providing opportunities for the authors to exchange their know-how with industrial experts. Advice and resourcefulness of technical teams from Fuchs Lubricants and Rhodia Inc. are sincerely acknowledged as well.

REFERENCES

1. W.J. Bartz, Ecotribology: Environmentally acceptable tribological practices, *Tribol. Int.*, 39, 728–733 (2006).
2. D.N. Garkunov, *Scientific Discoveries in Tribo-Technologies. No-Wear Effect Under Friction. Hydrogen Wear of Metals*, p. 383, MAA Publishing House, Moscow, Russia (2007).
3. P. Kula and R. Pietrasik, Hydrogen's interaction with hardened surface layers in lubricated frictional couples, *Tribol. Int.*, 40, 1613–1618 (2007).
4. J.C. Childers, The chemistry of metalworking fluids, in *Metalworking Fluids*, J.C. Childers (ed.), pp. 165–190, Marcel Dekker, New York (1994).
5. J.R. Mieczkowski and S.J. Asadauskas, Spreading of aqueous vegetable oil emulsions on metal surfaces, in *Proceedings of 17th International Symposium on Surfactants in Solution*, Berlin, Germany (2008).
6. J.M. Perez and S.J. Asadauskas, Utilization of vegetable oils in metalworking fluids, in *Biobased Industrial Fluids and Lubricants*, S.Z. Erhan and J.M. Perez (eds.), pp. 59–64, AOCS Press, Champaign, IL (2002).
7. C. Watkins, Tall oil fatty acids, *INFORM: Int. News Fats Oils Related Mater.*, 11, 580–588 (2000).
8. Y.P. Zhao, R. Turay, and L. Hundley, Monitoring and predicting emulsion stability of metal working fluids by salt titration and turbiscan, *Tribol. Trans.*, 49, 117–123 (2006).
9. A. Cambiella, J.M. Benito, C. Pazos, J. Coca, M. Ratoi, and H.A. Spikes, The effect of emulsifier concentration on the lubricating properties of oil-in-water emulsions, *Tribol. Lett.*, 22, 53–65 (2006).
10. M.S. Briceño, M. Ramirez, J. Bullón, and J.-L. Salager, Customizing drop size distribution to change emulsion viscosity, in *Proceedings of 2nd World Congress on Emulsion*, Paper No. 2-1-094, Bordeaux, France (1997).
11. B.P. Binks and P.D.I. Fletcher, Particles adsorbed at the oil-water interface: A theoretical comparison between spheres of uniform wettability and "Janus" particles, *Langmuir*, 17, 4708–4710 (2001).
12. P.M. Kruglyakov, Hydrophile-lipophile balance of surfactants and solid particles, *Stud. Interface Sci.*, 49, 404 (2000).
13. L. Dong and D.T. Johnson, Adsorption of acicular particles at liquid–fluid interfaces and the influence of the line tension, *Langmuir*, 21, 3838–3849 (2005).
14. E.H. Lucassen-Reynders and M. van den Tempel, Stabilization of water-in-oil emulsions by solid particles, *J. Phys. Chem.*, 67, 731–734 (1963).
15. B.P. Binks and S.O. Lumsdon, Catastrophic phase inversion of water-in-oil emulsions stabilized by hydrophobic silica, *Langmuir*, 16, 2539–2547 (2000).
16. M.J. Garvey, I.C. Griffith, and I. Thompson, Lubricant comprising an oil-in-water emulsion, a process for the preparation thereof and the use of the lubricant, U.S. Patent No. 4995995 (1991).
17. J.J. Benner, F. Sadeghi, M.R. Hoeprich, and M.C. Frank, Lubricating properties of water in oil emulsions, *J. Tribol.*, 128, 296–312, (2006).
18. R. Varadaraj, Oil-in-water-in-oil emulsion, U.S. Patent No. 7338924 (2008).
19. A. Jonsson, J. Bokstrom, A.-C. Malmvik, and T. Warnheim, Preparation and characterization of surface active erucic acid derivatives, *J. Am. Oil Chemists Soc.*, 67, 733–738 (1990).
20. D. Atwood and A.T. Florence, *Surfactants Systems: Their Chemistry, Pharmacy and Biology*, Chapman and Hall, New York (1983).
21. C. Tanford, Thermodynamics of micelle formation: Prediction of micelle size and size distribution, *Proc. Natl. Acad. Sci. USA*, 71, 1811–1815 (1974).

22. B.L.V. Prasad and S.I. Stoeva, Colloid systems: Micelles and nanocrystals, in *Dekker Encyclopedia of Nanoscience and Nanotechnology*, Vol. 2, pp. 856–862 (2004).
23. E.W. Kaler, A.K. Murthy, B.E. Rodriguez, and J.A. Zasadzinski, Spontaneous vesicle formation in aqueous mixtures of single-tailed surfactants, *Science*, 245, 1371–1374 (1989).
24. Ho-Young Cheon, Noh-Hee Jeong, and Hong-Un Kim, Spontaneous vesicle formation in aqueous mixtures of cationic gemini surfactant and sodium lauryl ether sulfate, *Bull. Korean Chem. Soc.*, 26, 107–114 (2005).
25. R. Zana, Micelles and vesicles, in *Encyclopedia of Supramolecular Chemistry*, J.L. Atwood and J.W. Steed (eds.), Marcel Dekker, New York, pp. 861–866 (2004).
26. S. Hollinger, J.-M. Georges, D. Mazuyer, G. Lorentz, O. Aguerre, and N. Du, High-pressure lubrication with lamellar structures in aqueous lubricant, *Tribol. Lett.*, 9, 143–151 (2000).
27. D.A. Sabatini, A. Witthayapanyanon, E.J. Acosta, and J.H. Harwell, Formulation of ultralow interfacial tension systems using extended surfactants, *J. Surfactants Deterg.*, 9, 331–339 (2006).
28. A. Graciaa, J. Lachaise, C. Cucuphat, M. Bourrel, and J. L. Salager, Improving solubilization in microemulsions with additives. 2. Long chain alcohols as lipophilic linkers, *Langmuir*, 9, 3371–3374 (1993).
29. T.C. Whetzel and S. Rodman, Improved lubrication in cold strip rolling, *Iron Steel Eng.*, 36,123–132 (1959).
30. H. Yang, S.R. Schmid, T.J. Kasun, and R.A. Reich, Elastohydrodynamic film thickness and tractions for oil-in-water emulsions, *Tribol. Trans.*, 47, 123–129 (2004).
31. M. Ratoi-Salagean, H. Spikes, and R. Hoogendoorn, The design of lubricious oil-in-water emulsions, *Proc. Inst. Mech. Eng. Part J: J. Eng. Tribol.*, 211, 195–208 (1997).
32. Y. Kimura and K. Okada, Lubricating properties of oil-in-water emulsions, *Tribol. Trans.*, 32, 524–532 (1989).
33. U. Chia Chen, Y. Shi Liu, C.C. Chang, and J. Fin Lin, The effect of the additive concentration in emulsions to the tribological behavior of a cold rolling tube under sliding contact, *Tribol. Int.*, 35, 309–320 (2002).
34. H.L. Rieffe, M. Ratoi-Salagean, and H.A. Spikes, Optimizing film formation by oil-in-water emulsions, *Tribol. Trans.*, 40, 569–578 (1997).
35. L. Ma, J. Luo, Ch. Zhang, Sh. Liu, X. Lu, D. Guo, J.B. Ma, and T. Zhu, Film forming characteristics of oil-in-water emulsion with super-low oil concentration, *Colloids Surf. A*, 340, 70–76 (2009).
36. G. Verberne, A. Schroeder, G. Halperin, Y. Barenholz, and I. Etsion, Liposomes as potential biolubricant additives for wear reduction in human synovial joints, *Wear*, 268, 1037–1042 (2009).
37. S. Sivan, A. Schroeder, G. Verberne, Y. Merkher, D. Diminsky, A. Priev, A. Maroudas, G. Halperin, D. Nitzan, I. Etsion, and Y. Barenholz, Liposomes act as effective biolubricants for friction reduction in human synovial joints, *Langmuir*, 26, 1107–1116 (2010).
38. A.-M. Trunfio-Sfarghiu, Y. Berthier, M.-H. Meurisse, and J.-P. Rieu, Role of nanomechanical properties in the tribological performance of phospholipid biomimetic surfaces, *Langmuir*, 24, 8765–8771 (2008).
39. R. Nagarajan, Molecular theory for mixed micelles, *Langmuir*, 1, 331–341 (1985).
40. E. Bormashenko, Why does the Cassie–Baxter equation apply? *Colloids Surf. A*, 324, 47–50 (2008).
41. P.G. de Gennes, X. Hua, and P. Levinson, Dynamics of wetting: Local contact angles, *J. Fluid Mech.*, 212, 55–63 (1990).
42. M.J. Rosen and M. Dahanayake, *Industrial Utilization of Surfactants. Principles and Practice*, p. 83, AOCS Press, Champaign, IL (2000).

43. M.J. Rosen and H. Liu, Mechanism of the enhanced spreading of some mixtures of anionic and cationic hydrocarbon chain surfactants on a highly hydrophobic polyethylene surface, *J. Surfactants Deterg.*, 8, 157–163 (2005).
44. Y. Wu and M.J. Rosen, Synergism in the spreading of hydrocarbon-chain surfactants on polyethylene films—Anionic and cationic mixtures by a two-step procedure, *Langmuir*, 21, 2342–2348 (2005).
45. Q. Zhou, Y. Wu, and M.J. Rosen, Surfactant-surfactant molecular interactions in mixed monolayers at a highly hydrophobic solid/aqueous solution interface and their relationship to enhanced spreading on the solid substrate, *Langmuir*, 19, 7955–7962 (2003).

8 Surfactant Effects on Biobased Emulsions Used as Lubrication Fluids

Kenneth M. Doll and Brajendra K. Sharma

CONTENTS

ABSTRACT

The successful formulation of a lubricating emulsion requires carefully balancing the mixture of base oil, water, and a plethora of additives. The factors that affect the performance of lubricating emulsions are both macroscopic, such as stability, and microscopic, such as surface adsorption. Presented here is a review of some of the large body of literature devoted to this subject. Also

173

presented is a short study on the use of chemically modified vegetable oil-based aqueous emulsions and their performance in a variety of lubrication tests. Of note is that their coefficient of lubrication, a measure of how well a lubricating emulsion compares to that of its corresponding base oil, is at or above 1.0 for the biobased systems and only ~0.5 for a petroleum-based lubricant.

8.1 INTRODUCTION

Materials science is a rapidly changing and exciting field [1]. The use of more environmentally friendly materials has become a goal of manufacturers in almost all aspects of modern society, including chemistry [2,3], energy [4], polymers, and, especially, surfactants [5].

There is also an overall emphasis on utilizing renewable resources whenever possible [6–9] as dictated in U.S. Executive Orders 13101 and 13134 [10]. Surfactants are one area where nature has supplied tremendous diversity [11] and appears to offer an opportunity to increase biobased material utilization. This has been a focus of the research of many groups over the years [12–15].

The physical properties of vegetable oil and chemically modified vegetable oils are well suited for a range of lubricant products [16–23], including stamping fluids [24], greases [25–29], hydraulic oils [30,31], and, potentially, motor oils. Developments in genetic modifications, such as high oleic soybean oil (SBO) or castor oil equivalents [32], or new crop oils, such as meadowfoam and lesquerella [33–35], will undoubtedly increase these applications.

A lubricating emulsion is a complicated formulation of components (Figure 8.1) blended in order to accomplish a variety of functions, including reducing friction, reducing temperature, removing excess material, and preventing welding and seizing, all to help the manufacturer increase production rate and tool lifetime. Typically, emulsions are aqueous although, in some cases, supercritical carbon dioxide or air dispersions are also used [36]. The base oil can be petroleum based, synthetic oil, or vegetable oil. Differences in these base oils necessitate different

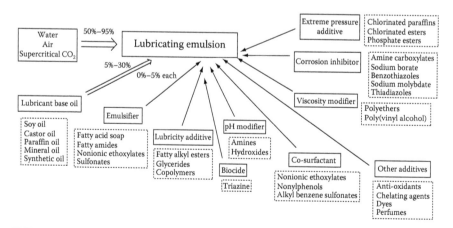

FIGURE 8.1 Typical components of a lubricating emulsion and some examples.

types and concentrations of emulsifiers to stabilize the mixture. Cosurfactants and pH modifiers are commonly added to enhance and preserve emulsion stability, and viscosity modifiers and lubricity additives help the base oil to perform its friction-reducing tasks. Finally, other additives, such as extreme pressure additives, perfumes, dyes, and corrosion inhibitors, help tune the emulsion for specific applications. Rounding off the mixture is a required biocide [37], especially important to use in biodegradable materials.

The surfactant is arguably one of the most important elements of the formulation. In theory, any emulsifier may work, but it has been found that controlling the oil droplet size is critical for lubricant performance. There is a conundrum that must be balanced as smaller drops tend to form more stable emulsions [38], but larger drops help the formation of thicker films [39]. After optimizing these parameters [40–42], a third factor, foaming, must also be considered [43,44]. Fortunately, this problem can sometimes be overcome with the addition of further additives [43].

Theories for emulsification are plentiful [45,46], but models of emulsions in lubrication applications are more scarce. A brief review of some of these theories is presented in Sections 8.2–8.5 below.

8.2 EMULSOLS AND EARLY DIFFERENTIAL SCANNING CALORIMETRY

In the 1980s, an interesting study of oil emulsions, termed "emulsols," was performed [47]. It determined the physical and chemical properties required of the oil in order to form these emulsions. The researchers' hope was that more readily available base oils could be used without sacrifice in performance. The overall conclusion was that formulations utilizing medium viscosity mineral base oil and a sulfonate-type emulsifier could be used with extreme pressure and antiwear additives to form relatively stable emulsions. The researchers divided the emulsols into two groups. The first group contained 5%–10% lubricating oil and were only suitable for severe cutting conditions. Generally, they were not completely stable for 24 h. The second group had 3%–10% oil content and was recommended for a wider range of applications due to their lower corrosion tendency.

This study led to a few conclusions. First, the observed pH value of the formulations was between 8.8 and 9.6 and formulations with higher pH values tended to be more stable. Next, the observed surface tensions of the emulsions were from 33.3 to 42.2 mN m^{-1}. As expected, the samples of the lowest values were stable but, surprisingly, the sample with the highest observed surface tension was also stable. Finally, the area of emulsion-based lubrication research was, and still is, suffering from a lack of fundamental information. Fortunately, the authors of the study were able to recommend an available feedstock.

Later in the 1980s, the analytical technique of differential scanning calorimetry was brought into the biobased emulsion lubricants world [48]. The group studied oil samples that had been used in a steel rolling mill for varying time periods between 6 h and 104 days. Initially, the results were surprising in that fresh emulsion had an oxidation onset temperature of 164°C, whereas 104-day-old emulsion was stable up to 204°C. A look at the data, however, showed that this apparent instability was simply caused by a higher content of volatile components in the original emulsion that were not present after oil use. The overall conclusion that there was the potential for rapid change in a lubricating emulsion in the initial stages of use was found to be valid.

8.3 MODELS FOR LUBRICATION

Fundamental studies followed in the early 1990s. Sakaguchi et al. [39] published their dynamic concentration model for lubrication, which claimed that conventional elastohydrodynamic analysis tended to underestimate film thickness. This error was caused by treating the effective viscosity of an emulsion in terms of the viscosity of water and a correction factor. This caused an underestimation of film thickness by up to a factor of five, especially in situations where the droplet size of the emulsion was greater than the film thickness. The authors presented [39] a series of equations where they modeled the film-forming behavior of tallow oil-in-water emulsions. The method basically derives a relationship that showed that the surface coverage increased as the effective concentration of the oil in the emulsion increased, up to the point where behavior of an inverted emulsion, water-in-oil, was reached.

Szeri [63] took a different approach to emulsion lubrication, developing models to explain the observed phenomenon that lubricant mixtures do not exhibit Newtonian behavior, even if each of the components behaves individually as a Newtonian fluid. This was an interesting theory where sentences like "The notion of overlapping continua is a hypothesis about materials in much the same way as is the notion of a continuum in itself" lead to a model where a series of 10 equations must be solved for 10 unknowns utilizing 13 variables that must be determined experimentally. The one thing this theory did demonstrate, although not something that can be easily applied, was that there is a considerable dependence on the system, where specific "loads" applied to a bearing can form either oil-in-water or water-in-oil emulsions from the same lubricant composition.

8.4 STRUCTURES

Under certain pressure conditions, emulsion systems can take advantage of the formation of various microstructures, such as lamellar dispersions and vesicular dispersions, which can be highly effective in lubricants in wire drawing applications [49]. A system where an aqueous solution of amines and phosphate esters is mixed with an oil phase containing fatty acids and fatty acid phosphates was used. The vesicular solution is quite effective in reducing friction, but the lamellar solution, formed by the addition of copper and zinc ions to the vesicular solution, is even more effective. It retains an interfacial film at loads >4000 N as opposed to ~2800 N for the vesicular system. Although this system was quite specific, it points to some significant new possibilities for lubricant formulators and designers.

Other work on the relationship between lubrication efficacy and the inherent ability of the formulation to produce liquid crystalline structures [50] has been performed. It was concluded that surfactants having a double-chain structure were more prone to forming these effective structures. This behavior was predicted by a model from Kabalnov and Wennerstrom [66] in which the surfactant geometry and formation of a balanced monolayer were found to be important in the stability of these structures.

An interesting natural surfactant for production of these structures is soy lecithin [51]. Several different types of flow behaviors were observed in these dispersions. At concentrations below 120 g L^{-1}, Newtonian behavior was observed if the dispersions were sonicated. If a different dispersion preparation method was used, or higher concentrations were used in the sonicated system, then deviations from Newtonian behavior were observed, which were attributed to vesicle fusion.

Another system utilizes lecithin stabilization augmented by an ordinary anionic surfactant—sodium dodecyl sulfonate (SDS). Still other emulsions were made to compare bovine serum albumin, which functions as a protein-based emulsification stabilizer, with lecithin and SDS. In all the systems where SDS was incorporated before emulsification, it tended to dominate the emulsion droplet size and zeta potential values. This effect was diminished, especially in the bovine serum albumin system, if the SDS was added after the emulsion had been preformed with one of the other stabilizers. No other differences between the lecithin and bovine serum albumin systems were observed.

There are a multitude of possibilities for utilizing surfactant properties to control the formation of these structures. Sometimes, they enhance lubricity, as mentioned above, whereas sometimes these structures are considered undesirable [52], i.e., in a detergent. What is clear is that the use of natural surfactants [11,53], such as proteins, carbohydrates, and phospholipids, in combination with synthetics [54–56] may be necessary, along with appropriate tools and models [52,57–61], to formulate truly superior products.

8.5   HYDROPHILE–LIPOPHILE BALANCE

A simple parameter called the hydrophile–lipophile balance (HLB) is used to compare surfactant properties [62,63]. It is simply an index that represents the ratio of the hydrophilic and hydrophobic portions of a surfactant. This simple relationship has been a staple of the surfactant industry for half a century, especially in the comparison of nonionic surfactants [64–66]. A corresponding set of values, called the required HLB, has been developed for many hydrocarbons and other organic liquids by observation of the phase inversion temperature of emulsions of these liquids stabilized with different surfactants [67]. A very simple method utilizing this value in lubrication formulation has been recently reported [68]. The process involves simply matching the required HLB of a specific lubricant component, or blend of lubricant components, with a surfactant system of the same corresponding value. The study shows that if a match within ±0.5 units is obtained, an emulsion with the optimum performance and cost relationship is formed. This is especially important when selecting surfactant systems for vegetable oil–based lubricants, considering the required HLB of SBO is 6, whereas that of paraffinic mineral oil is 10 [67].

8.6 PERFORMANCE

All these theories and processes are useful, but evaluation of a lubricant emulsion must also involve examining performance under lubricating conditions. Some of the important parameters that must be addressed are emulsion stability, friction and wear reduction, and fluid stability.

8.6.1 EMULSION STABILITY

Many types of tests have been performed on emulsions that are specifically designed to be lubricants. The test methods used for lubricants are the same as those used for conventional emulsions and include surface tension [69], zeta potential [70–72], interfacial tension [73,74], and simple phase separation observation.

In one study [75], surface tension measurements were made at 25°C on solutions of 17 different surfactants. A trend was observed where the surfactants that exhibited the maximum reduction in surface tension also formed the best cutting oil emulsions. Other studies used turbidity values [76] or laser light readings [77] to rapidly determine the stability of metalworking fluid emulsions where destabilization of emulsion could be detected much sooner than by visual inspection for phase separation. However, it was noted that these methods would be inapplicable to some formulations where background signals could overwhelm the emulsion droplet signal.

More sophisticated and systematic studies of emulsion stability as a function of surfactant structure [78] or water hardness [77,79] have also been reported. These studies show that a three-phase diagram, termed a "formulation triangle," was necessary to discover the trends. For each three-component system, the triangle has 10 areas representing different concentrations. Various nonionic surfactants and anionic cosurfactants were used to stabilize three representative base oils, canola oil, SBO, and a fatty acid trimethylolpropane ester. Droplet diameters between 10 and 1000 nm were observed, and increased anionic surfactant concentrations tended to result in smaller drops. More interestingly, the differences in oils were illustrated by comparison of the 10 areas in their formulation triangles. SBO and canola oil were found to form stable emulsions under similar conditions, 84% of the time. On the other hand, comparison of trimethylolpropane to SBO and canola oil gave similar results only 70% and 60% of the time, respectively. This is another indication that vegetable oil stabilization is fundamentally different from mineral oil stabilization.

The most complete study [80] not only determined the emulsion stability by measuring surface tension and drop size but also measured film thickness by ultrathin film interferometry. The surprising conclusion was that the adhesion of the oil droplet to the metal surface was the dominant factor in lubricant performance, whereas the macroscopic properties, such as drop size, had little effect. Researchers from the same group [81] have reported that zeta potentials were useful in monitoring lubrication emulsions. A zeta potential of −10 mV or stronger is required to stop the emulsion from destabilizing and being rendered completely useless.

8.6.2   Friction and Wear Results

Tribological tests on lubricating emulsions are also available in the literature. Two recent examples are studies where aqueous emulsions were tested by the four-ball test method [82] and the grinding wheel wear test [83]. The four-ball test method showed that by using ethoxylated fatty acid ester lubricants, friction could be reduced sixfold and wear could be cut in half. The general trend was that if the degree of ethoxylation increased, the lubricant performed at a higher level. This was attributed to a stronger surface activity caused by the increase of the polar ethoxylate groups.

In the grinding test [99], the performance of the fluid was determined by analyzing grinding wheel wear and substrate surface roughness. A quantity called the "G ratio" represents the amount of substrate removed divided by the loss of mass of the wheel. The values for the optimized biobased fluid were over 1500, nearly as large as for the petroleum-based product. Experiments using the biobased fluid also showed lower surface roughness results compared to the experiments using petroleum product.

8.6.3   Fluid Maintenance

Deterioration of the lubricating emulsion is another parameter that significantly affects the performance of a fluid [79]. Oxidation is one of the key areas of concern when biobased lubricating oils are used. The oxidation of oils is a well-studied, though somewhat complicated, phenomenon [84]. A recent study [85] has demonstrated further complication in emulsion systems by uncovering a drop size dependence of oxidation rate. Oxidation studies of aqueous emulsions of methyl linoleate, with drop sizes ranging from 41 to 5900 nm, showed that reactions were found to proceed with a shorter induction time on the smaller drops. This could be related to the larger total surface area of the small oil drops. A surprising result was that the effective rate of the overall oxidation was slightly slower for the smaller drops. This was attributed to changes in drop size during the oxidation.

Microbial contamination is probably a bigger problem than oxidation or emulsion stability. There are at least 52 bacterial and 11 fungal organisms [86] that cause concern in emulsion lubricants. This issue not only affects production but is also a health concern [87]. Conditions such as hypersensitivity pneumonitis and granulomatous allergic lung disease [88] have been diagnosed in workers exposed to metalworking fluids. Membrane filtration has been used [89,90] for controlling these problems, and considerable ongoing effort is underway to improve membrane efficiency.

Even though there are some means of regeneration of fluids [81], where additional additives can be added to the lubricant emulsion in order to rejuvenate its properties, the emulsion lubricant will have to be disposed of at some point [91]. In order to accomplish this, the emulsion, which has been carefully designed to be stable, must be destabilized into separate oil and water phases. A new technique called pervaporation [92], involving the use of a special membrane to remove oil, is one area of promise. Another study mentions a potential advantage in using a supercritical carbon dioxide–based rapeseed oil emulsion system [93]. This product was shown to be more environmentally friendly over its lifecycle.

8.7 SYNTHESIS OF CHEMICALLY MODIFIED FATTY ACID METHYL ESTERS

It has been shown that chemical modification of vegetable oils can improve their physical properties such as surface adsorption [94–97], oxidative stability [98,99], and viscosity [100,101]. Prior to this report, it had not been determined if these oils were suitable for emulsion-based lubrication. A detailed study of the emulsification properties of these oils and the resultant stability is available elsewhere [102]. For comparison, the emulsions in this study were stabilized with an ordinary alcohol ethoxylate surfactant at 1% concentration and with no other additives.

8.7.1 MATERIALS

Five biobased oils and a petroleum-based oil were used in this study. SBO (refined, bleached, deoderized grade, KIC Chemicals, New Paltz, NY), castor oil (Refined, KIC Chemicals, New Paltz, NY), epoxidized SBO; Vikoflex® 7170, Arkema, Philadelphia, PA), 2-ethylhexylsoyate (Vikoflex® 4050, Arkema, Philadelphia, PA), methyl soyate (SoyGold 1100, Ag. Environmental Products, Omaha, NE), petroleum lubricant basestock (ZX-7R, Castrol, Wayne, NJ), and Brij 30 surfactant (Sigma-Aldrich, St. Louis, MO) were all used as received.

8.7.2 EMULSION PREPARATION

The emulsion samples were prepared in a 400 mL beaker using a high-shear mixer (Silverson, East Longmeadow, MA) equipped with 5/8″ microtubular mixing assembly and general purpose disintegrating head. The mixer was brought up to full speed, ~9900 rpm, in the beaker that contained 100.0 mL of deionized water. The concentrations of additives were measured by volume. A 1.0 mL syringe was used to deliver 1.0 mL of Brij 30 surfactant and was followed by 1.0, 2.0, or 5.0 mL of oil addition using a 10.0 mL syringe. The blending was continued for 10 min.

8.7.3 FRICTION TESTING

Friction testing was conducted on a four-ball test geometry (Model Multi-Specimen, Falex Corporation, Sugar Grove, IL) at ambient temperature. The balls (52100 steel, 12.7 mm diameter, 64–66 Rc hardness) were thoroughly cleaned successively with dichloromethane and hexane before each experiment. Test fluid, 10 mL of emulsion, was poured into the test cup to cover the stationary balls. Tests were conducted at a normal load of 40 kg (392 N), 1200 rpm, at room temperature, for 60 min. After the test, the wear scar diameters on the lower three balls were measured using a microscope (6SD with L2 fiber-optic light, Leica, Bannockburn, IL) and a digital microscopic camera (PAXcam, Villa Park, IL) with Pax-it 6.4 software. Two wear scar measurements, perpendicular to each other, were recorded for each ball and the average of the six measurements was calculated. Duplicate tests were always conducted using a new set of balls. The standard deviation of six measurements was less than 0.04 mm in all experiments.

8.7.4 Performance Properties

Emulsions of the five biobased oils, SBO, ESO, castor oil, methyl soyate (Soy-Gold 1100), 2-ethylhexyl epoxy soyate (Vikoflex 4050), and one petroleum-based oil, Castrol ZX-7R (7R), were made at 1%, 2%, and 5% (v/v) oil concentrations. Each emulsion as well as a sample of un-emulsified base oil was used in friction measurements. Replicate experiments were performed and the following data were measured: coefficient of friction (CoF), emulsion temperature change, and wear scar diameter. The results of each of these measurements will be discussed next.

8.7.4.1 Coefficient of Friction

CoF can give insight into the effectiveness of a lubricant. A CoF value of 0.0 would mean that no torque is needed to spin the top ball. This, of course, is only achievable in theory, so the goal of a lubricant is to approach that value. The CoF values (shown in Figure 8.2) are averages of duplicate runs for each respective concentration of oil.

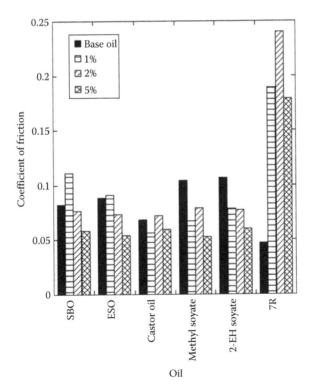

FIGURE 8.2 CoF observed in testing of neat base oils and 1%, 2%, and 5% aqueous emulsions. Note that pure water produced a CoF of 0.48 under these conditions.

It can be seen that for the base oil tests, the petroleum-based 7R oil had the lowest CoF by a wide margin. However, upon emulsification, its effectiveness was greatly reduced. Among the biobased oils, there were much less dramatic changes between the base oils and the emulsions. When using the triacylglyceride lubricants, SBO, ESO, castor oil, or the ester lubricants, SoyGold 1100 and Vikoflex 4050, as the base oil in emulsions, their performance was an improvement over the experiments using base oil alone in most cases.

8.7.4.2 Sample Temperature Change

The friction tests were conducted over 60 min with constant monitoring of the temperature. Emulsion temperatures are summarized (Table 8.1) and show values quite consistent with the friction results. The samples with lowest friction results, petroleum-based 7R oil, SoyGold 1100 5%, ESO 5%, Castor 1%, all had final temperatures of 40°C–42°C, the lowest observed. The worst friction results, 7R emulsions, SBO 1%, SoyGold 1100 base oil, Vikoflex base oil, all had final temperatures of 49°C–55°C, the highest observed. This finding is quite consistent with the expected temperature increase that is positively correlated with friction.

8.7.4.3 Wear Scar Diameter

Wear scar diameter is perhaps the most relevant value for many emulsion lubricating applications. Using wear scar data, two criteria were evaluated. The first was how large the wear scar diameter was, and the second was how efficiently an emulsion lubricated compared to neat base oil. The wear scar data (Figure 8.3) show results with some of the same trends as the friction data. The neat petroleum-based 7R oil run had the smallest scar of any of the base oils, but the wear nearly doubled when that oil was used in an emulsion. All the other oils, except SBO, actually gave smaller wear scar diameter as emulsions than as base oils. Of the materials

TABLE 8.1
Final Temperature (°C) after 60 min of Friction Testing Performed at Ambient Temperature

Oil	Temperature (°C)			
	Base Oil	1% Emulsion	2% Emulsion	5% Emulsion
SBO	44	46	45	44
ESO	46	42	42	40
Castor	43	41	45	43
Soy-Gold	49	45	44	42
Vikoflex4050	50	43	45	46
Petroleum-based oil (7R)	41	55	54	52

Note: An accurate value for a test using pure water could not be obtained due to instrument limitations. However, it would have been significantly higher than any of these values.

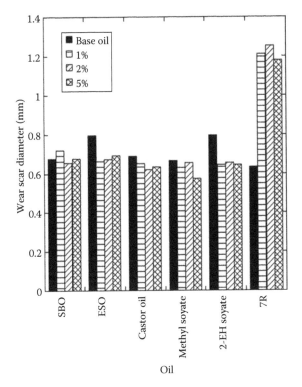

FIGURE 8.3   Wear scar diameter observed in testing of neat base oils and 1%, 2%, and 5% aqueous emulsions. Note that pure water produced a wear scar diameter of 2.4 mm under these conditions.

investigated, the best wear properties were observed for the methyl soyate (SoyGold 1100) 5% emulsion.

Another way of analyzing the data is to calculate the coefficient of lubricity (L). This value is calculated from the simple ratio shown in Equation 8.1.

$$L = \frac{\text{Wear scar diameter using base oil alone}}{\text{Wear scar diameter using emulsion}} \tag{8.1}$$

According to Equation 8.1, the lower the value of L, the poorer the performance of the emulsion. The values of the coefficient of lubricity (Table 8.2) show that the petroleum-based 7R oil is a poor performer in an emulsion, whereas most of the biobased oils have values above 1.0 and are very efficient emulsion lubricants.

8.8   CONCLUSION

There are no lubricating emulsion formulations suitable that work for all the applications needed in industry. The factors that determine if a lubricating emulsion will be effective are numerous and include both microscopic and macroscopic properties. An emulsion will have to be optimized for cost and performance. In this study, we

TABLE 8.2
Coefficient of Lubricity (L in Equation 8.1) of Emulsions
Used in This Study

Oil	Coefficient of Lubricity		
	1% Emulsion	2% Emulsion	5% Emulsion
SBO	0.91	1.02	1.00
ESO	1.21	1.18	1.16
Castor	1.04	1.11	1.08
Soy-Gold	1.06	1.02	1.18
Vikoflex4050	1.27	1.23	1.26
Petroleum-based oil (7R)	0.52	0.50	0.53

Note: The values indicate that the petroleum-based 7R emulsions are quite poor performers (L~0.5) compared to emulsions of biobased oils (L≥1).

observed that emulsions with as little as a 1% biobased oil concentration can give many of the same friction-reducing properties as the neat base oil. On the other hand, a petroleum-based oil did not display such behavior. This indicates that perhaps biobased fluids have superior potential for use in areas where emulsions-based lubricants are advantageous.

ACKNOWLEDGMENTS

We would like to thank Cynthia M. Ruder, Eric S. Johnsen, and Richard H. Henz for assistance in the work presented here. This work was part of the in-house research of the Agricultural Research Service of the United States Department of Agriculture. Mention of trade names or commercial products in this publication is solely for the purpose of providing specific information and does not imply recommendation or endorsement by the U.S. Department of Agriculture. USDA is an equal opportunity provider and employer.

REFERENCES

1. S. Lee and N. D. Spencer, Materials Science: Sweet, Hairy, Soft, and Slippery, *Science* 319, 575–576 (2008).
2. J. C. Warner, A. S. Cannon, and K. M. Dye, Green Chemistry, *Environ. Impact Assess. Rev.* 24, 775–799 (2004).
3. P. T. Anastas and M. M. Kirchhoff, Origins, Current Status, and Future Challenges of Green Chemistry, *Acc. Chem. Res.* 35, 686–694 (2002).
4. G. Knothe, The History of Vegetable Oil-Based Diesel Fuels. In: *The Biodiesel Handbook*, G. Knothe (ed.) AOCS Press, Champaign, IL, pp. 4–16 (2005).
5. E. S. Stevens, *Green Plastics: An Introduction to the New Science of Biodegradable Plastics*. Princeton University Press, Princeton, NJ (2002).

6. R. M. Navarro, M. A. Pena, and J. L. G. Fierro, Hydrogen Production Reaction from Carbon Feedstocks: Fossil Fuels and Biomass, *Chem. Rev.* 107, 3952–3991 (2007).
7. A. Corma, S. Iborra, and A. Velty, Chemical Routes for the Transformation of Biomass into Chemicals, *Chem. Rev.* 107, 2411–2502 (2007).
8. S. Z. Erhan, B. K. Sharma, and K. M. Doll, Production of Agricultural Commodities in the United States, *Chem. Oggi/Chem. Today* 27, 34–36 (2009).
9. S. Z. Erhan, B. K. Sharma, and K. M. Doll, Opportunities for Industrial Uses of Agricultural Products, *Chem. Oggi/Chem. Today* 27, 38–41 (2009).
10. Guideline for Designating Biobased Products for Federal Procurement, *Fed. Regist.* 70, 1792–1812 (2005).
11. V. M. Dembitsky, Astonishing Diversity of Natural Surfactants: 4. Fatty Acid Amide Glycosides, Their Analogs and Derivatives, *Lipids* 40, 641–660 (2005).
12. I. Gawrilow, Vegetable Oil Usage in Lubricants, *Inform* 15, 702–705 (2004).
13. A. R. Lansdown, *Lubrication and Lubrication Science.* ASME Press, New York (2004).
14. A. P. Pratap, A. S. Kadam, and D. N. Bhowmick, Modified Oils and Fats as Biolubricants, *INFORM* 16, 282–285 (2005).
15. M. Wood, Ars Scientists Just Say "No" to Petroleum, *Agric. Res.* 57, 14–16 (2009).
16. S. Z. Erhan, A. Adhvaryu, and Z. Liu, Chemically Modified Vegetable Oil-Based Industrial Fluid, U.S. Patent 6583302 (2003).
17. B. K. Sharma, A. Adhvaryu, Z. Liu, and S. Z. Erhan, Chemical Modification of Vegetable Oils for Lubricant Applications, *J. Am. Oil Chem. Soc.* 83, 129–136 (2006).
18. S. Z. Erhan, A. Adhvaryu, and B. K. Sharma, Chemically Functionalized Vegetable Oils. In: *Synthetics, Mineral Oils, and Bio-Based Lubricants Chemistry and Technology,* L. R. Rudnick (ed.) CRC Press, Boca Raton, FL, pp. 14–30 (2005).
19. K. M. Doll, B. K. Sharma, and S. Z. Erhan, Synthesis of Branched Methyl Hydroxy Stearates Including an Ester from Bio-Based Levulinic Acid, *Ind. Eng. Chem. Res.* 46, 3513–3519 (2007).
20. B. R. Moser, B. K. Sharma, K. M. Doll, and S. Z. Erhan, Diesters from Oleic Acid: Synthesis, Low Temperature Properties, and Oxidation Stability, *J. Am. Oil Chem. Soc.* 84, 675–680 (2007).
21. B. K. Sharma, Z. Liu, A. Adhvaryu, and S. Z. Erhan, One-Pot Synthesis of Chemically modified Vegetable Oils, *J. Agric. Food Chem.* 56, 3049–3056 (2008).
22. A. Biswas, B. K. Sharma, J. L. Willett, S. Z. Erhan, and H. N. Cheng, Soybean Oil as a Renewable Feedstock for Nitrogen-Containing Derivatives, *Energy Environ. Sci.* 1, 639–644 (2008).
23. S. Z. Erhan, B. K. Sharma, Z. Liu, and A. Adhvaryu, Lubricant Base Stock Potential of Chemically Modified Vegetable Oils, *J. Agric. Food Chem.* 56, 8919–8925 (2008).
24. A. C. Carcel, D. Palomares, E. Rodilla, and M. A. P. Puig, Evaluation of Vegetable Oils as Pre-Lube Oils for Stamping, *Mater. Des.* 26, 587–593 (2005).
25. J. W. Hagemann and J. A. Rothfus, Thermal Behavior of Prospective Hydroxy Acid Grease Thickeners, *J. Am. Oil Chem. Soc.* 68, 139–143 (1991).
26. P. R. Grives, The Manufacture of Biodegradable Nontoxic Lubricating Grease, *NLGI Spokesman* 63, 25–29 (2000).
27. B. K. Sharma, K. M. Doll, and S. Z. Erhan, Biobased Greases: Soap Structure and Composition Effects of Tribological Properties. In: *Surfactants in Tribology,* G. Biresaw and K. L. Mittal (eds.) CRC Press, Boca Raton, FL, pp. 309–323 (2008).
28. B. K. Sharma, A. Adhvaryu, J. M. Perez, and S. Z. Erhan, Biobased Grease with Improved Oxidation Performance for Industrial Application, *J. Agric. Food Chem.* 54, 7594–7599 (2006).
29. B. K. Sharma, A. Adhvaryu, J. M. Perez, and S. Z. Erhan, Soybean Oil Based Greases: Influence of Composition on Thermo-Oxidative and Tribochemical Behavior, *J. Agric. Food Chem.* 53, 2961–2968 (2005).

30. L. Tocci, Rethinking Hydraulic Fluids, *Lubes 'N' Greases* 12, 26 (2006).
31. J. Suszkiw, Statue of Liberty Goes Green with Soy-Based Elevator Fluid, *Inform* 15, 705 (2004).
32. T. A. McKeon, Castor Plant Gene for Castor Oil Biosynthesis, *Ind. Bioprocess.* 29, 7 (2007).
33. J. W. Goodrum and D. P. Geller, Influence of Fatty Acid Methyl Esters from Hydroxylated Vegetable Oils on Diesel Fuel Lubricity, *Bioresource Technol.* 96, 851–855 (2005).
34. B. R. Moser, S. C. Cermak, and T. A. Isbell, Evaluation of Castor and Lesquerella Oil Derivatives as Additives in Biodiesel and Ultralow Sulfur Diesel Fuels, *Energy Fuels* 22, 1349–1352 (2008).
35. S. C. Cermak and T. A. Isbell, Pilot-Plant Distillation of Meadowfoam Fatty Acids, *Ind. Crops Prod.* 15, 145–154 (2002).
36. S. J. Skerlos, K. F. Hayes, and A. F. Clarens, Metal Working Lubricant Formulations Based on Supercritical Carbon Dioxide, U.S. Patent 2006/0247139 (2006).
37. M. A. Wright, The Application of Biocides in Metalworking Fluids. In: *Industrial Biocides: Selection and Application*, D. R. Karsa and D. Ashworth (eds.) The Royal Society of Chemistry, Cambridge, U.K., pp. 111–118 (2002).
38. J. Piao and S. Adachi, Stability of O/W Emulsions Prepared Using Various Monoacyl Sugar Alcohols as an Emulsifier, *Innovat. Food Sci. Emerg. Tech.* 7; 211–216 (2006).
39. W. R. D. Wilson, Y. Sakaguchi, and S. R. Schmid, A Dynamic Concentration Model for Lubrication with Oil-in-Water Emulsions, *Wear* 161, 207–212 (1993).
40. I. Roland, G. Piel, L. Delattre, and B. Evrard, Systematic Characterization of Oil-in-Water Emulsions for Formulation Design, *Int. J. Pharm.* 263, 85–94 (2003).
41. S. Tcholakova, N. D. Denkov, and T. Banner, Role of Surfactant Type and Concentration for the Mean Drop Size During Emulsification in Turbulent Flow, *Langmuir* 20, 7444–7458 (2004).
42. S. Tcholakova, N. D. Denkov, D. Sidzhakova, I. B. Ivanov, and B. Campbell, Interrelation between Drop Size and Protein Adsorption at Various Emulsification Conditions, *Langmuir* 19, 5640–5649 (2003).
43. R. R. Schnelle and O. Klocker, Selecting Defoamers, *Polym. Paint Colour J.* 194, 22–24 (2004).
44. R. J. Pugh, Foaming, Foam Films, Antifoaming and Defoaming, *Adv. Colloid Interface Sci.* 64, 67–142 (1996).
45. M. J. Rosen and M. Dahanayake, *Industrial Utilization of Surfactants: Principles and Practice.* AOCS Press, Champaign, IL (2000).
46. M. J. Rosen, *Surfactants and Interfacial Phenomena.* John Wiley & Sons, Inc., Hoboken, NJ (2004).
47. V. A. Serov, A. I. Maksimova, and S. B. Dorfman, Emulsols with Antiwear and Extreme-Pressure Additives, *Chem. Technol. Fuels Oils* 18, 34–35 (1982).
48. B. O. Haglund and T. Luks, Characterisation of the Ageing of Cold Rolling Emulsions by Differential Scanning Calorimetry, *Thermochim. Acta* 72, 51–54 (1984).
49. S. Hollinger, J.-M. Georges, D. Mazuyer, G. Lorentz, O. Aguerre, and N. Du, High-Pressure Lubrication with Lamellar Structures in Aqueous Lubricant, *Tribol. Lett.* 9, 143–151 (2001).
50. K. Boschkova, B. Kronberg, J. J. R. Stalgren, K. Persson, and M. R. Salagean, Lubrication in Aqueous Solutions Using Cationic Surfactants a Study of Static and Dynamic Forces, *Langmuir* 18, 1680–1687 (2002).
51. M. Manconi, J. Aparicio, A. O. Vila, J. Pendas, J. Figueruelo, and F. Molina, Viscoelastic Properties of Concentrated Dispersions in Water of Soy Lecithin, *Colloids Surf. A: Physicochem. Eng. Asp.* 222, 141–145 (2003).

52. N. Komesvarakul, M. D. Sanders, E. Szekeres, E. J. Acosta, J. F. Faller, T. Mentlik, L. B. Fisher, G. Nicoll, D. A. Sabatini, and J. F. Scamehorn, Microemulsions of Triglyceride-Based Oils: The Effect of Co-Oil and Salinity on Phase Diagrams, *J. Cosmet. Sci.* 57, 309–325 (2006).

53. D. K. Allen and B. Y. Tao, Carbohydrate-Alkyl Ester Derivatives as Biosurfactants, *J. Surfact. Deterg.* 2, 383–390 (1999).

54. F. Comelles, J. Sanchez-Leal, and J. J. Gonzalez, Soybean Oil Microemulsions with Oleic Acid/Glycols as Cosurfactants, *J. Surfact. Deterg.* 8, 257–262 (2005).

55. C. Rodríguez, D. P. Acharya, S. Hinata, H. Kunieda, and M. Ishitobi, Effect of Ionic Surfactants on the Phase Behavior and Structure of Sucrose Ester/Water/Oil Systems, *J. Colloid Interface Sci.* 262, 500–505 (2003).

56. S. Dima, R. Cretu, and T. Florea, Emulsification of Vegetable Oils with Nonionic Surfactants and Polymers, *Revue Roumaine de Chimie* 53, 399–403 (2008).

57. K. Aramaki and H. M. Khalid, Effect of Addition and Molecular Size of Triglyceride Oils on Phase Behavior and Surfactant Self-Assemblies, *J. Oleo Sci.* 53, 557–563 (2004).

58. P. A. Winsor, Binary and Multicomponent Solutions of Amphiphilic Compounds. Solubilization and the Formation, Structure, and Theoretical Significance of Liquid Crystalline Solutions, *Chem. Rev.* 68, 1–40 (1968).

59. T. Doan, E. Acosta, J. F. Scamehorn, and D. A. Sabatini, Formulating Middle-Phase Microemulsions Using Mixed Anionic and Cationic Surfactant Systems, *J. Surfact. Deterg.* 6, 215–224 (2003).

60. C. Tongcumpou, E. J. Acosta, L. B. Quencer, A. F. Joseph, J. F. Scamehorn, D. A. Sabatini, S. Chavadej, and N. Yanumet, Microemulsion Formation and Detergency with Oily Soils: I. Phase Behavior and Interfacial Tension, *J. Surfact. Deterg.* 6, 191–203 (2003).

61. C. Tongcumpou, E. J. Acosta, L. B. Quencer, A. F. Joseph, J. F. Scamehorn, D. A. Sabatini, S. Chavadej, and N. Yanumet, Microemulsion Formation and Detergency with Oily Soils: II. Detergency Formulation and Performance, *J. Surfact. Deterg.* 6, 205–214 (2003).

62. W. C. Griffin, Classification of Surface-Active Agents by Hlb, *J. Soc. Cosmet. Chem.* 1, 312–326 (1949).

63. W. C. Griffin, Calculation of Hlb Values of Non-Ionic Surfactants, *J. Soc. Cosmet. Chem.* 5, 249–256 (1954).

64. J. Wu, Y. Xu, T. Dabros, and H. Hamza, Development of a Method for Measurement of Relative Solubility of Nonionic Surfactants, *Colloids Surf. A: Physicochem. Eng. Asp.* 232, 229–237 (2004).

65. M. Fukuda, The Importance of Lipophobicity in Surfactants: Methods for Measuring Lipophobicity and Its Effect on the Properties of Two Types of Nonionic Surfactant, *J. Colloid Interface Sci.* 289, 512–520 (2005).

66. J.-P. Hsu and A. Nacy, Behavior of Soybean Oil-in-Water Emulsion Stabilized by Nonionic Surfactant, *J. Colloid Interface Sci.* 259, 373–381 (2003).

67. K. Shinoda and H. Kunieda, Phase Properties of Emulsions: PIT and HLB. In: *Encyclopedia of Emulsion Technology*, P. Bechler (ed.) Marcel Dekker, Inc., New York, pp. 354–361 (1983).

68. N. Canter, Hlb: A New System for Water-Based Metalworking Fluids, *Tribol. Lubric. Tech.* 60, 12–13 (2004).

69. I. Langmuir, The Constitution and Fundamental Properties of Solids and Liquids. II. Liquids, *J. Am. Chem. Soc.* 39, 1848–1906 (1917).

70. L. Kong, J. K. Beattie, and R. J. Hunter, Electroacoustic Study of Concentrated Oil-in-Water Emulsions, *J. Colloid Interface Sci.* 238, 70–79 (2001).

71. H. Parra-Barraza, D. Hernandez-Montiel, J. Lizardi, J. Hernandez, R. Herrera Urbina, and M. A. Valdez, The Zeta Potential and Surface Properties of Asphaltenes Obtained with Different Crude Oil/N-Heptane Proportions, *Fuel* 82, 869–874 (2003).
72. Á. Cambiella, J. M. Benito, C. Pazos, and J. Coca, Interfacial Properties of Oil-in-Water Emulsions Designed to Be Used as Metalworking Fluids, *Colloids Surf. A: Physicochem. Eng. Asp.* 305, 112–119 (2007).
73. I. Kobayashi, S. Mukataka, and M. Nakajima, Effects of Type and Physical Properties of Oil Phase on Oil-in-Water Emulsion Droplet Formation in Straight-through Microchannel Emulsification, Experimental and Cfd Studies, *Langmuir* 21, 5722–5730 (2005).
74. R. Nagarajan and E. Ruckenstein, Molecular Theory of Microemulsions, *Langmuir* 16, 6400–6415 (2000).
75. A. M. Al Sabagh, N. A. Maysour, N. M. Nasser, and M. R. Sorour, Some Cutting Oil Formulations Based on Local Prepared Emulsifiers Part I: Preparation of Some Emulsifiers Based on Local Raw Materials to Stabilize Cutting Oil Emulsions, *J. Dispersion Sci. Technol.* 27, 239–250 (2006).
76. J. Deluhery and N. Rajagopalan, A Turbidimetric Method for the Rapid Evaluation of Mwf Emulsion Stability, *Colloids Surf. A: Physicochem. Eng. Asp.* 256, 145–149 (2005).
77. Y. P. Zhao, R. Turay, and L. Hundley, Monitoring and Predicting Emulsion Stability of Metalworking Fluids by Salt Titration and Laser Light Scattering Method, *Tribol. Trans.* 49, 117–123 (2006).
78. F. Zhao, A. Clarens, A. Murphree, K. Hayes, and S. J. Skerlos, Structural Aspects of Surfactant Selection for the Design of Vegetable Oil Semi-Synthetic Metalworking Fluids, *Environ. Sci. Technol.* 40, 7930–7937 (2006).
79. J. B. Zimmerman, K. F. Hayes, and S. J. Skerlos, Influence of Ion Accumulation on the Emulsion Stability and Performance of Semi-Synthetic Metalworking Fluids, *Environ. Sci. Technol.* 38, 2482–2490 (2004).
80. A. Cambiella, J. Benito, C. Pazos, J. Coca, M. Ratoi, and H. Spikes, The Effect of Emulsifier Concentration on the Lubricating Properties of Oil-in-Water Emulsions, *Tribol. Lett.* 22, 53–65 (2006).
81. E. Fernandez, J. M. Benito, C. Pazos, J. Coca, I. Ruiz, and G. Rios, Regeneration of an Oil-in-Water Emulsion after Use in an Industrial Copper Rolling Process, *Colloids Surf. A Physicochem. Eng. Asp.* 263, 363–369 (2005).
82. M. W. Sulek and A. Bocho-Janiszewska, The Effect of Ethoxylated Esters on the Lubricating Properties of Their Aqueous Solutions, *Tribol. Lett.* 24, 187–194 (2006).
83. S. M. Alves and J. F. G. de Oliveira, Development of New Cutting Fluid for Grinding Process Adjusting Mechanical Performance and Environmental Impact, *J. Mater Process Technol.* 179, 185–189 (2006).
84. E. N. Frankel, *Lipid Oxidation.* The Oil Press, Bridgewater, England (2005).
85. H. Imai, T. Maeda, M. Shima, and S. Adachi, Oxidation of Methyl Linoleate in Oil-in-Water Micro- and Nanoemulsion Systems, *J. Am. Oil Chem. Soc.* 85, 809–815 (2008).
86. L. H. G. Morton, J. W. Gillatt, E. O. Warrilow, and M. Greenhalgh, A Potential Method for the Recognition of Metalworking Fluid Spoilage Organisms, *Int. Biodeterior. Biodegrad.* 48, 162–166 (2001).
87. K. Suuronen, M.-L. Henriks-Eckerman, R. Riala, and T. Tuomi, Respiratory Exposure to Components of Water-Miscible Metalworking Fluids, *Ann. Occup. Hyg.* 52, 607–614 (2008).
88. M. Veillette, P. S. Thorne, T. Gordon, and C. Duchaine, Six Month Tracking of Microbial Growth in a Metalworking Fluid after System Cleaning and Recharging, *Ann. Occup. Hyg.* 48, 541–546 (2004).

89. J. Deluhery and N. Rajagopalan, Use of Paramagnetic Particles in Membrane Integrity Testing, *J. Memb. Sci.* 318, 176–181 (2008).
90. K. R. Januszkiewicz, A. R. Riahi, and S. Barakat, High Temperature Tribological Behaviour of Lubricating Emulsions, *Wear* 256, 1050–1061 (2004).
91. C. Cheng, D. Phipps, and R. M. Alkhaddar, Treatment of Spent Metalworking Fluids, *Water Res.* 39, 4051–4063 (2005).
92. A. Z. Hadj-Ziane and S. Moulay, Microemulsion Breakdown Using the Pervaporation Technique: Application to Cutting Oil Models, *Desalination* 170, 91–97 (2004).
93. A. F. Clarens, J. B. Zimmerman, G. A. Keoleian, K. F. Hayes, and S. J. Skerlos, Comparison of Life Cycle Emissions and Energy Consumption for Environmentally Adapted Metalworking Fluid Systems, *Environ. Sci. Technol.* 42, 8534–8540 (2008).
94. T. L. Kurth, B. K. Sharma, K. M. Doll, and S. Z. Erhan, Adsorption Behavior of Epoxidized Fatty Esters via Boundary Lubrication Coefficient of Friction Measurements, *Chem. Eng. Commun.* 194, 1065–1077 (2007).
95. T. L. Kurth, J. A. Byars, S. C. Cermak, B. K. Sharma, and G. Biresaw, Non-Linear Adsorption Modeling of Fatty Esters and Oleic Estolide Esters via Boundary Lubrication Coefficient of Friction Measurements, *Wear* 262, 536–544 (2007).
96. K. M. Doll, B. K. Sharma, and S. Z. Erhan, Friction Reducing Properties and Stability of Epoxidized Oleochemicals, *CLEAN—Soil, Air, Water* 36, 700–705 (2008).
97. A. Adhvaryu, B. K. Sharma, H. S. Hwang, S. Z. Erhan, and J. M. Perez, Development of Biobased Synthetic Fluids: Application of Molecular Modeling to Structure-Physical Property Relationship, *Ind. Eng. Chem. Res.* 45, 928–933 (2006).
98. B. K. Sharma, K. M. Doll, and S. Z. Erhan, Oxidation, Friction Reducing, and Low Temperature Properties of Epoxy Fatty Acid Methyl Esters, *Green Chem.* 9, 469–474 (2007).
99. B. K. Sharma, K. M. Doll, and S. Z. Erhan, Ester Hydroxy Derivatives of Methyl Oleate: Tribological, Oxidation and Low Temperature Properties, *Bioresource Technol.* 99, 7333–7340 (2008).
100. B. K. Sharma, A. Adhvaryu, S. K. Sahoo, A. J. Stipanovic, and S. Z. Erhan, Influence of Chemical Structures on Low-Temperature Rheology, Oxidative Stability, and Physical Properties of Group II and III Base Oils, *Energy Fuels* 18, 952–959 (2004).
101. Z. Liu, B. K. Sharma, and S. Z. Erhan, From Oligomers to Molecular Giants of Soybean Oil in Supercritical Carbon Dioxide Medium: 1. Preparation of Polymers with Lower Molecular Weight from Soybean Oil, *Biomacromolecules* 8, 233–239 (2006).
102. K. M. Doll and B. K. Sharma, Emulsification of Chemically Modified Vegetable Oils for Lubricant Use, *J. Surfact. Deterg.* In press DOI 10.1007/s11743-010-1203-x (2010).

9 Hydrolyzed Fatty Oil Surfactants as Friction Modifiers at Water/Oil Emulsion Interface

Ahmed M. Al-Sabagh, Rasha A. El-Ghazawy,
Mahmoud R. Noor El-Din, and Nadia G. Kandile

CONTENTS

ABSTRACT

This chapter provieds a comprehensive description of hydrolyzed fatty oil surfactants as friction modifiers at water-in-oil emulsion interface. It focuses on tribological principles and their applications in crude oil fields to develop more economical demulsifiers. It discusses the crude oil emulsion, demulsification of crude oil emulsion, its rheological properties in relation to demulsification efficiency, and kinetics and mechanism of demulsification process.

9.1 INTRODUCTION

Tribology, including friction, is the science and technology of interacting surfaces in relative motion and of related subjects and practices. Tribology is crucial to modern machinery that uses sliding and rolling surfaces. One of the most effective means of controlling friction and wear is by proper lubrication, which provides smooth running and satisfactory lifetime for machine elements. Lubricants can be liquid or solid. The role of surface roughness, mechanisms of adhesion, friction and wear, and physical and chemical interactions between the lubricant and the interacting surfaces must be understood for optimum performance and reliability. Friction is usually subdivided into several varieties including dry, lubricated, skin, internal, and fluid frictions. Fluid friction is used to describe the force resisting relative lateral motion of two solid surfaces separated by a layer of gas or liquid or the friction between layers within a fluid that are moving relative to each other [1,2]. The latter description is suitable to define interacting water and crude oil surfaces found in crude oil emulsions.

The increased problem of crude oil fields that are now producing both crude oil and water pushes researchers to develop more economical chemical demulsifiers. Such demulsifiers alter the oil–water interfacial properties and water droplets size, thus modifying the fluid friction at crude oil–water interface. Attempts have been made to correlate demulsifier performance and certain properties such as molecular structure, hydrophile–lipophile balance (HLB), interfacial tension, interfacial viscosity, partition coefficient, dynamic interfacial tension, and relative solubility number [3–11], but no generalized conclusions have been reached. The selection of a suitable demulsifier for a certain oil emulsion is still a practical art. The high molecular weight components of crude oil, especially asphaltenes and resins (wrongly named natural surfactants), are responsible for crude oil emulsion stability. They do not develop very high surface pressure but the steric stabilization in combination with stabilization by small particles is the most plausible mechanism for the stabilization of such emulsions. On adding an effective demulsifier, an acceleration of flocculation and film rupture processes occurs. This acceleration is a result of the replacement of natural surfactant molecules with demulsifier molecules having low interfacial tension [12]. On the other hand, the dramatic changes in rheological properties of the emulsion upon adding the demulsifier is one of the important factors that should be studied. Generally, oil with a high viscosity has the ability to hold up more and larger water droplets than oil with a lower viscosity. Thus, lowering the oil viscosity increases the mobility of water droplets and thereby leads to collisions, coalescence, and a further increase in the rate of water separation.

Fatty oils are one of the main raw materials that are easily subjected to hydrolysis and esterification. Demulsifiers based on the available raw materials such as soybean, linseed, and castor oils are reported here and the mechanism of demulsification was followed up through studying demulsification kinetics using simple light microscope. The demulsifiers produced were first evaluated with the conventional bottle test using a stable crude oil emulsion containing 7.14 wt% asphaltene. The best demulsifier was then subjected to further analysis for effect of water and asphaltene contents, some surface active properties, rheological properties, kinetics of demulsification, and photomicrographic studies. Photomicrography was found to be a good indicator for the demulsifier impact on modifying water-in-oil (w/o) interface friction [3,13].

9.2   CRUDE OIL EMULSION

An emulsion is a significantly stable dispersion of droplets of a liquid with a certain size in a second immiscible liquid [14,15]. For such a stable dispersion, a third component must be present to stabilize this thermodynamically unstable system. Surfactants play a central role in such systems through adsorption at liquid–liquid interface as an oriented interfacial film, creating sufficient kinetic stability.

In crude oil production, the presence of water in the aquifer layer beneath crude oil zone or pumping water to enhance oil production of old wells results in water coproduction with the crude oil. Although water may be in a discrete phase as it flows up at the well connections to pipelines, it becomes emulsified as it passes through the region of very high shear at the wellhead forming w/o emulsion. Such an emulsion is stabilized by a variety of indigenous surfactants and solid particles found originally in crude oil [16,17]. The coproduction of water with crude oil gives rise to a variety of problems including extra-expense of pumping and transportation, associated corrosion problems, required additional production equipments, poisoning of downstream refinery catalysts, and increased oil viscosity.

9.3   DEMULSIFICATION OF CRUDE OIL EMULSION

9.3.1   Demulsification Techniques

In early days, the general practice was to fill a tank with emulsified oil, introduce the reagent, agitate the mixture by means of gas or air or even by hand, heat it by circulation through a heater as required, and then allow water to separate from oil. Current practice is to employ a continuous process "flow-line treatment," introducing the reagent continuously into the well "downhole," heating the mixture during passage through the flow line, if required, and then leaving the emulsion to break in a special tank (settling tank). When continuous or flow-line treatment is used, the demulsifier may be injected at the desired point. It is usually injected as near as possible to the wellhead, to take advantage of the natural heat and of agitation arising from the turbulence of flow in the system [18].

Several methods are used to break petroleum w/o emulsions like gravity settling, distilling at atmospheric or elevaterd pressure, use of chemicals, applying ultrasound or microwave radiation etc. However, chemical demulsification process is so far the most widely used method in petroleum industry. This method is based on introducing an agent that counteracts the stabilizing influence of naturally present emulsifying agents. These emulsifying agents are high molecular weight compounds namely asphaltenes and resins. When two water droplets approach due to gravitational forces, thermal convection, or agitation, the intervening film of the continuous phase will drain. The shear stresses that exist in the film associated with drainage tend to concentrate the natural surfactant molecules outside the film and lower their concentration inside the film. This will set up an interfacial tension gradient along that film. Demulsifier molecules and natural surfactants then compete with each other for adsorption onto the voids created in that film [19]. When the demulsifier molecules, which lower interfacial tension much more than the natural surfactants, are adsorbed at the interface, an interfacial tension gradient is set up leading to again drainage of the film of the continuous phase; in other words, an assissted film drainage upon replacing natural surfactant molecules with demulsifier molecules. Ultimately, the film becomes very thin and ruptures due to close proximity of adjacent dispersed phase surfaces and local absence of natural and demulsifier surfactant molecules [20]. This process may be seen under a microscope where the interface of approaching droplets becomes flat [21]. Photomicrography as the demulsification process procceeds would show how the demulsifier facilitates sliding of water droplets. This facile sliding is a result of decreased o/w interfacial roughness and the physicochemical role of the demulsifier.

9.3.2   Mechanism of Demulsification

It is an unfortunate fact that there exists no single coherent theory regarding the mechanism of demulsification. Demulsification is believed to be accompanied by inversion followed by coagulation. It is important to note that the coagulation of the dispersed phase occurs as a two-stage process: flocculation and coalescence. In the second stage, termed coalescence, aggregates combine to form a single drop. This is an irreversible process, leading to a decrease in the number of water droplets and finally to complete demulsification [22]. A third new stage was proposed [23] as a final stage in completion of demulsification process and will be discussed later.

9.3.3   Polyoxyethylene and Poly (Oxyethylene-co-Oxypropylene) Nonionic Surfactants as Demulsifiers

Surfactants show characteristic molecular structure (amphipathic structure) by combining two structurally dissimilar groups with opposing solubility tendencies. The opposing groups within the surfactant molecule are usually designated as hydrophobic and hydrophilic groups. The hydrophobic moiety in synthetic surfactants may be produced from petroleum-derived feedstocks or natural raw materials (vegetable- and animal-derived oils and fats, fatty acids, carbohydrates, etc.). Oil-bearing seeds are used here as the main source for hydrophobic moieties.

The hydrophilic part of the surfactant is considered as the main feature that classifies surfactants into cationic, anionic, nonionic, and amphoteric categories. Nonionic surfactants bear no apparent ionic charge on the molecule. Hydrophilicity of this class is provided by hydrogen bonding of oxygen atoms, hydroxyl groups, etc., with water molecules [24]. Polyoxyethylene [poly(ethylene glycol)] can be easily attached to a variety of hydrophobic organic molecules by ethoxylation of molecules containing active hydrogen atoms or direct reaction with polyoxyethylene if possible. The produced structures can be solubilized in aqueous medium through hydrogen bonding with water molecules. To obtain complete miscibility with water at room temperature, ca. 60–75 wt% of polyoxyethylene content is needed in most nonionic molecules. Also, poly (oxyethylene-co-oxypropylene) P(EO-PO) are used as hydrophilic moiety; these block polymers (BP) are not strongly surface active but they exhibit commercially useful surfactant properties. A great variety of P(EO-PO) nonionic surfactants are marketed under the trademark Pluronic polyol. Synthesis of P(EO-PO) block copolymers varies according to the required end product. However, polyethylene oxide (PEO) moiety is considered as a hydrophilic part, while polypropylene oxide (PPO) is the hydrophobic one. The hydrophobicity of PPO is ascribed to little hydrophilic contribution of an oxygen atom that constitutes only a small proportion of the whole molecule.

First, soybean, linseed, and castor oils were hydrolyzed separately using KOH [24]. Then, the resulting fatty acids were individually reacted with equimolar quantity of maleic anhydride (MA). Thus, 1 mole of each adduct was esterified with 1 mole of different molecular weights poly(ethylene glycol) (PEG) (PEG 2000, PEG 4000) or P(EO-PO) block copolymer (BP 3000, BP 6400 and BP 14500) using xylene as a solvent and 1% p-toluene sulphonic acid as a catalyst. The reaction was carried out until the theoretical amount of water was collected azeotropically forming hemiesters. The progress of the reaction was followed using Fourier Transform Infrared Spectroscopy (FTIR) spectroscopy. Samples were coded as XY_n, where X: represents the name of the hydrolyzed oil (soybean, S; linseed, L; or castor, C), Y: represents PEG or BP (P or B, respectively) and n represents the molecular weight of PEG or BP (viz., 2 for 2000, 3 for 3000, 4 for 4000, 6 for 6400 and 14 for 14,500 mol^{-1}). The general chemical structures of the prepared demulsifiers based on P(EO-PO) and PEO are shown in Figure 9.1.

9.3.4   FACTORS AFFECTING THE DEMULSIFICATION PROCESS

9.3.4.1   Effect of Demulsifier Chemical Structure

The formulations of demulsifiers are selected to provide specific properties including HLB, stability, rate of diffusion to the interface, and effectiveness of destabilizing the interface. However, the selection of the demulsifier can be made by applying standard testing method for water and sediment in crude oil (Basic Sediment and Water [BS&W]) (American Society for Testing and Materials [ASTM] D 4007) [25] or using the most common bottle test.

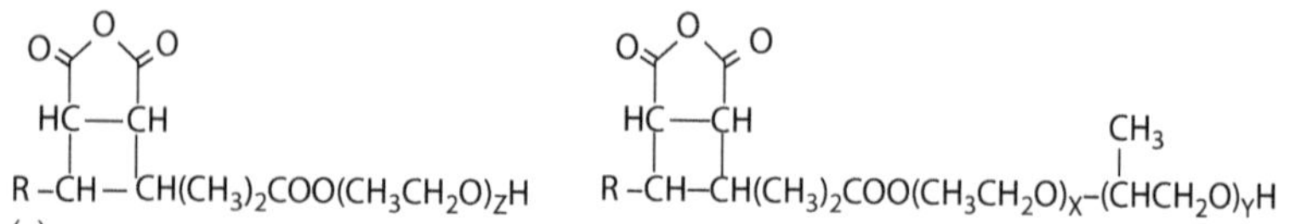

(a)

(b)

where: R represents

i. $CH_3CH_2CH=CHCH_2CH=CHCH_2$ For hydrolyzed linseed oil (52%)

    OH
    |
ii. $CH_3(CH_2)_5CHCH_2$ For hydrolyzed castor oil (87%)

iii. $CH_3(CH_2)_4CH=CHCH_2$ For hydrolyzed soybean oil (51%)

(a) represents polyethylene oxide (PEO) based demulsifiers, where:

z denotes ethylene oxide repeating units in PEO

for mol. wt. 2000 mol^{-1}: z = 45.4

mol. wt. 4000 mol^{-1}: z = 90.9

(b) represents poly (ethylene oxide-co-propylene oxide) (P(EO-PO)) based demulsifiers, where:

x and y represent numbers of repeating EO and PO units along P(EO-PO) with different molecular weights as follows

for mol. wt. = 3,000 mol^{-1}: x = 24 and y = 34

mol. wt. = 6,400 mol^{-1} : x = 44 and y = 76.9

mol. wt. = 14,500 mol^{-1} : x = 44 and y = 216.6

FIGURE 9.1 Main chemical structures for (a) PEO- and (b) P(EO-PO)-based demulsifiers.

It is well known that when a surfactant is brought into contact with a water–oil system, surfactant molecules have the opportunity to pass into both phases. For a system at equilibrium, a state of lower energy for the surfactant molecule would be the one in which it straddles the interface, with the polar group remaining in the water and the hydrophobic tail in the oil. As the concentration of surfactant at the interface becomes high, the adjacent molecules would tend to orient themselves to be substantially parallel to each other [26]. Lu et al. [27] have focused on the relative positions of alkyl chain and oxyethylene chain of octaethylene glycol monododecyl ether and the widths of their distributions normal to the air–water interface. For this purpose, they have used neutron specular reflection in combination with isotopic labeling. When surfactant molecules pack together at the interface forming a mono-layer, they do not act independently and intermolecular interactions are often quite strong between neighboring molecules [26]. The hydrophobic interactions between adjacent alkyl groups provide a force tending to pack the molecules closer together. The hydrophilic groups, on the other hand, have a strong affinity for water that there is tendency for them to be spaced out to allow as much water as possible to solvate each head group.

Separation times required for complete demulsification by the prepared demulsifiers based on soybean (S) and linseed (L) oils at 300 ppm for 30:70 water: crude oil (Type X (where X is II)) emulsions are shown in Figures 9.2 and 9.3, respectively. From these figures, it is obvious that the demulsifiers containing only ethylene oxide groups (PEG) as a hydrophilic moiety are less effective than those containing

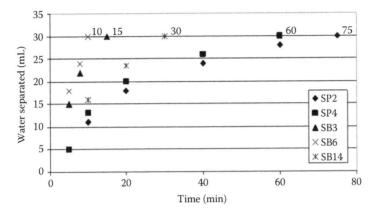

FIGURE 9.2 Demulsification (bottle test) by hydrolyzed soybean oil-based demulsifiers at 300 ppm for 30% water content in type X (where X is II) oil emulsion at 60°C. Numbers next to plotted symbols are the measured time, in minutes, for complete (100%) demulsification.

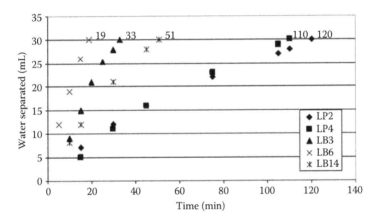

FIGURE 9.3 Demulsification (bottle test) by hydrolyzed linseed oil-based demulsifiers at 300 ppm for 30% water content in type X (where X is II) oil emulsion at 60°C. Numbers next to plotted symbols are the measured time, in minutes, for complete (100%) demulsification.

both EO and PO groups (BP). This is probably due to the presence of repeated PEO chains through each block copolymer (BP) molecule that facilitates collecting more water droplets in close proximity to each other—along each polymer chain—than the PEG-ended demulsifier molecule can do. A proposed mechanism is depicted in Scheme 9.1. In the proposed mechanism, the BP structure may provide opportunity for dispersed water droplets to form clusters more readily.

Also, it can be seen that the increase in (EO) content is accompanied by an increase in the demulsification efficiency. This behavior may be due to more suitable amphiphilic structure of the demulsifier molecule that permits effective partitioning at w/o interface and hence more effective adsorption. As a result of efficient adsorption,

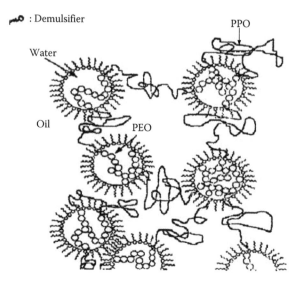

SCHEME 9.1   Schematic diagram of a PEO–PPO block copolymer based demulsifier show-ing accumulation of water around PEO (hydrophilic part) while PPO (hydrophobic part) is located in the oil phase.

better demulsification performance is expected. This observation is in agreement with that reported by Shetty et al. [28], that is, good performance of a demulsifier is related to high percentage of hydrophilic content. On the other hand, increasing the molecular weight of BP in the demulsifier molecule from 3000 to 6400 as in LB_3 and LB_6, respectively, shows an increase in the demulsification efficiency. This behavior may be ascribed to the increasing probability of water droplet numbers that coalesce forming channels and terminate demulsification of the emulsified water much faster. A further increase in the BP molecular weight up to 14,500 as in LB_{14} causes a decrease in the demulsification efficiency (see Figure 9.3). This is probably due to the fact that high increase in the BP molecular weight causes more coiling of the polymer chain. This will lead to a decrease in surface concentration and thus a less demulsification efficiency. It is worth mentioning that LB_6 gives 100% efficiency at 300 ppm in 19 min (see Figure 9.3). This fast demulsification at reasonable concen-tration is considered as a desirable step in preparing an effective demulsifier based on available crude materials (without purification).

Figure 9.4 shows the demulsification results of the demulsifiers based on castor oil (C). It reveals that these demulsifiers show the same performance trend as those based on soybean (S) and linseed oils (L) regarding the hydrophilic moiety of the demulsifier. This similarity is due to similar structures. Comparing the time required for complete demulsification, castor-based demulsifiers are less effective than the others (Figures 9.2 through 9.4). In this case, the composition of the hydrolyzed fatty acids plays an important role, as the main fatty acid with highest constituent percent is considered as the effective component. Hydrolyzed soybean oil contains 51% lin-oleic acid, while hydrolyzed linseed oil contains 52% linolenic acid and hydrolyzed castor oil contains 87% ricinoleic acid [29].

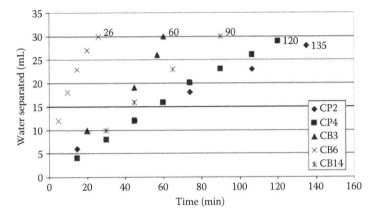

FIGURE 9.4 Demulsification (bottle test) by hydrolyzed castor oil-based demulsifiers at 300 ppm for 30% water content in type X (where X is II) oil emulsion at 60°C. Numbers next to plotted symbols are the measured time, in minutes, for complete (100%) demulsification.

This behavior is explained on the basis of aqueous solubility, functional group, and its polarity. As well known, a surfactant molecule with its amphipathic structure has a property of adsorbing onto w/o interfaces till reaching a certain critical concentration. However, demulsifiers are designed to reduce emulsion stability by displacing or destroying the effectiveness of the original stabilizing agents (such as asphaltenes) at the interfaces. The polarity of the demulsifier is an important parameter affecting the adsorption of surfactants at the interface since demulsifiers are usually added to the continuous phase (crude oil) within which they must diffuse to the interface and disrupt the stabilizing interfacial film. The polar functional groups (here, the hydroxyl groups of the hydrolyzed castor oil backbone containing ricinoleic acid as a main component) should aid adsorption at the interface. In fact, the position of the side hydroxyl group (in the hydrolyzed castor oil backbone) away from the remaining hydrophilic part can cause disturbance in the mode of demulsifier packing on the interface. Consequently, low demulsification efficiency is observed for this series. Although (C)-based demulsifiers are of low activity, CB_6 shows a fast 100% demulsification at 300 ppm after 26 min (see Figure 9.4), that is, it is an efficient demulsifier that can be used in practical applications.

9.3.4.2 Effect of Emulsion Water Content

Water content in a w/o emulsion is one of the important factors affecting demulsification efficiency or stability of an emulsion in general. The persisting emulsions not only depend on asphaltene, resin, and paraffin but also on water content [5]. However, at low water content, the external pressure of the continuous phase (oil) is greater than the internal pressure of water droplets [22]. This leads to an increase in the mechanical stability of the interface film, so that breaking of this strong interface film will need high temperature and a strong demulsifier. On the other hand, at increasing water content, the external pressure of water phase becomes less than the internal pressure [22]. In a regular w/o emulsion, the maximum stability of the emulsion will occur

TABLE 9.1
Effect of Water Content on Demulsification Efficiency of LB$_6$ Demulsifier for Type X (Where X Is II) Emulsions at 60°C

Water Content (%)	LB$_6$ Concn. (ppm)	Coalescence Rate Constant (min mL^{-1})	Separation Time (min)	Demulsification Efficiency (%)
30	25	0.4745	61	100
	50	0.5301	53	100
	100	0.7817	37	100
	200	1.1266	24	100
50	25	1.3426	40	100
	50	1.6119	29	100
	100	2.2419	21	100
	200	2.5931	17	100
70	25	2.2091	34	100
	50	3.5696	21	100
	100	4.0799	16	100
	200	4.5135	14	100

at a particular ratio of water to oil. Typically, this maximum is found at low water contents as water droplets have only a small chance of colliding and coalescing with each other. Increasing the water percentage may destroy the stability of the emulsion.

This behavior is evident upon investigating a selected demulsifier, namely LB$_6$ with 30%, 50%, and 70% water-to-oil ratio. Values of separation rate constant, time for maximum separation, and demulsification efficiency in presence of different concentrations of LB$_6$ at the selected water contents are listed in Table 9.1.

The data in Table 9.1 indicate that the time taken for complete water separation and the required demulsifier dose generally decreased with increasing water content for all demulsifier concentrations. For example, Table 9.1 shows that 100% demulsification efficiency was exhibited after 37, 21, and 16 min with 30%, 50%, and 70% water contents, respectively, at 100 ppm, whereas the same efficiency was obtained at 53, 29, and 21 min with 30%, 50%, and 70% water contents, respectively, at 50 ppm. This may be explained by the reduction of the interfacial film pressure in the presence of the demulsifier or at least the film is no longer as rigid as in its absence.

9.3.4.3 Effect of Asphaltene Content

The stability of the emulsion is controlled by the type and amount of indigenous surfactants and/or finally divided solids that act as emulsifying agents. These emulsifiers form interfacial films around the droplets of the dispersed phase and create a barrier that slows down or prevents coalescence of the droplets. The displacement of the asphaltenic film at the w/o interface by the demulsifier is an important factor in destabilizing this film.

The effect of asphaltene content (9.45%, 7.14%, and 0.723%) on the stability of w/o emulsion in the absence and presence of LB$_6$ is illustrated in Table 9.2. The

TABLE 9.2

Effect of Asphaltene Content on Demulsification Efficiency of LB$_6$ Demulsifier (Type X (Where X Is I, II or III)) Oil Emulsion with 30% Water Content) at 60°C

Demulsifier	Asphaltene Content (%)	Concn (ppm)	Coalescence Rate Constant (min mL^{-1})	Separation Time (min)	Demulsification Efficiency (%)
Control	9.45 (Type I)	0	—	106,560	81
LB$_6$		100	0.3843	91	100
		200	0.5515	61	100
		300	0.4923	57	100
Control	7.18 (Type II)	0	—	61,920	30
LB$_6$		100	0.7817	37	100
		200	1.1266	24	100
		300	1.3025	19	100
Control	0.723 (Type III)	0	0.2295	165,600	90
LB$_6$		10	0.5516	30	80
		20	0.9171	19	86.6
		50	1.3638	10	93.3

control samples show incomplete demulsification even at very low asphaltene content (0.723%). Only 30% of the emulsified water was separated after 43 days for a control sample of type X (where X is II) emulsion (7.18% asphaltene). The data in Table 9.2 show that the time taken for complete water separation increases with increasing asphaltene content. On adding 100 ppm of LB$_6$ to type X (where X is I) emulsion (9.45% asphaltene), complete demulsification of the emulsified water is observed after 91 min, while on its addition on type X (where X is II) emulsion (7.14% asphaltene) the separation completes after 37 min. Generally, the hydrodynamic force between the water droplets and asphaltenes present at the interface governs the eventual rupture of w/o emulsion. The asphaltene is responsible for increasing the stability of emulsion by forming this stable mechanical interface and decreasing the hydrodynamic force between water droplets and asphaltene [30]. Consequently, the ability of the demulsifier to migrate through the oil phase to adsorb on the water droplet interface will be hindered. Accordingly, the rupture of the interface and draining of water become more difficult. Interestingly, one can notice that crudes with low asphaltene content generally have a lower viscosity (see Table 9.3). Wax does have a large impact on the solution viscosity, but it has a rather low melting point. Thus working at this high temperature through the demulsification (60°C, an equivalent temperature to that of the heater used at process area) would cancel the effect of wax content on viscosity. However, the increase in viscosity of the oil phase on using crudes with higher asphaltene content would consequently hinder the demulsification process. The increase in concentration of the injected demulsifier leads to better demulsification. For example, type X (where X is II) emulsion show decreased separation time on increasing LB$_6$ demulsifier concentration from 100 to 300 ppm (see Table 9.2).

TABLE 9.3

Physicochemical Properties of Crude Oils Used

Specification	Method	Crude Oil Types		
		I	II	III
Specific gravity at 60°F	IP[a] 160	0.793	0.874	0.811
API gravity at 60°F	IP 160	36.29	40.96	42.6
Kinematic viscosity at 40°C in centistokes	IP 71	295.0	222.8	9.2
Asphaltene content (wt%)	IP 143	9.45	7.14	0.723
Paraffin wax (wt%)	UOP[b] 46	2.86	3.1	16.1
Water content (%)	IP 74/70	0.5	0.5	0.4

[a] Institute of Petroleum (IP) testing methods for petroleum and fuel products.
[b] Universal Oil Products testing methods.

9.3.4.4 Effect of Demulsifier Solvent

The demulsifier is a viscous material and is diluted with an organic solvent. This organic solvent is seen as having a viscosity-reducing role, yet it can also have a positive benefit to emulsion breaking if selected correctly. There is little reason to assume that the solvent affects the interface between the crude oil and the emulsified water droplets in an emulsion but there is strong evidence that the solvent affects the demulsifier molecules. Even if the solvents were equally soluble in the crude oil, it is possible to predict how the surface active behavior of the demulsifiers can differ. The highly aggregated demulsifier molecules will not be as surface active as those of low aggregation. Xylene is more compatible with crude oil than propanol. We can, therefore, explain how xylene, which is more soluble in crude oil, gives a more uniform distribution of low or unaggregated demulsifier molecules, which are able to adsorb readily at a crude oil–water interface. Furthermore, solvents that are added to chemical demulsifiers to reduce their viscosity for easy handling/delivery can have a very significant effect on the final efficiency of the demulsifier through facilitating the latter solubility in the continuous phase of w/o emulsion [31].

9.3.4.5 Effect of Using Demulsifier Blends

It was observed that sometimes mixtures of demulsifiers were more effective than when used individually [32]. Practically, formulations of different types of demulsifiers having varied functions such as decreasing crude oil salinity and viscosity and removing asphaltenes are used. However, synergism between demulsifiers used in a formulation should be examined well. Also, compatibility between the demulsifiers used in the formulation enhances demulsification efficiency. For example, demulsifiers with varied HLB or wide distribution of molecular weights are formulated to provide specific properties including solubility, rate of diffusion into the interface, and effectiveness in destabilizing the interface. In this case, each component of the demulsifier possesses a different partitioning ability and a different interfacial activity due to various chemical structures or types. On the other hand, conflict within a formulation can lead to less demulsification efficiency. For example, blending demulsifiers with bulky structures may lead to competition during diffusion step to the interface and altered packing geometry of the adsorbed molecules.

9.3.4.6 Effect of Demulisifer Surface Activity

The surface active properties of the demulsifier used determine how fast the surfactant would orient itself at the interface, which is an important factor in determining its effectiveness.

9.3.4.6.1 Micelle Formation

Micelle formation is a property of surfactants that may be as fundamental as their property of being adsorbed at interfaces. It is the property of forming colloidal sized clusters in solution. Micelle formation, or micellization, is an important phenomenon not only because a number of important interfacial phenomena, such as detergency and solubilization that depend on the existence of micelles in solution, but also because it affects other interfacial phenomena, such as surface or interfacial tension reduction that do not directly involve micelles [33].

Changes in temperature, concentration of surfactant, additives in the liquid phase, and structural groups in the surfactant all may cause change in the size, shape, and aggregation number of the micelle, with a structure varying from spherical through rod- or disklike to lamellar in shape [34].

The physicochemical properties of surfactants vary markedly above and below a specific surfactant concentration called critical micelle concentration (cmc) [30,35].

Below the cmc value, the physicochemical properties (e.g., conductivities, electromotive force, etc.) of ionic surfactants like sodium dodecylsulfate, resemble those of a strong electrolyte. Above the cmc value, these properties change dramatically, indicating occurrence of highly cooperative association process.

The nature and limits of applicability of specific methods for determining cmc vary widely. Fluorescence spectroscopy, nuclear magnetic resonance spectroscopy, conductivity, calorimetry, surface tension, and foaming are the most common cmc determination methods. Many of these methods have been reviewed by Mukerjee and Mysels [36]. The determination of cmc at elevated temperature and pressure is experimentally much more difficult than for ambient conditions and comparatively little work has been done in this area. For example, cmc values at up to 166°C have been reported by Evens and Wightman [37]. Some cmc values using calorimetry up to 200°C have been reported by Noll [38] and using fluorine-19 nuclear magnetic resonance (^{19}FNMR) up to 180°C by Shinoda et al. [39]. Surface tension measurement is the classical method for determining cmc but many surface tension methods are not suitable for use with aqueous solutions at elevated temperatures [40,41]. Conducting cmc determination at elevated pressures and temperatures is even more difficult and only a few studies have been reported, mostly employing conductivity methods [42,43]. Morrison et al. [44] describe a dynamic foam height method for the determination of cmc, which is suitable for use at high temperature and pressure. This method is much more rapid than the surface tension method.

Different amounts of LB_6 were dissolved in double distilled water and their surface tension was determined at 25°C using the Du Nouy tensiometer. The cmc of LB_6 was determined by the method adopted by Larinov [45]. The surface tension versus concentration (SVC) curves were plotted for the prepared surfactants at different temperatures. The cmc values were determined from the abrupt change in the slope of SVC curves.

9.3.4.6.2 Thermodynamics of Micellization and Adsorption of Demulsifiers

The values of cmc together with surface tension versus the bulk concentration were utilized to calculate Gibbs free energy of micellization (ΔG_{mic}) and adsorption (ΔG_{ad}) for LB_6 (see Table 9.4).

The ability for micellization depends on the thermodynamic parameter (standard free energy, ΔG_{mic}). Most information on the free energy of micellization has been obtained indirectly using Equation 9.1 through determination of cmc value for the required system [40].

$$\Delta G_{mic} = 2.3 \ RT \ (1-\alpha) \log (cmc), \tag{9.1}$$

where

ΔG_{mic} is the standard free energy of micellization in kJ mol^{-1}
R is the universal gas constant (R=3.418 J mol^{-1} K^{-1})
T is the absolute temperature in °K (T=t°C+273)
α is the fraction of counterions bound by the micelle of ionic surfactants (α=0 for nonionic surfactants) and cmc is the critical micelle concentration in g mol^{-1}

Many investigators have dealt with the thermodynamics of surfactant adsorption at interfaces [46]. ΔG_{ad} values were calculated via Equation 9.2

$$\Delta G_{ad} = \Delta G_{mic} - 0.6023\pi_{cmc}A_{min}, \tag{9.2}$$

where

ΔG_{ad} is the standard free energy of adsorption in kJ mol^{-1}
π_{cmc} is the surfactant adsorption effectiveness in mN m^{-1}
A_{min} is the minimum area per surfactant molecule at the oil–water interface in nm

TABLE 9.4
Some Surface Active Properties of LB_6

Property	Value
IFT (mN m^{-1})	3×10^{-3}
K_p (partition coefficient)	0.612
ΔG_{mic} (kJ mol^{-1})	−25.08
ΔG_{ad} (kJ mol^{-1})	−25.783

Some surface active properties (cmc, interfacial tension (IFT) and partition coefficient (K_p), ΔG_{mic}, ΔG_{ad}) of LB_6 are listed in Table 9.4. It is noticed that this demulsifier favors adsorption over micellization. More negative $\Delta G_{ad} = -25.783$ kJ mol^{-1} compared to -25.08 kJ mol^{-1} for ΔG_{mic} indicates more feasible adsorption of the demulsifier at o–w interface. This behavior supports its efficiency as a demulsifier regarding its ability to displace natural emulsifiers. Also, there is a significant reduction in IFT of crude oil/water emulsion by adding 100 ppm of LB_6, reaching 3×10^{-3} mN m^{-1} (see Table 9.4). This low IFT leads to a decrease in the stability of the interfacial film. Consequently, the drainage of water would be easier and hence promoting coalescence.

Spinning drop interfacial tensiometer equipped with a built-in occulometer (an ocular lens having a two dimensional scale meter) was used for measuring the interfacial tension between crude oil and demulsifier. The demulsifiser was dissolved in distilled or seawater at a concentration of 100 ppm and 30°C. The denser phase (oil phase) was injected by means of microliter syringe to a spinning capillary tube containing the surfactant solution at 30°C and 3000 rpm speed of rotation. As the oil drop elongated, its diameter was measured by the occulometer. The interfacial tension γ was calculated from Equation 9.3.

$$\gamma = 3.427 \times 10^{-7} (0.7d)^3 \times n^3 \times \Delta\rho, \tag{9.3}$$

where
  γ is the interfacial tension in mN m^{-1}
  d is the diameter of the drop in mm
  n is the speed of rotation in rpm
  $\Delta\rho$ is the density difference between the oil phase and the surfactant aqueous solution in g/cm^3

9.3.4.6.3 Hydrophile–Lipophile Balance Number of Demulsifier

The HLB concept, first proposed by Griffin [47], is considered to be an important parameter for selecting suitable demulsifier for breaking w/o emulsions [48,49]. Nonionic surfactants offer certain advantages over conventional ionic ones. The most important one is the ability to control the hydrophilicity of the surfactant. This is particularly true for polyoxyethylene compounds in which there is a possibility of varying the length of the polyoxyethylene chain and consequently modifying the HLB.

Attempts have been made to arrive at some correlation between demulsifier performance and properties such as molecular structure, HLB, interfacial tension, interfacial viscosity, partition coefficient, dynamic interfacial tension, and relative solubility number [4,5]. But no generalized conclusions have been reached.

There have been numerous attempts to formulate simple rules concerning emulsion stability. Historically, the first one was Bancroft's rule [50], which states that a stable emulsion is obtained when the surfactant is soluble in the continuous phase. A more sophisticated criterion was proposed by Griffin [47] who introduced the concept of HLB.

It has been found that the efficiency of a demulsifier may be related to its HLB value [48]. Commercially used demulsifiers for w/o emulsions are oil soluble, that is, having low HLB (4–10). However, HLB is just an indicator of the preferentiality of a demulsifier but not its efficiency. Different formulas for calculating HLB numbers are available. For example, Davies and Rideal give the following expression [51]:

$$HLB = 7 + (\text{number of hydrophilic groups}) - 0.475 n_c, \qquad (9.4)$$

where n_c is the number of $-CH_2$ groups in the lipophilic part of the molecule.

HLB numbers for normal nonionic surfactants were determined by simple calculation [14].

A commonly used formula for nonionics is

$$HLB = \frac{M_H}{M_H + M_L} \times 20, \qquad (9.5)$$

where

M_H is molecular weight of hydrophilic portion of the molecule
M_L is molecular weight of lipophilic portion of the molecule

For alcohol ethoxylates and alkyl phenol ethoxylates

$$HLB = \frac{e.o.\%}{5}, \qquad (9.6)$$

where e.o. % is the percent of ethylene oxide in the molecule
For fatty acid esters of polyol,

$$HLB = 20 \left[1 - (\text{saponification number/acid number}) \right] \qquad (9.7)$$

When the demulsifier is initially introduced to the w/o emulsion, it will be thermodynamically stable at the interface of the water droplets. In this respect, the surfactants possessing high HLB migrate faster to the interface than those having low HLB. As a result of such enhanced migration toward the interface, the surfactant forms a continuous aqueous environment (like channels) between the dispersed water droplets. This leads to rupture of the interfacial oil film surrounding the water droplets.

9.3.4.6.4 Partition Coefficient

In certain cases, some components of an oil-soluble demulsifier can be dissolved (partitioned) into the water drop phase (dispersed phase); this process is known as partitioning. When the demulsifier components exist in the dispersed (droplet) phase, the partitioning of demulsifier components in the droplet phase can strongly affect the w/o dynamic interfacial properties, such as the interfacial tension [37].

Partition coefficient (K_P) is defined as the equilibrium ratio of the demulsifier concentration in the water phase (C_W) to the demulsifier concentration in the oil phase (C_O) [47]—that is,

$$K_P = \frac{C_W}{C_O},\tag{9.8}$$

where
  C_W is the surfactant concentration in water phase in mg L^{-1}
  C_O is the surfactant concentration in oil phase in mg L^{-1}

Krawczyk et al. [30] identified partitioning of the demulsifier as an important parameter in the mechanism of demulsification. They concluded from experimental evidence that a demulsifier with a partition coefficient close to unity has a better performance. Bhardwaj and Hartland [52] found that partitioning of demulsifier effectively reduces interfacial tension and increases demulsification performance by rapid adsorption of the demulsifier at the w/o interface. It is generally believed that the most effective demulsifier is the one that partitions equally into the water and oil phases. This balance would lead to a maximum in the surface adsorption of demulsifier and a minimum in the interfacial tension [53]. Shetty et al. [28] found that a water-soluble demulsifier could effectively destabilize the w/o emulsions. They studied the effects of demulsifiers with varying HLB on the destabilization of w/o emulsions and concluded that a demulsifier could have a very good performance when the demulsifier contains a high percentage of hydrophilic groups (high HLB number) [22].

Usually the demulsifier adsorption at oil–water interface is diffusion controlled, and the diffusivity of the demulsifier can be described as a mean value of diffusivities in the oil and water phases. If the demulsifier is soluble in both oil and water phases (i.e., the demulsifier is partitioned), then the diffusivity (D) can be expressed as (see Equations 9.9 through 9.11) [54]

$$D = \left(\sqrt{\frac{C_o}{C_o + C_w}} D_o + \sqrt{\frac{C_w}{C_o + C_w}} D_w \right)^2,\tag{9.9}$$

where
  D is the diffusion coefficient of the demulsifier in cm^2 s^{-1}
  C_o and C_w are the demulsifier concentrations (mg L^{-1}) in oil and water phases, respectively
  D_o and D_w are the diffusion coefficients (cm^2 s^{-1}) in oil and water phases, respectively

$$D = \frac{1}{1 + K_p}\left(D_o + K_p D_w + \sqrt{K_p D_o D_w} \right)\tag{9.10}$$

$$D \approx \frac{1}{1+K_p}\left(D_o + K_p D_w\right) \qquad (9.11)$$

When the demulsifier is partitioned it means that it can transfer onto the interface from both water and oil phases with relatively high diffusion rate [4,52,55]. Good demulsifier partitioning at an oil–water interface indicates its ability to orient itself in either water or oil. K_p for a demulsifier approaching unity is hypotheosized by Berger et al. [53] to optimize demulsifier performance. However, if a demulsifier is insoluble in the water phase, the coefficient K_p is small and the demulsifier adsorption rate will depend on the oil phase diffusivity (D_o) in the diffusion-controlled adsorption process.

LB$_6$ shows K_p value of 0.612 a value near unity (see Table 9.4). This value indicates the ability of the demulsifier to partition well at the w/o interface.

9.4 RHEOLOGICAL PROPERTIES IN RELATION TO DEMULSIFICATION EFFICIENCY

Rheology of injected emulsion samples (with demulsifiers) indicates the effect of demulsifier surfactant on the flow properties of the continuous phase (crude oil). A high bulk oil viscosity hinders drop settling velocities and limits the rate of coagulation and coalescence [56]. Droplet coalescence requires coagulation followed by the drainage of the dividing continuous phase film. Lowering the viscosity by adding a demulsifier leads to favorable collisions, coalescence, and thus increasing the rate of water separation. Jones et al. [57] surmise that in the early stages of demulsification process, the rate of film drainage will be primarily governed by the bulk phase viscosity but as the film thins, the interfacial viscosity becomes increasingly influential. For a certain asphaltene content (7.14%), the effect of bulk viscosity of the emulsion on the emulsion stability was also studied using different concentrations of LB$_6$. Table 9.5 shows yield value (τ_b), apparent viscosity (η_{app}), and dynamic viscosity (η_d) for 30% w/o emulsion at different concentrations of LB$_6$ at varying temperature. These data show that apparent viscosity (η_{app}) and dynamic viscosity (η_d) are affected strongly by temperature, that is, they decrease with increasing temperature. For emulsions samples without demulsifier (Table 9.5), a significant drop in both η_{app} values from 350 to 50.6 and η_d values from ~2.048 to ~0.403 was observed on increasing temperature from 30°C to 60°C. The decrease in the emulsion viscosity is mainly due to a decrease in the continuous phase viscosity. On adding different concentrations of LB$_6$ to the control type X (where X is I) emulsion, a sudden decrease in τ_b, η_{app}, and η_d was observed. Only 100 ppm of LB$_6$ results in a decrease in τ_b by fourfold (see Table 9.5). By injecting the emulsions with increased concentrations of LB$_6$ it was found that τ_b, η_{app}, and η_d decreased slightly with increasing the concentration of the added demulsifier at the same temperature. This decrease in the viscosity may be explained on the basis of droplet size of the dispersed phase (water) [58,59]. In other words, it will depend on the viscosity of dispersion medium.

The mechanism of decreasing the continuous phase viscosity is related to demulsification progress as shown in Scheme 9.2. According to Scheme 9.2, in phase 1

TABLE 9.5

Rheological Parameters for Type X (Where X Is II) Crude Oil and Its 30% w/o Emulsion at 60°C without and with Different Concentrations of LB$_6$

Sample	Temp. (°C)	LB$_6$ Concn. (ppm)	τ_b^a	η_{app}^b	η_d^c
Control (crude oil)	30	—	16.4	111	0.7944
	40		10.6	71.67	0.522
	50		6.76	45	0.327
	60		4.18	26	0.172
30% w/o emulsion	30	—	56	350.0	2.048
	40		32	203.1	1.312
	50		19	125.9	0.88
	60		17	50.6	0.403
30% w/o emulsion with LB$_6$	30	100	13.52	179.49	0.135
	40		7.406	97.747	0.105
	50		4.733	59.777	0.099
	60		2.898	37.436	0.047
	30	200	10.62	138.67	0.346
	40		7.019	91.188	0.204
	50		4.347	54.711	0.124
	60		1.867	24.716	0.143
	30	300	10.30	134.37	0.588
	40		6.214	80.407	0.206
	50		4.121	52.598	0.085
	60		2.576	34.282	0.022

^{a} τ_b yield value.
^{b} η_{app} apparent viscosity.
^{c} η_d dynamic viscosity.

(emulsion with 50% water content at the initial stage of treating the emulsion with demulsifier), the droplets size is nearly equal and as the time progresses water droplets coalesce to reach coalescence stage, phase 2. By dynamic rotation, the draining of water droplets occurs, which is accompanied by a decrease in the dynamic viscosity and thus the yield value. This behavior continues until water clusters form (phase 3). Clusters combine to form channels and a complete phase separation occurs (phase 4). At this point, the viscosity of the system drops to nearly the original water-free crude oil viscosity (pay attention to viscosity values of injected emulsions with LB$_6$ and control samples in Table 9.5). From the tribology point of view, adding a demulsifier decreases the friction between water and crude oil droplets, which prevents aggregation of water and thus demulsification. Water separation out of the system enhances the flow of oil layers as seen through rheology parameters.

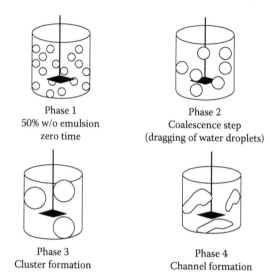

Phase 1
50% w/o emulsion
zero time

Phase 2
Coalescence step
(dragging of water droplets)

Phase 3
Cluster formation

Phase 4
Channel formation

SCHEME 9.2 Schematic diagram showing the three steps of demulsification (phase 2–4). In the last phase the viscosity equals that of pure oil due to complete separation of water.

9.5 KINETICS AND MECHANISM OF DEMULSIFICATION PROCESS

Kinetics of different demulsification stages were investigated using photomicrography of emulsion samples at different time intervals. Micrographs for blank emulsion (30% water content) and emulsions injected with 300 ppm of CB_{14} or LB_6 were taken under high power resolving microscope. Water droplets were sized per micrograph giving information about interfacial area of the dispersed phase. A specific surface area of the dispersed phase (S) was calculated using Equation (9.12). As coalescence of water droplets leads to a decrease in surface area of the dispersed phase, the variation of S with time was used to quantitatively study the kinetics of demulsification. The relation between interfacial area per gram of the dispersed phase (S) and time was followed and the mechanism of demulsification was determined using the following first-order rate equation (see Equation 9.13) [60]:

$$S = \frac{A}{M},$$

(9.12)

where
   A is the area of the dispersed phase
   M is the mass of the dispersed phase

$$\log (S) = -Kt + C,$$

(9.13)

where
   K is the rate constant
   t is the time
   C is a constant

Micrograph of a control emulsion sample was taken every 5 days (see Figure 9.5). These micrographs show small variation in water droplets size during 35 days. This may be due to the mechanical stability character of asphaltenes and resins on the interface film that hinders the drainage of the neighboring water droplets. This explanation is in agreement with the demulsification efficiency data obtained using the bottle test, Figures 9.1 through 9.3. CB_{14} and LB_6 were selected as models for studying kinetics and mechanism of demulsification as their separation times for complete demulsification are different (90 min for CB_{14} and 19 min for LB_6; see Figures 9.3 and 9.4). However Figures 9.6 and 9.7 show micrographs of 30% w/o emulsion samples injected with 300 ppm of CB_{14} and LB_6, respectively. For CB_{14}, Figure 9.6, the radius of water droplets generally starts to increase after 30 min, whereas the good LB_6 demulsifier shows a fast increase in the water droplet radius from 3 to 6 min. It should also be mentioned that no distinct water separation was observed for CB_{14} after 80 min (see Figure 9.5, which actually shows 100% efficiency after 100 min at 300 ppm dose). On the other hand, a significant water separation is shown by LB_6 after 24 min, a result that is comparable with that obtained by the bottle test, Figure 9.2.

Plot of log (S) versus time of demulsification for 30% water emulsions of a control sample and those injected with 300 ppm of CB_{14} and LB_6 are shown in Figures 9.8 through 9.10. The curves in these figures can be divided into three parts A, B, and C, which describe the demulsification mechanism. Part A refers to the rate of coalescence, part B indicates the rate of cluster formation, and part C manifests to water channel formation where at the end of this stage a complete water separation occurs. This mechanism can also be observed clearly under microscope. The increase in the volume of water droplets (as seen in slides 1–7 in Figure 9.6 and slides 1–6 in Figure 9.7) show coalescence and cluster formation. Whereas slides 8 and 9 in Figure 9.6 and slides 7 and 8 in Figure 9.7 show channel formation until complete demulsification is achieved.

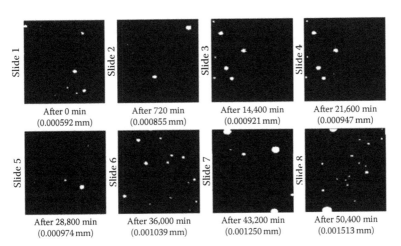

FIGURE 9.5 Micrographs for a control sample of type X (where X is II) oil emulsion at 30% water and 60°C (Photo/5 days). Numbers written at bottom left are to indicate average diameter of water droplet.

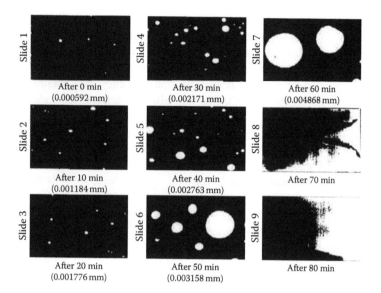

FIGURE 9.6 Micrographs for type X (where X is II) oil emulsion (at 30% water and 60°C) injected by 300 ppm of CB_{14}. Numbers written at bottom left are to indicate average diameter of water droplet.

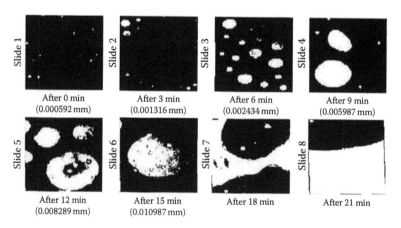

FIGURE 9.7 Micrographs for type X (where X is II) crude oil emulsion (at 30% water and 60°C) injected by 300 ppm of LB_6. Numbers written at bottom left are to indicate average diameter of water droplet.

The values of rate constants (k) for each stage (viz. coalescence, cluster, and channel formation) are listed in Table 9.6. The control emulsion sample exhibited very slow rates of coalescence (3.1×10^{-4} μm^2 mg^{-1} min^{-1}), cluster formation (3.9×10^{-3} μm^2 mg^{-1} min^{-1}), and channel formation (3.6×10^{-3} μm^2 mg^{-1} min^{-1}) after 35 days of incubation at 60°C. Emulsion sample injected with CB_{14} shows an increase in the rate constants for all three stages with a magnitude of the order

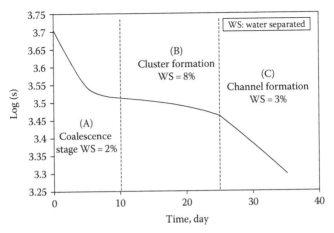

FIGURE 9.8 Variation of specific surface area of dispersed phase (S) with time for a blank sample (type X (where X is II) crude oil emulsion with 30% water at 60°C). The dashed lines separate the three stages of the demulsification process.

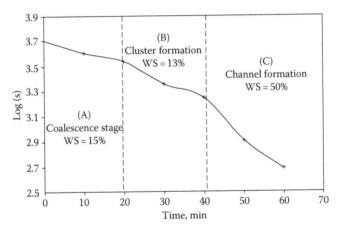

FIGURE 9.9 Variation of specific surface area of dispersed phase (S) with time for a type X (where X is II) crude oil emulsion (at 30% water and 60°C) injected by 300 ppm of CB_{14}. The dashed lines separate the three stages of the demulsification process.

of 10^2. LB_6 sample shows only two parts, that is, cluster and channel formation; in this instance, the coalescence and cluster formation overlapped in one fast stage. This may be attributed to the efficient demulsification power of LB_6. These observations can be translated into application in oil fields. By knowing the required time for complete demulsification, it is possible to specify the time needed for crude oil emulsion settling in settling tank (after which crude oil should be nearly free from any water traces).

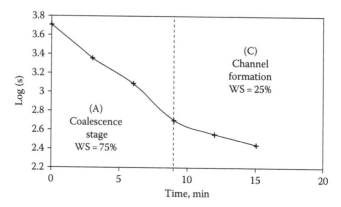

FIGURE 9.10 Variation of specific surface area of dispersed phase (S) with time for a type X (where X is II) crude oil emulsion (at 30% water and 60°C) injected by 300 ppm of LB_6. The dashed lines separate the stages of the demulsification process. (A) coalescence stage and (C) channel formation.

TABLE 9.6
Rate Constants of Demulsification Process for Control, CB_{14} and LB_6 Using 30% Water in Type X (Where X Is II) Oil Emulsions at 60°C

	Rate Constants (μm^2 mg^{-1} min^{-1})		
Sample	Coalescence Stage (A)	Cluster Stage (B)	Channel Formation (C)
Control	3.1×10^{-4}	3.9×10^{-3}	3.6×10^{-3}
CB14	2.4×10^{-3}	3×10^{-2}	1×10^{-2}
LB6	1.5×10^{-2}		1.2×10

9.6 CONCLUSIONS

The following conclusions are made from the present study:

1. Efficient demulsifiers for asphaltenic crude oil in water emulsions can be synthesized using commercial raw (without purification) materials and available techniques.
2. The increase in water content of the emulsion leads to fast demulsification and only low doses of the demulsifier are required.
3. The increase in the asphaltene content of the emulsion is one of the factors that is responsible for the stability of the w/o emulsion.
4. Low interfacial tension and low Gibbs free energy of adsorption are required for a good demulsifier.

5. The addition of the demulsifier may cause a decrease in the viscosity of the emulsion.
6. Kinetics of the demulsification process can be followed using photomicrography of the samples.
7. The demulsification process can be divided into three stages: coalescence, cluster formation, and channel formation.
8. Demulsifiers play an important role in friction modification at an o/w interface.

ACKNOWLEDGMENTS

The authors would like to gratefully acknowledge exploration laboratory—Egyptian Petroleum Research Institute (EPRI), Egypt—for photomicrographic analysis support.

REFERENCES

1. F.P. Beer and E.R. Johanston Jr., *Vector Mechanics for Engineers: Statics and Dynamics*, 6th edn., McGraw-Hill, New York (1996).
2. J.L. Meriam and L.G. Kraige, *Engineering Mechanics*, 5th edn., John Wiley & Sons, New York (2002).
3. I.C. Sharma, I. Haque, and S.N. Srivastava, Chemical demulsification of natural petroleum emulsions of Assam (India), *Colloid Polym. Sci.*, 260, 616–626 (1982).
4. A. Bhardwaj and S. Hartland, Study of demulsification of water-in-crude oil emulsion, *J. Dispersion Sci. Technol.*, 14, 541–557 (1993).
5. N.N. Zaki and A.M. Al-Sabagh, Efficiency of polyalkylphenols-polyalkylenepolyamines-formaldehyde ethoxylates as de-emulsifiers for water-in-crude oil-emulsions, *Tenside Surf. Det.*, 34, 12–17 (1997).
6. A.M. Al-Sabagh, S.A. Nehal, M.N. Amal, and A.M. Gabr, Synthesis and evaluation of some polymeric surfactants for treating crude oil—Part II. Destabilization of naturally occurring water-in-oil emulsions by polyalkylphenol formaldehyde amine resins, *Polym. Adv. Technol.*, 13, 346–352 (2002).
7. A.M. Al-Sabagh, A.M. Badawi, and M.R. Noor El-Din, Breaking of water-in-crude oil emulsions by novel demulsifiers based on maleic anhydride-oleic acid adduct, *Petroleum Sci. Technol.*, 20, 887–914 (2002).
8. A.M. Al-Sabagh, S.A. Nehal, M.N. Amal, and A.M. Gabr, Synthesis and evaluation of some polymeric surfactants for treating crude oil emulsions—Part I: Treatment of sandy soil polluted with crude oil by monomeric and polymeric surfactants, *Colloids Surf. A: Physicochem. Eng. Asp.*, 216, 9–19 (2003).
9. A.M. Al-Sabagh, M.R. Noor El-Din, and H.M. Mohamed, Effect of demulsifiers on surface active and rheological properties of water-in-oil emulsion in relation to the demulsification efficiency, *Egypt. J. Petroleum*, 15, 49–60 (2006).
10. A.M. Al-Sabagh, N.E. Maysour, M.N. Naser, and M.R. Noor El-Din, Synthesis and evaluation of some modified polyoxyethylene—polyoxypropylene block polymer as water-in-oil emulsion breakers, *J. Dispersion Sci. Technol.*, 28, 537–545 (2007).
11. A.M. Al-Sabagh, N.E. Maysour, and M.R. Noor El-Din, Investigate the demulsification efficiency of some novel demulsifiers in relation to their surface active properties, *J. Dispersion Sci. Technol.*, 28, 547–555 (2007).

12. A.M. Al-Sabagh, M.R. Noor El-Din, R.E. Morsi, and M.Z. Elsabee, Demulsification efficiency of some novel styrene-maleic anhydride esters copolymers, *J. Appl. Polym. Sci.*, 108, 2301–2311 (2008).

13. P. Rajinder, Rheology of blends of suspensions and emulsions, *Ind. Eng. Chem. Res.*, 35, 5005–5010 (1999).

14. P. Becher and M.J. Schick, *Macroemulsions in Nonionic Surfactants*, Surfactant Science Series, 23, Marcel Dekker Inc., New York (1987).

15. K.J. Liasant (ed.), *Emulsions and Emulsion Technology, Part III*, Surfactant Science Series, Vol. 6, Marcel Dekker, New York (1984).

16. J.D. McLean and P.K. Kilpatrick, Effects of asphaltene solvency on stability of water-in-crude-oil emulsions, *J. Colloid Interface Sci.*, 189, 242–253 (1997).

17. I.B. Ivanov, R.K. Jain, P. Somasundaran, and T.T. Traykov, The role of surfactants on the coalescences of emulsion droplets, In: *Solution Chemistry of Surfactants*, K.A. Mittal, Plenum Press, New York, Vol. 2, pp. 817–840 (1979).

18. W. Clayton, *The Theory of Emulsions and Their Technical Treatment*, 15th edn., Chemical Publishing Co., New York (1954).

19. L. Xia, S. Lu, and G. Cao, Stability and demulsification of emulsions stabilized by asphaltenes or resins, *J. Colloid Interface Sci.*, 271, 504–506 (2004).

20. O.P. Strausz, T.W. Mojelsky, and E.M. Lown, The molecular structure of asphaltene: An unfolding story, *Fuel*, 71, 1355–1363 (1992).

21. J. Djuve, X. Yang, I.J. Fjellanger, J. Sjoblom, and E. Pelizzetti, Chemical destabilization of crude oil based emulsions and asphaltene stabilized emulsions, *Colloid Polym. Sci.*, 279, 232–239 (2001).

22. Y.K. Kim and D.T. Wasan, Effect of demulsifier partitioning on the destabilization of water-in-oil emulsions, *Ind. Eng. Chem. Res.*, 35, 1141–1149 (1996).

23. R.A. El-Ghazawy, A.M. Al-Sabagh, N.G. Kandile, and M. Reyad, Synthesis and preliminary demulsification efficiency evaluation of new demulsifiers based on fatty oils, *J. Dispersion Sci. Technol.*, 31, 1423–1431 (2010).

24. L.N. Ferguson, The water solubilities of ethers, *J. Am. Chem. Soc.*, 77, 5288–5289 (1955).

25. Annual book of ASTM standards, American Chemical Society for Testing and Materials, *ASTM D4007-08 Standard Test Method for Water and Sediment in Crude Oil by the Centrifuge Method (Laboratory Procedure)*, Philadelphia, Vol. 5.02 (2008).

26. H.C. John, *Surfactant Aggregation*, Blackie, Chapman and Hall, New York (1992).

27. J.R. Lu, Z.X. Li, R.K. Thomas, E.J. Staples, L. Thompson, I. Tucker, and J. Penfold, Neutron reflection from a layer of monododecyl octaethylene glycol adsorbed at the air-liquid interface: The structure of the layer and the effects of temperature, *J. Phys. Chem.*, 98, 6559–6567 (1994).

28. C.S. Shetty, A.D. Nikolov, D.T. Wasan, and B.R. Bhattacharyya, Demulsification of water in oil emulsions using water soluble demulsifiers, *J. Dispersion Sci. Technol.*, 13, 121–133 (1992).

29. Z.W. Wicks, F.N. Jones, and S.P. Pappas, *Organic Coatings: Science and Technology, Vol 1: Film Formation, Components and Appearance*, Wiley Interscience, New York (1992).

30. M.A. Krawczyk, D.T. Wasan, and C. Shetty, Chemical demulsification of petroleum emulsions using oil-soluble demulsifiers, *Ind. Eng. Chem. Res.*, 30, 367–375 (1991).

31. B.P. Singh, Performance of demulsifiers: Prediction based on film pressure-area isotherms and solvent properties, *Energy Sources, Part A: Recov. Util. Environ. Effects*, 16, 377–385 (1994).

32. B.P. Singh and B.P. Pandey, Emulsification and demulsification study of crude oil-water system, *Res. Ind.*, 36, 203–207 (1991).

33. P.C. Hieiemenz and R. Rajagopalan, *Principles of Colloid and Surface Chemistry*, 3rd edn., Marcel Dekker, New York (1997).
34. S. Ikeda, S. Ozeki, and S. Hayashi, Size and shape of charged micelles of ionic surfactants in aqueous salt solutions, *Biophysical Chem.*, 11, 417–423 (1980).
35. D. Myers, *Surfactant Science and Technology*, VCH Publishers, New York (1988).
36. P. Mukerjee and K.J. Mysels, *Critical Micelle Concentrations of Aqueous Surfactant Systems*, National Bureau of Standards, NSRDS-NBS36, Washington, DC (1971).
37. D.F. Evans and P.J. Wightman, Micelle formation above 100°C, *J. Colloid Interface Sci.*, 86, 515–524 (1982).
38. L.A. Noll, The effect of temperature, salinity, and alcohol on the critical micelle concentration of surfactants, In: *Proceeding SPE International Symposium on Oilfield Chemistry, Society of Petroleum Engineering*, Richardson, Anaheim, CA (1991).
39. K. Shinoda, M. Kobayashi, and N. Yamaguchi, Effect of "iceberg" formation of water on the enthalpy and entropy of solution of paraffin chain compounds: The effect of temperature on the critical micelle concentration of lithium perfluorooctanesulfonate, *J. Phys.Chem.*, 91, 5292–5294 (1987).
40. L.L. Schramm, *Surfactants: Fundamental and Application in the Petroleum Industry*, Cambridge University Press, Cambridge, U.K. (2000).
41. E.N. Stasiuk and L.L. Schramm, The temperature dependence of the critical micelle concentrations of foam-forming surfactants, *J. Colloid Interface Sci.*, 178, 324–333 (1996).
42. C. La Mesa, B. Sesta, M.G. Bonicelli, and G.F. Ceccaroni, Phase diagram of the binary system water-(dodecyldimethylammonio) propanesulfonate, *Langmuir*, 6, 728–731 (1990).
43. G. Sugihara and P. Mukerjee, High-pressure study of micelle formation in aqueous solutions of sodium perfluorooctanoate, *J. Phys. Chem.*, 85, 1612–1616 (1981).
44. C. Morrison, L.L Schramm, and E.N. Stasiuk, A new method for the estimation of critical micelle concentrations at elevated temperature and pressures, *J. Petroleum. Sci. Eng.*, 15, 91–100 (1996).
45. L.I.O. Larinov, *Surface Chemistry*, Reinhold, New York (1962).
46. T.N. Castro Dantas, E.Ferreira Moura, H. Scatena Júnior, A.A. Dantas Neto, and A. Gurgel, Micellization and adsorption thermodynamics of novel ionic surfactants at fluid interfaces, *Colloids Surf. A: Physicochem. Eng. Asp.*, 207, 243–252 (2002).
47. W.C. Griffin, Calculation of HLB values of non-ionic surfactants, *J. Soc. Cosmet. Chem.*, 5, 249–256 (1949).
48. D.G. Cooper, J.E. Zajig, E.J. Cannel, and J.W. Wood, The relevance of "HLB" to de-emulsification of a mixture of heavy oil, water and clay, *Can. J. Chem. Eng.*, 58, 576–580 (1980).
49. J. Sjoblom, *Emulsions—A Fundamental and Practical Approch*, Kluwer Academic Publisher, Norway (1992).
50. W.D. Bancroft, The theory of emulsification, V, *J. Phys. Chem.*, 17, 514–519 (1913).
51. J.T. Davies and E.K. Rideal, *Interfacial Phenomena*, 2nd edn., Academic Press, New York (1963).
52. A. Bhardwaj and S. Hartland, Dynamics of emulsification and demulsification of water in crude oil emulsions, *Ind. Eng. Chem. Res.*, 33, 1271–1279 (1994).
53. P.D. Berger, C. Hsu, and J.P. Arendell, Designing and selecting demulsifiers for optimum field performance based on production fluid characteristics, *SPE Prod. Eng.*, 3, 522 (1987).
54. J.V. Hunsel and P. Joos, Stress-relaxation experiments at the oil/water interface, *Colloids Surf.*, 25, 251–261 (1987).
55. A. Bhardwaj and S. Hartland, Kinetics of coalescence of water droplets in water-in-crude oil emulsion, *J. Dispersion Sci. Technol.*, 15, 133–146 (1994).

56. R.C. Little, Chemical demulsification of aged, crude oil emulsions, *Environ. Sci. Technol.*, 15, 1184–1190 (1981).

57. T.J. Jones, E.L. Neustadter, and K.P. Whittingham, Water-in-crude oil emulsion stability and emulsion destabilization by chemical demulsifiers, *J Cand. Petroleum. Technol.*, 17, 100–108 (1978).

58. J.S. Chong, E.B. Christiansen, and A.D. Bear, Rheology of concentrated suspensions, *J. Appl. Polym. Sci.*, 15, 2007–2021 (1971).

59. A.B. Metzner, Rheology of suspensions in polymeric liquids, *J. Rheol.*, 29, 739–775 (1985).

60. A. Anil and S. Hartland, Kinetics of coalescence of water droplets in water-in-crude oil emulsions, *J. Dispersion Sci. Technol.*, 15, 133–146 (1994).

10 Antiwear and Antiseizure Properties of Aqueous Solutions of Alkyl Polyglucosides

Marian W. Sulek and Tomasz Wasilewski

CONTENTS

ABSTRACT

The objective of this study to investigate the antiwear and antiseizure properties of aqueous solutions of alkyl polyglucosides (APGs). The solutions, used
as lubricating substances, were prepared from two kinds of APGs differing in
the length of the alkyl chain. Antiwear properties were evaluated by means of
a ball-on-disk tribometer (T-11) and a four-ball machine (T-02). The measurements were conducted at constant loads of 5, 10, 20, 30, 40, and 50 N (T-11) and
1.5, 2.0, 2.5, 3.0, and 3.5 kN (T-02). The quantities characterizing antiseizure
properties, namely, scuffing load (P_t), seizure load (P_{oz}), and limiting pressure
of seizure (p_{oz}), were determined using the T-02 device and during the test the
load was increased from 0 to 7.2 kN at a steady rate of 409 N/s. The effects of
load, friction time, and the length of the alkyl chain in the APG molecules on
the tribological quantities measured were analyzed. The results obtained were
interpreted in terms of physicochemical properties of APG solutions, primarily their surface activity and aggregate-forming capability in the bulk phase
and in the surface phase.

10.1 INTRODUCTION

Water is a substance commonly found in nature. It is cheap, is readily available, and
does not have a harmful effect on living organisms; it is practically not hazardous
in any way to the natural environment. These factors have instigated us to consider
application of water as a base lubricant. Water, however, has a number of disadvantages: corrosive action, inadequate lubricating properties, a narrow range of working
temperatures, and low viscosity are the most important ones. These shortcomings
may be eliminated by a proper selection of friction pair materials or introduction of
appropriate additives. A number of publications on tribological properties of water
and water solutions of active additives can be found in the literature [1–25]. We
have also been doing intensive research on water-based lubricants for several years
[26–34]. An improvement in tribological properties of water can be achieved by
using, as additives, surface active agents from the following groups: anionic (alkyl
ether sulfates [26]), cationic (quaternary ammonium salts [27]), and nonionic (ethoxylated sorbitan esters [28,29], ethoxylated alcohols [30], ethoxylated oils [31], and
alkyl polyglucosides [APGs] [32–34]). We have focused primarily on the effect of the
structure and concentration of compounds in a lubricating substance and the effect of
the kind of the mesophases formed on various tribological characteristics: resistance
to motion, wear, and antiseizure properties [26–34].

Additives used in water-based lubricating substances must also fulfill ecological
criteria. They should not pose any hazard at the stages of production, application, and
utilization. Therefore, it is practically impossible to use traditional additives, which
are usually insoluble in water and frequently toxic. Moreover, the compounds should
also be cheap and, preferably, produced on an industrial scale. Another important
factor is that the additives should be efficient at low concentrations. These features
ensure profitability of using water as a lubricant base and, at the same time, increase
the probability of using it in industrial practice.

From among the compounds mentioned above, particular attention was paid to APGs, mainly because of the fact that they do not have a harmful effect on humans or the natural environment. Their physicochemical properties are also of interest. They exist in the form of monomers in aqueous solutions at low concentrations and form micelles above the critical micelle concentration (CMC). Aqueous solutions of APGs appear in the form of lyotropic liquid crystals (LLCs) at concentrations of about 50% or more [35,36]. By analogy to graphite or molybdenum disulfide, one may expect to see an effect of these structures on tribological properties. Another advantage of APGs is their high surface activity. This ensures a significant increase in the concentration of this type of additives in the surface phase and, consequently, a modification of the structures of a lubricating film by possible formation of mesophases.

In this work, an attempt has been made to show how addition of APGs to water affects their antiwear and antiseizure properties. Results of tribological tests will be correlated with physicochemical properties of aqueous solutions of APGs with different alkyl chain lengths.

10.2 CHARACTERISTICS OF ALKYL POLYGLUCOSIDES AND THEIR AQUEOUS SOLUTIONS

APGs (Figure 10.1) are surfactants. Their hydrophilic group usually consists of one to five condensed glucoside parts (degree of polymerization [DP] = 1–5). The hydrophobic group in APG molecules is an alkyl chain containing from 6 to 18 carbon atoms. Depending on the structure of the molecule, APGs exhibit different solubility in different solvents. The compounds with the DP from 1 to 5 containing alkyl chains whose length is from 6 to 14 carbon atoms are water soluble. The molecules whose chain length is 16 to 22 carbon atoms are oil soluble. An increase in the DP increases solubility in a polar medium, while lengthening of the alkyl chain increases solubility in a nonpolar medium [35,36].

From the point of view of fundamental research, pure chemical entities should be used as additives. Taking into consideration the application of the results obtained, we aim to ensure that the substance tested is inexpensive and has appropriate performance features. Therefore, we decided to use mixtures of APGs of various structures, which are commercial products prepared by industrial synthesis from the reaction of fatty alcohols and carbohydrates (sugars). The reactants used in the synthesis are naturally occurring substances such as coconut oil or palm oil and glucose derived from maize or potatoes [35,36]. Such products are mixtures of compounds

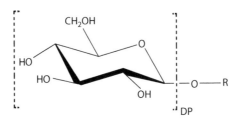

FIGURE 10.1   Alkyl polyglucoside (APG) (R—alkyl chain, DP—degree of polymerization).

with various alkyl chain lengths and various degrees of polymerization of the glucoside part.

Since industrially produced commercial products are used in this investigation, there is a need to verify and supplement physicochemical data. In view of the application of APGs as lubricant components, it is particularly important to determine the surface activity of these compounds and the range of concentrations at which micelles form (CMC) in the bulk phase of the solutions. Viscosity coefficient values obtained for various compounds at various concentrations are also important.

10.2.1 SURFACE TENSION AND WETTABILITY OF STEEL

Two APGs were used in this investigation: APGs containing C12 and C14 alkyl chains (the average alkyl chain length in this mixture is C 12.5) and APGs containing C8 and C12 alkyl chains (the average alkyl chain length is C 9.1).

Changes in the surface tension and wettability of steel surfaces as a function of concentration were used as a measure of surface activity of APGs. Added to water, APGs cause a reduction in surface tension (σ) [33]. As expected, surface activity of longer chain compounds (APG C12.5) is higher than surface activity of shorter chain compounds (C9.1). Over a twofold decrease in the surface tension of water ($\sigma = 30$ mN/m) was already observed at 0.05 wt% concentration. The solutions of APG C9.1 reached a value of 30 mN/m at a concentration of about 1 wt%.

Higher surface activity of long-chain APGs was confirmed during examination of wettability of a steel surface [33]. The solutions of APG C12.5 caused a significant reduction in the contact angle on steel from about 90°–60° already at a concentration of 0.05 wt%. In the case of APG C9.1 solutions, such a significant reduction in the contact angle occurred at a concentration of 0.5 wt%.

Taking into account surface activity of the compounds based on measurements of surface tension and the contact angle on steel, it seems that the optimal concentration of APGs in lubricant compositions should exceed 1%. At such additive concentrations, the observed values of surface tension and contact angles were about half the values for pure water [33].

10.2.2 CMC AND FORMATION OF MICELLES IN BULK PHASE

APGs exist in aqueous solutions in the form of monomers up to the value of the CMC. Micelles form in solutions above this concentration. The CMC depends on the length of the alkyl chain and the DP (Table 10.1) [36]. APGs can form micelles of various shapes. The compounds with alkyl chains C8–C10 form spherical or disklike micelles at concentrations above 0.007 wt%. With an increase in concentration the aggregation number increases, and the micelles undergo deformation and become cylindrical at a concentration of about 0.05 wt%. The increase in the hydrophobic part of the molecules of APGs leads to an increased tendency of formation of cylindrical micelles. The aggregation numbers and sizes of micelles are given in Table 10.2 [35].

The change of a micelle shape from spherical to cylindrical corresponds to an increase in the aggregation number (number of surfactant molecules in one micelle).

TABLE 10.1
Effect of Alkyl Polyglucoside Molecule Structure on Critical Micelle Concentration Value

	CMC (wt%)
APG C$_8$–C$_{10}$	0.056
APG C$_{10}$–C$_{12}$	0.027
APG C$_{12}$–C$_{14}$	
DP = 1.2	0.0092
DP = 1.5	0.01
DP = 1.8	0.011

Source: Nickel, D. et al., In: *Alkyl Polyglucosides*, Hill, K., von Rybiński, W., and Stoll, G., (eds), pp. 39–70, VCH, New York, 1996.

TABLE 10.2
Aggregation Numbers and Sizes of Micelles for Various Alkyl Polyglucosides

	Aggregation Number (N)	Lenghts of Micelle (nm)
APG C8–C10 (DP = 1.2)	160	28.8
APG C10–C12 (DP = 1.2)	250	37.0
APG C12–C14 (DP = 1.2)	450	51.2
APG C12–C14 (DP = 1.6)	920	119.0

Source: Balzer, D., In: *Nonionic Surfactants: Alkyl Polyglucosides*, Balzer, D. and Luders, H., (eds), pp. 85–278, Marcel Dekker, New York, 2000.

The largest size of micelles was observed for APGs with alkyl chain length of C12–C14 and DP = 1.6. The length of such cylindrical micelles was about 120 nm [35].

10.2.3 VISCOSITY OF AQUEOUS SOLUTIONS OF ALKYL POLYGLUCOSIDES

The viscosity (η) of aqueous solutions of APGs was investigated as a function of concentration at 25°C. The shape and number of aggregates formed have a direct effect on viscosity (η) of solutions. Changes in viscosity as a function of concentration of APGs with C8–C10 chain lengths are not really large. One to thirty percent solutions show a η value of about 10 mPa · s. The viscosity of the solutions begins to rise quite significantly beyond 30% concentration, reaching a η value of 100 mPa·s at 50%, the highest concentration examined.

Considerably higher viscosity values were observed for aqueous solutions of APGs with C12–C14 chains. The viscosity of 4% concentration of APG C12-C14 solutions was tenfold higher than the value for APG C8-C10 solutions. Above 4% concentrations the viscosity of APG C12-C14 solutions arose steadily with increasing concentration and reached a value of about 10,000 mPa·s at 50% concentration [32].

10.3   EXPERIMENTAL DETAILS

10.3.1   MATERIALS

Two polyglucoside materials produced by Cognis (Germany)—Glucopon 600CSUP and 215CSUP—were used in this investigation. Glucopon 600CSUP is a 50% aqueous solution of APGs containing C12 and C14 alkyl chains with an average DP = 1.4. The average alkyl chain length in this mixture is C 12.5. The other material, Glucopon 215CSUP, is a 63% aqueous solution of APGs containing C8 and C12 alkyl chains and with an average DP = 1.5. The average alkyl chain length in the mixture is C 9.1. The two types of APGs are being referred to as APG C12.5 and APG C9.1, respectively. The two designations unequivocally describe the average alkyl chain length. The composition of the commercial products used in this investigation is shown in Table 10.3 [36].

TABLE 10.3
Composition of Alkyl Polyglucosides Commercial Products

	Glucopon 215CSUP	Glucopon 600CSUP
	APG C9.1	APG C12.5
Hydrophilic Group	(wt%)	(wt%)
Monoglucosides	44	63
Diglucosides	14	16
Triglucosides	9	8
Tetraglucosides	7	3.5
Pentaglucosides	3	1.5
Hexaglucosides	1.5	1.5
Higher glucosides	21	6.5
C Chain Distribution		
C8	45	—
C10	55	—
C12	—	75
C14	—	25

Source: Nickel, D. et al., In: *Alkyl Polyglucosides*, Hill, K., von Rybiński, W., and Stoll, G., (eds), pp. 39–70, VCH, New York, 1996.

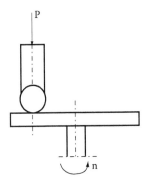

FIGURE 10.2 Schematic diagram of a ball-on-disk friction pair (T-11 device).

Based on the analysis of physicochemical test results, 4% aqueous solutions of the two types of APGs, APG C9.1 and APG C12.5, were selected for this investigation. At this concentration, which exceeds the CMC value, the surface phase is fully saturated. In the bulk phase, spherical micelles were mostly formed in the APG C9.1 solutions and cylindrical micelles were mostly formed in the APG C12.5 solutions. It is also important to note that liquid crystalline structures do not yet appear in the bulk phase solutions at a 4 wt% concentration of either additives.

The physicochemical properties of aqueous solutions of APGs strongly depend on the length of the alkyl chain. Lubricant compositions in the form of mixtures of APG C9.1 and APG C12.5 at weight ratios of 9:1, 7:3, 5:5, 3:7, and 1:9 were prepared in order to analyze more thoroughly the effect of the average alkyl chain length on the tribological quantities being measured. As a result, APG compositions whose average alkyl chain lengths were C9.44, C10.12, C10.8, C11.48, and C12.16 were obtained. The compositions were used to make aqueous solutions in which the concentration of APGs was 4 wt%. Redistilled water was used to prepare the solutions in all cases.

All the tribological tests were carried out using a steel–steel friction pair. Friction pair elements were made of 52100 bearing steel. Surface roughness was $R_a = 0.032$ μm, hardness 60–65 HRC. Before the tests, all components of the friction couples were thoroughly chemically cleaned. An ultrasonic cleaner was employed. The elements were cleaned in *n*-hexane, acetone, ethyl alcohol, and distilled water. After the cleaning process all friction-couple components were dried (50°C, 30 min).

10.3.2 METHODS

Two types of testers were used to conduct tribological tests: a ball-on-disk apparatus and a four-ball machine.

10.3.2.1 Ball-on-Disk Apparatus

A T-11 ball-on-disk tribological tester produced by ITeE-PIB (Radom, Poland) was used. Figure 10.2 shows a schematic diagram of the friction pair used in the apparatus. The steel balls used had a diameter of 6.35 mm and the diameter of the steel disks was 25 mm. The duration of the test was 900 s and the sliding speed was 0.1 m/s. The loads applied were 5, 10, 20, 30, 40, and 50 N. Based on friction force measurements, the coefficient of friction (μ) was calculated from the formula

$$\mu = \frac{F_T}{P},\tag{10.1}$$

where
F_T is the friction force [N]
P is the applied load [N]

The friction coefficient values for the entire 900 s of test were averaged. Three independent measurements were conducted; the averages are presented in the figures. The measurement error is the standard deviation of the arithmetic mean at an assumed confidence level of 95%.

10.3.2.2 Four-Ball Machine

A T-02 tribological tester produced by ITeE-PIB (Radom, Poland) was used. Figure 10.3 shows a schematic diagram of the friction pair used in the apparatus. The steel balls used in the tests had a diameter of 12.7 mm.

The T-02 four-ball apparatus was used in two kinds of tests: tests at a constant load (stationary tests) and tests during which the load on the friction pair was increased linearly (this method makes it possible to determine quantities characterizing seizure). The tests were carried out according to the procedure described in the literature [37].

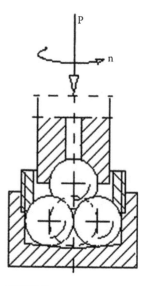

FIGURE 10.3 Schematic diagram of a friction pair of a T-02 four-ball machine.

The tests at a constant load were performed at 200 rpm rotational speed of the spindle, test duration was 0.25, 1, 2 or 3 h, friction pair loads were 1.5, 2.0, 2.5, 3.0, and 3.5 kN. The coefficient of friction (μ) was determined from the measured friction torque (M_T), using the following equation:

$$\mu = 222.47 \frac{M_T}{P},\tag{10.2}$$

where

M_T is the friction torque [N·m]

P is the applied load [N], the coefficient 222.47 results from the distribution of forces in the four-ball machine

The tests at variable loads were conducted at a 500 rpm of rotational speed. The applied load was increased from 0 to 7.2 kN at a rate of 409 N/s. The dependence of friction torque on load was obtained for each lubricating substance. The data was used to determine: (a) the scuffing load, P_t, at which boundary layer rupture occurs (it manifests itself by a sudden increase in the friction torque), and (b) the seizure load, P_{oz}, which corresponds to the friction torque of 10 N·m. After the tests the limiting pressure of seizure, p_{oz}, was calculated as follows:

$$p_{oz} = 0.52 \frac{P_{oz}}{d^2} [\text{kN/mm}^2],\tag{10.3}$$

where d is the mean wear scar diameter in mm.

10.3.2.3 Reflection Microscopy

In order to evaluate the antiwear properties of the test solutions, wear scar diameters (d) were measured both parallel and perpendicular to the direction of friction after each test. The arithmetic mean of the data was calculated and reported here. A polar reflection microscope produced by PZO (Warsaw, Poland) was used to measure the wear scar diameters.

10.3.2.4 Profilometry
Wear scar profiles were measured after each friction test. The device used for the investigation was Profilometer TOPO 01 manufactured by The Institute of Advanced Manufacturing Technology (Krakow, Poland). In this investigation, the scanning needle traced a 4 mm long wear scar at a speed of 0.5 m/s. The resulting profilograms were used to determine the maximum depth of the scar. Measurements were conducted perpendicular to the line connecting the edges of the scar. Surface roughness was also measured before and after individual tests. The roughness measurement parameters were measurement path, 0.4 mm, and measurement speed, −0.1 m/s. The roughness data were processed using "Profilometer 2D" and "Topografia" software.

10.3.2.5 Atomic Force Microscopy
The behavior of aqueous solutions of APGs at the steel–liquid interface was analyzed using atomic force microscopy (AFM). A Nanoscope III AFM was used. During the test, the scanning needle was immersed in the solution and set in the tapping mode. Measurements were conducted at 22°C and the area of the scanned surfaces was 1×1 μm. The test results are presented in the form of profilograms. Values of R_a and R_{max} were also determined from the resulting profilograms.

10.3.2.6 Viscosity Measurement
A Brookfield RV DVI + viscometer was used for the viscosity measurements. The measurements were carried out at 25°C and at a rotational speed of the spindle of 100 rpm.

10.4 RESULTS AND DISCUSSION

10.4.1 ANTIWEAR PROPERTIES OF AQUEOUS SOLUTIONS OF ALKYL POLYGLUCOSIDES

10.4.1.1 Effect of Applied Load
The effect of 4% aqueous solutions of APG C9.1 and APG C12.5 on wear of the friction elements at various loads on the friction pairs was investigated as a function of load. A load range of 5–50 N was used in the T-11 tests, whereas a load range of 1.5–3.5 kN was used in the case of the T-02 device.

The wear observed for the ball-on-disk pair for the loads of 5 and 10 N differs from the wear for 30, 40, and 50 N (Figure 10.4). In the case of APG C9.1 solutions, wear scar diameter varied in the 0.42–0.43 mm range at lower loads and in the 0.34–0.38 mm range at higher loads. A different tendency was observed in the case of APG C12.5 solutions. The d value was in the 0.26–0.27 mm range for low loads, while for higher loads, it was in the range of 0.34–0.38 mm. However, no significant effect of load on the coefficient of friction of either material was observed (Figure 10.5). The μ values ranged from 0.13 to 0.16 at various loads.

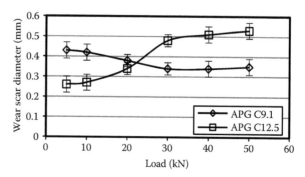

FIGURE 10.4 Dependence of wear scar diameter on load for 4% aqueous solutions of APG C9.1 and C12.5. Tribometer with a steel ball–steel disk pair (T-11), load 5–50 N, linear velocity 0.1 m/s, friction path 90 m, test duration 900 s.

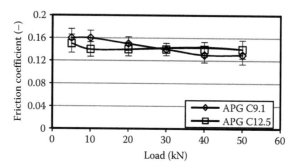

FIGURE 10.5 Dependence of the coefficient of friction on load for 4% aqueous solutions of APG C9.1 and C12.5. Tribometer with a steel ball–steel disk pair (T-11), load 5–50 N, linear velocity 0.1 m/s, friction path 90 m, test duration 900 s.

The four-ball machine test results indicate that wear scar diameter increases with increasing applied load (Figure 10.6). Relatively lower values of d were obtained for APG C12.5 solutions compared to APG C9.1 solutions. The largest difference in the values of wear scar diameters between these two materials was up to 0.2–0.3 mm and was observed at the highest loads tested (3.0 and 3.5 kN).

The effect of the type of APG on coefficient of friction is considerably larger than on wear (Figure 10.7). For example, the μ value obtained for solutions of APGs with longer alkyl chains at the load of 2.5 kN was about 0.07, whereas it was as much as 0.17 for APGs with shorter chains. The differences were also quite significant for the loads of 2.0, 3.0, and 3.5 kN, and they amounted to about 0.03–0.06.

Summing up the effect of load on antiwear properties of aqueous solutions of APGs, one can notice a lack of a definite trend for changes. During tests employing a ball-on-disk pair, there was an increase in wear as a function of load for APG 12.5 solutions, while for APG C9.1 solutions wear decreased significantly at the highest of the loads examined. This tendency was not confirmed by four-ball machine tests. In this case, the solutions of both kinds of APGs did not reduce wear with an increase in load. It is interesting that compared to APG C9.1, APG C12.5

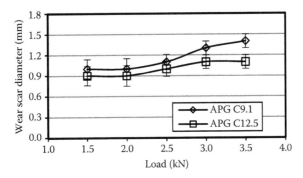

FIGURE 10.6   Dependence of wear scar diameter on load for 4% aqueous solutions of APG C9.1 and C12.5. Four-ball machine (T-02), rotational speed of 200 rpm, test duration 900 s.

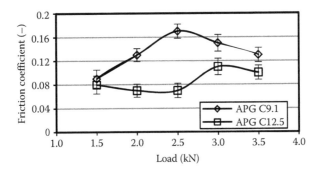

FIGURE 10.7   Dependence of the coefficient of friction on load for 4% aqueous solutions of APG C9.1 and C12.5. Four-ball machine (T-02), rotational speed of 200 rpm, test duration 900 s.

exhibits slightly better antiwear properties and considerably higher resistance to motion.

10.4.1.2   Effect of Friction Time

The effect of friction time on wear of friction pair elements was evaluated. Four percent aqueous solutions of APG C9.1 and APG C12.5 were used as lubricating substances. The tests were carried out with the four-ball machine at a load of 2.5 kN and a rotational speed of 200 rpm. Four test time durations (15 min, 1, 2, and 3 h) were investigated for each lubricating substance. The diameters and depths of the wear scars of the balls as well as surface roughness of the scars were evaluated after individual measurements and the average value of the coefficient of friction was calculated. The results are summarized in Figures 10.8 through 10.10. Examples of scar profiles after various friction test time durations are presented in Figures 10.11 and 10.12.

It has been found that for both APGs tested the wear scar diameter of the balls increased with an increase in friction test time. However, the increase of scar diameter was much larger for APG C9.1 than for APG C12.5. Considerable differences in the size of wear scar diameter (d) appeared already after 1 h tests. The wear scar

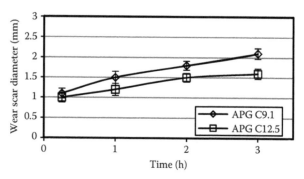

FIGURE 10.8 Dependence of wear scar diameter on friction test time for 4% aqueous solutions of APG C9.1 and C12.5. Four-ball machine (T-02), rotational speed of 200 rpm, load 2.5 kN, test duration 0.25, 1, 2, and 3 h.

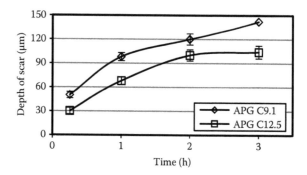

FIGURE 10.9 Dependence of wear scar depth on friction test time for 4% aqueous solutions of APG C9.1 and C12.5. Four-ball machine (T-02), rotational speed of 200 rpm, load 2.5 kN, test duration 0.25, 1, 2, and 3 h.

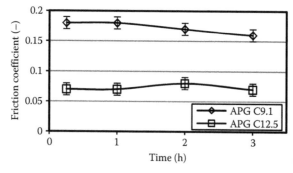

FIGURE 10.10 Dependence of the coefficient of friction on friction test time for 4% aqueous solutions of APG C9.1 and C12.5. Four-ball machine (T-02), rotational speed of 200 rpm, load 2.5 kN, test duration 0.25, 1, 2, and 3 h.

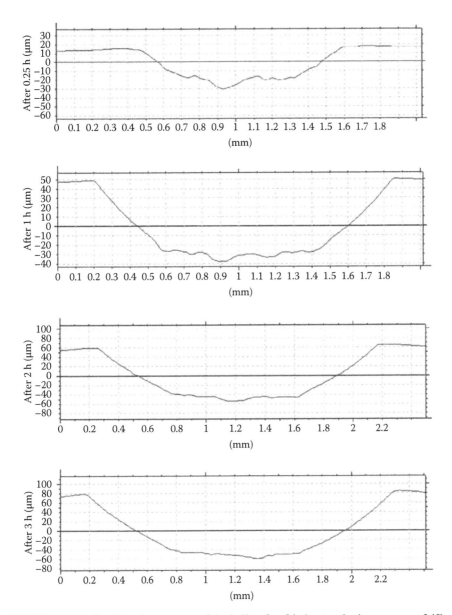

FIGURE 10.11 Profiles of wear scars of the balls, after friction test in the presence of 4% aqueous solution of APG C9.1. Four-ball machine (T-02), rotational speed of 200 rpm, initial temperature 25°C, load 2.5 kN, test duration 0.25, 1, 2, and 3 h.

diameter for APG C9.1 solutions was 1.5 mm, whereas the value obtained for APG C12.5 solutions was 1.2 mm. This difference increased after 2 and 3 h tests. The d values were 1.8 and 2.1 mm for APGs with shorter alkyl chains and 1.5 and 1.6 mm for APGs with longer chains. Similar differences were observed in the wear scar depths of the balls for these two solutions. Depth values of 50, 98, 120, and 142 μm

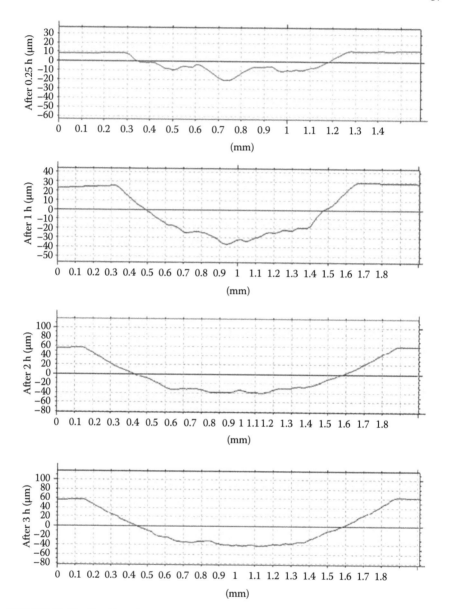

FIGURE 10.12 Profiles of wear scars of the balls, after friction test in the presence of 4% aqueous solution of APG C12.5. Four-ball machine (T-02), rotational speed of 200 rpm, initial temperature 25°C, load 2.5 kN, test duration 0.25, 1, 2, and 3 h.

were obtained for APG C9.1 after 0.25, 1, 2, and 3 h tests, whereas the corresponding values for APG C12.5 the values were 30, 68, 100, and 104 μm, respectively. The results indicate that the chain length in APG molecules significantly affect their antiwear performance. After 3 h of test, the 4% aqueous solution of APG C12.5 gave lower scar depth results (by about 25%) than 4% APG C9.1 solutions.

Relatively large differences were also observed in the ability of individual solutions to reduce resistance to motion. The averaged values of the coefficient of friction (μ) for shorter chain APGs ranged from 0.16 to 0.18, whereas for longer chain APGs they ranged from 0.07 to 0.08. The μ values were thus 2–2.5 times lower in the case of APG C12.5 solutions than those of APG C9.1 solutions.

In this investigation, 4% aqueous solutions of APGs were used. At this concentration of the additives, according to literature data [35,36], micelles predominate both in the surface phase and in the bulk phase. However, it is known that different kinds of aggregates are formed by APGs with different alkyl chain lengths. Based on this, it can be expected that the lubricant film formed during friction will have a different aggregate structures. In the light of such results, it is possible to formulate a hypothesis that at high loads of the order of several kN the compounds capable of forming cylindrical micelles show a better antiwear action. Analogous to LLCs formed in the bulk phase, they can form mesophases with a hexagonal structure in the surface phase.

These conclusions can be confirmed using the results from the examination of roughness of the steel surface covered with water or aqueous solutions of APGs. This study was conducted using an AFM. The results are given in Figure 10.13.

Roughness of the steel surface is relatively quite considerable. The value of R_a is about 3.6 nm. If the steel surface is covered with an aqueous solution of APG C9.1 (1% concentration), roughness is reduced to the R_a value of about 1.2 nm. It may be expected that spherical micelles which are present in the surface phase fill a part of microasperities. This results in smoothing of the surface and, hence, the area of real contact increases, which leads to a considerable reduction in pressure in the friction area. This contributes obviously to a reduction in motion resistance.

In the case of the profile obtained for the steel surface covered with a 1% APG C12.5 solution, one can notice a significant reduction in roughness compared to the surface covered with water or an APG C9.1 solution. In this case, the value of R_a was 0.25 nm (Figure 10.13). Such a significant reduction in roughness might indicate that microasperities were filled with aggregates of considerably large sizes.

Of course, AFM studies do not provide information on the thickness of a lubricant film. However, relatively low values of the coefficient of friction observed during tests with APG C12.5 solutions indicate that the lubricant film layer was not broken. In the case of APG C9.1 solutions, the values of μ are considerably higher. This may suggest less efficient protection by these types of compounds or lower stability of the lubricant film formed by spherical micelles.

10.4.1.3 Effect of Alkyl Chain Length

In tests at constant load on T-02 tester, it has been shown that long-chain APGs (APG C12.5) exhibit lower resistance to motion and lower wear than short-chain APGs (APG C9.1). In order to confirm the direction of these changes, solutions differing in average lengths of the alkyl chain were prepared. Four percent aqueous solutions of APG compositions with average alkyl chain lengths of C9.1, C9.44, C10.12, C10.8, C11.48, C12.16, and C12.5 were used as lubricating substances. Tribological tests

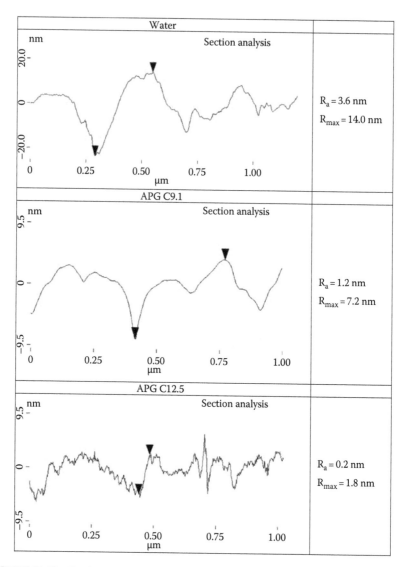

FIGURE 10.13 Profilograms obtained using AFM for steel surface immersed in water and aqueous solutions of APGs. Scanning area: 1×1 μm.

were carried out with of a four-ball machine at a load of 2.5 kN and a rotational speed of 200 rpm. Duration of each test was 3 h.

It has been found that increasing the length of the alkyl chain in APG molecules leads to a reduction in wear and resistance to motion (Table 10.4) [34]. A 1.9 mm wear scar diameter (d) was observed for APGs with C9.44–C10.8 chains. Two other solutions—containing APGs C11.48 and C12.16—produced a wear scar diameter of 1.7 mm, whereas the value of d was 1.6 mm for APG with the longest chain. Considerably larger differences were observed in the coefficients of friction. The

TABLE 10.4
Wear Scar Diameters and Coefficients of Friction from Four-Ball Friction
Tests in 4% Aqueous Solutions of Alkyl Polyglucosides

	APG C9.1	APG C9.44	APG C10.12	APG C10.8	APG C11.48	APG C12.16	APG C12.5
Wear scar diameter* (mm)	2.1	1.9	1.9	1.9	1.7	1.7	1.6
Coefficient of friction* (−)	0.16	0.14	0.11	0.11	0.10	0.08	0.07

Source: Sulek, M.W. and Wasilewski, T., *Tribologie und Schmierungstechnik*, 51, 9, 2004.
*Four-ball machine (T-02), rotational speed of 200 rpm, load 2.5 kN, test duration 3 h.

value of μ reached the level of 0.16 for APG C9.1. Slight lengthening of chains of APG molecules, to 9.44, led to a reduction of μ to 0.14. Increasing the fraction of long-chain molecules in APG compositions effected a further reduction in resistance to motion. The μ value obtained for APGs C10.12, C10.8, and C11.48 was in the range of 0.1–0.11. The lowest values of μ of 0.07 and 0.08 were observed for molecules with the longest alkyl chains [34].

Viscosity measurements were performed on the solutions tested. It has been found that viscosity of solutions of APGs with alkyl chain lengths of 9.1–11.48 is around 10 mPa·s and changes slightly with chain length. However, a significant increase in viscosity—up to 80–100 mPa·s—was recorded for APGs with C12.16 and C12.5 chains. The results confirm that long and cylindrical micelles which form in such systems cause a considerable increase in the viscosity of the system [34].

The results obtained confirm the assumption that an important role in effective wear prevention is played by the length of the alkyl chain in APG molecules and hence by the micellar structure of the lubricant film. Wear reduction and a very distinct decrease in resistance to motion were recorded for APG C12.16 and C12.5 solutions whose viscosity was 80 and 100 mPa·s, respectively. The presence of cylindrical micelles in those lubricating substances effectively protects rubbing surfaces.

10.4.2 ANTISEIZURE PROPERTIES OF AQUEOUS SOLUTIONS OF ALKYL POLYGLUCOSIDES

Antiseizure properties of 4% aqueous solutions of APGs were analyzed. APG compositions differing in the average length of the alkyl chain (C9.1, C9.44, C10.12, C10.8, C11.48, C12.16, and C12.5) were used. The quantities characterizing antiseizure properties (P_t, P_{oz}, and p_{oz}) were determined using a T-02 tester according to the procedure described in the literature [37] (Section 10.3.2). The results obtained are presented in Figures 10.14 and 10.15.

Scuffing parameter (P_t) is a measure of the stability of a lubricant film. Application of load higher then P_t leads to a direct contact of the friction pair materials in microareas and, consequently, to an increase in wear and resistance to motion. The lubricating substance should have the highest possible value of P_t [26,37]. The 4% solutions

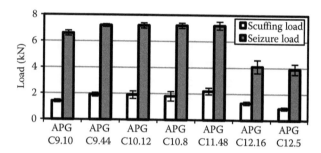

FIGURE 10.14 Dependence of scuffing load (P_t) and seizure load (P_{oz}) on alkyl chain length of APG molecules. Lubricating medium—4 wt% aqueous solutions of mixtures of APGs. Tribological tests: four-ball machine (T-02), rotational speed of 500 rpm, load increment 409 N/s.

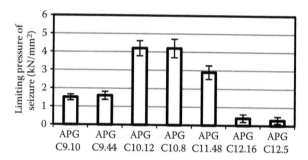

FIGURE 10.15 Dependence of limiting pressure of seizure (p_{oz}) on alkyl chain length of APG molecules. Lubricating medium—4 wt% of aqueous solutions of APG mixtures. Tribological study: four-ball machine (T-02), rotational speed of 500 rpm, load increment 409 N/s.

of APGs analyzed here differed in their antiscuffing capabilities. Lengthening of the average alkyl chain length in APG molecules from C9.1 to C11.48 leads to an increase of P_t from 1.4 to 2.2 kN. A dramatic decrease in scuffing load was recorded for APG C12.16—the value obtained was 1.3 kN. A further drop in P_t, to the value of 0.9 kN, was observed for APG C12.5.

The trends in changes in seizure load as a function of the composition of APGs is similar to changes in scuffing load (Figure 10.14). P_{oz} had the maximum value in this test for the solutions of APGs whose average alkyl chain length was C9.44–C11.48. The values of P_{oz} were lower for the chain lengths of C9.1, C12.1, and C12.5. The compositions of APGs with the longest chains for which the P_{oz} values were about 4.0 kN exhibited particularly low antiseizure performance.

The effect of the length of the alkyl chains of APGs is particularly important for the observed changes in limiting pressure of seizure (p_{oz} in Equation 10.3) as a function of the composition of the lubricating substance (Figure 10.15). Solutions of APGs with C9.1 and C9.44 chains reached a value of 1.5–1.6 kN/mm². Two other APG compositions (APG C10.12 and C10.8) reached very high p_{oz} values of 4.2 kN/mm². A lower value, 2.9 kN/mm², was obtained for the APG C11.48 solution, whereas definitely the lowest values of limiting pressure of seizure were recorded for the solutions of APGs with the longest chains C12.16 and C12.5–0.4 and 0.3 kN/mm^2, respectively.

The analysis of the results obtained points to a different influence of chain length in APG molecules on antiseizure properties than on antiwear properties. The APG compositions which were the most effective ones in antiwear action protected friction pair elements against seizure only to a small extent. The mechanism can be explained as follows: short-chain APGs produce a surface lubricating film containing spherical micelles. Such a film exhibits low stability and ineffectively protects surfaces against wear. Under extreme temperature and pressure conditions this film does not protect surfaces against direct contact. However, when the peaks of micro-asperities are in contact, relatively small spherical micelles may fill the spaces formed and remain in the friction zone. This phenomenon leads to a decrease in the intensity of the process. Long-chain APGs, in turn, form a relatively well-ordered and, in effect, stable lubricant film on the surface which effectively prevents wear and reduces resistance to motion. However, under conditions of extreme temperatures and pressures, the film is broken and relatively large APG micelles are not able to remain in the friction area and to prevent seizure, even to a small degree.

This assumption can be confirmed by the results observed for limiting pressure of seizure—the parameter which combines, in fact, antiwear and antiseizure actions (p_{oz} in Equation 10.3). The APG mixtures with average alkyl chain lengths of C10.12 and C10.8 reached very high values, several times higher than the values obtained for other solutions. This may indicate that in this way the structure of APG compositions has been optimized for various applications.

10.5 SUMMARY

The issue of using surfactants as additives to modify lubricating properties is not only interesting from the theoretical point of view but it also has practical applications [1–15]. APGs play a key role in these applications due to their high surface activity, lack of environmental hazard, and high degree of biodegradability.

The results discussed in this chapter available indicate that APGs used as additives improve lubricating properties of water [32–34]. In view of the physicochemical studies carried out so far [32–34], APG solutions have shown useful tribological properties due to their surface activity and formation of micelles both in the surface phase and in the bulk phase. The tribological properties depend primarily on the length of the alkyl chain. The results obtained generally confirm this thesis. It turns out, however, that the correlations obtained are not monotonic, which would indicate a complicated lubricating film formation process under friction. Based on the results presented here, it is possible to optimize the structure and compositions of APGs for applications under any desired friction conditions.

ACKNOWLEDGMENTS

This work was supported by The National Centre for Research and Development during the years 2008–2010—Grant No. RZ 15 0023 04.

REFERENCES

1. M. Graca, J. Bongaerts, J. Stokes, and S. Granick, Friction and adsorption of aqueous polyoxyethylene (Tween) surfactants at hydrophobic surfaces, *J. Colloid Interface Sci.*, 315, 662–670 (2007).
2. S.K. Mistra and R.O. Skold, Lubrication studies of aqueous mixtures of inversely soluble components, *Colloids Surf. A*, 170, 91–106 (2000).
3. D. Kumar and S.K. Biswas, Microscopic frictional response of sodium oleate self-assembled on steel, *Tribol. Lett.*, 30, 199–204 (2008).
4. K. Boschkova, A. Feiler, B. Kronberg, and J.J.R. Stalgren, Adsorption and frictional properties of gemini surfactants at solid surfaces, *Langmuir*, 18, 7930–7935 (2002).
5. K. Boschkova, B. Kronberg, M. Rutland, and T. Imae, Study of thin surfactant films under shear using the tribological surface force apparatus, *Tribol. Int.*, 34, 815–822 (2001).
6. M. Ratoi and H.A. Spikes, Lubricating properties of aqueous surfactant solutions, *Tribol. Trans.*, 42, 479–486 (1999).
7. H.A. Spikes, Mixed lubrication—An overview, *Lubr. Sci.*, 9, 221–253 (1997).
8. G. Guangteng, P.M. Cann, A.V. Olver, and H.A. Spikes, An experimental study of film thickness between rough surfaces in EHD contacts, *Tribol. Int.*, 33, 183–189 (2000).
9. M. Ratoi, V. Anghel, C. Bovington, and H.A. Spikes, Mechanisms of oiliness additives, *Tribol. Int.*, 33, 241–247 (2000).
10. S.M. Hsu and R.S. Gates, Boundary lubricating films: Formation and lubrication mechanism, *Tribol. Int.*, 38, 305–312 (2005).
11. S.M. Hsu, Molecular basis of lubrication, *Tribol. Int.*, 37, 553–559 (2004).
12. S.M. Hsu, Nano-lubrication: Concept and design, *Tribol. Int.*, 37, 537–545 (2004).
13. Ch. Zhang, Research on thin film lubrication: State of the art, *Tribol. Int.*, 38, 443–448 (2005).
14. Y. Zhang, Boundary lubrication—An important lubrication in the following time, *Mol. Liquids*, 128, 56–59 (2006).
15. S. Zhang and H. Lan, Developments in tribological research on ultrathin films, *Tribol. Int.*, 35, 321–327 (2002).
16. R.M. Matveevskii, Problems of boundary lubrication, *Tribol. Int.*, 28, 51–54 (1995).
17. H. Yoshizawa, Y.L. Chen, and J. Israelachvili, Recent advances in molecular level understanding of adhesion, friction and lubrication, *Wear*, 168, 161–166 (1993).
18. Y.K. Cho, L. Cai, and S. Granick, Molecular tribology of lubricants and additives, *Tribol. Int.*, 30, 889–894 (1997).
19. Cz. Kajdas, Importance of the triboemission process for tribochemical reaction, *Tribol. Int.*, 38, 337–353 (2005).
20. A. Morina, A. Neville, M. Priest, and J.H. Green, ZDDP and MoDTC interactions in boundary lubrication—The effect of temperature and ZDDP/MoDTC ratio, *Tribol. Int.*, 39, 1545–1557 (2006).
21. J. Vicente, J.R. Stokes, and H.A. Spikes, Soft lubrication of model hydrocolloids, *Food Hydrocolloid.*, 20, 483–491 (2006).
22. J.H. Choo, H.A. Spikes, M. Ratoi, R. Glovnea, and A. Forrest, Friction reduction in low-load hydrodynamic lubrication with a hydrophobic surface, *Tribol. Int.*, 40, 154–159 (2007).
23. Ch. Gao and B. Bhushan, Tribological performance of magnetic thin-film glass disks: Its relation to surface roughness and lubricant structure and its thickness, *Wear*, 190, 60–75 (1995).
24. Y.L. Wu and B. Dacre, Effect of lubricant-additives on the kinetics and mechanism of ZDDP adsorption on steel surfaces, *Tribol. Int.*, 30, 445–453 (1997).

25. N.K. Myshkin, M.I. Petrokovets, and A.V. Kovalev, Tribology of polymers: Adhesion, friction, wear and mass transfer, *Tribol. Int.*, 38, 910–921 (2005).
26. M.W. Sulek and T. Wasilewski, Antiseizure properties of aqueous solutions of compounds forming liquid crystalline structures, *Tribol. Lett.*, 18, 197–205 (2005).
27. M.W. Sulek, T. Wasilewski, and M. Zieba, Tribological and physical-chemical properties of aqueous solutions of cationic surfactants, *Ind. Lubr. Tribol.*, 62, 279–284 (2010).
28. T. Wasilewski, Aqueous solutions of the mixtures of nonionic surfactants as modern ecological lubricants, in: *Surfactants in Tribology*, G. Biresaw and K.L. Mittal (Eds.), pp. 355–388, CRC Press (Taylor & Francis), Boca Raton, FL, 2008.
29. T. Wasilewski and M.W. Sulek, Paraffin oil solutions of the mixture of sorbitan monolaurate—ethoxylated sorbitan monolaurate as lubricants, *Wear*, 261, 230–234 (2006).
30. M.W. Sulek, Aqueous solutions of oxyethylated fatty alcohols as model lubricating substances, in: *Surfactants in Tribology*, G. Biresaw and K.L. Mittal (Eds.), pp. 325–353, CRC Press (Taylor & Francis), Boca Raton, FL, 2008.
31. M.W. Sulek and A. Bocho-Janiszewska, The effect of ethoxylated esters on the lubricating properties of their aqueous solutions, *Tribol. Lett.*, 24, 187–194 (2006).
32. M.W. Sulek and T. Wasilewski, Tribological properties of aqueous solutions of alkyl polyglucosides, *Wear*, 260, 193–204 (2006).
33. M.W. Sulek and T. Wasilewski, Influence of critical micelle concentration (cmc) on tribological properties of aqueous solutions of alkyl polyglucosides, *Tribol. Trans.*, 52, 12–20 (2009).
34. M.W. Sulek and T. Wasilewski, Einfluss der Kettenlänge von Alkylpolyglukosiden auf die tribologischen Eigenschaften ihrer wässrigen Lösungen, *Tribol. Sch.*, 51, 9–13 (2004) (in German).
35. D. Balzer, Surfactant properties, in: *Nonionic Surfactants: Alkyl Polyglucosides*, D. Balzer and H. Luders (Eds.), pp. 85–278, Marcel Dekker, New York, 2000.
36. D. Nickel, T. Forster, and W. von Rybinski, Physicochemical properties of alkyl polyglucosides, in: *Alkyl Polyglucosides*, K. Hill, W. von Rybiński, and G. Stoll (Eds.), pp. 39–70, VCH, New York, 1996.
37. W. Piekoszewski, M. Szczerek, and W. Tuszynski, The action of lubricants under extreme pressure conditions in a modified four-ball tester, *Wear*, 240, 183–193 (2001).

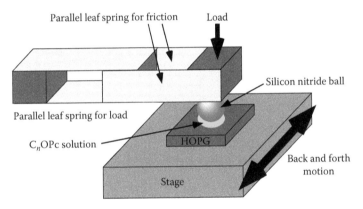

FIGURE 4.1 Schematic model of pin-on-plate tribometer. (From Miyake, K. et al., *Tribol. Lett.*, 31, 9, 2008. With permission.)

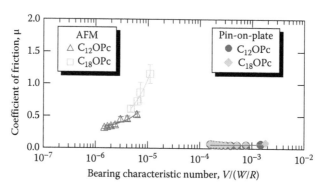

FIGURE 4.9 Stribeck curve with a parameter of $V/(W/R)$, where V is the sliding velocity, W is the normal load, and R is the radius of the curvature of the tip. Open triangles and squares correspond to the data obtained using AFM for $C_{12}OPc$ and $C_{18}OPc$, respectively. Filled circles and diamonds correspond to the data obtained using pin-on-plate tribometer for $C_{12}OPc$ and $C_{18}OPc$, respectively. The thick lines indicate the fitted curve by linear correlation.

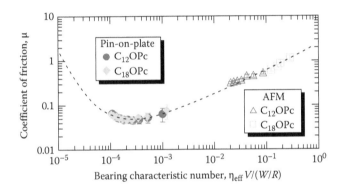

FIGURE 4.10 Stribeck curve with a parameter of $\eta_{eff}\, V/(W/R)$, where η_{eff} is the effective viscosity, V is the sliding velocity, W is the normal load, and R is the radius of the curvature of the tip. Open triangles and squares correspond to the data obtained using AFM for $C_{12}OPc$ and $C_{18}OPc$, respectively. Filled circles and diamonds correspond to the data obtained using pin-on-plate tribometer for $C_{12}OPc$ and $C_{18}OPc$, respectively. The dashed line traces the data on the basis of the shape of the Stribeck curve.

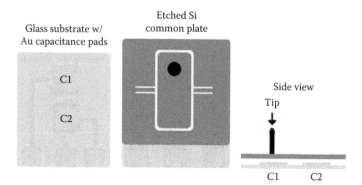

FIGURE 6.5 Schematic of the IFM sensor. Gold capacitance pads (C1 and C2) are bonded to a glass substrate (a), to which an etched Si common plate is bonded (b). Also shown are three Au leads on the bottom of the glass substrate to which electrical connections are made. The left and right leads connect to capacitance pads, while the center is bonded to the common plate. Etched W tips are bonded to the common plate above one of the capacitance pads. (c) Side view of the common plate with tip suspended above the substrate.

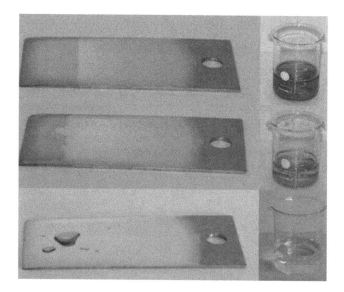

FIGURE 7.10 Wetting of a steel plate by water-based lubricant emulsions with 10%, 2%, and 0% solids (top to bottom). *Left*: coated plates (actual size). *Right*: beakers with the "synthetic fluid" microemulsions (reduced image size).

FIGURE 11.1 Examples of catastrophic failures due to tribological problems in oil well construction: (a) erosion failures by drilling fluids, (b) sliding and fretting wear on thrust-bearing racetrack, (c) sliding and fatigue wear on drill pipe by metal-metal contact, (d) impact and sliding wear at *PDC* bits, (e) two-body abrasion wear at mud pump liners, and (f) fatigue and two-body abrasion on mud pump piston.

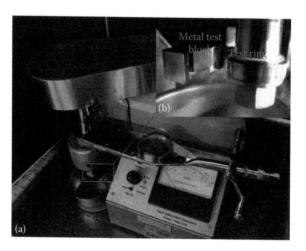

FIGURE 11.3 Fann-type lubricity tester (a). Detailed block-on-ring configuration (b).

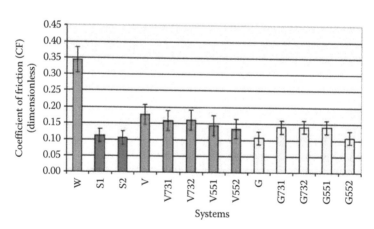

FIGURE 11.4 CF of different systems evaluated. The abbreviation code is presented in Table 11.5.

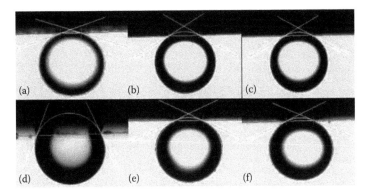

FIGURE 11.5 Contact angles of two different oils in ASS on metal surface: (a) NAM-oil/water, (b) NAM-oil/1% ASS, (c) NAM-oil/2% ASS, (d) diesel/water, (e) diesel/1% ASS, and (f) diesel/2% ASS.

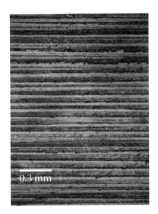

FIGURE 11.6 Optical image of the center of an unused metal test block.

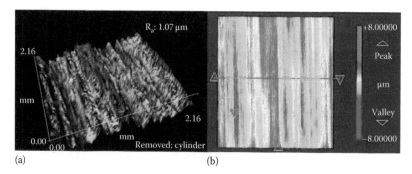

(a)   (b)

FIGURE 11.7 Optical surface profilometry images of an unused metal test block: (a) areal view and (b) profile.

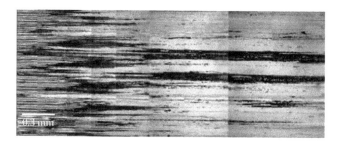

FIGURE 11.8 Metal surface after the lubricity test for lubricant V731.

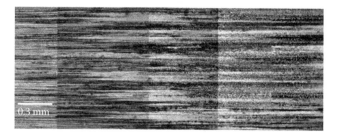

FIGURE 11.9 Metal surface after the lubricity test for lubricant V732.

FIGURE 11.10   Metal surface after the lubricity test for lubricant V551.

FIGURE 11.11   Metal surface after the lubricity test for lubricant V552.

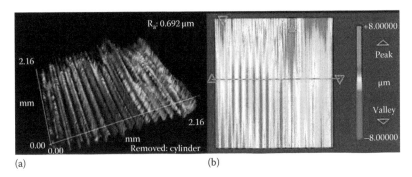

(a)   (b)

FIGURE 11.12   Optical surface profilometry images of the metal block after the lubricity test for lubricant V731: (a) areal view and (b) profile.

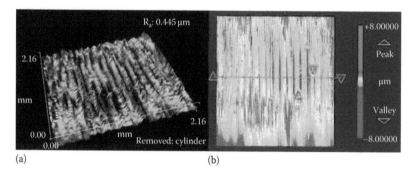

(a)   (b)

FIGURE 11.13   Optical surface profilometry images of the metal block after the lubricity test for lubricant V732: (a) areal view and (b) profile.

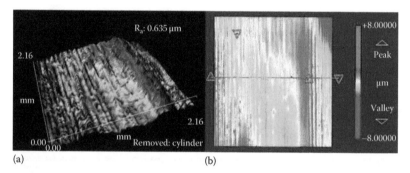

(a)   (b)

FIGURE 11.14   Optical surface profilometry images of the metal block after the lubricity test for lubricant V551: (a) areal view and (b) profile.

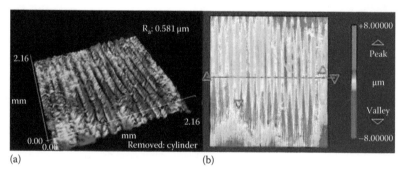

(a)   (b)

FIGURE 11.15   Optical surface profilometry images of the metal block after the lubricity test for lubricant V552: (a) areal view and (b) profile.

FIGURE 11.16 Metal surface after the lubricity test for lubricant W.

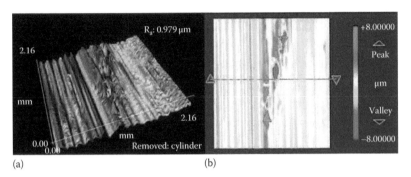

(a)   (b)

FIGURE 11.17 Optical surface profilometry images of the metal block after the lubricity test for lubricant W: (a) areal view and (b) profile.

FIGURE 11.18 Metal surface after the lubricity test for lubricant S1.

FIGURE 11.19 Metal surface after the lubricity test for lubricant S2.

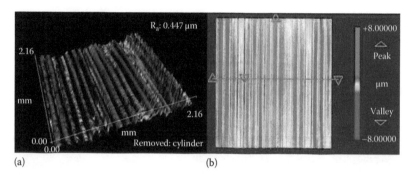

(a)   (b)

FIGURE 11.20 Optical surface profilometry images of the metal block after the lubricity test for lubricant S1: (a) areal view and (b) profile.

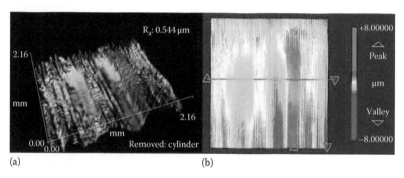

(a)   (b)

FIGURE 11.21 Optical surface profilometry images of the metal block after the lubricity test for lubricant S2: (a) areal view and (b) profile.

FIGURE 11.22 Metal surface after the lubricity test for lubricant V.

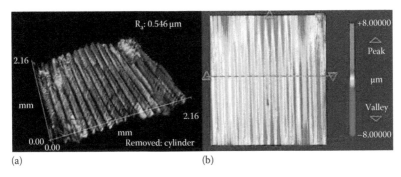

(a) (b)

FIGURE 11.23 Optical surface profilometry images of the metal block after the lubricity test for lubricant V: (a) areal view and (b) profile.

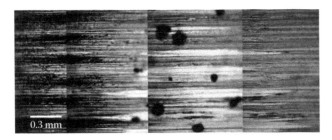

FIGURE 11.24 Metal surface after the lubricity test for lubricant G731.

FIGURE 11.25 Metal surface after the lubricity test for lubricant G732.

FIGURE 11.26 Metal surface after the lubricity test for lubricant G551.

FIGURE 11.27 Metal surface after the lubricity test for lubricant G552.

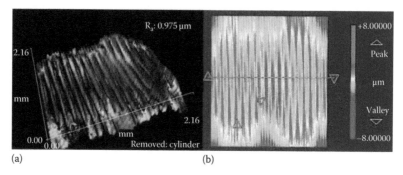

(a)   (b)

FIGURE 11.28 Optical surface profilometry images of the metal block after the lubricity test for lubricant G731: (a) areal view and (b) profile.

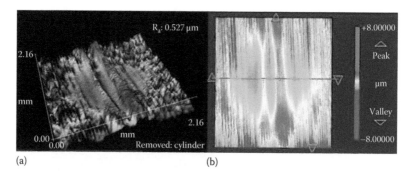

(a)   (b)

FIGURE 11.29 Optical surface profilometry images of the metal block after the lubricity test for lubricant G732: (a) areal view and (b) profile.

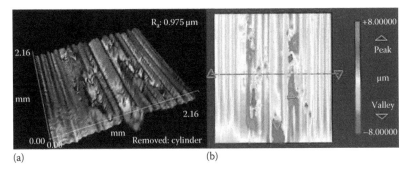

(a)   (b)

FIGURE 11.30 Optical surface profilometry images of the metal block after the lubricity test for lubricant G551: (a) areal view and (b) profile.

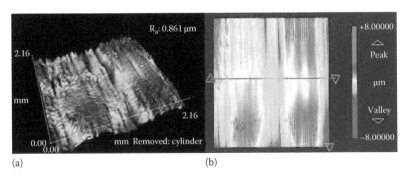

(a)   (b)

FIGURE 11.31 Optical surface profilometry images of the metal block after the lubricity test for lubricant G552: (a) areal view and (b) profile.

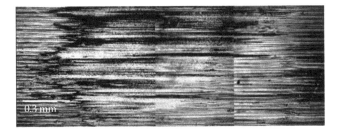

FIGURE 11.32 Metal surface after the lubricity test for lubricant G.

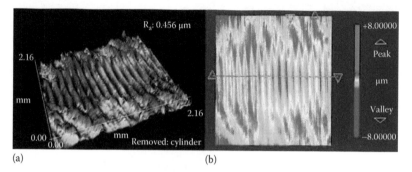

R_a: 0.456 μm

2.16

mm

2.16

mm

0.00

0.00

Removed: cylinder

+8.00000

Peak

μm

Valley

−8.00000

(a)

(b)

FIGURE 11.33 Optical surface profilometry images of the metal block after the lubricity test for lubricant G: (a) areal view and (b) profile.

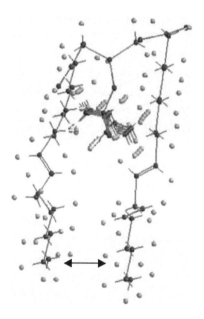

FIGURE 13.3 Minimum energy conformation of oleic-linoleic acid pairs forming the double-end side of "tuning fork" structure, with the third oleic chain forming the stem in the bent conformation.

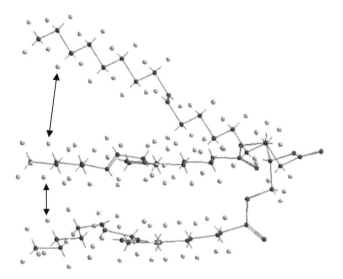

FIGURE 13.4 Minimum energy conformation of 2-linoleic acid pairs forming the double-end side of "tuning fork" structure, with the third oleic chain forming the stem in the bent conformation.

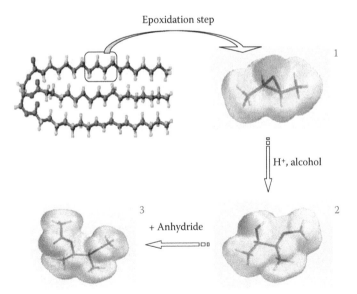

FIGURE 13.6 Formation of epoxy (**1**), hydroxyl-ether (**2**), and ester-ether (**3**) compounds from triacylglycerol molecule.

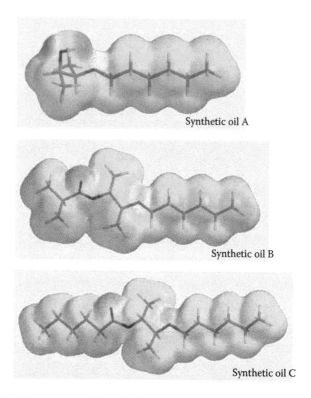

Synthetic oil A

Synthetic oil B

Synthetic oil C

FIGURE 13.7 Electron cloud distribution of synthetic fluids A, B, and C shown in Scheme 13.1.

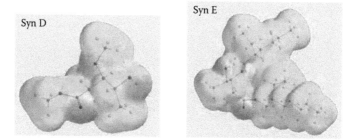

Syn D

Syn E

FIGURE 13.8 Electron cloud distribution of synthetic fluids D and E shown in Scheme 13.2.

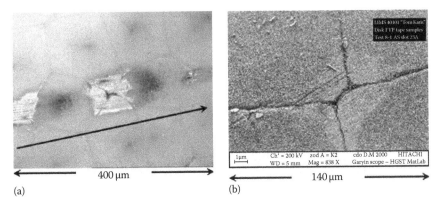

400 μm		140 μm	
(a)		(b)	

FIGURE 18.27 Micrographs of disk polishing tape after an extended period of sliding on a disk. (a) Optical micrograph of worn region. The arrow shows the sliding direction. (b) SEM micrograph of worn region. (Courtesy of Stone, G., Hitachi Global Storage Technologies, San Jose Materials Analysis Laboratory, San Jose, CA.)

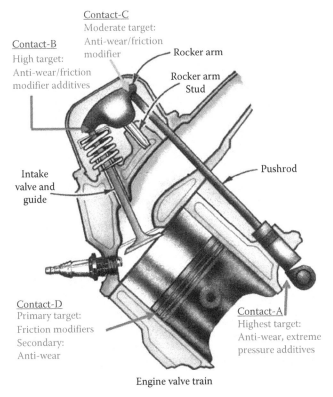

FIGURE 19.1 Illustration of a valve train typical of a small block Chevrolet V-8 engine. (Courtesy of Joseph N. Valentine.)

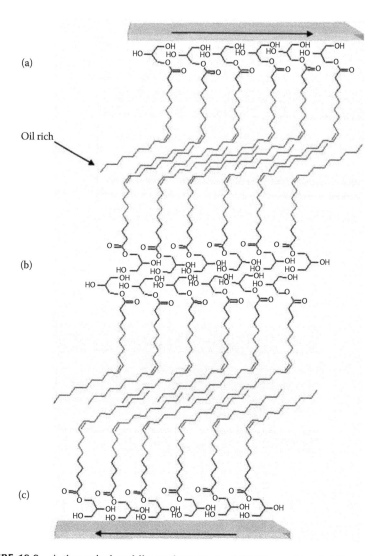

FIGURE 19.9 A theoretical multilayered structure of GMO, capable of deformation and reforming during a dynamic state, set up between friction surfaces in motion. (a) The polar portion of (GMO) in the first bilayer is bound to the top metal surface, while the nonpolar (oil soluble) chains interact. (b) The polar hydrogen bonding interactions between the ester and hydroxyl functional groups holding the two bilayers together. (c) The bonding of the polar portion of (GMO) to the lower metal surface.

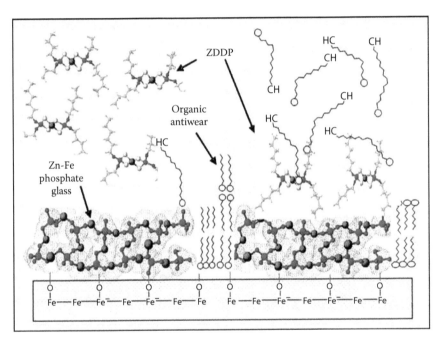

FIGURE 19.15 A combined ZDDP-OAW additive film formation model. While ZDDP inorganic glass protective films form on the metal surface bound to iron oxides, the smaller polar OAW molecules help to fill in the gaps formed in the glass under normal wear action. The resulting film may be reinforced and stronger when both additives are combined.

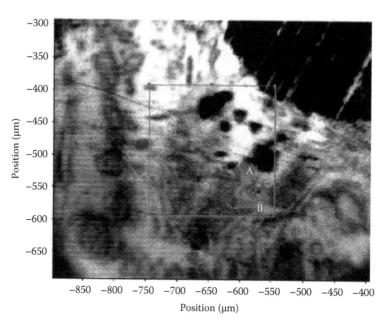

FIGURE 19.17 Microscopic image near the end of the wear scar above identifying two points (A and B) and identified with marks where FTIR spectra were collected.

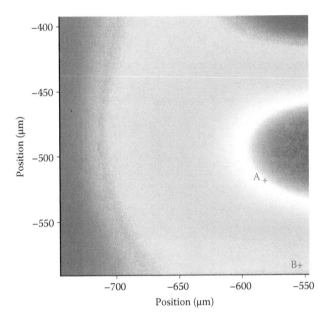

FIGURE 19.19 Red regions indicate higher absorbance (more C=O concentration), and blue is lower absorbance (less C=O). The map is based on a $40 \times 40\,\mu m$ spot size over the area given.

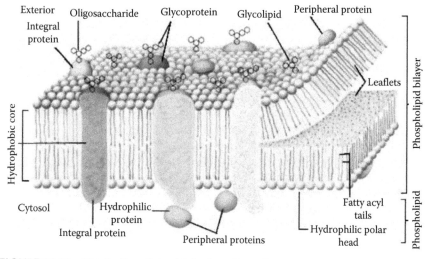

FIGURE 20.18 Distribution of steroids in the cell membrane.

11 Formulation Effects on the Lubricity of O/W Emulsions Used as Oil Well Working Fluids

J.M. González, F. Quintero, R.L. Márquez,
S.D. Rosales, and G. Quercia

CONTENTS

ABSTRACT

In oil well drilling, completion, and maintenance operations, the rotating pipe bears against the side of the hole at numerous points, giving rise to two main friction manifestations known as torque and drag. Torque refers to the pipe resistance to rotation and drag to hoisting and lowering. Excessive torque and drag can cause unacceptable loss of power making oil well operations less efficient, especially in high-angle and extended-reach wells. In these cases, lubricity becomes one of the main functions of the fluid. In the oil industry, there are oil well working fluids of different nature, classified according to the external phase as water-based fluids (WBFs), oil-based fluids, and pneumatic or gas-based fluid systems. Within WBFs, there are oil-in-water (O/W) emulsions, developed as a technological solution for oil well operations in low-pressure reservoirs. In this work, tribological properties of O/W emulsions have been studied as a function of their physicochemical formulation, especially oil type (nonaromatic mineral oil [NAM-oil], diesel) and surfactant concentration (1, 2% w/v) along with the oil/water ratio (70/30, 50/50) as formulation variables. The lubrication performance was established by measuring the coefficient of friction (CF), and optical microscopy imaging in conjunction with optical surface profilometry was used to evaluate antiwear properties. Additionally, contact angle measurements were performed to correlate the wettability phenomenon with the lubricity of O/W emulsions. Based on the results, it was established that with the surfactants mixture used in this study, the oil type does not have a significant effect on the CF of O/W emulsions, due to the similar wettability behavior observed at the metal surface. However, NAM-oil/W emulsions have better antiwear properties than the diesel/W emulsions. Also, the lubricity performance and antiwear properties of O/W emulsions are affected by oil/water ratio and surfactants mixture concentration, showing a systemic interaction between these two parameters.

11.1 INTRODUCTION

11.1.1 Tribological Phenomena in Well Construction

In oil well construction and maintenance processes, particularly during drilling, all the equipment and fluid systems present different tribological phenomena and related problems. Table 11.1 summarizes the principal and the particular tribological events encountered in the oil well operation system. The dominant wear modes include impact wear, abrasion, and slurry erosion [1], and when not controlled or predicted, they could cause catastrophic failures (Figure 11.1) of the equipment and the wellbore, with the ultimate loss of the hole.

The main frictional events in oil well operations are torque and drag since they are caused by the rotation and sliding of pipe inside the well, affecting a large surface (3 km of pipe in metal-to-metal contact), and can be minimized by lubricity with a large volume (600 m³) of the circulating working fluid.

TABLE 11.1
Main Tribological Problems in Oil Well Construction and Maintenance Operations

Equipment	Main Tribologically Related Problems
Engines	Adhesion, abrasion, fretting, corrosion
Shearing, mixing, and product storage facilities	Abrasion (two and three body), erosion (dry and wet), stamping
Flowlines, stand pipe, valves, and hoses	Erosion (dry and slurry), abrasion, stamping, corrosion
Mud and centrifugal pumps	Abrasion (two and three body), erosion (slurry), adhesion, fretting, corrosion
Hoisting systems (derrick, crown block, traveling block, thread ropes, etc.)	Impact, fatigue, abrasion (two and three body)
Rotary table or top drive	Abrasion (two and three body), fretting, impact, fatigue, stamping
Pipes and accessories (heavy wates, centralizers, mud motors, directional controls tools, etc.)	Abrasion (two and three body), fatigue, erosion (slurry), impact, corrosion
Drill bit	Abrasion (two and three body), fatigue, erosion (slurry), impact, torque and drag, corrosion, oxidation
Blow-out preventer and kill system	Abrasion (two and three body), fatigue, erosion (slurry), impact, corrosion
Solid control systems (shakers, hydrocyclones, cutting lines, etc.)	Fretting, erosion (slurry), abrasion (two and three body), fatigue, impact

(a)   (b)   (c)

(d)   (e)   (f)

FIGURE 11.1 (See color insert.) Examples of catastrophic failures due to tribological problems in oil well construction: (a) erosion failures by drilling fluids, (b) sliding and fretting wear on thrust-bearing racetrack, (c) sliding and fatigue wear on drill pipe by metal-metal contact, (d) impact and sliding wear at *PDC* bits, (e) two-body abrasion wear at mud pump liners, and (f) fatigue and two-body abrasion on mud pump piston.

11.1.1.1 Torque and Drag

Torque and drag are two frictional forces that appear during many oil well operations (drilling, completion, and maintenance), produced by the resistance to rotation (torque) and to raising and lowering (drag), in contact with either the wellbore (metal-to-rock) or the casing (metal-to-metal) [2–4] (Figure 11.2).

Management of torque and drag is a crucial part of well design and well operations, for complex well architectures and extended reach wells (ERWs). For example, rig limits can be compromised by excessive torque surpassing topdrive capacity and excessive drags that can lead to the inability to slide pipe for oriented drilling or failure to land a casing or completion string. Similarly, high overpulls can exceed derrick lifting capacity. In addition, downhole frictional forces can compromise pipe limits, leading to problems arising from pipe failures (i.e., twistoffs, collapse, buckling, and fracture) or stuck pipe [5].

For this reason, it is important to understand the mechanisms related to torque and drag frictional forces and lubrication by different working fluids, and how these affect the wellbore stability and components in the oil well operational system, in relative rotation or sliding movements.

11.1.1.2 Lubricity by Oil Well Working Fluids

One of the functions of oil well working fluids is to cool and lubricate the string pipe. Conventionally, the lubricity coefficient is used to quantify the lubricating properties of fluids, and it is measured as the coefficient of friction (CF) of a steel test block pressed against a test rotating ring by a torque arm, to simulate metal-to-metal friction between the pipe and the casing. Lubricity coefficient values are on a scale of 0.01–0.50 and low coefficient denotes good lubricity by a fluid [6].

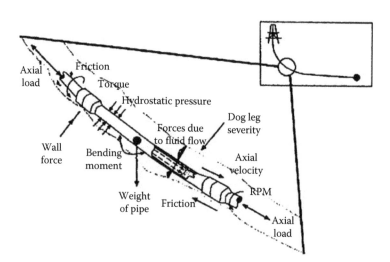

FIGURE 11.2 Friction forces in oil well construction. (Reproduced Aston, M.S. et al., Techniques for solving torque and drag problems in today's drilling environment, Paper # SPE 48939, In: *SPE Annual Technical Conference and Exhibition Proceedings*, 1998. With permission.)

It is known that the behavior of a lubricant can be classified into three separate regimes depicted in the Stribeck curve as follows: (a) the hydrodynamic regime, suitable at high speeds, in which the surfaces are fully separated and the lubrication is governed by bulk rheological properties of the lubricant; no direct physical contact interaction between surfaces occurs, so wear process cannot take place except surface fatigue wear, cavitation, or fluid erosion. (b) The boundary lubrication regime is present at low sliding speeds and high loads, where friction is determined by both surface–surface asperities interaction and ability of the lubricant to adsorb chemically onto the surface and form an interfacial film. Different sliding wear mechanisms may occur in this regime such as corrosive, fatigue, and adhesive wear depending on the dynamic conditions. Increasing sliding speeds, loads, and operating temperature will establish the extreme pressure (EP) lubrication regime, which is based on the concept of a sacrificial film generated by the reaction between lubricant additives and exposed metal surface, preventing metallic contact and severe wear of the surface. (c) Between these two regimes, a mixed regime can be recognized, where the pressure of the fluid carries only part of the load, while the other part is sustained by the surface asperities [7,8].

The lubricity test simulates load conditions present in oil well operations (pipe weight against casing), corresponding to EP lubrication mechanism [3] and it is currently limited to steel-on-steel testing. It is believed that this is adequate for most lubricant screening purposes, as the cased hole section constitutes 85% of the total hole length [9].

Although oil-based fluids (OBFs) have lower CF values, there are wellbore conditions and environmental requirements that will limit their use. This is the case of oil well operations in low-pressure reservoir that require fluids with density lower than that of oil and adequate rheological properties, especially for critical high-angle and ERWs, where lubricity is also a main design criterion.

Lubricants are often added to the oil well working fluids to reduce friction and minimize torque and drag. They are additives generally available as film-producing liquids or solid beads, powders, or fibers. Liquid additives include glycols, oils, esters, fatty acid esters, surfactants, and polymer-based lubricants and solid additives, such as graphite, calcium carbonate flakes, glass, and plastic beads [5].

11.1.2 OIL-IN-WATER EMULSIONS USED AS OIL WELL WORKING FLUIDS

O/W emulsions have been specifically developed to drill, complete, and maintain oil wells in low-pressure or depleted reservoirs [10], where water-based low-density fluids (lower than water's) are required.

O/W emulsions, used as oil well working fluids in low-pressure reservoirs, are designed with a high internal oil phase concentration (HIPC:50%–90%) to both lower the density by substituting water by oil mass, and impart the required rheological properties to the emulsion. The HIPC-O/W emulsion stability and viscosity greatly depend on the type and concentration of the surfactant system used as emulsifier [11,12]. Other additives such as salts potassium chloride (KCl) and pH modifiers (monoethanolamine [MEA]) are incorporated in oil well working emulsions for specific purposes like clay swelling inhibition.

Most studies published on lubricating behavior of O/W emulsions have been done under metal working, rolling-cooling, drilling, and cutting conditions, where they have been specifically designed as lubricant fluids, with a low emulsified oil concentration (3%–5%) [13–16]. Combination of oil lubricity and the latent heat of water provides the optimum fluid for this application. Mining and petroleum machinery is also lubricated by water-based fluids (WBFs) to minimize the risk of fire from leakage of lubricants [7]. The most severe limitation of these lubricants is the temperature range in which they can be successfully applied [17].

The main objective of this work was to study the influence of oil type, dispersed oil fraction, and surfactant concentration on the tribological behavior of O/W emulsions, designed to drill, complete, and maintain high-angle and extended reached wells, located in low-pressure reservoirs, where high lubricity and low density are primary design criteria. No additives were added to the O/W emulsion formulations, assuming that the dispersed oil and/or the surfactant will provide the lubricity required, without increasing the cost of the formulation.

11.2 EXPERIMENTAL DETAILS

11.2.1 MATERIALS

11.2.1.1 Surfactants

The emulsifier system was a hydrophilic anionic/nonionic mixture at a specific mass ratio of two commercial surfactants: an alkyl ether sulfate sodium salt and an alkyl ethoxylated alcohol (made by Clariant, Venezuela); and they were used as received. The critical micelle concentration (CMC) of the aqueous surfactant solutions (ASS) and interfacial tension between the base oil and the aqueous surfactant solution were determined for the individual surfactants and the mixture, using a Dataphysics DCAT11 tensiometer, employing the Wilhelmy plate technique (Table 11.2). Deionized water was used for preparing the solutions.

11.2.1.2 Oils and Other Chemical Additives

A nonaromatic mineral oil (NAM-oil) and diesel were used as the dispersed phase in the O/W emulsions. They were used as received. Table 11.3 shows the characteristics of the dispersed phases used. The salt evaluated was KCl (Sigma-Aldrich, 99%

TABLE 11.2
Characteristics of Surfactants Used

Surfactant Description	HLB	Interfacial Tension (mN/m, 25°C)	CMC (ppm, 25°C)	Active Material (%)	Molecular Weight (g/mol)
Alkyl ether sulfate sodium salt	18	28.4	86.4	27–30	432
Alkyl ethoxylated alcohol	16	42.0	8.0	99.0	1564
Surfactant mixture	—	34.9	49.7	20.0	—

TABLE 11.3
Characteristics of Oils Used as Dispersed Phase

Oil	Saturated (%)	Aromatics (%)	Density (g/cm³, 20°C)	Viscosity (cP, 180 s⁻¹, 49°C)
NAM-oil[a]	>99	<1	0.813	2.09
Diesel	75	25	0.867	2.55

[a] Nonaromatic mineral oil.

purity), and MEA (Sigma-Aldrich, 98% purity) was used to adjust the pH in a range from 8 to 10.

11.2.2 PROCEDURES

11.2.2.1 O/W Emulsion Preparation

The corresponding amount of oil (50% or 70%) was slowly added to the aqueous surfactant solution (1% or 2%), while stirring at 10,000 rpm, room temperature, for 10 min. To characterize the emulsions, average drop diameter ($D_{0.5}$) was determined by laser-scattering technique with a Malvern Mastersizer Hydro 2000G and viscosity by a Physica MCR 301 rheometer, Anton Paar, at 180 s⁻¹ and 49°C. Table 11.4 shows the characterization results of the emulsions with surfactant concentration of 1% and 2%.

11.2.2.2 Lubricity Test

This device is based on the Amontons friction law [7], where the CF is defined as the ratio of parallel frictional force F and the normal force W applied to the surface.

TABLE 11.4
Characteristics of O/W Emulsions

	Oil/Water Ratio 70/30		Oil/Water Ratio 50/50	
Dispersed Oil	Droplet Diameter ($D_{[0.5]}$, μm)	Viscosity (cP, 49°C)	Droplet Diameter ($D_{[0.5]}$, μm)	Viscosity (cP, 49°C)
Surfactant concentration 1%				
NAM-oil[a]	2.331	36.85	2.914	5.91
Diesel	2.261	67.89	2.646	4.72
Surfactant concentration 2%				
NAM-oil[a]	1.795	37.19	2.188	6.51
Diesel	1.773	68.27	2.041	8.27

[a] Nonaromatic mineral oil.

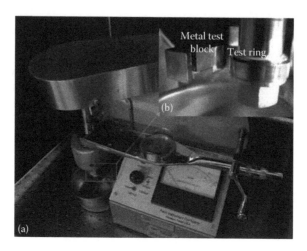

FIGURE 11.3 (See color insert.) Fann-type lubricity tester (a). Detailed block-on-ring configuration (b).

In order to evaluate the effect of O/W emulsion composition on its lubricating property, the lubricity coefficient was measured as the CF or friction factor, using a block-on-ring tribometer (Fann Lubricity Tester, model 212 EP, Figure 11.3). The following recommended procedure was used: apply a constant load (W) 444.8 N by means of the torque arm, adjust the rotational speed of the ring at 60 ± 10 rpm, after 600 s take the ampere reading on the meter, which is converted to the CF. Measurements were performed every 60 s after an equilibration period of 300 s.

The test blocks had 12.32 ± 0.10 mm wide and 19.05 ± 0.41 mm long test surfaces. Each block was supplied with four flat faces and the roughness was between 0.51 and 0.76 µm. The test rotating rings had a width of 13.06 ± 0.05 mm, a perimeter of 154.51 ± 0.23 mm, and a diameter of 49.22 ± 0.125. Both test blocks and rotating rings were made of carburized steel, having a Rockwell hardness C scale number of 58–62 or a Vickers hardness number 653–746 [18,19].

11.2.2.3 Contact Angle Measurements
An optical contact angle device (OCA Dataphysics) was used to evaluate the wetting properties of the oils and ASS on the steel test block surface, at room temperature. The metal surface was immersed in the surfactant solution and then a 2 µL oil droplet was placed on the metal surface with a syringe and allowed to spread until no further change in the contact angle was observed. Images of the solid/liquid/liquid (S/L/L) system were captured by a high-resolution CCD camera.

11.2.2.4 Optical Microscope Images and Profilometry Analysis
Pictures of metal test block surface were taken with an optical microscope, OLYMPUS BX51 objective 10×, in order to describe the wear pattern on the surface after completing the lubricity test on an unused block. No cleaning process was performed, they were dried in an oven at 50°C for 5 min. Several consecutive pictures were taken, which represent an entire or part of a scar on the metal surface. Due to

the curvature on the metal test block, the edges or part of the pictures may appear blurry; however, this does not have a significant effect on describing qualitatively the wear suffered by the surface.

Additionally, the roughness and texture of the metal test block were analyzed using a Zygo NewView 6000 optical surface profilometer. The studied square area (2.16×2.16 mm) in all cases was in the center of the block, assumed as the major friction zone. All metal blocks were cleaned with acetone prior to the profilometry analysis to remove any deposits on the surface. The images were analyzed using the cylindrical filter that allows to flatten the surface.

11.3 RESULTS AND DISCUSSION

O/W emulsions used in oil well operations have been specifically designed for low-pressure or depleted reservoirs, by dispersing a high oil proportion (>50%) to obtain density lower than water, and still being a WBF. These O/W emulsions must be stable under pressure, temperature, shear, and contamination conditions present during operations. To guarantee stability, a high-surfactant concentration is required (>200 CMC). If a low-pressure reservoir is accessed by a high-angle or extended reached well, then lubricity by the O/W emulsions is also a crucial property.

In a previous work [20], hydrophilic surfactants were systematically selected to obtain very stable O/W emulsions. In this study, the effects of oil type, oil/water ratio, and surfactant mixture concentration were evaluated on the tribological properties of the O/W emulsions. The oils and ASS were also evaluated to discriminate the role of each one in the lubricity by the O/W emulsions. Table 11.5 presents nomenclature of the systems used.

Tribological behavior of the systems was studied by determining the CF and wear of the steel block surface subjected to friction in the lubricity test, following the procedures described in the previous section.

TABLE 11.5
Systems Nomenclature

W	water
S1	1% w/v aqueous surfactant solution
S2	2% w/v aqueous surfactant solution
V	Nonaromatic mineral oil (NAM-oil)
V731	70/30 NAM-oil/water ratio and 1% w/v surfactant concentration
V732	70/30 NAM-oil/water ratio and 2% w/v surfactant concentration
V551	50/50 NAM-oil/water ratio and 1% w/v surfactant concentration
V552	50/50 NAM-oil/water ratio and 2% w/v surfactant concentration
G	diesel
G731	70/30 diesel/water ratio and 1% w/v surfactant concentration
G732	70/30 diesel/water ratio and 2% w/v surfactant concentration
G551	50/50 diesel/water ratio and 1% w/v surfactant concentration
G552	50/50 diesel/water ratio and 2% w/v surfactant concentration

The mechanisms involved in lubrication using O/W emulsions are far more complex than those involving single-phase hydrocarbon solutions. According to the literature, during O/W emulsions lubrication process, water is partially excluded from the loaded contacts due to an oil pool forming at the interface, and as a result, the performance of an O/W emulsion is close to that of the pure oil at mild operational conditions [7,15,21].

The results presented in this investigation correspond to a boundary and EP lubrication regimes, imposed by the mechanical design, and the medium-to-high pressure operational conditions of the Lubricity Tester, that simulates metal–metal friction between pipes, during oil well operations [6,22,23].

11.3.1 Coefficient of Friction Studies

11.3.1.1 O/W Emulsions

Figure 11.4 shows the CF values for all studied systems: O/W emulsions, as well as water, ASS and oils, as their main components. The results presented are arithmetic means of CF for the last 5 min of the experiment (total experimental time of 10 min). The error bars shown in Figure 11.4 represent standard deviations of three independent tests. The formulation variables considered were oil type (NAM-oil, diesel) and surfactants mixture concentration (1%, 2% w/v) along with the oil/water ratio (70/30, 50/50).

11.3.1.1.1 NAM-Oil/W Emulsions

O/W emulsions formulated with NAM-oil have similar or lower CF as that of pure oil (dispersed internal phase), but higher than that of ASS, and show considerable CF reduction, up to three times compared to pure water (Figure 11.4). If it is assumed that water is partially excluded from the metal contact zone [7,21], then surfactant adsorption on the metal surface and its interaction with NAM-oil are

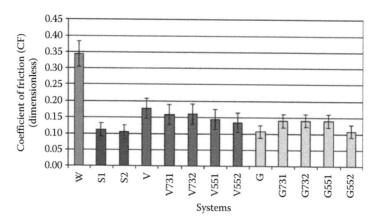

FIGURE 11.4 (See color insert.) CF of different systems evaluated. The abbreviation code is presented in Table 11.5.

responsible almost totally for the lubrication properties of emulsions, as it has been well established in the literature [24–26].

The CF values of the NAM-oil/W emulsion show a small increment when the NAM-oil content is raised from 50% to 70%. Now, the liquid/liquid interfacial area is higher and requires more surfactant to adsorb and stabilize the emulsion suggesting the possibility that less surfactant is available to adsorb and lubricate the metal surface. However, this effect is not so important because the surfactant concentration exceeds 200 times the CMC; hence, there are enough surface aggregates to lubricate the metal surface.

Likewise, increasing surfactant concentration from 1% to 2% does not have a remarkable effect on CF because at 1% the surfactant concentration is already 200 times the CMC, meaning that the interfaces are saturated and the additional surfactant will remain in the bulk.

From these observations, it can be inferred that in NAM-oil/W emulsion, lubricity is provided by interaction between the surfactant aggregates adsorbed at metal surface and the oil molecules present in the system [15,27].

11.3.1.1.2 Diesel/W Emulsions

Even though diesel has environmental restrictions, there are occasions, like completion and maintenance operations, where diesel/W emulsions can be safely used, because they will be displaced from the well and treated together with the produced oil. In this case, safety and health precautions should be taken to prepare and manage the diesel/W emulsion. For this reason, the tribological properties of the O/W emulsions were also evaluated using diesel as the dispersed phase.

When diesel is added to the ASS, an increase in emulsion CF is observed with respect to both the ASS and pure diesel. The emulsions prepared with 50% of dispersed oil phase show a CF that depends on the surfactant concentration (G551, G552): increasing surfactant concentration from 1% to 2% will render a lower CF. In general, the emulsions prepared with either NAM-oil or diesel show similar CF values (0.14–0.16).

The difference in CF behavior between NAM-oil and diesel, when they are emulsified, could be explained in terms of oil's wettability, i.e., ability of each oil to wet the metal surface.

11.3.2 WETTABILITY STUDIES

Contact angle measurement is a common technique to determine wettability of a solid surface by liquids. Table 11.6 shows the variation of contact angle of NAM-oil and diesel surrounded by water with different surfactant concentrations, measured from inside the oil drop to the oil/water interface. For NAM-oil/ASS, a slight reduction of contact angle was observed, compared with NAM-oil/water system, implying that NAM-oil surface affinity is higher in the presence of the evaluated surfactants. For diesel, an inverse tendency was observed indicating a decrease of diesel wettability on the metal surface with the ASS. Increasing surfactant concentration from 1% to 2% in both systems does not change oil surface wettability.

TABLE 11.6
Contact Angles of Two Different Oils in Aqueous Surfactant Solutions on Metal Surface

System Oil/Aqueous Surfactant Concentration	Contact Angle (θ)		
	Left (°)	Right (°)	Average
NAM-oil	160.3	160.4	160.4
NAM-oil/1%	152.3	152.5	152.4
NAM-oil/2%	151.0	151.1	151.1
Diesel	114.9	113.0	114.0
Diesel/1%	156.8	156.9	156.9
Diesel/2%	155.3	155.6	155.5

Influence of surfactant concentration on the oil's wetting properties.

Two important concepts that need to be considered are oil polarity and orientation and packing of the adsorbed surfactant [11]. Oil polarity will affect oil-surface affinity, as observed in Figure 11.5. The nonpolar NAM-oil, a mixture of saturated hydrocarbon compounds, practically does not wet the metal surface when surrounded by water (Figure 11.5a); but diesel, that contains a high quantity of polar aromatic compounds, shows a high affinity for the metal surface, due to adsorption by polarization of aromatic π electrons [11].

Concerning orientation and packing of the absorbed surfactant, it is known that above the CMC surfactants will aggregate in different geometrical forms (spherical, cylindrical and others) called micelles, and when adsorbed at a metal surface these have been designated as surface aggregates to distinguish them from micelles in solution [11].

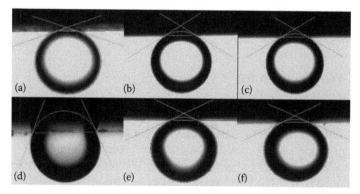

FIGURE 11.5 (See color insert.) Contact angles of two different oils in ASS on metal surface: (a) NAM-oil/water, (b) NAM-oil/1% ASS, (c) NAM-oil/2% ASS, (d) diesel/water, (e) diesel/1% ASS, and (f) diesel/2% ASS.

Since the highest surfactant proportion is anionic, the surface aggregates will have a negative charge, which will be imparted to the metal surface. If the metal surface is now negatively charged, diesel will have less metal surface affinity (Figure 11.5d and f) due to repulsion between aromatic π electrons and the negative charge of surface aggregates. In the case of NAM-oil, which contains a high percent of saturated organic compounds, the addition of surfactant slightly increases the oil surface affinity (Figure 11.5b and c). It is believed that this occurs due to the possible interaction between oil molecules and the hydrophobic groups of the surface aggregates adsorbed on the metal surface.

According to the literature, the wettability phenomenon is related to the tribological properties of O/W emulsions [28]. In this study, both oils achieve similar metal surface wettability (151.1°–156.9°), when surfactants mixture is present in the oil–water–metal system. This wettability behavior is in correspondence with the similarity of CF values obtained for O/W emulsions (Table 11.6 and Figure 11.4).

Improving oil wettability will decrease the CF of emulsions. Compared with pure oils as lubricants, similar to lower CF values would be obtained for NAM-oil/W emulsions as NAM-oil wettability slightly increases (Figure 11.5). Likewise, oil wettability decreases in the diesel–metal–ASS system, then similar to higher diesel/W emulsions CF values would be obtained (Figures 11.4 and 11.5).

11.3.3 WEAR EVALUATION

Wear was qualitatively evaluated by analyzing and comparing several consecutive pictures of the metal block surface wear scar, taken with an optical microscope, after completing the lubricity test. Additionally, the roughness and texture of the metal test block were analyzed using an optical surface profilometer.

Optical profilometry test includes a roughness index (R_a), which is an arithmetic mean roughness. In this study, roughness index was not considered as a parameter because steel blocks had a rough surface in the original state (before lubricity test) (Figure 11.6) and comparison could be misleading or ambiguous. After the test, what is relevant is the scar or wear pattern characteristics, i.e., whether it is homogeneous, symmetrical, flat, or with crest and valleys.

Figure 11.6 is the optical microscope image of the center of an unused metal test block, which is presented as a reference pattern of the metal surface. The surface profilometry images are also shown (Figure 11.7).

11.3.3.1 NAM-Oil/W Emulsions

Figures 11.8 through 11.11 are the optical microscope pictures of the different steel test blocks after performing the lubricity test, where 50/50 and 70/30 NAM-oil/W emulsions stabilized with 1% and 2% surfactant concentrations were used as lubricants. Pictures with less amount of dispersed NAM-oil (50/50) show that

FIGURE 11.6 (See color insert.) Optical image of the center of an unused metal test block.

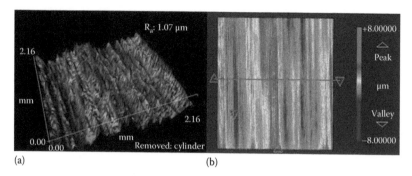

(a)   (b)

FIGURE 11.7 (See color insert.) Optical surface profilometry images of an unused metal test block: (a) areal view and (b) profile.

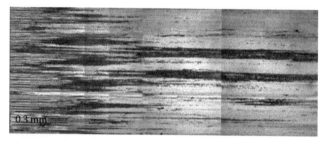

FIGURE 11.8 (See color insert.) Metal surface after the lubricity test for lubricant V731.

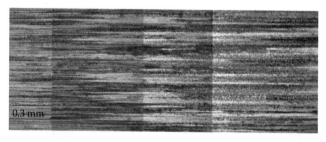

FIGURE 11.9 (See color insert.) Metal surface after the lubricity test for lubricant V732.

FIGURE 11.10 (See color insert.) Metal surface after the lubricity test for lubricant V551.

FIGURE 11.11 (See color insert.) Metal surface after the lubricity test for lubricant V552.

an oxidation process occurred with deposition of debris on the steel test block surface, possibly due to metal asperities contacts; in these emulsions, increasing surfactant concentration from 1% to 2% reduces surface's oxidation (Figures 11.10 and 11.11). For emulsion V731 (more oil, less surfactant) lower oxidation and no debris are observed, but when raising surfactant concentration from 1% to 2% (V732), the oxidation product is higher and homogeneously distributed along the surface (Figure 11.9).

Figures 11.12 through 11.15 are the optical surface profilometry images of the steel test block surface used with the NAM-oil/W emulsions. In general, low wear is observed in these pictures, the most damaged surface being the one that uses emulsions with less oil and less surfactant (V551) and more oil and more surfactant (V732). This analysis is in agreement with the optical microscopy results previously discussed. No metal mass transfer is noted with NAM-oil/W emulsions.

A very interesting observation is that, even though there is low plastic deformation in all the samples, surfaces that were lubricated with V731 and V552 present a symmetrical wear pattern, with crest and valleys, even more homogeneous than the reference pattern (Figures 11.6 and 11.7). It seems that the V731 and V552 lubricant emulsions uniformly distribute the load applied.

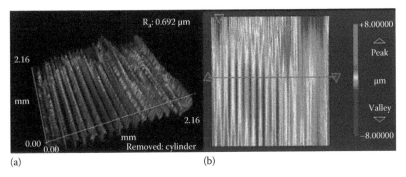

(a)   (b)

FIGURE 11.12 (See color insert.) Optical surface profilometry images of the metal block after the lubricity test for lubricant V731: (a) areal view and (b) profile.

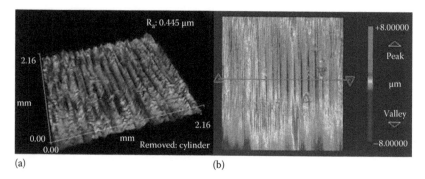

(a)   (b)

FIGURE 11.13 (See color insert.)   Optical surface profilometry images of the metal block after the lubricity test for lubricant V732: (a) areal view and (b) profile.

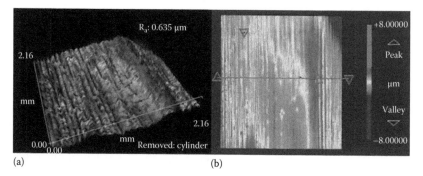

(a)   (b)

FIGURE 11.14 (See color insert.)   Optical surface profilometry images of the metal block after the lubricity test for lubricant V551: (a) areal view and (b) profile.

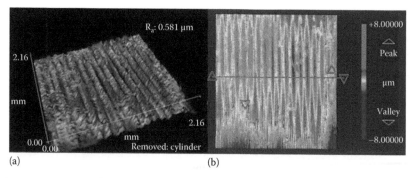

(a)   (b)

FIGURE 11.15 (See color insert.)   Optical surface profilometry images of the metal block after the lubricity test for lubricant V552: (a) areal view and (b) profile.

FIGURE 11.16 (See color insert.) Metal surface after the lubricity test for lubricant W.

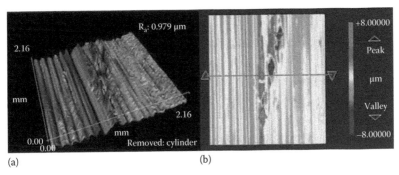

(a)   (b)

FIGURE 11.17 (See color insert.) Optical surface profilometry images of the metal block after the lubricity test for lubricant W: (a) areal view and (b) profile.

To explain the difference in wear behavior as a function of emulsion formulation, the individual performance of all the systems was studied: water (Figures 11.16 and 11.17), the ASS (Figures 11.18 through 11.21), and pure NAM-oil (Figures 11.22 and 11.23).

The ASS with 1% surfactant concentration has the best performance of all evaluated systems (Figures 11.18 and 11.19). A higher surfactant concentration has a detrimental effect on antiwear properties (Figures 11.19 and 11.20).

FIGURE 11.18 (See color insert.) Metal surface after the lubricity test for lubricant S1.

FIGURE 11.19 (See color insert.) Metal surface after the lubricity test for lubricant S2.

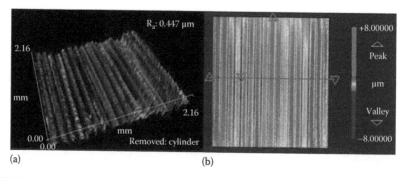

FIGURE 11.20 (See color insert.) Optical surface profilometry images of the metal block after the lubricity test for lubricant S1: (a) areal view and (b) profile.

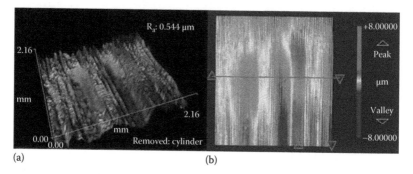

FIGURE 11.21 (See color insert.) Optical surface profilometry images of the metal block after the lubricity test for lubricant S2: (a) areal view and (b) profile.

FIGURE 11.22 (See color insert.) Metal surface after the lubricity test for lubricant V.

Figure 11.16 shows the optical microscope picture taken of the steel test block using water (W) as the lubricant. In the middle of the picture, part of the scar can be seen caused by possible metal–metal contact, and between the block and the ring the metal looks polished. In the outer area of the scar, the metal suffers an oxidation process with deposition of fine wear debris on the surface.

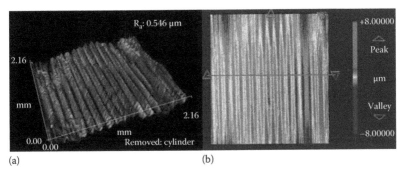

(a)   (b)

FIGURE 11.23 (See color insert.) Optical surface profilometry images of the metal block after the lubricity test for lubricant V: (a) areal view and (b) profile.

The profilometry shows severe wear on the center of the metal block evidencing metal–metal asperities contact (Figure 11.17); such pattern could be generated by the existence of adhesive wear mechanism initiated by lubricant film failure due to the poor load-carrying properties of pure water [7].

The optical microscope picture of the steel test block surface after using 1% ASS (S1) as the lubricant (Figure 11.18) shows debris and fine oxide particles dispersed over the surface. Also, a colored film (light blue with patterns on light yellow and red) can be seen in the friction zone. This colored film could be a result of a possible chemical degradation of the sulfate group of surfactant molecules adsorbed onto the metal surface [8,26,29]. This chemical reaction may allow the formation of a sacrificial film on the surface. It is known that additives with sulfur, chlorine, or phosphorus and others may react with exposed metallic surfaces creating protective, low shear strength surface films, which reduce friction and wear; these additives are termed as EP additives [7,30]. If this presumption is true, the surfactant evaluated acts like a boundary and EP additive in water, under the experimental conditions evaluated [7].

Figure 11.20 shows surface plastic deformation in an even more homogeneous and very symmetrical wear pattern, than the one produced with V731 and V552 NAM-oil emulsions (Figures 11.12 and 11.15). It is known that surfactants may adsorb onto the surface, depending on the surface affinity, whereas the monomers and micelles will fill up microasperities generating a film. This film strongly bonded with the surface may enlarge the real area of contact distributing the load and thus, improving the load-carrying capacity of the system [7,24].

When the surfactant concentration was increased in the aqueous solution (Figure 11.19), the colored film was still present, but in this case, a significant area of wear was noted. Increasing surfactant concentration (Figures 11.19 and 11.21) has a detrimental effect on the wear reduction properties of the ASS, probably due to a change in the aggregation size and geometry of the adsorbed surface aggregates, producing a weaker unstable film, with a loss in load-carrying properties allowing contact between the sliding surfaces [7,24,25].

Likewise, metal surface tested with pure NAM-oil (Figures 11.22 and 11.23) shows no debris, low-to-moderate wear, and no colored film. This last observation is in correspondence with the absence of S-, Cl-, and P-compounds in the NAM-oil, which would react with the metal surface [8]. Profilometry analysis (Figure 11.23) presents a symmetrical and homogeneous wear pattern, similar to S1, implying that NAM-oil also possesses good load-carrying capacity under the experimental conditions evaluated.

These results allow to propose that the film with the best performance is an ordered structure formed by the adsorbed surface aggregates and oil molecules, which is determined by oil and surfactant concentrations [11,27,28].

11.3.3.2 Diesel/W Emulsions

Figures 11.24 through 11.27 are the optical microscope pictures of the different metal blocks, where diesel/W emulsions stabilized with 1% and 2% surfactant concentrations used as lubricants. In all pictures, fine oxidation particles are observed

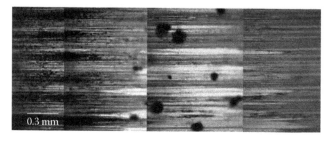

FIGURE 11.24 (See color insert.) Metal surface after the lubricity test for lubricant G731.

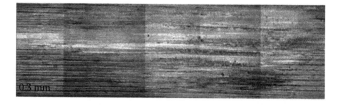

FIGURE 11.25 (See color insert.) Metal surface after the lubricity test for lubricant G732.

FIGURE 11.26 (See color insert.) Metal surface after the lubricity test for lubricant G551.

FIGURE 11.27 (See color insert.) Metal surface after the lubricity test for lubricant G552.

distributed along the surface. Except for system G731, a colored film (light blue with patterns on light yellow and red) is formed, similar to the ASS systems (Figures 11.18 and 11.19).

Formulation G731 permits generation of a lubricant film strongly bonded to the surface, which distributes more effectively the load applied than the other diesel/W formulations, avoiding metal–metal contacts and hence the formation of sacrificial film is obviated.

Figures 11.28 through 11.31 are the optical surface profilometry images of the metal blocks after the test where diesel/W emulsions were used as lubricants. High to severe wear process was observed in these pictures. Adhesive wear with metal mass loss was present with system G551 as lubricant (Figure 11.30), like what was observed when water was used as lubricant (Figure 11.17), which is an evidence of lubricant film failure [7].

Similarly to NAM-oil/W emulsions, the surface deformation depends on the evaluated formulation: low plastic deformation was observed with higher diesel/W ratio (G731) and lower surfactant concentration (Figure 11.28), while high to severe plastic deformation was observed in other formulations (G732, G551, and G552) (Figures 11.29 through 11.31).

Pure diesel was also evaluated and the optical microscope image (Figure 11.32) shows the oxidation process suffered by the metal surface after using diesel (G) as lubricant, producing a large quantity of fine oxide particles deposited on the metal surface. Although, diesel has traces of sulfur compounds and other impurities, the

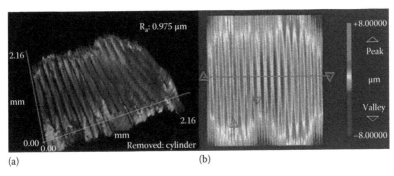

(a)   (b)

FIGURE 11.28 (See color insert.) Optical surface profilometry images of the metal block after the lubricity test for lubricant G731: (a) areal view and (b) profile.

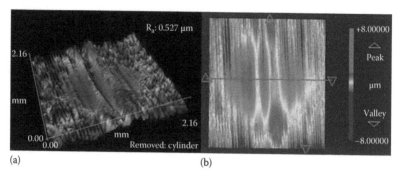

(a)   (b)

FIGURE 11.29 (See color insert.) Optical surface profilometry images of the metal block after the lubricity test for lubricant G732: (a) areal view and (b) profile.

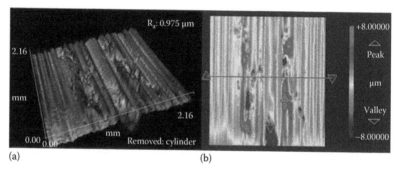

(a)   (b)

FIGURE 11.30 (See color insert.) Optical surface profilometry images of the metal block after the lubricity test for lubricant G551: (a) areal view and (b) profile.

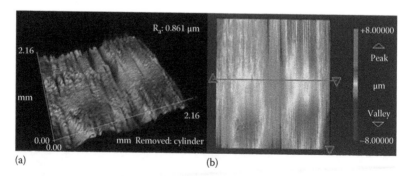

(a)   (b)

FIGURE 11.31 (See color insert.) Optical surface profilometry images of the metal block after the lubricity test for lubricant G552: (a) areal view and (b) profile.

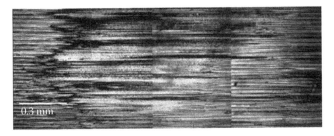

FIGURE 11.32 (See color insert.) Metal surface after the lubricity test for lubricant G.

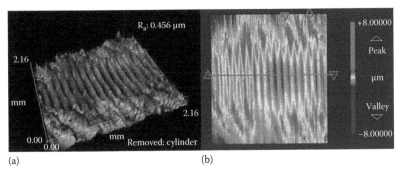

(a)   (b)

FIGURE 11.33 (See color insert.) Optical surface profilometry images of the metal block after the lubricity test for lubricant G: (a) areal view and (b) profile.

colored film was not observed, meaning that the colored (sacrificial) film is produced by the possible reaction of the adsorbed surfactant molecules with surface metal atoms, producing colored metallic sulfur salts [7,8,29].

Optical surface profilometry image (Figure 11.33) of the metal block tested with diesel (G) presents high wear accompanied by high plastic deformation bordering the friction zone.

11.4 CONCLUSIONS

With the surfactants mixture used in this study, the oil type does not have a significant effect on the CF of O/W emulsions due to the similar wettability behavior observed at the metal surface. However, NAM-oil/W emulsions have better antiwear properties than the diesel/W emulsions. This implies that conventional CF measurements are not enough to evaluate lubricating properties of oil well working fluids. Surface analysis must be done to compare antiwear performance.

Under the evaluated test conditions and with the specified surfactants mixture, the lubricity performance and antiwear properties of O/W emulsions are affected by oil/water ratio and surfactants mixture concentration, showing a systemic interaction between these two parameters. Being under a boundary and EP lubrication regimes, emulsion viscosity and droplet size do not have any effect on the lubricating properties.

The film formed at metal surface with all NAM-oil/W emulsions and diesel (70)/W(30) emulsion with 1% surfactant concentration can withstand loads higher than 488 N before entering in the EP lubrication regime. All other evaluated systems fall in the EP lubrication regime at lower loads and where they will form a sacrificial film if surfactant is present.

The 1% surfactants mixture studied could be used as a lubricant additive for pure water-based oil well working fluids.

REFERENCES

1. J.J. Truhan, R. Menon, and P.J. Blau. The evaluation of various cladding materials for down-hole drilling applications using the pin-on-disk test. *Wear* 259, 1308–1313 (2005).
2. R. Caenn and G.V. Chillingar. Drilling fluids: State of art. *Petrol. Sci. Eng.* 14, 221–230 (1996).
3. M. Zamora and M. Stephens. Drilling fluids. In: *Petroleum Well Construction.* Chapter 5: pp. 119–142, John Wiley & Son Ltd., England (1998).
4. W.E. Foxenberg, S.A. Ali, T.P. Long, and J. Vian. Field experience shows that new lubricant reduces friction and improves formation compatibility and environmental impact. Paper # SPE 112483. In: *SPE International Symposium and Exhibition on Formation Damage Control Proceedings* (2008).
5. M.S. Aston, P.J. Hearn, and G. McGhee. Techniques for solving torque and drag problems in today's drilling environment. Paper # SPE 48939. In: *SPE Annual Technical Conference and Exhibition Proceedings* (1998).
6. J.D. Kercheville, A.A. Hinds, and W.R. Clements. Comparison of environmentally acceptable materials with diesel oil for drilling mud lubricity and spotting fluid formulations. Paper # IADC/SPE 14797. In: *IADC/SPE Drilling Conference Proceedings* (1986).
7. G. Stachowiak and A.W. Batchelor. *Engineering Tribology*, 2nd edn., Chapter 1: pp. 1–7, Chapter 2: pp. 11–45, Chapter 3: pp. 51–95, Chapter 8: pp. 357–404, Chapter 12: pp. 533–550, Butterwort-Heinemann Inc., Boston, MA (2001).
8. S.M. Hsu. Molecular basis of lubrication. *Tribol. Int.* 37, 553–559 (2004).
9. J. Holand, S.A. Kvamme, T.H. Omland, A. Saasen, and K. Taugbol. Lubricants enabled completion of ERD well paper # IADC/SPE 105730. In: *IADC/SPE A Drilling Conference Proceedings* (2007).
10. J. Blanco, Inventor; L. Quintero, Inventor. PDVSA Intevep, S.A., assignee. Well stimulation with hydrophilic fluids. United States of America, U.S. patent 5783525 (July 21, 1998).
11. M.J. Rosen. *Surfactants and Interfacial Phenomena*, 3rd edn., Chapter 2: pp. 35–95, Chapter 6: pp. 243–268, Chapter 8: pp. 306–320, John Wiley & Sons, Hoboken, NJ (2004).
12. L. Bécu, P. Grondin, A. Colin, and S. Manneville. How does a concentrated emulsion flow? Yielding, local rheology, and wall slip. *Colloids Surf.* 263, 146–152 (2004).
13. Z. Pawlak, B.E. Klamecki, T. Rauckyte, G.P. Shpenkov, and A. Kopkowski. The tribochemical and micellar aspects of cutting fluids. *Tribol. Int.* 38, 1–4 (2005).
14. A.K. Tieu, P.B. Kosasih, and A. Godbole. A thermal analysis of strip-rolling in mixed-film lubrication with O/W emulsions. *Tribol. Int.* 39, 1591–1600 (2006).

15. A. Cambiella, J.M. Benito, C. Pazos, J. Coca, A. Hernandez, and J.E. Fernandez. Formulation of emulsifiable cutting fluids and extreme pressure behavior. *J. Mater. Proc. Technol.* 184, 139–145 (2007).
16. U.C. Chen, Y.S. Liu, C.C. Chang, and J.F Lin. The effect of the additive concentration in emulsions to the tribological behavior of a cold rolling tube under sliding contact. *Tribol. Int.* 35, 309–320 (2002).
17. K.R. Januszkiewicz, A.R. Riahi, and S. Barakat. High temperature tribological behavior of lubricating emulsions. *Wear* 256, 1050–1061 (2004).
18. ASTM D 2509-03. Standard Test Method for Measurement of Load-Carrying Capacity of Lubricating Grease (Timken Method). 326/83 (88). Reapproved 2008.
19. ASTM D 2782-02. Standard Test Method for Measurement of Extreme-Pressure of Lubricating Fluids (Timken Method). 240/84. Reapproved 2008.
20. F. Quintero, R.L. Márquez, J.M. González, and S.D. Rosales Anzola. Formulation effects on the stability of drilling oil/water emulsions based on interfacial dilatational rheology and light scattering techniques. Paper presented at the *17th International Symposium on Surfactants in Solution (SIS)*, Berlin, Germany (August 2008).
21. A. Chojnicka, G. Sala, C.G. Kruif, and F. Van de Velde. The interactions between oil droplets and gel matrix affect the lubrication properties of sheared emulsion-filled gels. *Food Hydrocolloids* 23, 1038–1046 (2009).
22. EP/Lubricity Tester, *Manual of Operation*. Baroid Division NL Industries, Inc., Houston, TX (1983).
23. R.H. Sifferman, T.M.M. Muijs, G.F. Fanta, F.C. Felker, and S.M. Erhan. Starch-lubricant compositions for improved lubricity and fluid loss in water-based drilling muds. Paper # SPE 80213. In: *SPE International Symposium on Oilfield Chemistry Proceedings* (2003).
24. M.W. Sulek and T. Wasilewski. Tribological properties of aqueous solutions of alkyl polyglucosides. *Wear* 260, 193–204 (2006).
25. M. Graca, J. Bongaerts, J.R. Stokes, and S. Granick. J. Friction and adsorption of aqueous polyoxyethylene (Tween) surfactants at hydrophobic surfaces. *Colloid Interface Sci.* 315, 662–670 (2007).
26. S.K. Misra and R.O. Sköld. Lubrication studies of aqueous mixtures of inversely soluble components. *Colloids Surf. A: Physicochem. Eng. Asp.* 170, 91–106 (2000).
27. T. Wasilewski and M.W. Sulek. Paraffin oil solution of the mixture of sorbitan monolaurate-ethoxilated sorbitan monolaurate as lubricants. *Wear* 261, 230–234 (2006).
28. A. Cambiella, J.M. Benito, C. Pazos, and J. Coca. Interfacial properties of oil-in-water emulsions designed to be used as metalworking fluids. *Colloids Surf. A: Physicochem. Eng. Asp.* 305, 112–119 (2007).
29. B.K. Sharma, A. Adhvaryu, and S.Z. Erhan. Friction and wear behavior of thioether hydroxy vegetable oil. *Tribol. Int.* 42, 353–358 (2009).
30. O. Furlong, F. Gao, P. Kotvis, and W.T. Tysoe. Understanding the tribological chemistry of chlorine-, sulphur- and phosphorus-containing additives. *Tribol. Int.* 40, 699–708 (2008).

Part III

Biobased Lubricants

12 Estolides: Biobased Lubricants

Steven C. Cermak

CONTENTS

ABSTRACT

Estolides were originally developed as a cost-effective derivative from vegetable oil sources to overcome the problems associated with standard vegetable oils as lubricants. Classic estolides, estolide fatty-acids (FAs) and estolide esters, are formed by the formation of a carbocation at the site of unsaturation that can undergo nucleophilic addition by another FA, with or without carbocation migration along the length of the chain, to form an ester linkage. The secondary ester linkages of the estolide are more resistant to hydrolysis than those of triglycerides (TGs), and the unique structure of the estolide results in materials that have far superior physical properties for lubricant applications than vegetable and mineral oils.

A second type of estolide, TGs estolides are from lesquerella and castor TGs. TG-estolides have been developed directly from the TG of the vegetable oil. The TG-estolide synthesis utilizes the hydroxyl group already on the molecule for a simple esterification process to form these new estolides. The unique structures of these TG-estolides allow for the materials that have far superior physical properties for lubricant applications than vegetable and mineral oils.

All the different types of estolides and estolide esters to date from oleic acid, saturated FAs, meadowfoam FAs, tallow FAs, castor and lesquerella oil or FAs, or any source of hydroxy FAs have shown promise in cosmetics, biodiesel additives, coatings, and biodegradable lubricants. The different types of estolides and esters usually outperform the commercially available industrial products such as petroleum-based hydraulic fluids, soy-based fluids, and petroleum oils.

12.1   INTRODUCTION

The use and investigation of biobased industrial oils and derivatives have increased dramatically in past years. The United States has recently seen crude oil prices topping $150/barrel compared to <$20/barrel in the late 1990s [1]. At these prices, we have entered the price zone that makes biobased lubricants not only economically feasible but a necessity for future economic growth. During the summer of 2008, we entered into the price range where biobased materials are profitable but still most companies and investors are concerned about the return of very low crude prices that history has shown [1]. In 2009, the world has seen new Green Energy or Renewable Energy companies start up and government leaders pushing for these technologies.

Our world economies were pushed to financial breaking points at the end of 2008 as our demand for petroleum outpaced our ability to pay for this high-priced energy source. With the fall of world markets, countries going under financially, and the price of crude oil falling to as low as ~$35/barrel since the peak in July 2008, the real question will be: will we still push toward biobased materials or return to our old habits? Past trends have shown that peaks in the crude oil markets have been followed by long periods of reduced prices, which have driven biobased competitors out of the market and out of business. Current data and experts suggest that increased prices for crude oil are likely to return in the near future as our economies recover and our global demands increase, which should help propel the demand for biobased materials.

There has also been much debate on global warming and what is being done to the planet "Earth." The consumers of today are aware of their effects on the environment and are becoming willing to replace petroleum products with biobased "Green" materials. These recent world events have given researchers the go-ahead to explore biobased materials as potential fuels, lubricants, and base material for industrial Green products. In particular, basic questions such as "what makes a biobased material a good lubricant?" or "how does structure impact certain physical properties?" are being investigated and reported in the literature [2].

Vegetable oils are one source of oil being investigated as a possible biobased lubricant. Oils from sunflower, soybean, castor, rapeseed, and coconut are a few examples of oils currently being studied. Vegetable oils and their derivatives show superior biodegradability as compared to petroleum oils [3] and synthetic esters in actual field studies [4,5]. Vegetable oils and synthetic esters also have superior viscosity indices [6] and lubricating properties compared to mineral oils. The coefficient of friction for vegetable oils is nearly one-half of that of refined mineral oils [7]. Vegetable oil-based lubricants and derivatives also have excellent lubricity and biodegradability properties for which they are being more closely examined as a base stock for lubricants and functional fluids [7–11].

Two major problems with vegetable oils as functional fluids are their low resistance to thermal oxidative stability [12] and poor low-temperature properties [13,14]. However, with the addition of additives, these properties can sometimes be improved but only at the sacrifice of biodegradability, toxicity, and increased costs.

Originally, development of a cost-effective derivative from a vegetable oil source was envisioned, which would overcome the problems associated with standard vegetable oils as lubricants. Estolides (Figure 12.1) are one example of synthetic biobased materials that may help lessen our demand on foreign oils as well as limit our greenhouse emissions. Estolides have been found in nature [15] and have been synthesized [16–18] in the laboratory and have led to the development of new products from industrial new crops [17–22].

Estolides are a class of esters, based on vegetable oils [23] made by the formation of a carbocation at the site of unsaturation that can undergo nucleophilic addition by another fatty-acid (FA), with or without carbocation migration along the length of the chain, to form an ester linkage (Figure 12.1). These ester linkages are used to help characterize the structure of the estolide, since the estolide number (EN) is defined as the average number of FAs added to a base FA (Figure 12.1, EN$=n+1$). The secondary ester linkages of the estolide are more resistant to hydrolysis than those of triglycerides (TG), and the unique structure of the estolide results in materials that have far superior physical properties for lubricant applications than vegetable and mineral oils [24,25].

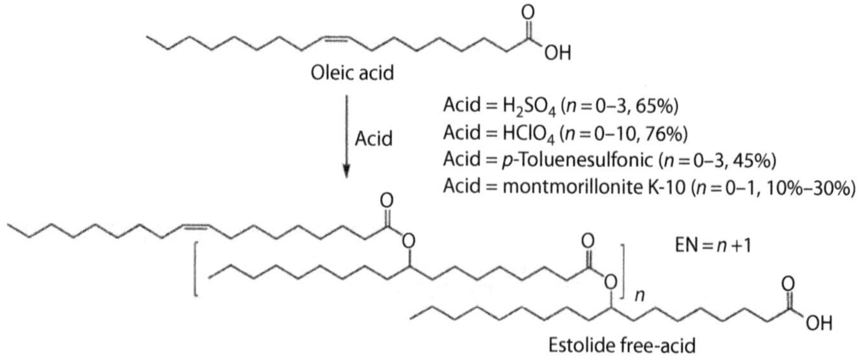

FIGURE 12.1 General oleic estolide free-acid synthesis.

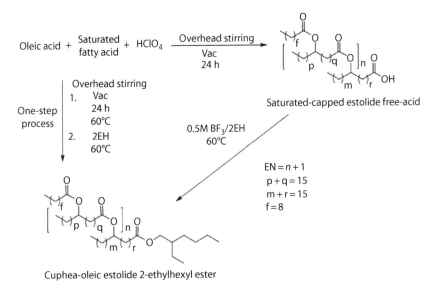

Saturated-capped estolide free-acid

$EN = n+1$
$p+q = 15$
$m+r = 15$
$f = 8$

Cuphea-oleic estolide 2-ethylhexyl ester

FIGURE 12.2   General saturated capped estolide 2EH ester synthesis.

In the early 2000, Cermak and Isbell [18,24] developed a series of saturated estolides (Figure 12.2) in which oleic acid and saturated FAs ranging from butyric through stearic were treated with 0.4 equivalents of perchloric acid at either 45°C or 55°C to produce complex estolides, saturated-oleic estolide 2-ethylhexyl esters (O-EST-2EHs). The shorter-chain, saturated FAs, that is, butyric, and hexanoic acid provide materials with higher degrees of oligomerization ($EN = 3.3$) than stearic acid ($EN = 1.4$). These saturated-capped estolide free-acids and estolide esters showed very different physical properties in terms of cold temperature properties. Cermak and Isbell [24] theorized that, by varying the capping material on the estolide, the crystal lattice structure of the material was disrupted as it approached its pour point, which led to estolide esters with excellent low-temperature properties, pour point of −36°C, and cloud point of −41°C. These saturated-capped estolide 2EH esters have eliminated common problems associated with the use of vegetable oils as functional fluids.

Estolides have been developed from both FAs and directly from the vegetable oil (Figure 12.3). In order to obtain a wide variety of different estolides, new crops were used to explore the full range of possibilities. Meadowfoam (*Limnanthes alba*) is a promising new crop that is grown mainly near the Pacific Coast of North America. Meadowfoam, a winter annual, is establishing a role as an alternative crop for seed grasses currently grown in this region. The seed oil is unique in that the TG consists of a >95% mixture of FAs with carbon chain lengths greater than 18. The composition of the FAs is (~66%) 5-eicosenoic acid, (~16%) 5,13-docosadienoic acid, and (~11%) 5- and 13-docosenoic acids [26].

Meadowfoam oil and its FAs have been utilized in the development of novel materials with potential use as industrial agents. The oil has been vulcanized and

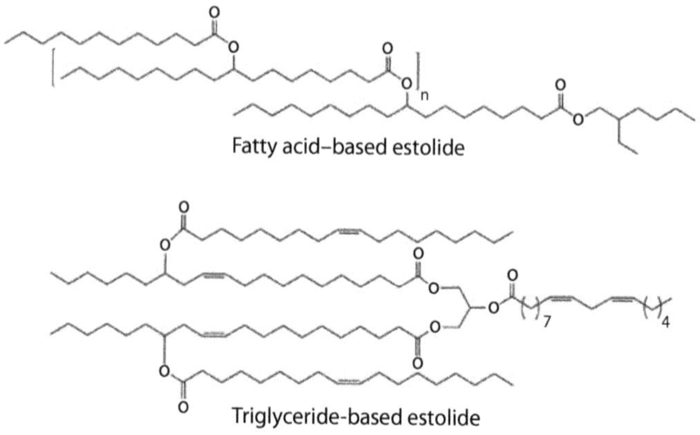

Fatty acid–based estolide

Triglyceride-based estolide

FIGURE 12.3   Estolide 2-ethylhexyl esters versus TG-estolides.

exhibited good properties in rubber applications [27–29]. The FAs have been converted to amides [30,31], lactones [32], dimer acids [33], and estolides [34,35]. The last are unique oligomeric FAs that contain a secondary ester linkage on the alkyl backbone of the FA. The high-pressure clay technique with meadowfoam FAs provided modest yields (<30%) of monoestolide. In contrast, the mineral acid-catalyzed condensation of oleic acid provided good yields of estolide (70%) with the formation of many ester oligomers (EN 2.65). The mineral acid-catalyzed process used for oleic acid estolide synthesis proved useful in the development of meadowfoam estolide. In addition, the proximal location of the Δ-5 unsaturation contributes to the physical properties of this estolide.

Another new crop that was utilized to produce FA-based estolides was from cuphea [21]. Cuphea (*Lythraceae*) is an annual plant that produces a small seed rich in saturated medium-chain TG. The initial oil characterization of a number of cuphea species was done at the United States Department of Agriculture (USDA) Research Center in Peoria, IL, in the early 1960s [36]. To address the need for higher seed yields, higher oil content, and less seed shattering, Steve Knapp [37] at Oregon State University began developing promising cuphea crosses. One of these new crosses, PSR-23 (where PSR refers to partial seed retention), was developed at the University of Oregon and planted in the Midwest. This cross has been mechanically harvested with conventional farm equipment by researchers at the USDA in Morris, Minnesota, and Peoria, Illinois, for the past 9 years. PSR-23 is a cross that is high in C10, decanoic acid (65%; Table 12.1) while C12, lauric, is the major FA found in coconut oil.

Both TG and FA-based estolides are possible from the new crop Lesquerella [20,22]. *Lesquerella fendleri* is a winter annual seed oil crop native to the desert southwestern United States and is currently undergoing an intensive research effort for its successful introduction into agriculture. Lesquerella produces a small seed that contains about >30% oil that is composed of 55%–64% hydroxy FA [38,39]. The hydroxy FAs of lesquerella (Table 12.2) [39] are lesquerolic (55%–60%, 14-hydroxy-*cis*-11-eicosenoic acid) and auricolic (2%–4%, 14-hydroxy-cis-17-eicosenoic acid),

TABLE 12.1

FA Compositions of Cuphea and Coconut Oils[a]

Percent FA (Mass %)

FAME[b]	8:0	10:0	12:0	14:0	16:0	18:0	18:1	18:2
Cuphea[c]	0.6	65.6	3.2	6.5	7.2	0.8	9.1	5.9
Coconut	5.1	4.9	48.3	21.7	10.6	7.0	2.4	—

[a] Determined by GC (SP-2380, 30 m × 0.25 mm inner diameter [i.d.]).
[b] FA methyl ester.
[c] PSR-23.

TABLE 12.2

Chemical Composition of Lesquerella and Castor Oils[a]

FAME[b]	Lesquerella Oil (Mass %)	Castor Oil (Mass %)
16:0	1.1	1.0
16:1	0.7	—
18:0	1.8	—
18:1	15.4	3.7
18:2	6.9	4.4
18:3	12.2	—
20:0	0.2	—
20:1	1.0	—
20:2	0.2	—
18:1 Hydroxy	0.6	89.0
20:1 Hydroxy	55.4	1.1
20:2 Hydroxy	3.8	—

[a] Determined by GC (SP-2380, 30 m × 0.25 mm i.d.).
[b] FA methyl ester.

a homologue of ricinoleic acid obtained from castor oil. Lesquerella production in the past few years has averaged 16.2 ha grown each year. In addition, seed yield of 2016 kg/ha compared to past releases of 1344 kg/ha [40] will jointly allow lesquerella to be competitive in the hydroxy oil market.

Unlike either estolide FAs or estolide esters (Figure 12.1) that are formed when the carboxylic acid functionality of one FA links to the site of unsaturation of another FA to form oligomeric esters, the FA esters of lesquerella and castor have a hydroxy functionality that provides a site for simple esterification to take place to produce TG-based estolides (Figure 12.4). Hydroxy FAs, such as lesquerella, can be readily converted into oil-based estolides [41,42] either as TGs in the presence

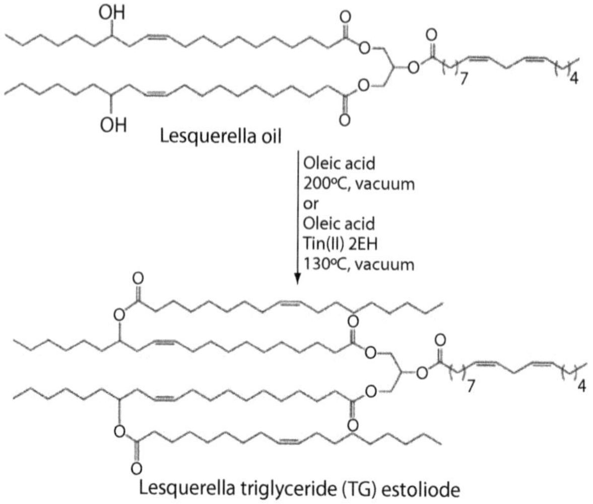

FIGURE 12.4 General lesquerella TG-estolide synthesis.

of free FA or from homopolymerization of the split FAs. The synthesis of some estolides from castor oil and FAs has been reported [43,44], but not as a complete set with the physical properties (where as Cermak and Isbell [20] recently published the missing data).

Lesquerella and castor TG estolides synthesized directly from the TG were recently reported in a detailed study on the synthesis of TG-estolides [22]. Hayes and Kleiman [42] synthesized estolides from the free lesquerolic and oleic acids using a lipase catalyst. However, no reports were available on either estolide types from lesquerella and castor capped with different FAs or their physical properties until reports either by Isbell or Cermak [20,45].

There are many challenges when introducing a new crop such as meadowfoam, cuphea, or lesquerella into the marketplace. Not only must farmers and policy makers be willing to try new things, but the market must also be able to support these new feedstocks or products. Most products from new crops must be so-called intermediate products to meet the economic challenges all new crops demand. For example, in the case of cuphea, for cuphea, to meet detergent manufacturers' demands for short-chain FA, farmers would need to plant several million acres. Since it would be unfeasible to go from the research plots (<10 ac) today to millions of acres in the next season, farmers must become interested in growing new crops, and there must be a guaranteed economic advantage in doing so. One way this is accomplished is with the use of high-value end products that do not require large initial plantings. The cuphea crop of the future will have a FA profile similar to those of coconut and palm kernel oils (Table 12.1). The most promising intermediate products from the cuphea crop are estolide esters, which would be used as a biobased lubricant.

All estolide free acids and estolide esters to date from oleic acid [23], saturated FAs [19,24], meadowfoam FAs [46], tallow FAs [47], castor and lesquerella oil or FAs [20,22], or any source of hydroxy FAs [48] have shown promise in cosmetics [49], bio-diesel additives [50], coatings, and biodegradable lubricants. The estolide free acids and esters usually outperform the commercially available industrial products such as petroleum-based hydraulic fluids, soy-based fluids, and petroleum oils [51].

12.2 MATERIALS

Oleic acid (90%), potassium hydroxide, 2,6-di-*tert*-butyl-4-methylphenol (BHT), palladium 0.25 wt.% on activated carbon, and N,O-*bis*(trimethylsilyl)acetamide (derivatization grade) were purchased from Sigma-Aldrich (Milwaukee, WI). Ethyl acetate and hexanes (for extractions), acetone (for high-performance liquid chromatography, HPLC), perchloric acid (70%), concentrated sulfuric acid (98%), methanol, castor oil, and 2EH alcohol were purchased from Fisher Scientific Co. (Fairlawn, NJ). Acetonitrile and acetic acid (both for HPLC), charcoal, sodium hydrogenphosphate, and sodium dihydrogenphosphate were obtained from EM Science (Gibbstown, NJ). Ethanol was purchased from AAPER Alcohol and Chemical Company (Shelbyville, KY). Pyridine was purchased from Mallinckrodt (Paris, KY). Filter paper was obtained from Whatman (Clifton, NJ). Lesquerella oil was obtained from cold-pressed *Lesquerella fendleri* seed and then subsequently alkali refined, bleached, and deodorized. Cuphea oil was obtained from pressed cuphea germplasm line PSR 23 (*Cuphea viscosissima* × *Cuphea lanceolata*) obtained from cuphea seeds harvested from USDA plots in Morris, MN, and Peoria, IL. The crude cuphea oil was then subsequently alkali refined, bleached, and deodorized. The FA methyl ester (FAME) standard mixtures were obtained from Alltech Associates, Inc. (Deerfield, IL). Petroleum oils: Valvoline® 5W-30, 10W-30, 10W-40, 20W-50, SAE-30, Mobil® 10W-30, and synthetic oils: Castrol Synthetic® 10W-30, Mobil Synthetic® 10W-30, and Penzoil Synthetic® 10W–30 were obtained from Wal-Mart Department Store® (Peoria, IL). Soy-based oil: Biosoy® and Soylink® were obtained as free samples from the University of Northern Iowa (Cedar Falls, IA). Hydraulic fluid: Traveler Premium Universal Hydraulic Fluid® IVG-46 and Traveler All Season Hydraulic Fluid® IVG-46 were obtained from Tractor Supply Company® (Peoria, IL). Meadowfoam oil was supplied by the Fanning Corp. (Chicago, IL). Environlogic 132, 146, and 168 were obtained as free samples from Terresolve Technologies Ltd. (Eastlake, OH). Aeroshell 15W-50 Aviation oil obtained as a free sample from the Central Illinois Aviators (Galesburg, IL). Lubrizol Additive 7652 obtained as a free sample from the Lubrizol Corporation (Wickliffe, OH). Alkylated diphenyl amine (Vanlube NA) and a mixture of alkylated diphenyl amines (Vanlube SL) antioxidants were obtained from Vanderbilt Corporation (Norwalk, CT) and were used as supplied.

12.3 ESTOLIDE SYNTHESIS

As of 2010, two different types of estolides have been synthesized: FA-based estolides and TG-based estolides (Figure 12.3). The estolides are synthesized [16,17] by the formation of a carbocation at the site of unsaturation on a FA, which can undergo

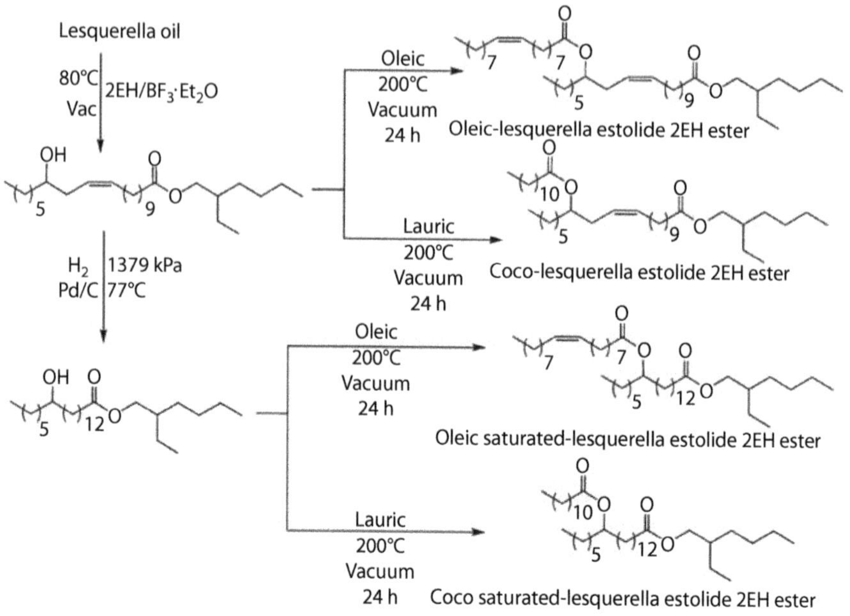

FIGURE 12.5 General lesquerella and castor estolide 2EH ester synthesis.

nucleophilic addition by another FA, with or without carbocation migration along the length of the chain, to form an ester linkage (Figure 12.1). The simple estolide structure has been easily modified by the addition of a saturated FA (Figure 12.2) using this same technology [18]. On the other hand, the TG estolides require that the vegetable oil has a hydroxy functionality present that provides a site for simple ester-ification to take place with a FA to produce TG-estolides (Figure 12.4). Estolides have also been synthesized by taking advantage of these two different technologies where the starting material is a hydroxy FA but have used the TG-estolide technol-ogy to produce the final estolides (Figure 12.5). Finally, derivates of the latter have also been explored where the unsaturation was moved to different locations in the estolide (Figure 12.5).

12.3.1 ESTOLIDE FREE-ACIDS

Acid-catalyzed condensation reactions were conducted without solvent in a 500 mL, baffled, jacketed reactor with a three-necked reaction kettle cover. The reactor was connected to a recirculating constant temperature bath maintained at either 45°C or 55°C ± 0.1°C. The reactor was equipped with an overhead stir motor using a glass shaft and a Teflon blade. The reactions were conducted at atmospheric pressure in a sealed flask. Reactions were performed under the general conditions described above while varying the type of saturated FA as reported in Table 12.3. In most cases, oleic acid (100.0 g, 354.0 mmol) and saturated FAs, i.e., lauric (35.5 g, 177.0 mmol), were combined together and heated to either 45°C or 55°C. Once the temperature was reached, perchloric acid (0.40 equivalents) was added and the flask was stoppered.

TABLE 12.3
Estolide Free-Acids Condensation Reaction (Figure 12.2) with Oleic Acid and Varying Saturated FAs[a]

Estolide[b]	Unsaturated FA	Saturated FA	Temp. (°C)	Percentage Estolides	GC EN	GC IV	% Capped[c]	Gardner Color
A	Oleic	Butyric	45	53.9	3.31	15.3	33.1	8
B	Oleic	Butyric	55	47.4	2.57	19.3	39.0	11
C	Oleic	Caproic	45	57.0	3.27	13.9	34.4	9
D	Oleic	Caproic	55	51.7	3.13	15.5	30.9	11
E	Oleic	Octanoic	45	58.8	2.89	14.2	42.4	10
F	Oleic	Octanoic	55	48.9	2.60	17.1	33.7	12
G	Oleic	Decanoic	45	64.8	2.68	12.0	53.3	18
H	Oleic	Decanoic	55	56.4	2.50	11.8	57.2	18
I	Oleic	Lauric	45	63.7	2.20	12.2	58.2	7
J	Oleic	Lauric	55	60.1	2.11	12.5	59.5	11
K	Oleic	Myristic	45	64.0	1.81	11.7	64.6	6
L	Oleic	Myristic	55	54.6	1.81	11.7	64.7	10
M	Oleic	Palmitic	45	58.5	1.92	10.1	68.4	8
N	Oleic	Palmitic	55	58.6	1.67	14.0	62.6	10
O	Oleic	Stearic	45	48.7	1.43	20.9	42.7	11
P	Oleic	Stearic	55	44.5	1.36	13.7	64.5	11

[a] Reactions were run for 24 h with overhead stirring in a 2:1 ratio, oleic:saturated FAs with 0.4 equivalents of perchloric acid.
[b] Continued on Table 12.10.
[c] Ratio of estolide capped with saturated FAs as determined by GC (SP 2380, 30 m × 0.25 mm i.d.).

Product distribution was monitored by HPLC, GC, and/or GC/MS as described below. Completed reactions were quenched by the addition of 0.5 M Na_2HPO_4 (212.4 mmol, 424.8 mL) to the reaction vessel. The reactor was disconnected from the circulating bath, and the solution was allowed to cool with stirring for 30 min. The material was transferred to a separatory funnel followed by the addition of 200 mL of a 2:1 ethyl acetate:hexanes solution. The pH of the organic layer was adjusted to 5.3–6.0 with the aid of a pH 5 buffer (NaH_2PO_4, 519 g in 4 L H_2O, 2×50 mL) followed by brine (2×50 mL). The organic layer was dried over sodium sulfate and filtered. Unreacted materials and by-products were removed using a Kügelrohr-distillation unit with a vacuum of 6–13 Pa and a temperature of 160°C–190°C to remove any lactones, saturated, and unsaturated FAs.

12.3.2 ESTOLIDE 2-ETHYLHEXYL ESTERS

The distilled, estolide free-acids were combined with a 0.5 M BF_3/2EH alcohol solution (3×estolide wt, w/v) in a 500 mL round-bottomed flask. The reactions were conducted at 60°C with magnetic stirring and monitored hourly by normal phase HPLC as described below (Table 12.4). Esterification reactions were run until >99%

2.84

TABLE 12.4

Esterification (Figure 12.2) of Estolide Free-Acids (Table 12.3) with 2EH Alcohol[a]

Estolide Ester[b]	GC EN	GC IV	Gardner Color	Decolorized Gardner Color	Gardner Color Improvement	Acid Value (mg/g)
A-2EH	2.84	18.8	11	7	4	0.91
B-2EH	2.95	17.8	13	11	2	1.56
C-2EH	3.46	13.7	11	10	1	0.94
D-2EH	2.69	17.9	13	11	2	0.87
E-2EH	2.96	14.6	11	8	3	1.16
F-2EH	3.07	15.5	14	12	2	1.19
G-2EH	2.69	11.9	18	18	0	1.05
H-2EH	2.30	12.4	18	18	0	1.46
I-2EH	2.16	12.7	12	11	1	0.96
J-2EH	1.92	14.0	15	13	2	0.90
K-2EH	1.98	10.6	11	8	3	0.78
L-2EH	1.77	11.8	14	11	3	1.03
M-2EH	1.35	27.5	18	15	3	0.12
N-2EH	1.13	17.2	17	12	5	1.42
O-2EH	1.09	25.0	12	10	2	0.80
P-2EH	1.13	14.8	14	11	3	0.60

[a] Esterification reactions were run with magnetic stirring and 0.5 M BF$_3$.
[b] Continued on Table 12.11.

complete and then were transferred to a separatory funnel and washed with brine (2 × 75 mL). The pH of the organic layer was adjusted to 5.3–6.0 with the aid of pH 5 buffer (NaH$_2$PO$_4$, 519 g in 4 L H$_2$O, 2 × 50 mL). The oil was dried over sodium sulfate and filtered. Unreacted materials were removed using a Kügelrohr-distillation unit with a vacuum of 6–13 Pa and a temperature of 100°C–120°C to remove any excess 2EH alcohol.

12.3.3 Estolide 2-Ethylhexyl Esters (One Step)

Acid-catalyzed condensation reactions were conducted without solvent in a 500 mL, baffled, jacketed reactor with a three-necked reaction kettle cover. The reactor was connected to a recirculating constant temperature bath maintained at ±0.1°C of the set point. The reactor was equipped with an overhead stir motor using a glass shaft and a Teflon blade. In most cases, oleic acid (100.0 g, 354.0 mmol) and saturated FAs, coco FAs (35.5 g, 177.0 mmol) were combined together and heated to 60°C under house vacuum, as shown in Table 12.5. Once the desired temperature was reached, perchloric acid (0.05 eq., 26.5 mmol, 2.3 mL; Table 12.5) was added, and the flask was placed under vacuum and stirred for 24 h. After 24 h, 2EH alcohol (59.6 g, 457.6 mmol, 67.5 mL) was added to the vessel, vacuum was restored, and

TABLE 12.5
Physical Properties of O-EST-2EH Esters[a] (See Figure 12.2 for Synthetic Scheme)

Estolide[b]	Oleic to Coconut Ratio	GC EN[c]	GC IV	Capped %	Pour Point (°C)	Cloud Point (°C)	Vis. at 40°C (cSt)	VI	Gardner Color
CO-EH-AD	1:3	1.49	7.33	82.0	−21	−18	149.5	138	17
CO-EH-BD	1:1	1.91	7.33	49.5	−24	−25	58.4	175	12
CO-EH-CD	1:2	1.46	5.81	58.0	−27	−22	61.1	164	13
CO-EH-DD	2:1	1.94	12.8	35.6	−33	−33	92.8	170	12
CO-EH-ED	3:1	1.96	14.2	40.9	−33	−32	86.3	232	12

[a] Estolides Kügelrohr-distilled at 160°C–190°C at 6–13 Pa to remove monomer.
[b] (Coco-oleic)-(2EH ester)-(sample # distilled).
[c] Estolide number.

the mixture was stirred for three additional hours. The completed reactions were quenched by the addition of KOH (22.3 mmol, 1.25 g, 1.2 eq. based on $HClO_4$) in 90% ethanol/water (10 mL) solution. The reactor was disconnected from the circulating bath, and the solution was allowed to cool with stirring for 30 min. The material was filtered through a Buckner funnel with Whatman #1 filter paper. The organic layer was dried over sodium sulfate and filtered. Unreacted materials and by-products were removed using a Kügelrohr-distillation unit with a vacuum of 6–13 Pa and a temperature of 160°C–190°C to remove any 2EH alcohol, lactones, saturated, and unsaturated FAs.

12.3.4 LESQUERELLA AND CASTOR MONO/FULLY CAPPED TRIGLYCERIDE ESTOLIDES

The hydroxy TGs of lesquerella and castor oil were converted into their corresponding mono (one capping group per TG molecule) and fully capped (all hydroxy functionalities capped per TG molecule) TG-estolides according to the reaction depicted in Figure 12.4. TG estolides were synthesized from a series of even-numbered carbon saturated FAs from acetic (C2) to stearic (C18) plus oleic. Due to the physical properties of the capping FAs, the TG-estolides required a number of different procedures in order to utilize the wide range of capping materials (C2–C18).

12.3.4.1 C2–C4

For the synthesis of monocapped TG-estolides, refined lesquerella oil (63.4 g, 65.8 mmol) or castor oil (61.2 g, 65.9 mmol) was combined with an equimolar amount of the corresponding acid anhydride as listed in Tables 12.6 and 12.7. Pyridine (0.98 g, 12.4 mmol) was added as a catalyst. Reactants were combined in

TABLE 12.6
Physical Properties of Lesquerella TG-Estolides (See Figure 12.4 for Synthetic Scheme)

TG Estolides	FA	EN[a]	Pour Point (°C)	Cloud Point (°C)	Vis. at 40°C (cSt)	Vis. at 100°C (cSt)	VI
Lesquerella	NA	NA	−21	−22	127.7	15.2	123
L2-M[b]	C2:0	0.90	−21	>r.t.	92.8	14.6	164
L2-F[b]	C2:0	1.70	−30	−18	79.7	14.2	186
L4-M	C4:0	0.88	−27	<−27	86.7	14.4	173
L4-F	C4:0	1.50	−33	−30	74.7	13.9	194
L6-M	C6:0	0.78	−33	>r.t.	103.0	15.9	165
L6-F	C6:0	1.23	−36	>r.t.	87.9	15.2	183
L8-M	C8:0	0.75	−27	−25	115.7	15.1	187
L8-F	C8:0	1.41	−33	−27	76.0	14.8	205
L10-M	C10:0	0.66	−27	−26	118.5	17.2	159
L10-F	C10:0	1.51	−30	−17	99.9	16.7	182
L12-M	C12:0	1.00	−27	−23	110.8	17.4	173
L12-F	C12:0	1.60	−18	−28	101.0	17.2	186
H₂-L12-M	H[c] L12-M	0.97	mp 28–38	NA	NA	NA	NA
H₂-L12-F	H[c] L12-M	1.61	mp 20–32	NA	NA	NA	NA
L14-M	C14:0	1.29	−18	1	129.9	19.2	168
L14-F	C14:0	1.46	3	21	118.6	18.4	174
L16-M	C16:0	0.83	0	15	135.2	20.5	176
L16-F	C16:0	1.75	6	27	114.2	18.7	184
L18-M	C18:0	1.46	9	28	137.4	20.5	173
L18-F	C18:0	1.75	24	45	Solid	34.3	NA
L18:1-M	C18:1	0.97	−27	−16	119.6	18.7	176
L18:1-F	C18:1	1.56	−27	−16	95.1	17.0	195

[a] EN determined by NMR.
[b] M means mono-capped and F means fully capped.
[c] Hydrogenated material.

a three-necked round-bottomed flask equipped with magnetic stirrer, cold water condenser, and temperature probe. For the synthesis of fully capped lesquerella estolides, refined lesquerella oil (63.5 g, 65.9 mmol) was combined with a 2.5 molar equivalent of the corresponding acid anhydride as listed in Tables 12.6 and 12.7. For the synthesis of fully capped castor TG-estolides, castor oil (60.2 g, 64.8 mmol) was combined with a 4 molar equivalent of the corresponding acid anhydride as listed in Tables 12.6 and 12.7. Pyridine (0.98 g, 12.4 mmol) was added as a catalyst. The round-bottomed flask was set up as previously described. All acetic anhydride reactions were run at 50°C for 24–26 h and then dissolved in hexane and washed with 5% H_2SO_4(aq.), brine, and NaH_2PO_4 (pH 5) solutions. Butyric anhydride reactions were run at 60°C for 4–6 h and then dissolved in hexane and washed with 1 M KOH,

TABLE 12.7
Physical Properties of Castor TG-Estolides (See Figure 12.4 for a General Synthetic Scheme)

TG Estolides	FA	EN[a]	Pour Point (°C)	Cloud Point (°C)	Vis. at 40°C (cSt)	Vis. at 100°C (cSt)	VI
Castor	NA	NA	−15	−34	260.4	20.1	89
C2-M[b]	C2:0	0.82	−24	<−24	147.4	16.8	122
C2-F[b]	C2:0	2.59	−27	<−27	110.0	15.6	150
C4-M	C4:0	0.93	−27	<−27	117.4	15.4	138
C4-F	C4:0	2.70	−33	−30	82.2	14.0	177
C6-M	C6:0	0.80	−36	<−36	133.0	19.4	167
C6-F	C6:0	1.67	−45	<−45	79.0	17.0	234
C8-M	C8:0	0.98	−21	<−21	203.0	21.1	123
C8-F	C8:0	2.54	−36	<−36	105.9	16.8	172
C10-M	C10:0	1.04	−27	<−27	183.8	21.0	135
C10-F	C10:0	2.34	−36	<−36	91.8	16.2	191
C12-M	C12:0	0.92	−27	<−27	193.1	21.4	132
C12-F	C12:0	2.36	−33	<−33	120.0	19.1	181
H₂-C12-M	H[c] C12-M	1.19	24–36	NA	NA	NA	NA
H₂-C12-F	H[c] C12-M	2.35	−3	21	161.2	22.1	164
C14-M	C14:0	1.11	−24	−17	223.4	25.6	146
C14-F	C14:0	2.69	−18	−7	155.9	23.2	179
C16-M	C16:0	0.81	−18	−3	220.6	24.1	137
C16-F	C16:0	2.69	3	21	177.8	26.1	182
C18-M	C18:0	2.10	9	27	226.8	26.6	151
C18-F	C18:0	2.42	18	24	174.5	25.7	182
C18:1-M	C18:1	1.55	−33	<−50	186.9	23.5	154
C18:1-F	C18:1	2.69	−27	−27	131.8	21.9	195

[a] EN determined by NMR.
[b] M means mono-capped and F means fully capped.
[c] Hydrogenated material.

brine, NaH_2PO_4 (pH 5), and Na_2HPO_4 (pH 9) solutions. Ethanol (85%) was added to break the emulsions formed.

12.3.4.2 C6–C12

For the monocapped TG-estolides, refined lesquerella oil (1.33 kg, 1.38 mol) or castor oil (1.25 kg, 1.35 mol) and an equimolar amount of the corresponding FA listed in Tables 12.6 and 12.7 were combined in a three-necked round-bottomed flask. For the synthesis of fully capped lesquerella estolides, refined lesquerella oil (1.33 kg, 1.38 mol) and a 2.5 molar equivalent of the corresponding FA listed in Table 12.6 were combined in a three-necked round-bottomed flask. For the synthesis of fully capped castor TG-estolides, castor oil (1.25 kg, 1.35 mol) and a 4 molar equivalent of the corresponding FA listed in Table 12.7 were combined in a three-necked round-bottomed flask of sufficient size. In all of the reactions, the flask was equipped with a magnetic stir bar, temperature probe, and two condensers connected in series; this condenser setup was necessary to recirculate all the unreacted FAs back into the flask. The first condenser was connected to a recirculating water bath and maintained at a temperature slightly above the melting point of the FA. The second condenser was connected to the first with a 75°C distillation head and cooled with cold tap water. The outlet of the second condenser was fitted with a vacuum distillation adapter and round-bottomed flask to collect the water of reaction. The reaction was catalyzed with 0.1 wt.% tin(II) 2-ethylhexanoate (1.30 g, 3.21 mmol) per gram of oil. Reactants and catalyst were heated to 130°C under vacuum (12–18 Pa) for 24 h. At the conclusion of the reaction, brine was added followed by ethanol (85%) to break any emulsion that formed. Unreacted FAs were removed using a Kügelrohr-distillation unit with a vacuum of 6–13 Pa and a temperature of 140°C–160°C; distillation required 3–5 h for complete removal of excess FAs. When precipitated salts were observed in the finished product, the material was taken up in hexane and gravity filtered.

12.3.4.3   C14–C18

For monocapped TG-estolides, lesquerella oil (1.33 kg, 1.38 mol) or castor oil (1.25 kg, 1.35 mol) and an equimolar amount of the corresponding FA listed in Tables 12.6 and 12.7 were placed in a three-necked round-bottomed flask. For the synthesis of the fully capped lesquerella TG-estolides, refined lesquerella oil (1.33 kg, 1.38 mol) and a 2.5 molar equivalent of the corresponding FA listed in Table 12.6 were combined in a three-necked round-bottomed flask. For the synthesis of fully capped castor TG-estolides, castor oil (1.25 kg, 1.35 mol) and a 4 molar equivalent of the corresponding FA listed in Table 12.7 were combined in a three-necked round-bottomed flask of sufficient size. In all of the reactions, the flask was fitted with a temperature probe, vacuum adapter, and stopper. The reaction was performed under vacuum (20 Pa) and held at 200°C for 24 h using a heating mantle controlled by a J-Kem Gemini-2 temperature controller utilizing a temperature probe immersed below the liquid surface in the flask. When the reaction time was reached, the solution was allowed to cool to room temperature under vacuum. Unreacted FAs were removed using a Kügelrohr-distillation unit with a vacuum of 6–13 Pa and a temperature of 150°C–170°C. A colorless distillate of FA was obtained with a yellow residue of TG-estolide.

12.3.4.4   Hydrogenation of Lauric-Lesquerella and Lauric-Castor TG-Estolides

Hydrogenation was performed by combining neat TG-estolide (Tables 12.6 and 12.7) and 0.25 wt.% Pd on activated carbon into a stainless steel pressure reactor (Pressure Products Industries, Warminster, PA). The reactor was charged to 1379 kPa hydrogen after purging with hydrogen. A temperature of 80°C was maintained, and the reaction was stirred for 2.5–5 h until hydrogen consumption ceased. The product was separated from the catalyst by vacuum filtration through Celite, calcinated diatomaceous earth diatomite, and #50 Whatman filter paper in a jacketed funnel heated with steam. A 10 mg sample was converted into its methyl ester as described below and injected onto the GC to confirm saturation of the estolide.

12.3.5 CASTOR AND LESQUERELLA 2-ETHYLHEXYL ESTERS

Acid-catalyzed transesterification reactions were conducted with 2-ethylhexanol in a three-necked round-bottomed flask under vacuum. The flask was evacuated by vacuum (20 Pa) and maintained at 80°C for 24 h using a heating mantle controlled by a J-Kem Gemini-2 temperature controller utilizing a temperature probe immersed below the liquid level in the flask. For the synthesis of both lesquerella 2EH esters and castor 2EH esters, a solution of boron trifluoride dimethyl etherate (0.5 M, 126.0 mL) was added to lesquerella oil (1.00 kg, 1.04 mol) and 2-ethylhexanol (1.56 kg, 11.99 mol, 1.84 L). After 24 h, the mixtures were allowed to cool, transferred to a separatory funnel followed by the addition of 200 mL of a 1:1 ethyl acetate:hexane solution. The pH of the organic layer was adjusted to 5.3–6.0 with a pH 5 buffer (NaH_2PO_4, 519 g in 4L H_2O, 3 × 100 mL) followed by brine (2 × 50 mL). The organic layer was dried over sodium sulfate and filtered. Unreacted materials were removed using a Kügelrohr-distillation unit with a vacuum of 6–13 Pa and a temperature of 90°C–110°C to remove excess 2-ethylhexanol. The residue then underwent a second Kügelrohr-distillation under vacuum (6–13 Pa) at 180°C–200°C to yield the purified 2EH esters (Figure 12.5) as a light yellow liquid.

12.3.6 HYDROGENATION OF LESQUERELLA AND CASTOR FATTY ESTERS

Hydrogenation was performed by combining hydroxy FA ester (Figure 12.5), that is, lesquerella 2EH ester (311.0 g, 710.0 mmol), hexane (200 mL), and 0.25 wt.% Pd (1.24 g) on activated carbon into a stainless steel pressure reactor (Pressure Products Industries, Warminster, Pennsylvania). The reactor was charged to 1379 kPa hydrogen after first purging with hydrogen. A room temperature setting was maintained, and the reactions (Table 12.8) were mixed for 2–5 h until hydrogen consumption ceased. The product was separated from the catalyst by vacuum filtration through silica gel and #50 Whatman filter paper. The saturated fatty ester was then dried over sodium sulfate and filtered. All products were concentrated in vacuo. A 10 mg sample was converted into its methyl ester and injected onto the GC to confirm saturation of the fatty ester.

12.3.7 SATURATED AND UNSATURATED CASTOR AND
LESQUERELLA ESTOLIDE ESTERS

The saturated and unsaturated lesquerella and castor esters were combined with the corresponding FAs as listed in Table 12.8. In most cases, 1.0 equivalent of lesquerella 2EH ester (200.0 g, 456.6 mmol) or castor 2EH ester (200.0 g, 487.8 mmol) and 1.5 equivalents of a capping material, i.e., oleic acid (195.9 g, 684.9 mmol) as

TABLE 12.8

Physical Properties of Unsaturated and Saturated Castor and Lesquerella-Based Estolide Esters (See Figure 12.5 for Synthetic Scheme)

Estolide	Capping FA	Hydroxy FA Ester[a]	Product Name	Pour Point (°C)	Cloud Point (°C)	Vis. at 40°C (cSt)	Vis. at 100°C (cSt)	VI	Gardner Color
1	Oleic	Castor	Oleic-Cas	−54	<−54	34.5	7.6	196	2+
2	Stearic	Castor	Stea-Cas	3	23	41.7	8.6	191	8−
3	Coco[b]	Castor	Coco-Cas	−36	−30	29.0	6.5	186	6−
4	2-Ethylhexanoic	Castor	2EH-Cas	−51	<−51	70.6	11.8	164	13−
5	Oleic	Lesquerella	Oleic-Les	−48	−35	35.4	7.8	200	3+
6	Stearic	Lesquerella	Stea-Les	3	12	38.6	8.2	195	4−
7	Coco[b]	Lesquerella	Coco-Les	−24	<−24	40.4	8.4	192	17
8	2-Ethylhexanoic	Lesquerella	2EH-Les	−54	<−54	51.1	10.1	189	8−
9	Oleic	Sat. Castor	Oleic-H-Cas	−36	<−36	68.3	12.2	178	16+
10	Stearic	Sat. Castor	Stea-H-Cas	6	r.t.	43.6	8.7	186	7+
11	Oleic	Sat. Lesquerella	Oleic-H-Les	−12	−6	37.0	7.9	196	9+
12	Stearic	Sat. Lesquerella	Stea-H-Les	6	31	45.7	9.1	187	6+

[a] 2EH ester.

[b] FAs from coconut oil.

listed in Table 12.8 were combined in a three-necked round-bottomed flask. The flask was fitted with temperature probe, vacuum adapter, and a stopper. The reaction was performed under vacuum (20 Pa) and maintained at 200°C for 24 h using a heating mantle controlled by a J-Kem Gemini-2 temperature controller utilizing a temperature probe immersed below the liquid level in the flask. When the reaction time was reached (24 h), the solution was allowed to cool to room temperature under vacuum. All reactions underwent Kügelrohr-distillation under vacuum (6–13 Pa) at 180°C–200°C to yield the purified saturated and unsaturated castor and lesquerella estolide 2EH esters (Figure 12.5).

12.4 ESTOLIDE ANALYSIS AND IDENTIFICATION

Chemical identification and analytical analysis are very important techniques that make it possible to adequately describe the different series of estolide compounds. Under different synthesis conditions, the estolides can vary greatly in size, and the ability to characterize these structures is very important to understanding the physical properties of the estolides. The estolides can be easily analyzed by analytical equipment (HPLC) without chemical modifications to the molecules. The estolides can also be easily chemically modified to allow for chemical characterization, which will help determine the size and connectivity of the estolides.

12.4.1 HPLC ANALYSIS OF FA-BASED ESTOLIDES

Reversed-phase HPLC analyses were performed on a Thermo Separation Spectra System AS1000 autosampler/injector (Fremont, CA) with a P2000 binary gradient pump from Thermo Separation Products coupled to an Alltech ELSD 500 evaporative light scattering detector (ELSD). A C-8 reverse-phase analysis used to separate reaction mixtures was carried out with a Dynamax column (250 × 4.5 mm, 5 μm particle size) from Rainin Instrument Co. (Woburn, MA).

Two methods for reversed-phase analysis were used to separate the reaction mixtures. Method A (16 min run time) was used to follow the reaction. It provided information on the overall progress of the reaction. Method B (35 min run time) produced a more detailed separation of the reaction mixture, in particular, a separation of estolides, lactones, and hydroxy FAs.

Parameters for method A were as follows: flow rate of 1 mL/min; 0–4 min 80% acetonitrile/20% acetone; 6–10 min 100% acetone; and 11–16 min 80% acetonitrile/20% acetone. The ELSD drift tube was set at 55°C, with the nebulizer set at 1034.3 torr N_2 providing a flow rate of 2.0 standard liters per minute (SLPM). Retention times for eluted peaks were as follows: estolides, 9.8–13.0 min; methyl oleate, 6.3 min; oleic acid, 5.1 min; lactones, 4.8 min; and hydroxy acids, 4.1 min.

Parameters for method B were as follows: flow rate of 1 mL/min; 0–2 min 60% acetonitrile/40% acetone; 20–25 min 100% acetone; and 30–35 min 60% acetonitrile/40% acetone. The ELSD drift tube was set at 55°C, with the nebulizer set at 1034.3 torr N_2 providing a flow rate of 2.0 SLPM. Retention times for eluted peaks were as follows: estolides, 8.2–25.6 min; methyl oleate, 5.5 min; oleic acid, 4.8 min; γ-lactones, 4.5 min; δ-lactones, 4.1 min; and hydroxy acids, 3.8 min.

Normal-phase HPLC analyses were performed using a Spectra-Physics 8800 ternary pump (San Jose, CA) with a Spectra System AS3000 autosampler/injector from Thermo Separation Products coupled to a Varex ELSD III light scattering detector (Alltech Associates). A silica normal-phase analysis was carried out with a Dynamax column (250×4.6 mm, 8 μm) from Rainin Instrument Co. Components were eluted isocratically from the column with a 80:20 mixture of hexanes/acetone at a flow rate of 1 mL/min with the ELSD drift tube set at 45°C and nebulizer set at 517.1 torr N_2, flow rate 1.50 SLPM. Normal-phase HPLC was used to separate esterification reaction mixtures. Retention times for eluted peaks were as follows: 2EH estolide ester, 2.6–2.8 min; and estolide free-acid, 3.5–3.7 min depending on the capping FA.

12.4.2 HPLC Analysis of Oil-Based Estolides

Reverse-phased HPLC (C8) analyses were performed on a Spectra-Physics 8800 system with ternary pump (San Jose, CA) and Spectra System AS3000 autosampler/injector from Thermo Separation Products (Fremont, CA) coupled to an ELSD from Alltech Associates. A Dynamax (250 mm×4.6 mm, 60 Å, 8 μm) C8 column purchased from Rainin Instrument Co., a division of Varian (Walnut Creek, CA), was used to separate the mixtures. Components were eluted from the column using the following gradient program: 0–2 min hold, acetonitrile/acetone 60:40; gradient to 100% acetone up to 20 min; held at 100% acetone up to 25 min; reverse gradient to 60:40 acetonitrile/acetone at 30 min; the last composition held 30–35 min. A flow rate of 1 mL/min was used. The ELSD drift tube was set at 55°C with 1034.3 torr N_2 feed to the nebulizer to give a flow rate of 2.0 SLPM.

12.4.3 Acid Value

The acid values were measured on a 751 GPD Titrino from Metrohm Ltd. (Herisau, Switzerland). Acid values were determined by the AOCS Method Te 2a-64 [52] with ethanol substituted for methanol to increase the solubility of the estolide ester during the titration. All acid values were run in duplicate and average values were reported.

The acid numbers for the estolide esters show the ratio between how much of the estolide is an ester versus a free-acid. The acid number is one example of how the esterification reaction can be followed using a chemical analysis. The esters and acids do have different physical properties, and it is vital to know what the chemical structures of the estolides are when comparing their physical properties. Acid numbers of <3 mg/g for the estolide esters are deemed fully esterified and correspond well with the HPLC procedure.

12.4.4 GC Analysis of Hydroxy FAs

GC analysis of chemically modified estolides was conducted to quantify the ratio of hydroxy versus nonhydroxy fatty esters. The ratio is used to determine the EN for the individual estolide samples. Higher amounts/ratios of hydroxy fatty esters come

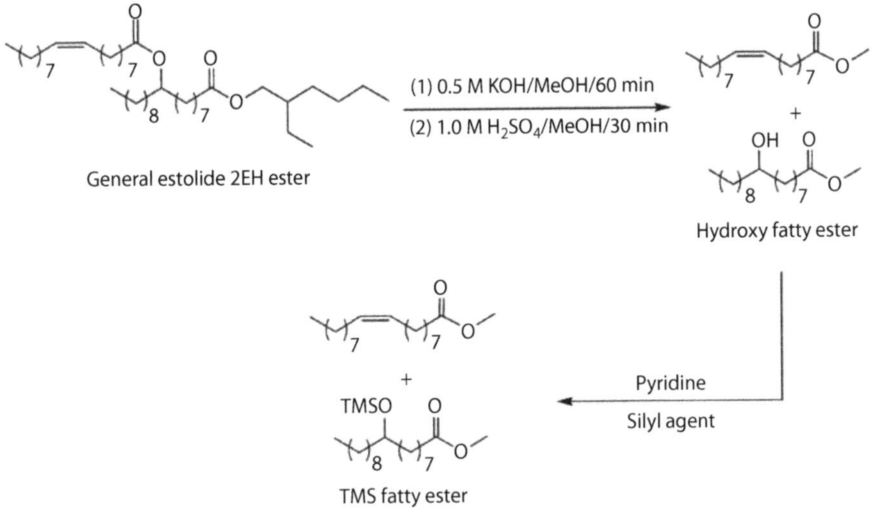

FIGURE 12.6   Estolide 2EH ester chemical derivatization.

from larger estolides. As the EN changes for a fixed series of estolides, the physical properties will also change. Usually, higher EN will have increased viscosities and results in higher pour points.

Analytical estolide samples for GC were prepared by heating a 10 mg sample of estolide free-acid or estolide 2EH ester in 0.5 mL of 0.5M KOH/MeOH to reflux on a heating block for 60 min in a sealed vial (Figure 12.6). After cooling to room temperature, 2 mL of 1 M H₂SO₄/MeOH was added to the vial; the vial was resealed and heated to reflux on a heating block for 30 min. The solution was transferred to a separatory funnel with water (1 mL) and washed with hexanes (2×2 mL), dried over sodium sulfate, gravity filtered, placed in a GC vial with hexanes, sealed, and injected into the GC and/or GC/MS.

12.4.5   TMS DERIVATIZATION OF HYDROXY FATTY ESTERS

Double-bond migration has been demonstrated under the acidic conditions at which the estolides are produced. The literature [23] has shown examples where the olefin position eventually spreads over the entire FA chain. When the olefin nears the carboxyl, we assume that estolide and lactone formation is in equilibrium at the γ- and δ-positions in classic ring-chain equilibrium typical of other lactones such as caprolactone [53].

Hydroxy FAMEs (10 mg; prepared as described above) were dissolved in 0.1 mL of pyridine and 0.05 mL N,O-bis(trimethylsilyl)acetamide (Figure 12.6). This solution was then placed in a sealed vial for 10 min at 100°C. After this time, hexanes (1 mL) were added, and the resulting solution was filtered though a silica gel plug. The filtered sample was placed in a sealed GC vial with hexanes and injected onto the GC and GC/MS. The main mass spectral features were m/e 371 (M+ −15, 1.2%),

TABLE 12.9

Sampling of Estolide Position as Determined by Gas Chromatography/Mass Spectrometry (GC/MS) of Trimethylsilyl Ether of Alkali-Hydrolyzed Estolide[a]

Estolide Fraction From	Hydroxy Position	MS Fragments Carbonyl	MS Fragments Alkyl	Total Abundance (%)
Table 12.3	5	203	285	7.7
Estolide H	6	217	271	19.9
	7	231	257	30.8
	8	245	243	35.9
	9	259	229	39.3
	10	273	215	40.0
	11	287	201	34.5
	12	301	187	27.5
	13	315	173	21.2

[a] The same relative results were obtained for all complex estolides.

73 (TMS+, 100%) and a Gaussian fragment representing cleavage at the C–C bond adjacent to the silyloxy positions (masses 173–315). The fragments and abundances are summarized in Table 12.9 relative to their respective positions on the fatty ester backbone; the same general relative results were obtained for all complex estolides. The estolide position was distributed from positions 5 to 13 with the original Δ9 and Δ10 positions having the largest abundances in the mass spectrum. This method demonstrated that, during the estolide reaction, there was migration of the double bond or of the carbocation. The intensity of a fragment in the mass spectrum is not solely dependent on concentration of a species in the mixture but more so on ionization energies and charge stabilizations. Thus, the apparent Gaussian distribution of hydroxy positions derived from the estolide, though likely due to structural similarities, might not have represented the amount of each position in the mixture, but merely indicated its presence.

12.4.6 ESTOLIDE NUMBERS AND IODINE VALUES

EN were determined by GC from the SP-2380 column analysis as previously described [23]. Iodine values (IV) were calculated from GC data using AOCS Method Cd 1c-85 [52].

Table 12.3 outlines a series of reactions that explore the formation of complex estolide free-acids at two different temperatures where a series of different saturated FAs, butyric through stearic, are used as the capping material to give the saturated-capped estolide free-acid (Figure 12.2). These new estolides have an oleic acid backbone with a terminal FA. Estolides are formed from the carbocationic homo-oligermization of unsaturated FAs [23] resulting from the addition of a FA carboxyl across the olefin. This condensation can continue resulting in oligomeric compounds where the average extent of oligomerization is defined as the EN (EN = n + 1, Figure 12.1). When saturated FAs are added to the reaction mixture, the oligomerization is terminated upon addition of the saturated FA to the olefin since the saturate provides no additional unsaturation to further the oligomerization. Consequently, the estolide is stopped at this point from further growth; thus, we term the estolide as being "capped."

In the synthesis of saturated-capped estolide free-acids (Figure 12.2), some unique structural possibilities were encountered. The EN in Table 12.3 demonstrated the amount of oligomerization for each set of estolides. As the chain length increased for the reactions at 45°C, the EN decreased most likely due to steric or pK_a effects. For example, the larger saturated FAs could have inhibited the reaction once they added to the estolide, limiting the propagation of estolide formation from the acid end. Solution acidity might also have affected EN. Since pK_a values increase slightly with chain length, shorter FAs have lower pK_a values and are, therefore, more acidic. Isbell [25] demonstrated that, as the amount of acid present in an estolide reaction increased, the ENs also increased. So, as the acidity of the solution increased, there should have been an increase in the EN of the estolides, which was observed. The EN was also found to be dependent on temperature. As the temperature increased to 55°C, EN decreased for the individual saturated-capped estolide free-acids, but the general trend observed at 45°C remained.

The estolide free-acid (Figure 12.2) is shown as a saturated-capped estolide with an IV of zero. In practice, however, the IVs for the reactions in Table 12.3 are in the teens. These higher-than-expected IVs suggest that not all the estolides are capped with saturated material; the estolide free-acids contain some unsaturated oleic materials. Isbell [25] reported that oleic estolide free-acids with comparable ENs have IVs around 35. An IV of 15 is close to an oleic estolide free-acid previously synthesized by Isbell after it was partially hydrogenated. The saturated-capped estolides have an advantage over these partially hydrogenated estolides. The capped estolide reaction involves one step and inexpensive reagents as compared to two steps and expensive reducing metals. Since fewer alkenes are present in the final capped estolides, the oxidative stability should be greater than standard oleic estolide free-acids if the trends described by Akoh [54] hold true. Akoh [54] reported that refined soybean oil had an oxidative stability index (OSI), [52] of 9.4 h at 110°C, but once the oil was partially hydrogenated, the OSI increased to 15.3 h at 110°C, an improvement of more than 60%.

12.4.7 NUCLEAR MAGNETIC RESONANCE

^{1}H and ^{13}C NMR spectra were obtained on a Bruker ARX-400 spectrometer (Karlsruhe, Germany) with a 5 mm dual proton/carbon probe (400 MHz ^{1}H/100.61 MHz ^{13}C) using CDCl$_3$ as a solvent in all experiments. The assignments of protons were not to the whole number. The representative NMR spectrum contained a compound, for example, that had an average EN 1.94 for the coco-oleic estolide 2-ethylhexyl esters (CO-EST-2EH; CO-EH-DD, Table 12.5), which made whole number assignment impossible. The data reported for the numbers of protons in the NMR spectrum reflected the actual numbers. The numbers could be multiplied by a factor to obtain whole numbers that corresponded to a whole number EN.

The proton spectra for the estolide free-acids in Figure 12.2, specifically estolide free-acid M (Table 12.3), show some key features of a typical estolide. The ester methine signal at 4.84 ppm is indicative of an estolide linkage. Other distinctive features are the α-methylene proton shift (2.32 ppm) adjacent to the acid and the α-methylene proton shift (2.25 ppm) adjacent to the ester. Integration of these signals provides a ratio for the number of ester bonds to acid functionalities. This ratio of α-ester/α-acid can be used as another means to determine the EN, that is, complementary to the GC method. The NMR indicates some presence of alkene in the estolide by the appearance of an alkene signal at 5.36 ppm. The alkene signal indicates that some of the estolide is capped with unsaturated material, i.e., oleic acid. The alkene signal in the ^{1}H NMR supports the IVs determined by GC, as the intensity of the NMR signals is comparable to the reported IVs.

The carbon NMR spectrum contains the expected estolide signals. There are two different carbonyl signals present at 179.2 ppm (acid) and 173.7 ppm (estolide ester). The other distinctive signal is the methine carbon at 74.1 ppm, which is common to esters. These major peaks in the carbon NMR are also confirmed by a Distortionless Enhancement by Polarization Transfer (DEPT) experiment. The alkene carbons are only slightly noticeable, as these signals are about the same as the signal-to-noise ratio.

The conversion to the 2EH estolide ester, M-2EH (Table 12.7) gives some predictable signal changes in the ^{1}H NMR. The α-carbonyl methylene protons have similar shifts, resulting in a multiplet from 2.27 to 2.22 ppm. As before, the alkene signal is noticeable at 5.35 ppm, confirming the IVs determined by GC. The carbon NMR signals are indicative of the 2EH ester and are confirmed by a DEPT experiment.

12.4.7.1 ^{1}H and ^{13}C NMR of Estolide Free-Acid (Table 12.3; Estolide M)

^{1}H NMR: δ 5.37–5.33 (m, 0.5H, –CH=CH–), 4.86–4.83 (m, 1.7H, –CH–OC=O–CH$_2$–), 2.32 (t, J = 7.4 Hz, 2H, –CH$_2$(C=O)–OH), 2.28–2.23 (m, 3.4H, –CH$_2$(C=O)–O–CH–), 1.96–1.20 (m, 74.3H), 0.88–0.84 (m, 8.3H, –CH$_3$). ^{13}C NMR: δ 179.7 (s, HO–C=O), 173.6 (s, –CH–O–(C=O)–CH$_2$–), 130.5 (d, –CH=CH–, very small signal, only a small amount of alkene present), 74.1 (d, –CH–O–C=O), 34.7 (t, –CH$_2$–), 34.2 (t, –CH$_2$–), 34.0 (t, –CH$_2$–), 32.5 (t, –CH$_2$–), 31.8 (t, –CH$_2$–), 31.8 (t, –CH$_2$–), 31.5 (t, –CH$_2$–), 29.8 (t, –CH$_2$–), 29.7 (t, –CH$_2$–), 29.6 (t, –CH$_2$–), 29.6 (t, –CH$_2$–), 29.5 (t, –CH$_2$–), 29.5 (t, –CH$_2$–), 29.4 (t, –CH$_2$–), 29.3 (t, –CH$_2$–), 29.2 (t, –CH$_2$–), 29.1 (t, –CH$_2$–), 29.0 (t, –CH$_2$–), 27.2 (t, –CH$_2$–), 25.3 (t, –CH$_2$–), 25.2 (t, –CH$_2$–), 24.9 (t, –CH$_2$–), 24.7 (t, –CH$_2$–), 24.6 (t, –CH$_2$–), 22.6 (t, –CH$_2$–), 13.9 (q, –CH$_3$).

12.4.7.2 ^{1}H and ^{13}C NMR of Estolide 2-Ethylhexyl Ester (Table 12.5; Estolide CO-EH-DD)

^{1}H NMR: δ 5.37–5.34 (m, 0.3H, –CH=CH–), 4.87–4.81 (m, 1.0H, –CH–OC=O–), 3.96 (d, J = 5.7 Hz, 1.5H, –OCH$_2$–CH(CH$_2$–)CH$_2$–), 2.29–2.24 (m, 4.1H, –CH$_2$(C=O)–O–CH$_2$–, –CH$_2$(C=O)–O–CH–), 1.96–1.24 (m, 55.7H), 0.89–0.85 (m, 10.7H, –CH$_3$). ^{13}C NMR: δ 174.0 (s, C=O), 173.5 (s, C=O), 130.0 (d, –CH=CH–, very small signals, only a small amount of alkene present), 73.9 (d, –CH–O–C=O), 66.5 (t, –O–CH$_2$–CH–), 38.6 (d, –CH$_2$–CH(CH$_2$–)–CH$_2$–), 34.3 (t, –CH$_2$–), 34.0 (t, –CH$_2$–), 31.8

(*t*, –C*H*₂–), 30.3 (*t*, –C*H*₂–), 29.6 (*t*, –C*H*₂–), 29.5 (*t*, –C*H*₂–), 29.5 (*t*, –C*H*₂–), 29.4 (*t*, –C*H*₂–), 29.4 (*t*, –C*H*₂–), 29.3 (*t*, –C*H*₂–), 29.2 (*t*, –C*H*₂–), 29.2 (*t*, –C*H*₂–), 29.1 (*t*, –C*H*₂–), 29.0 (*t*, –C*H*₂–), 28.8 (*t*, –C*H*₂–), 25.2 (*t*, –C*H*₂–), 24.9 (*t*, –C*H*₂–), 23.7 (*t*, –C*H*₂–), 22.8 (*t*, –C*H*₂–), 22.5 (*t*, –C*H*₂–), 14.0 (*q*, –C*H*₃), 13.9 (*q*, –C*H*₃), 10.9 (*q*, –C*H*₃).

12.4.7.3 ¹H and ¹³C NMR of Lesquerella TG-Estolide (Table 12.6; Estolide L18:1-F)

¹H NMR: δ 5.47–5.23 (*m*, 12.7H, –C*H*=C*H*– and (–O–C*H*₂)₂–CH–O–), 4.87 (*p*, *J*=6.2 Hz, 1.8H, –(CH₂)₂–C*H*–(O–CO–R), 4.28 (*dd*, *J*=4.3 and 11.9 Hz, 2H, (–O–C*H*₂–)₂–CH–O–), 4.13 (*dd*, *J*=6.0 and 11.9 Hz, 2H, (–O–C*H*₂–)₂–CH–O–), 2.78 (*m*, 2H, –(CH=CH)₂–C*H*₂–), 2.35–2.22 (*m*, 13.7H, –C*H*₂–CO₂R), 2.10–1.92 (*m*, 16.2H, –C*H*₂–CH=CH–) 1.70–1.45 (*m*, 14.5H, –C*H*₂–), 1.38–1.14 (*m*, 94H, –C*H*₂–), 0.87 (*t*, *J*=7.0 Hz, 15H, –C*H*₃). ¹³C NMR: δ 173.6 (*s*, –*C*=O), 173.3 (*s*, –*C*=O), 172.7 (*s*, –*C*=O), 132.6 (*d*, –*C*=C–CH₂CH–O–), 130.1 (*d*, –*C*=*C*–), 130.0 (*d*, –*C*=*C*–), 129.8 (*d*, –*C*=*C*–), 129.7 (*d*, –*C*=*C*–), 124.3 (*d*, –*C*=C–CH₂CH–O–), 73.7 (*d*, –*C*H–O–CO₂R), 68.9 (*d*, (–O*C*H₂)₂–CH–O–), 62.1 (*t*, (–O*C*H₂–O–)₂–CH–O–), 34.7 (*t*, –*C*H₂–), 34.2 (*t*, –*C*H₂–), 34.0 (*t*, –*C*H₂–), 33.7 (*t*, –*C*H₂–), 32.1 (*t*, –*C*H₂–), 32.0 (*t*, –*C*H₂–), 31.8 (*t*, –*C*H₂–), 29.8 (*t*, –*C*H₂–), 29.8 (*t*, –*C*H₂–), 29.7 (*t*, –*C*H₂–), 29.6 (*t*, –*C*H₂–), 29.5 (*t*, –*C*H₂–), 29.4 (*t*, –*C*H₂–), 29.4 (*t*, –*C*H₂–), 29.3 (*t*, –*C*H₂–), 29.2 (*t*, –*C*H₂–), 29.2 (*t*, –*C*H₂–), 29.1 (*t*, –*C*H₂–), 29.1 (*t*, –*C*H₂–), 27.4 (*t*, –*C*H₂–), 27.3 (*t*, –*C*H₂–), 27.2 (*t*, –*C*H₂–), 25.4 (*t*, –*C*H₂–), 25.2 (*t*, –*C*H₂–), 24.9 (*t*, –*C*H₂–), 22.7 (*t*, –*C*H₂–), 22.6 (*t*, –*C*H₂–), 14.2 (*q*, –*C*H₃), 14.1 (*q*, –*C*H₃).

12.4.7.4 ¹H and ¹³C NMR of Castor Estolide 2-Ethylhexyl Ester (Table 12.8; Estolide A)

¹H NMR: δ 5.49–5.37 (*m*, 3.8 H, –C*H*=C*H*–), 4.89–4.83 (*m*, 1.0 H, –C*H*–OC=O–), 3.98 (*d*, *J*=5.7 Hz, 1.8H, –OC*H*₂–CH(CH₂–)CH₂–), 2.32–2.24 (*m*, 6.0H, –C*H*₂ (C=O)–O–CH₂–, –C*H*₂ (C=O)–O–CH–), 2.06–1.95 (*m*, 5.6H, –C*H*₂–CH=CH–CH₂–), 1.65–1.19 (*m*, 51.6H, –C*H*₂–),0.90–0.85 (*m*, 11.8H, –C*H*₃). ¹³C NMR: δ 173.9 (*s*, –*C*=O, ester), 173.5 (*s*, –*C*=O, estolide), 132.4 (*d*, –*C*H=CH–), 129.9 (*d*, –*C*H=CH–), 126.6 (*d*, –*C*H=CH–), 124.2 (*d*, –*C*H=CH–), 73.5 (*d*, –*C*H–O–C=O), 66.5 (*t*, –O–*C*H₂–CH–), 38.6 (*d*, –CH₂–*C*H(CH₂–)–CH₂–), 34.6 (*t*, –*C*H₂–), 34.3 (*t*, –*C*H₂–), 33.5 (*t*, –*C*H₂–), 31.9 (*t*, –*C*H₂–), 31.6 (*t*, –*C*H₂–), 30.3 (*t*, –*C*H₂–), 29.7 (*t*, –*C*H₂–), 26.6 (*t*, –*C*H₂–), 29.4 (*t*, –*C*H₂–), 29.2 (*t*, –*C*H₂–), 29.1 (*t*, –*C*H₂–), 29.0 (*t*, –*C*H₂–), 28.8 (*t*, –*C*H₂–), 27.2 (*t*, –*C*H₂–), 27.1 (*t*, –*C*H₂–), 25.2 (*t*, –*C*H₂–), 25.0 (*t*, –*C*H₂–), 24.9 (*t*, –*C*H₂–), 23.6 (*t*, –*C*H₂–), 22.9 (*t*, –*C*H₂–), 22.5 (*t*, –*C*H₂–), 14.0 (*q*, –*C*H₃), 13.9 (*q*, –*C*H₃), 13.9 (*q*, –*C*H₃), 10.9 (*q*, –*C*H₃).

12.5 PHYSICAL PROPERTY TESTS AND RESULTS

Estolides, all types, have certain physical characteristics that could help eliminate common problems associated with the application of vegetable oils as functional fluids, such as low resistance to thermal oxidative stability [12] and poor low-temperature properties [13,14]. Simple oleic estolide esters (O-EST-2EHs), when formulated with a small amount of oxidative stability package (OSP), show better oxidative stability than both petroleum and vegetable oil-based fluids [55], but there is still room for improvement. There are a number of ways to improve the oxidative stability of an oil. Akoh [54] reported that refined soybean oil has an OSI of 9.4 h at 110°C, but once the oil is partially hydrogenated, the OSI increases to 15.3 h at 110°C, an improvement of more than 60%. The same approach was taken with the oleic estolides, where hydrogenation with 2% w/w of 10% palladium on activated carbon as catalyst gave completely saturated estolides [25]. The saturated oleic estolides are expected to be more oxidatively stable than the unsaturated estolides (assuming the same trend displayed by soybean oil holds true). However, in the cases examined by Isbell et al. [25], in which a new saturated estolide was synthesized via hydrogenation, the new saturated estolide had a high-pour point of −9°C, which is unsatisfactory for many functional fluids. Cermak and Isbell [19] envisioned a new class of saturated estolides with superior low-temperature properties.

Some of the estolides synthesized to date that best address cold temperature applications were made from a mixture of saturated and unsaturated FAs [20,21]. When saturated FAs are added to the estolide synthesis, a saturated-capped estolide is formed. These estolides have an oleic acid backbone with a terminal saturated tallow FA acting as a capping group. Cermak and Isbell [24] theorized that, by varying the capping material on the estolide, the crystal lattice structure of the material was disrupted as it approached its pour point, which led to estolide esters with excellent low-temperature properties: pour point of −36°C and cloud point of −41°C. These saturated-capped estolide 2EH esters thus have eliminated common problems associated with the use of vegetable oils as functional fluids. To date, all the different types of estolide FAs and estolide esters have compared favorably to commercially available industrial products such as petroleum-based hydraulic fluids, soy-based fluids, and petroleum oils [20,21].

12.5.1 Gardner Color

One of the most important physical properties to a consumer is the color of the oil or oil-based products. As a potential hydraulic fluid, the estolides need to meet the color requirements of currently used hydraulic fluids. Most consumers have become accustomed to the appearance of their oils, hydraulic fluids, and products. The measurement of the color of a material is designated as the Gardner color. The Gardner color scale is from 1 to 18 with 1 containing the least amount of color and 18 with the maximum amount of color.

Gardner color was measured on a Lovibond 3-Field Comparator from Tintometer Ltd. (Salisbury, England) using AOCS Method Td 1a-64 [52]. In many cases, the Gardner color of materials can be susceptible to the interpretation of the recorder; thus, the + and − notation was employed (as in Table 12.8) to designate samples that did not match one particular Gardner color numbers, with an upper limit of 18.

The estolide free-acids synthesized at 45°C had relatively low color (less color) on the Lovibond color scale (Table 12.3). As the reactions were repeated at 55°C, the colors of the estolides turned somewhat darker in every case, usually increased by two Gardner units. Esterification of the estolide to the 2EH esters caused additional problems, as the resulting Gardner colors were even darker (Table 12.4, high-Gardner color numbers) than the estolide free-acids (Table 12.3). This increase in Gardner number was expected as the estolide esters were subjected to continued acidic conditions and higher temperatures (60°C) than the nonesterified estolides. The estolide esters were decolorized with charcoal and the colors improved (lighter color) anywhere from one to five Gardner color units, Table 12.4. This decolorization step returned the estolide esters to nearly their original colors and made them commercially viable. The charcoal decolorization was used as a way to decolorize estolides, but our laboratory has had success with other methods as well [56].

In general, all the other estolides listed in the remaining tables had acceptable Gardner colors, and the color numbers could be improved further if necessary to meet certain applications requirements. Darkening of the estolides was caused by excessive heating under acidic conditions and with the use of crude starting materials.

12.5.2 Viscosity

Viscosity measurements were made using calibrated Cannon-Fenske viscometer tubes purchased from Cannon Instrument Co. (State College, PA). Viscosity measurements were made in a Temp-Trol (Precision Scientific, Chicago, IL) viscometer bath set at 40°C and 100°C. Viscosity and viscosity index (VI) were calculated using ASTM Methods D 445-97 [57] and ASTM D2270-93 [58], respectively. All viscosity measurements were conducted in duplicate runs and the average value is reported here.

12.5.2.1 Viscosities of Estolide Free-Acids and Estolide 2-Ethylhexyl Esters

The viscosities are presented as relatively average values for all of the estolide 2-ethylhexyl (2EH) esters and saturates (Tables 12.5 and 12.11) with an approximate range of 41–150 centistokes (cSt). The estolide free-acids (Table 12.10), as expected, had higher viscosities than the estolide esters (approximate range of 236–515 cSt), caused by hydrogen bonding of the carboxylate functionality. Viscosities of the estolide 2EH esters increased in an exponential manner with the EN (degree of oligomerization) (Tables 12.3 and 12.10) and (Tables 12.4 and 12.11). This same general trend was observed by Isbell et al. [25] with the simple O-EST-2EHs. The viscosity range of the estolide free-acids and estolide 2EH esters has proven to be useful for numerous applications.

In general, the estolides can have a wide range of viscosities depending on what the starting materials are and how large the estolides are allowed to form, through the use of reaction conditions. Longer reaction times will yield larger EN numbered estolides, thus, larger estolides with higher viscosities. With so many possible types and sizes of estolides, it is possible to almost any desired viscosity and other excellent physical properties. This will allow open opportunities for many potential industrial applications for both the estolide free-acids and estolide 2EH esters.

TABLE 12.10

Physical Properties of Estolide Free-Acids Condensation Reaction with Oleic Acid and Varying Saturated FAs[a] (See Figure 12.2 for Synthetic Scheme)

Estolide[b]	Pour Point (°C)	Cloud Point (°C)	Vis. at 40°C (cSt)	Vis. at 100°C (cSt)	VI
A	−27	−26	410.0	39.9	146
B	−18	−10	456.0	41.7	155
C	−24	−27	515.5	39.7	122
D	−21	−17	411.2	40.3	148
E	−24	−24	389.1	37.7	143
F	−18	−9	398.1	39.2	147
G	−21	—[c]	342.0	34.0	142
H	−21	—[c]	336.9	34.3	145
I	−25	−27	262.6	28.7	145
J	−16	−18	262.4	28.4	143
J	−18	−6	282.3	30.4	146
L	−9	7	290.5	30.0	140
M	−10	−12	267.1	28.7	143
N	−2	−2	236.4	26.5	145
O	−3	−2	296.5	31.0	143
P	3	19	296.6	30.6	141

[a] Reactions were run for 24 h with overhead stirring in a 2:1 ratio, oleic:saturated FAs with 0.4 equivalents of perchloric acid.

[b] Continued from Table 12.3.

[c] Material color too dark to determine accurate cloud point.

12.5.2.2 Viscosities of Castor and Lesquerella TG-Estolides

The two main factors in the TG-estolide series of homologous compounds that affect viscosity are hydrogen bonding by the hydroxyl groups and the steric bulk of the molecules. Intermolecular hydrogen bonding between the hydroxyl TG increases the steric bulk of the molecules by creating weakly associated dimers, trimers, etc., of the original TG, therefore increasing the viscosity of the oil. In all cases, the fully capped TG-estolides have lower viscosities for all chain-lengths than the mono-capped estolides (Tables 12.6 and 12.7). Capping of the hydroxyl moiety reduces or eliminates the hydrogen bonding and, consequently, reduces the viscosity of the oil. The viscosity trend across the chain-length series follows a linear relationship with increasing viscosity correlated to increasing chain length.

12.5.2.3 Viscosity Index

VI is a term used as a lubricating oil quality indicator, a measure of the change of kinematic viscosity with temperature. The viscosity of a lubricant is closely related to its ability to reduce friction. In all cases, the VI for the estolides is very high VI

TABLE 12.11
Properties of 2EH Estolide Esters (See Figure 12.2 for Synthetic Scheme)

Estolide Ester[a]	Pour Point (°C)	Cloud Point (°C)	Vis. at 40°C (cSt)	Vis. at 100°C (cSt)	VI
A-2EH	−30	−36	125.5	19.3	175
B-2EH	−19	−17	131.3	20.0	175
C-2EH	−30	−34	114.5	17.9	174
D-2EH	−27	−30	106.0	16.9	173
E-2EH	−36	−41	104.4	16.8	175
F-2EH	−24	−16	106.3	16.8	172
G-2EH	−39	—[b]	93.8	15.5	176
H-2EH	−24	—[b]	84.2	14.3	177
I-2EH	−36	−32	73.9	13.0	179
J-2EH	−27	−29	70.6	12.4	176
K-2EH	−25	−22	80.5	13.9	179
L-2EH	−18	−11	78.7	13.4	174
M-2EH	−12	−13	81.6	13.5	174
N-2EH	−12	−13	41.3	8.7	196
O-2EH	−15	4	81.8	14.0	177
P-2EH	−5	−1	77.1	13.4	178

[a] EH Ester, continued from Table 12.4.
[b] Material color too dark to determine accurate values.

(upper 100s to 200+). Most soy-based biolubricants have reported VI less than 100. The lower the VI number, the less desirable the material where viscosity may be an issue. VIs greater than 200 is outstanding and highly advantageous. The VI represents how close the viscosities of the material are at 40°C and 100°C; thus, the ideal lubricant would have the same viscosity at all temperatures.

12.5.3 POUR POINT AND CLOUD POINT

Pour points were measured by ASTM Method D97-96a [59] to an accuracy of ±3°C. The pour points were determined by placing a test jar with 50 mL of the sample into a cylinder submerged in a cooling medium. The sample temperature was reduced in 3°C increments until the material stopped pouring. The sample is cooled until it no longer flowed when the test jar is held in a horizontal position for 5 s. The temperature of the cooling medium was chosen based on the expected pour point of the material. Samples with pour points in the range +9°C to −6°C, −6°C to −24°C, and −24°C to −42°C were placed in baths of temperatures −18°C, −33°C, and −51°C, respectively. The pour point was defined as the coldest temperature at which the sample still poured. All pour points were determined in duplicate and average values are reported here.

Cloud points were determined by ASTM Method D2500-99 [60] to an accuracy of ±1°C. The cloud points were determined by placing a test jar with 50 mL of the sample into a cylinder submerged into a cooling medium. The sample temperature was reduced in 1°C increments until any cloudiness was observed at the bottom of the test jar. The temperature of the cooling medium was chosen based on the expected cloud point of the material. Samples with cloud points that ranged from ambient to 10°C, 9°C to −6, and −6°C to −24°C, −24°C to −42°C were placed in baths of temperatures 0°C, −18°C, −33°C, and −51°C, respectively. All cloud points were determined in duplicate and average values are reported.

12.5.3.1 Estolides Free-Acid and Estolide 2EH Esters Low-Temperature Properties

A series of estolide free-acids and estolide 2EH esters were synthesized from oleic acid and the appropriate saturated FAs with 0.4 mol equivalents of perchloric acid at either 45°C or 55°C for 24 h (Figure 12.2). Vacuum distillation removed any excess FAs and provided estolide samples. The alcohol portion of the ester functionality was determined by Isbell et al. [25] to have a significant role in pour point reductions because branched chain alcohols dramatically lower the pour point. Thus, these estolides were converted to their corresponding 2-ethylhexyl estolide esters to provide enhanced pour point capability. The reaction temperatures, saturated FAs, pour and cloud points, viscosity, VI, color, and EN for the free-acid estolides (Tables 12.3 and 12.10) and the estolide 2EH esters (Tables 12.4 and 12.11) are listed.

The pour points of saturated estolide free-acids and estolide 2EH esters synthesized at 45°C are displayed in Figure 12.7. As the chain length of the saturated FA component increased from C-4 to C-10, the pour points of the estolide 2EH esters decreased to −39°C, and then as the chain length increased to C-18, the pour points increased to −15°C. In every case, the estolide 2EH esters had better (lower) pour points than their corresponding estolide free-acids (Tables 12.10 and 12.11). As the

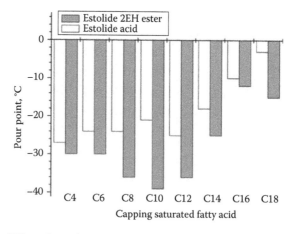

FIGURE 12.7 Effect of capping saturated FA chain length on pour point of estolide acid and estolide 2EH esters.

TABLE 12.12
Physical Properties of Cuphea-O-EST-2EH Esters (See Figure 12.2 for Synthetic Scheme)

Estolide[a]	Equiv HClO$_4$	GC EN[b]	Pour Point (°C)	Cloud Point (°C)	Estolide[c] Ester	Pour Point (°C)	Cloud Point (°C)	Acid Value (mg/g)
I	0.40	2.13	−3	15	I-2EH	−9	−7	1.60
II	0.20	2.02	−6	0	II-2EH	−12	−10	1.32
III	0.10	1.88	−27	−11	III-2EH	−21	−20	0.87
IV	0.05	1.75	−27	−23	IV-2EH	−33	−34	0.99
V	0.01	1.47	−27	−30	V-2EH	−42	−36	1.15

[a] Reactions were run for 24 h with overhead stirring under vacuum in a 2:1 molar ratio of oleic acid/cuphea FA at 60°C, followed by distillation at 180°C–200°C.
[b] Estolide number.
[c] Esterification reactions were run with magnetic stirring and 0.5 M BF^3 at 60°C.

chain length increased, the pour points did not vary much for the estolide 2EH esters until C-16 and C-18, when the pour point increased.

The cloud points of saturated estolide FAs and estolide 2EH esters synthesized at 45°C followed the same general trend as the pour points in Figure 12.7 (Tables 12.10 and 12.11). In general, all distilled estolides, either free-acids or esters, with low-pour points also have low-cloud points. The C-10 estolide 2EH ester (G-2EH, Tables 12.3 and 12.11) should have had the best (lowest) cloud point, but the material was much too dark to determine its cloud point. A Gardner color of 18 is a nontransparent black material.

Other mixtures of saturated FAs were explored using coconut (Table 12.5) and cuphea FAs (Table 12.12). The physical properties of various commercial materials and a CO-EST-2EH were compared with the best-performing cuphea-O-EST-2EH. Both of the estolide esters were completely unformulated, unlike the commercial products, which contain up to 40% additives designed to improve cold temperature properties. All the commercial products shown in Figure 12.8 had cold-weather-functional pour points except the soy-based oils such as Soylink. Soy-based products are known to have pour points that are too high and unsuitable for cold-weather climates [13].

12.5.3.2 Castor and Lesquerella Estolide 2EH Esters Pour Points

Castor and lesquerella estolide 2EH esters, where the castor and lesquerella base units were unsaturated (Figure 12.5), produced estolides that had the lowest pour points (pp) when capped with oleic acid (pp=−54°C and pp=−48°C, respectively) or with a branched material, 2-ethylhexanoic acid (pp=−51°C and pp=−54°C, respectively) (Table 12.8). By capping the FA esters of castor and lesquerella, the compounds no longer had the opportunity to undergo either intra- or intermolecular hydrogen-bonding interactions, thus yielding a lower pour point. As the capping

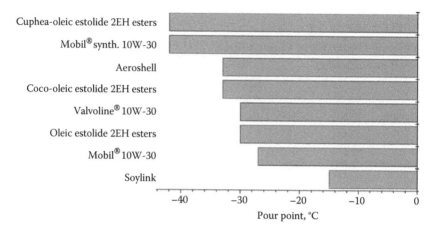

FIGURE 12.8 Pour-point comparison of unformulated estolides esters to commercial products.

material was changed to a saturated FA, dramatic changes occurred in the physical properties, stearic (pp=3°C castor and pp=3°C lesquerella) and coco (pp=−36°C castor and pp=−24°C lesquerella), respectively (Table 12.8). The stearic group had higher pour points than the base FA esters because the long saturated alkyl group allows for sufficient alkyl stacking. With shorter branched chains, the opposite trend was observed with both castor and lesquerella (Table 12.8, estolide entry 4 and 8). The shorter branched chains disrupt the stacking interactions and produce pour points that are considerably lower than the underivatized FA esters [20]. It has been previously demonstrated that coco-capped estolides have beneficial effects on the cold temperature properties of estolides [51].

As the base of the estolide was changed to a saturated unit, castor and lesquerella FA esters had pour points of 9°C and 15°C, respectively [20]. These saturated base unit esters were combined with either oleic or stearic acid. The oleic-capped estolide esters had reasonable pour points (pp=−36°C castor and pp=−12°C lesquerella), whereas the stearic-capped materials allowed for sufficient alkyl stacking, thus producing higher pour points (Table 12.8).

Some of the estolide esters synthesized from castor and lesquerella FA have outstanding (very low) pour points (<−50°C) (Figure 12.9). The best estolide to date has been a cuphea-O-EST-2EH with a pour point of −42°C [21]. With respect to the castor and lesquerella-based estolides, the general trend for the lowest pour point is that 2°C of unsaturation (oleic-castor and oleic-lesquerella) equals one estolide capped with some branching (2EH-lesquerella and 2EH-castor) followed by an estolide with one degree of unsaturation either as a capper or on the base unit (pp=−33°C, coco-castor) as shown in Table 12.8.

12.5.3.3 Castor and Lesquerella TG-Estolides Pour Points

The hydroxy TGs of lesquerella and castor oil were converted into their corresponding mono and fully capped estolides according to the reaction depicted in Figure 12.4.

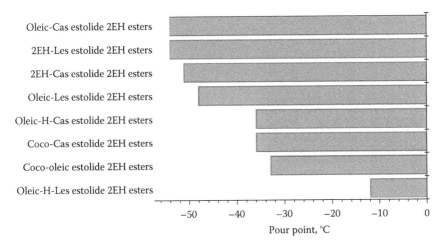

FIGURE 12.9 Pour-point comparison of hydrogenated (H) and nonhydrogenated lesquerella (Les) and castor (Cas) estolides 2EH esters.

Lesquerella and castor vegetable oils have pour points of −21°C and −15°C, respectively [45]. The fully capped lesquerella TG estolides had slightly lower pour points for all cases than the monocapped estolides. The lowest pour points were observed for both the mono (pp=−33°C) and fully capped lesquerella TG-estolides (pp=−36°C) when the capping group was hexanoic (Table 12.6, L6-M and L6-F). When the saturated capping chain was stearic, the materials were semisolids. The fully capped castor TG-estolides also had lower pour points than the monocapped materials for all saturated FA groups shorter than lauric. The lowest pour points observed for both mono (pp=−36°C) and fully capped castor TG-estolides (pp=−45°C) were also seen when the capping group was hexanoic (Table 12.7, C6-M and C6-F). The stearic fully capped castor TG-estolides were also solids at room temperature as well. The oleate-capped TG-estolides for both lesquerella and castor had pour points about −30°C.

The longer chain-capping groups provided sufficient alkyl stacking to yield pour points that were higher than the original oils. The shorter chain groups appear to disrupt the TG stacking interactions to yield pour points that were considerably lower than the underivatized oils. Interestingly, the acetate-capped TG-estolides with their two carbon units do not appear to significantly disrupt the stacking interactions in the oil that has the same pour point as lesquerella oil. Fully capping the lesquerella oil removed both the intra and intermolecular hydrogen-bonding interactions, yielding a lower pour point. When oleate was used as the capping FA, the unsaturation of the oleate improved the pour point, resulting in considerably lower pour-point temperatures, compared to the stearates. Thus, the pour point decreased from 24°C for stearate fully capped to −27°C for the oleate fully capped lesquerella TG-estolides (Table 12.6, L18-F and L18:1-F).

The lesquerella and castor TG-estolides compared favorably to previously synthesized estolides [45]. In terms of pour points, the TG-estolides were equal to the saturated-capped estolide 2EH ester series. Figure 12.10 shows the pour points of the TG-estolides, which are in the middle of estolides synthesized to date, with the

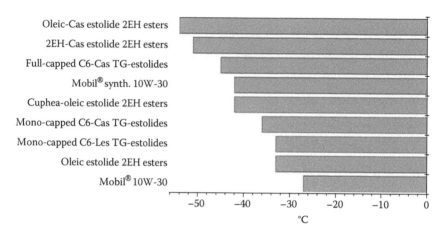

FIGURE 12.10 Pour-point comparison of various estolides and commercial engine oils.

lesquerella and castor estolide 2EH esters still being the estolides with the lowest pour points to date.

12.5.3.4 Estolides and Cloud Points

In general, all the estolides have cloud points that are very close to their individual pour points if all the excess monomers and fatty esters are removed. All materials containing monomers will have significantly higher cloud points. A high-cloud point could lead to filter clogging and poor pumpability in cold weather applications. Most petroleum-based motor oils and soybean-based products have cloud points near 0°C [51], which are unacceptable for most cold weather applications. The high-cloud point of commercially available base oils demonstrates a need for a better-performing cold weather oils. All synthesized estolides had cloud points and other physical properties that outperformed the base oil materials currently in the marketplace, even when the estolides are not formulated with low-temperature additives (Figure 12.11). Some of the best-performing estolides have cloud points <−50°C, which is not even achievable with conventional petroleum-based oils. Special niche markets as potential lubricants have developed for the estolides, since they have these cold temperature advantages.

12.5.4 ROTATING PRESSURIZED VESSEL OXIDATION TESTS

There are numerous ways by which the oxidative stability of an oil has been measured. Some of the common ways are as follows: OSI [54], rotating pressurized vessel oxidation test (RPVOT) [55], differential scanning calorimetry [61]; Indiana stirring oxidation test [62]; and the thin-film microoxidation test [63].

The original oleic-based estolide esters were developed as an industrial base oil or as an industrial oil additive, and so the material had to be evaluated under conditions commonly associated with industrial-based materials. These industrial oils will replace petroleum oils and by-products for which the recommended tests are

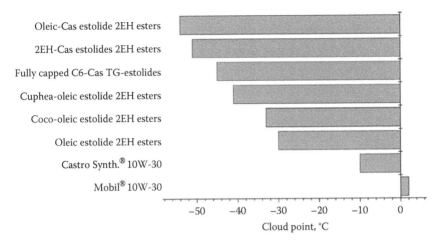

FIGURE 12.11 Cloud-point comparison of various estolide and commercial engine oils.

generally the microoxidation or RPVOT. The American Society for Testing Materials has developed detailed test procedures for these materials. For the RPVOT, the time to failure is reported in minutes. Failure identified as a pressure drop of 175 kPa from the maximum recorded pressure. The longer the RPVOT time the better the oxidative stability of the material. Typical commercial engine oils have RPVOT times in the range of 210–250 min (Table 12.13). The RPVOT test method calls for the oil to be tested with materials that would be present in most applications such as water, copper, and oxygen. The test has been accepted by biobased material producers as a suitable method to test the oxidative stability of these fluids. With vegetable-based materials, the RPVOT method tests both the thermal oxidative stability and hydrolytic stability [64].

Rotating bomb oxidation tests were conducted on a RPVOT apparatus manufactured by Koehler (Bohemia, NY) using the ASTM Method D 2272-98 [65]. Estolides and commercial products were tested at 150°C. Samples were measured to 50.0±0.5 g with 5.0 mL of reagent water added to the sample. The copper catalyst used was 3 m long and polished with 220 grit silicon carbide sandpaper produced by Abrasive Leaders and Innovators (Fairborn, OH) and was used immediately. The wire was wound to have an outside diameter of 44–48 mm and a weight of 55.6±0.3 g and a height of 40–42 mm. The bomb was assembled and slowly purged with oxygen twice. The bomb was charged with 90.0±0.5 psi (620 kPa) of oxygen and then tested for leaks by immersing in water. The test was complete after the pressure dropped more than 25.4 psi (175 kPa) from the maximum pressure. All samples were tested in duplicate runs and the average time is reported here.

12.5.4.1 Oxidative Stability: Estolide 2EH Esters

Some of the concerns with using vegetable-based materials are that they have poor oxidative stability [54,66] as well as below-standard cold weather properties [13]. Because estolides are derived from vegetable oils, one might assume they also have these same undesirable characteristics. However, cold weather properties of estolides

TABLE 12.13
RPVOT Values of Common Functional Fluids

Fluid	Average Time (min)
Aeroshell® 15W-50 Aviation oil	552
Biosoy®	28
Castrol® Synthetic 10W-30	246
Crambe oil[a]	13
Environlogic®-132 Terrsolve	67
Environlogic-146 Terrsolve	51
Environlogic-168 Terrsolve	71
Meadowfoam oil[a]—crude	20
Soybean oil[a]	13
Soylink®	83
Traveller® All Season H.F.[b] IVG-46	274
Traveller Premiun Universal H.F.[b] IVG-46	464
Valvoline® 5W-30	228
Valvoline 10W-30	223
Valvoline 10W-40	224
Valvoline 20W-50	214
Valvoline SAE-30	224

[a] Unformulated.
[b] Hydraulic fluid.

turned out to be surprisingly superior to materials currently on the market. Some concerns have also been raised regarding the oxidative stability of vegetable-based functional fluids [67]. A functional fluid is any material that can be used in any working device, hydraulic, crankcase, or lubricating fluid to name a few. A number of ways to improve the oxidative stability of an oil, including that of an estolide, exist such as through synthetic methods for oxidative stability improvements. Akoh [54] reported that refined soybean oil had an OSI [52] of 9.4h at 110°C, but once the oil was partially hydrogenated, the OSI increased to 15.3h at 110°C, an improvement of more than 60%. The same approach was taken with the oleic estolides, which underwent hydrogenation with 2% w/w of 10% palladium on activated carbon to obtain completely saturated estolides [25]. The saturated oleic estolides were expected to be more oxidatively stable than the unsaturated estolides by assuming that the same trend as displayed by soybean oil held true. However, the final product had undesirable cold weather properties that made it useless as a biobased functional fluid.

Saturated estolides were developed to eliminate the poor oxidative stability problems. In a one-step process, an estolide ester was synthesized with such excellent physical properties as pour and cloud points, viscosities, and oxidative stabilities. These saturated materials retained good pour point properties and had the potential for increased oxidative stability, in addition, to reduced production costs and chemical waste [19].

A series of formulation studies were conducted to explore their effects on oxidative stability and compare them with the stabilities of commercially available materials [55]. The RPVOT times were determined on a wide range of petroleum, vegetable-based, and synthetic materials at 150°C, and these results are listed in Table 12.13. Of the materials tested, those formulated with some sort of OSP performed the best. The normal petroleum and synthetic motor oils had acceptable RPVOT times greater than 200 min, whereas a premium hydraulic fluid had an RPVOT time of greater than 400 min. The best-performing material tested (Table 12.13) was a moderately priced aviation oil used for single engine, propeller planes with a RPVOT time of 552 min. All of the vegetable-based oils tested were unformulated and had very short RPVOT times, usually 20 min or less. Even crude meadowfoam oil, *Limnanthes alba*, which is the most oxidatively stable, crude vegetable oil [66] with an OSI of about 247 h, had an RPVOT time of only 20 min. This example demonstrates the harsh conditions that the RPVOT exerts on the fluids being tested. Other biobased materials listed in Table 12.13 had RPVOT times of less than 100 min. The average RPVOT times for these types of fluids are between 55 and 80 min. The two soybean-based materials listed in Table 12.13 were formulated with at least 40%–60% additives to make them perform at a marketable level. Akoh [54] and Isbell et al. [66] demonstrated that the unsaturation of soybean-based oils has a destructive effect on oxidative stability. Thus, Biosoy, a soybean derivative, had the shortest RPVOT time of 28 min of all the formulated materials tested. Two different estolide 2EH esters were synthesized from oleic acid (Figure 12.12) [16]. A third was synthesized from oleic and coconut FAs (Figure 12.12) [51] with 0.05 mol equivalents of perchloric acid at 60°C under vacuum followed by direct conversion to the corresponding ester by the addition

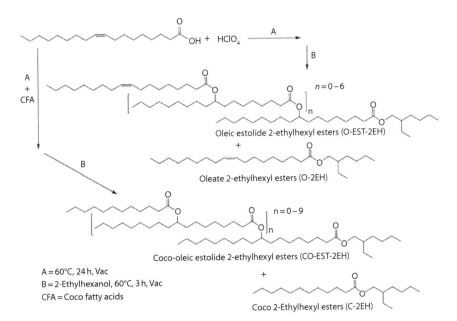

FIGURE 12.12 Synthesis of O-EST-2EH and CO-EST-2EH and C-2EH and O-2EH.

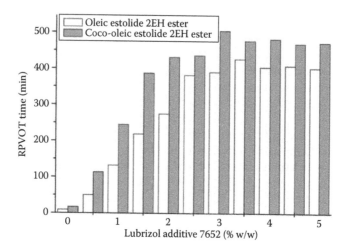

FIGURE 12.13 Estolide esters—RPVOT time versus the concentration of OSP.

of 2-ethylhexanol at 60°C for 3 h under vacuum (Figure 12.12). The two different estolide 2EH esters were vacuum distilled to remove any excess 2-ethylhexanol, whereas the CO-EST-2EH ester underwent additional vacuum distillation to remove any excess FAs [22].

RPVOT times were determined on the O-EST-2EH at 150°C while varying the amounts of an OSP, Lubrizol additive 7652 (Figure 12.13). The unformulated O-EST-2EH showed an expected low-RPVOT time of 8.5 min, as was typical for all vegetable oils. The OSP, Lubrizol additive 7652, was added to a concentration of 0.5% by weight. At 0.5% of the OSP, there was a fivefold increase in the RPVOT value to 50 min. Increasing the concentration of the OSP to 1.5% produced an RPVOT value of 219 min for the simple oleic estolide, which was comparable with the petroleum crankcase fluids [55]. This was an improvement of more than 25 times over the original stability time. The RPVOT values reached a maximum value at a concentration of 3.5% OSP, which produced a RPVOT time of 426 min. This time compared favorably with most premium petroleum-based hydraulic fluids. Further increase in concentration of the OSP from 4.0% to 10.0% showed no improvement on the overall RPVOT values (Figure 12.13).

The RPVOT times were also determined for the CO-EST-2EH as a function of the concentration of an OSP, Lubrizol additive 7652, at 150°C (Figure 12.13). The unformulated O-EST-2EH showed the expected low-RPVOT time of 17 min. In this case, the O-EST-2EH had RPVOT values almost twice that of the O-EST-2EHs. This could be accounted for in terms of IV. The simple O-EST-2EH had an IV of about 40, whereas the CO-EST-2EH had an IV of about 15. Oxidative stability and IV are known to be dependent on the amount of unsaturation in a molecule [66].

The OSP, Lubrizol additive 7652, was added at a concentration of 0.5% by weight in O-EST-2EH. At 0.5% of the OSP, there was a 6.5-fold improvement in the RPVOT value to 113 min. Increasing the OSP concentration to 1.0% increased the RPVOT time to 245 min, which exceeded the values for the petroleum crankcase oils listed in Table 12.13. Therefore, an RPVOT time of 200 min, common for most petroleum crankcase oils, could be easily achieved with less than 1.0% OSP. The result shows that the CO-EST-2EH could be easily and inexpensively formulated into a commercial crankcase formulation [55].

At a 2.0% concentration of the OSP, the RPVOT values for O-EST-2EH were similar to that of premium hydraulic fluids. The RPVOT values for O-EST-2EH reached a maximum with about 3.0% concentration of OSP, which produced an RPVOT time of 504 min. This value compares favorably with aviation oil for single-engine, propeller planes (Table 12.13), an improvement of more than 30-fold over the unformulated CO-EST-2EH. Further increase in the concentration of the OSP to 30%–10% showed no further improvement in oxidative stability (Figure 12.13).

The relative RPVOT values of O-EST-2EH and CO-EST-2EH with similar concentrations of Lubrizol additive 7652, commercial OSP, were not the same. There was a noticeable difference between the two estolide samples. The CO-EST-2EH generally gave longer RPVOT times at all concentrations of the oxidation stability package (Figure 12.13). The most noticeable difference was at the 1.5% concentration of OSP, where the CO-EST-2EH displayed almost twice the RPVOT value of the O-EST-2EH. O-EST-2EH and CO-EST-2EH displayed maximum RPVOT values at 2.5% and 3.0% of the OSP, respectively. The RPVOT values held somewhat steady or declined slightly with further increase of the concentration of the OSP (Figure 12.13).

12.5.4.2 Oxidative Stability: Lesquerella and Castor Estolide 2EH Esters

The oxidative stabilities of vegetable oils are very poor, whereas past estolide esters have increased the oxidative stability by reducing the amount of unsaturation in the molecule [55]. Estolide esters synthesized from hydroxy FAs (Figure 12.5) do not decrease the unsaturation in the molecule, and, therefore, improvements in the oxidative stability from the formation of estolide are unlikely. At the time of this study, lesquerella oil was only available in a limited supply, and so most larger scale destructive tests were conducted on castor versions. Castor versions have somewhat similar FA profiles as lesquerella (Table 12.2) in terms of effects on oxidative stabilities. This theory proved to be correct when pure castor-O-EST-2EH was tested for oxidative stability and displayed RPVOT time of 15 min (Figure 12.14). The RPVOT times for the estolides tested increased as the concentration of antioxidant package was increased (Figure 12.14). The unsaturated castor-based O-EST-2EH with antioxidant stability package concentration of 3.5% compared favorably with commercially available petroleum-based oils (Table 12.13). Increased RPVOT times were observed when the "capping" unit was replaced with saturated FAs (e.g., coco) (Figure 12.5). Addition of 3.5% OSP in coco-castor estolide 2EH ester resulted in RPVOT time that exceeded most commercially available oils (Figure 12.14). In addition, the cold temperature properties of these partially hydrogenated materials still exceed the commercially available materials (Table 12.8).

12.5.4.3 Oxidative Stability: Lesquerella and Castor TG-Estolides

The oxidative stabilities of vegetable oils are very poor due to the high amount of unsaturation inherently present in the molecule. The formation of estolides from oleic

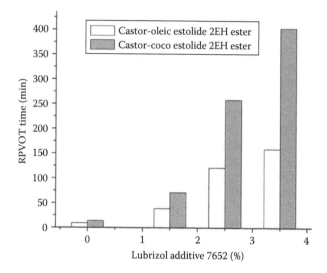

FIGURE 12.14 Castor estolide esters—RPVOT time versus the concentration of OSP.

acid reduces the amount of unsaturation in the molecule, which resulted in improved oxidative stability [55]. The TG-estolides synthesized, however, do not remove the unsaturation during the course of TG-estolide synthesis (Figure 12.4). Consequently, the RPVOT times for the TG-estolides are still short when completely unformulated, approximately 15 min (Table 12.14). The addition of an antioxidant package can improve the performance of the TG-estolides with respect to the RPVOT. However, even at 4% concentration of antioxidant, the TG-estolides displayed RPVOT times only in the 200 min range (RPVOT times of 200 min are typical of most mineral oil-based automotive crankcase fluids, Table 12.13).

Hydrogenation of the lauric-capped TG-estolides from both lesquerella and castor greatly improved their oxidation stability performance even with no antioxidant added, where the lesquerella monocapped TG-estolide had an RPVOT of 56 min. When 1% antioxidant was added to this material, the RPVOT time increased to 439 min.

The castor TG estolides performance was also improved at the same antioxidant concentration (1%), but the improvement in RPVOT time was smaller. Unfortunately, the pour points of these materials were negatively impacted because these materials are solids at room temperature.

12.5.4.4 Specially Designed Oxidative Package for Estolide 2EH Esters

Four different materials (Figure 12.12) were used to evaluated a new set of OSP for the estolide ester series. All the final products tested underwent simple vacuum distillation to remove excess 2EH alcohol. The simplest estolide, O-EST-2EH, underwent a further distillation at 180°C–200°C to remove the fatty esters (monomers), oleate 2EH esters (O-2EH), to yield the purified estolide ester (O-EST-2EH). A CO-EST-2EH was prepared under similar conditions to yield a distilled coco-based estolide ester sample. Then, a more complex mixture was examined; the complete CO-EST-2EH reaction was used as it is (no distillation), so the sample

TABLE 12.14
RPVOT[a] Values of Lesquerella and Castor TG-Estolides

TG Estolide[b]	Lubrizol 7652 Antioxidant Added (wt.%)				
	0	0.5	1	2	4
L12-M	13	17	29	66	118
L12-F	14	22	47	106	158
H₂L12-M	56	n.d.	439	559	541
H₂L12-F	37	n.d.	476	688	668
L18:1-M	16	18	29	34	121
L18:1-F	14	20	31	60	127
C12-M	18	30	44	102	241
C12-F	16	32	46	124	318
H₂C12-M	47	n.d.	142	289	396
H₂C12-L	28	n.d.	242	410	515
C18:1-M	15	23	52	90	197
C18:1-F	14	26	48	103	180

[a] All RPVOT times are reported in minutes.
[b] Estolides from Tables 12.6 and 12.7.
n.d., not determined.

contained both the coco 2EH ester (C-2EH) as well as the O-2EH ester in a ratio of 1:1.4 (C-2EH:O-2EH) as previously reported [51]. Finally, the ester fraction obtained as the distillate during the purification step (distillation) of the synthesis of CO-EST-2EH contained the esters O-2EH and C-2EH in the same ratio as reported above.

The two pure (distilled) estolides were analyzed for the amount of saturated capping material. The saturated-capped % values in Figure 12.12 were obtained from GC analysis of the estolide esters, which were saponified and esterified. The components of the GC were classified as one of the following: unsaturated, saturated, and hydroxy FAs. The percent saturated-capped estolide was calculated as follows:

$$\% \, \text{Saturated-capped} = \left[\frac{(\text{saturated FA})}{(100 - \text{hydroxy FA})} \right] \times 100 \qquad (12.1)$$

Analysis of GC data using Equation 12.1 showed that the CO-EST-2EH contained 42% saturated-capped estolide esters while the remaining material was capped with unsaturated FAs. Depending on the amount of the capped material, the physical properties of the estolides will vary, and the estolides interact differently, i.e., the more saturation tends to lead to better oxidative stability.

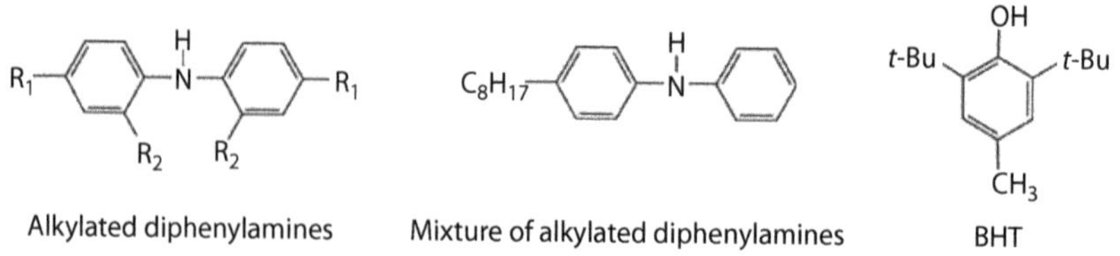

Alkylated diphenylamines   Mixture of alkylated diphenylamines   BHT

FIGURE 12.15 Arylamines and hindered phenol commercial oxidation inhibitors.

A series of OSP containing amines and hindered phenols (Figure 12.15) were examined. Both of these classes of antioxidant compounds act as radical scavengers through a hydrogen transfer mechanism [68]. Ten antioxidant packages were prepared by blending each amine (alkylated diphenyl amine or mixed alkylated diphenyl amine, Figure 12.15) antioxidant with a hindered phenol (BHT, Figure 12.15) antioxidant in the following ratios: 100:0, 75:25, 50:50, 25:75, and 0:100. Each antioxidant was blended into each of the estolide esters to a concentration of 0.5% or 1.0% (wt/wt). Each sample was used in duplicate RPVOT tests. The RPVOT values reported here were obtained following the RPVOT ASTM Method D 2272-98 exactly. Previously, soybean oil [69], which is known to have poor hydrolytic stability, was tested with water removed from the RPVOT ASTM method. This gave RVPOT times that were dramatically higher than the RPVOT times when the ASTM method was followed as published in the presence of water. This demonstrates that materials that display long RPVOT times based on unmodified ASTM Method D 2272-98 must be both thermally and hydrolytically stable. In our case, water was used in the RPVOT method; thus, the samples could potentially undergo hydrolysis that would yield unacceptable or lower RPVOT times. As most vegetable-based materials, the estolide-based samples tested did not suffer from this condition. The RPVOT times that are reported for the estolides represent a combination of both the thermal oxidative stability and hydrolytic stability of the samples tested. The RPVOT results for the individual samples from Figure 12.12 are displayed in Figures 12.16 through 12.19. Again, all samples without added OSP gave RPVOT times of less than 20 min, which was similar to data on vegetable oils without antioxidants (Table 12.13). In general, as the concentration of amine antioxidants increased, the RPVOT times increased for all samples. With 100% of a commercial hindered phenol (1% wt/wt), the RPVOT times, in general, increased from 16 min to about 100 min; this was not outstanding, but, in most cases, an improvement was visible and comparable with current commercial biobased materials (Table 12.13). In all cases, the samples with the highest RPVOT times had higher amounts of the amine packages added to the estolide-based materials.

In cases where the amines were the only antioxidant used, the alkylated diphenyl amine performed better than the mixed alkylated diphenyl amine. The simple O-EST-2EH (Figure 12.16) with 1.0% OSP, the 100% alkylated diphenyl amine had an RPVOT time of 248 min, whereas the mixed alkylated diphenyl amine displayed only 197 min. This trend held true with all four oil samples at both the 0.5% and 1.0% OSP concentrations. The most dramatic improvements were observed with two of the C-2EH samples (Figures 12.18 and 12.19). The mixture of CO-EST-2EH

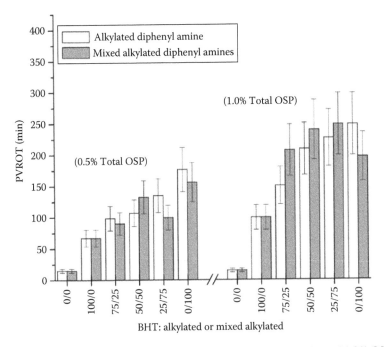

FIGURE 12.16   RPVOT times for O-EST-2EH (Figure 12.12) with 0.5% and 1.0% OSP.

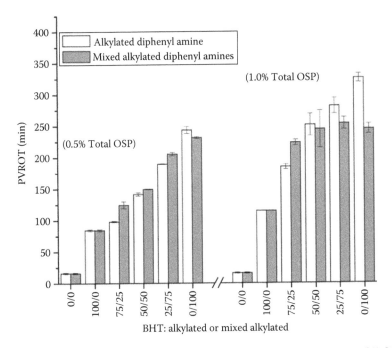

FIGURE 12.17   RPVOT times for CO-EST-2EH (Figure 12.12) with 0.5% and 1.0% OSP.

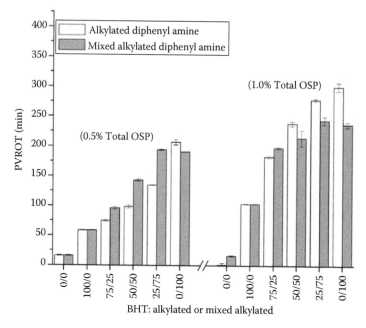

FIGURE 12.18 RPVOT times for CO-EST-2EHs containing CO-EST-2EH and C-2EH (Figure 12.12) with 0.5% and 1.0% OSP.

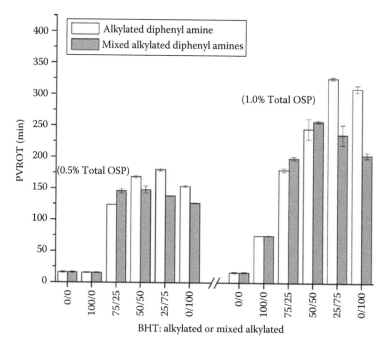

FIGURE 12.19 RPVOT times for coco 2EH esters (C-2EH, Figure 12.12) with 0.5% and 1.0% OSP.

and C-2EH (Figure 12.18) showed an improvement of more than 80 min or 33% in the RPVOT time at the 1.0% OSP. The C-2EH sample (Figure 12.19) showed an improvement of more than 105 min or 52% in the RPVOT time at the 1.0% OSP.

The simple O-EST-2EH (Figure 12.16) and the CO-EST-2EH (Figure 12.17) gave similar RPVOT times that were within experimental error of one another. Cermak and Isbell [21,55] previously showed that complex estolide esters had the longest RPVOT times with the addition of a differently formulated commercial antioxidant additive when capped with a very short chain, saturated FAs such as those present in cuphea and coconut. The coco-estolide ester had longer RPVOT times due to the saturation in the molecule. With the amine antioxidants, the estolides unexpectedly performed about the same with longer RPVOT times than under previous conditions. The similar RPVOT times between the two estolide samples show the effect of contributions of interactions of the estolides with the antioxidants.

The mixture of CO-EST-2EH and C-2EH (Figure 12.18) and C-2EH (Figure 12.19) had significantly longer RPVOT times than the pure (distilled) estolide esters with the amine additive packages. The CO-EST-2EH and C-2EH (Figure 12.18) had longer RPVOT times than the simple CO-EST-2EH (Figure 12.17). The 77 min increase in RPVOT time was attributed to the increased amounts of saturated esters present in the sample. The C-2EH (Figure 12.19) yielded a longer RPVOT time of 326 min, which was expected as the sample contained the highest amounts of saturation of all the samples analyzed.

12.5.5 Noack Evaporative Loss Test

Evaporative loss determinations were conducted on a Noack evaporative tester manufactured by Koehler (Bohemia, NY) using the ASTM Method D 5800-00a [70]. Estolides and commercial products were run at 250°C. Samples were measured to 65.0 ± 0.1 g to a precision of 0.01 g. The test was complete after 60 min at which time the extraction tube was disconnected within 15 s. The sample crucible was placed in a cold-water bath to a minimum depth of 30 mm for 30 min. All samples were run in duplicate, and the average evaporative losses are reported here.

An important fact in determining how well an oil will behave as a potential lubricant is to evaluate the oil volatility or evaporative loss. In most applications where oil is used as a lubricant, heat is generated due to friction. As the oil heats, it can and does become very volatile, as in the case of air-cooled gas engines (i.e., water pumps and lawnmowers). As the oil evaporates and escapes from the engine, the engine has less lubricating material, which increases the operating temperatures until engine damage or failure occurs. There are a number of ways to counter this volatility problem. The first would be to develop lubricating oils that would either dissipate the heat rapidly or create an excellent lubricant that would not generate heat. The simplest and most cost-effective method would be to produce oils that would have very low or no evaporative losses. Current oils have evaporative loss requirements of no more than 15% using the Noack method. Table 12.15 shows that all the commercial samples have evaporative losses very close to 15%. All the estolide esters have evaporative losses of only about 1%; this would allow the estolides to last longer in the very hot running engines, thus potentially extending engine life and reducing wear.

TABLE 12.15

Noack Value: Estolide Esters Comparison to Commercial Products

Sample	Loss (%)
Mobil® 10W-30[a]	14.1
Traveller® Premiun Universal H.F.[b] IVG-46	17.6
Penzoil Synthetic® 10W-30[c]	16.2
CO-EST-2EH (Table 12.5, Estolide CO-EH-DD)[d]	1.1
Oleic-Cas Estolide 2EH Ester (Table 12.8, Estolide 1)[d]	1.2
O-EST-2EH (Figure 12.12, Estolide O-EST-2EH)[d]	1.1

[a] Commercial petroleum oil.
[b] Commercial hydraulic fluid.
[c] Commercial synthetic oil.
[d] Unformulated.

12.5.6 ESTOLIDE ESTERS PILOT SYNTHESIS

The estolide reactions are very straightforward and high-yielding reactions [17,18]. Simple O-EST-2EHs and CO-EST-2EHs have been synthesized on the drum size quantity. Oleic sources included ~70% yellow oleic, >95% oleic (pharmaceutical grade), and oleic from a canola source. Oleic from all these sources has been successfully converted to estolides in the pilot plant. Figure 12.20 shows the general scheme for the synthesis of highly refined estolides with best pour, cloud points as well as oxidative stability. Once the reaction is quenched with base, the material is filtered

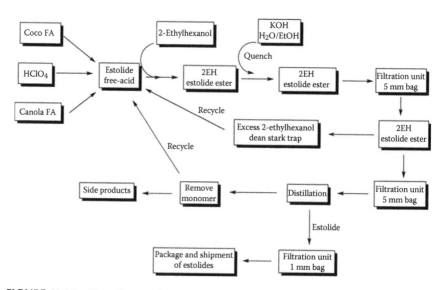

FIGURE 12.20   Flow diagram for the pilot scale (40 gal) synthesis of estolide EH ester.

through a 5 μm bag filter. The material is reheated to remove the ethanol, water, and finally excess 2-ethylhexanol. After cooling to room temperature, the estolide is filtered through another 5 μm bag filter and then distilled.

The most cost-effective estolide would be an undistilled estolide where the unreacted monomer remains with the base estolide material. The simple O-EST-2EH (Figure 12.2) would provide the most cost-effective and simplest estolide to produce. However, depending on final application, the other estolides would serve in extreme conditions where cold weather and high-oxidative stability requirements are a must.

12.5.6.1 Estolide Esters Pilot Distillation

A Myers Pilot-15 distillation unit (Myers Vacuum, Kittanning, PA) was used for pilot scale distillations [71]. It is a continuous, centrifugal 38.1 cm molecular still that contains all of the components needed for distilling raw feedstock (Figure 12.21). Raw feedstock is delivered with a metering valve and a gear pump. It first enters a degasser unit maintained, which is maintained at pressures between 0.72 and 0.91 Pa. It then enters the heated evaporator cone or distillation chamber, where the molecular distillation takes place. The distillate and residue are continuously removed by transfer pumps. The material is passed to and from each station through stainless steel transfer lines, which are traced with heating tapes. This unit was designed for use in extended or large scale distillations (~50 gal/day).

The major component of the 38.1 cm centrifugal molecular still is a concave, heated, evaporator cone, which rotates within the distillation chamber and is maintained at pressures as low as <0.009 Pa (Figure 12.22). Two vacuum, mechanical

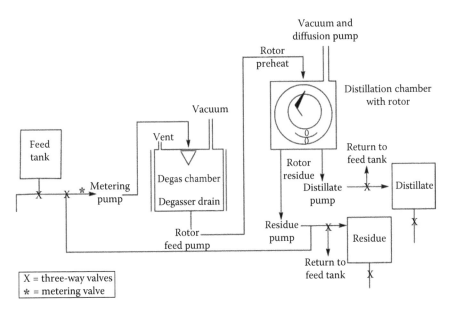

FIGURE 12.21 Operational flow diagram for the pilot scale (3 gal/h) molecular distillation unit for estolide production.

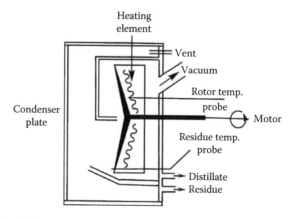

FIGURE 12.22 Side view of pilot scale (3 gal/h) distillation chamber used in estolide production.

vane pumps, and a high-vacuum diffusion pump supply both the degas and distillation chambers (Figure 12.22). Degassed and/or stripped feedstock is metered by the feed pump into the center of the spinning evaporator cone (1700 rpm) and is spread rapidly and evenly outward in a thin film over the entire surface by centrifugal forces. As the film spreads and is heated, part of the feedstock reaches a temperature at which it vaporizes and leaves the rotor surface to condense on the cooler surface of the condenser plate. With the help of centrifugal force, the unvaporized feedstock, the residue (estolide), is spun into a gutter, where it is removed from the still by a constant-speed transfer pump. The condensed vapor, the distillate (fatty esters), flows by gravity into a removal pipe located at the bottom of the distillation chamber and then pumped from the still by another constant speed transfer pump. The operational distillation unit has two temperature sensors in the distillation chamber (Figure 12.22). The first sensor, which monitors the temperature of the rotor, is located by the electric heating elements just beneath the 38.1 cm rotor. This first temperature corresponds to the temperature of the heating element and not the temperature of the rotor. The actual temperature of the rotor can be varied by varying the flow of feed material without affecting the element temperature. The second and most important temperature sensor measures the temperature of the residue. This sensor is located in the gutter where the residue is collected during distillation. The residue sensor provides the closest temperature of the rotor since it is a measurement of the temperature of the material that was just in contact with the rotor.

All the transfer lines and pumps are wrapped with heating tapes. The heating tape and electrical elements are controlled with digital controllers either based on percent power demand or actual measured temperature. Each region of the transfer line and pump has internal temperature probes that measure the temperature of the material as it passes the sensors.

The distillation unit allows for all estolides synthesized except for the TG-estolides to be distilled at levels acceptable for industrial purposes and formulated to meet those standards.

12.5.7 Conclusions

Estolides are very unique compounds that have potential as lubricants, edible applications, cosmetics, cooling fluids, hydraulic fluids, inks, etc. A wide range of estolides structures can be synthesized (Figures 12.1 through 12.5) through the use of various feedstocks such as new crops, animal, and traditional fats and oils. Estolide esters have outstanding cold weather and other desirable properties. They can be easily synthesized to meet various applications requirements. In terms of cold weather physical properties, castor and lesquerella estolide 2EH esters, where the castor and lesquerella base units are unsaturated-capped with 2-ethylhexanoic acid (Figure 12.5), have the lowest pour points (pp=−51°C and pp=−54°C), respectively (Table 12.8). In general, all the distilled saturated estolide esters had excellent (low) cloud points and performed well in low-temperature storage tests.

Viscosity is dictated by the extent of oligomerization. As the amount of oligomerization increased so did the viscosity. In general, the estolide free-acids were generally several hundred cSt more viscous than the corresponding esters. Estolides can be prepared in a very wide viscosity ranges as shown in Tables 12.5 through 12.8 and 12.10 through 12.12. Thus, there are many potential industrial applications for such materials.

All unformulated estolides outperformed commercial petroleum and biobased basestocks currently in the market with their outstanding physical properties and biodegradability. The estolide technologies have been licensed from USDA by Peaks and Prairies, LLC in Malta, MT, as a biolubricant for the period of 2007 through 2010.

ACKNOWLEDGMENTS

The author is extremely grateful to Girma Biresaw for his work and insight in the oxidative stability package for estolides. The author is also extremely grateful to Kendra B. Brandon, Jackie T. Chu, Stephanie S. DeKeyser, Amy B. Deppe, Amber L. Durham, Natalie A. LaFranzo, Benjamin A. Lowery, Jonathan L.A. Phillips, Rebekah M. Richardson, Alex L. Skender, and Melissa L. Winchell for their assistance with sample preparation, data collection, and synthesis. The author is grateful to Karl E. Vermillion and David Weisleder for performing all of the NMR experiments.

Names are necessary to report factually on available data; however, the USDA neither guarantees nor warrants the standard of the product, and the use of the name by USDA implies no approval of the product to the exclusion of others that may also be suitable.

REFERENCES

1. U.A. Energy Information Administration (2010), viewed November 16, 2010, http://tonto.eia.doe.gov/dnav/pet/PET_PRI_SPT_SI_D.htm
2. S.C. Cermak and T.A. Isbell, Synthesis and Physical Properties of Mono-Estolides with Varying Chain Lengths, *Ind. Crops Prod.*, 29, 205–213 (2009).

3. N.J. Novick, P.G. Metha, and P.B. McGoldrick, Assessment of the Biodegradability of Mineral Oil and Synthetic Ester Base Stock, Using CO_2 Ultimate Biodegradability Tests and CEC-L-33-T-82, *J. Synth. Lubr.*, 13, 19–30 (1996).

4. S.D. Haigh, Determination of Synthetic Lubricant Concentrations in Soil During Laboratory-Based Biodegradation Studies, *J. Synth. Lubr.*, 11, 83–98 (1994).

5. S.D. Haigh, Fate and Effects on Synthetic Lubricants in Soil: Biodegradation and Effect on Crops in the Field Studies, *Sci. Total Environ.*, 168, 71–83 (1995).

6. H. Noureddini, B.C. Teoh, and L.D. Clements, Viscosities of Vegetable Oils and Fatty Acids, *J. Am. Oil Chem. Soc.*, 69, 1189–1191 (1992).

7. T. Mang, Lubricants, in: *Lipids Technologies and Applications*, F.D. Gunstone and F.B. Padley (eds.), pp. 737–758, Marcel Deckker, New York (1997).

8. T. Mang, Biodegradable Lubricants and Their Future Markets, *Lipid Technol.*, 6, 139–143 (Nov./Dec. 1994).

9. F.A. Zaher and H.M. Nomany, Vegetable Oils and Lubricants, *Grasas Aceities (Seville)*, 39, 235–238 (1988).

10. J. Legrand and K. Durr, High-Performing Lubricants Based on Renewable Resources, *J. Agro. Food Ind. Hi-Tech.*, 9, 16–18 (1998).

11. N. Canter, It Isn't Easy Being Green, *Lubr. World*, 16–21 (September 2001).

12. R. Becker and A. Knorr, An Evaluation of Antioxidants for Vegetable Oils at Elevated Temperatures, *Lubr. Sci.*, 8, 95–117 (1996).

13. S. Asadauskas and S.Z. Erhan, Depression of Pour Points of Vegetable Oils by Blending with Diluents Used for Biodegradable Lubricants, *J. Am. Oil Chem. Soc.*, 76, 313–316 (1999).

14. G.R. Zehler, Performance Tiering of Biodegradable Hydraulic Fluids, *Lubr. World*, (September 2001).

15. R.D. Plattner, K. Payne-Wahl, L. Tjarks, and R. Kleiman, Hydroxy Acids and Estolide Triglycerides of *Heliophila amplexicaulis* l.f. Seed Oil, *Lipids*, 14, 576–579 (1979).

16. T.A. Isbell, K. Kleiman, and B.A. Plattner, Acid-Catalyzed Condensation of Oleic Acid into Estolides and Polyestolides, *J. Am. Oil Chem. Soc.*, 71, 169–174 (1994).

17. T.A. Isbell, T.P. Abbott, S. Asadauskas, and J.E. Lohr, Biodegradable Oleic Estolide Ester Base Stocks and Lubricants, U.S. Patent 6,018,063 (2000).

18. S.C. Cermak and T.A. Isbell, Biodegradable Oleic Estolide Ester Having Saturated Fatty Acid End Group Useful as Lubricant Base Stock, U.S. Patent 6,316,649 B1 (2000).

19. S.C. Cermak and T.A. Isbell, Synthesis of Estolides from Oleic and Saturated Fatty Acids, *J. Am. Oil Chem. Soc.*, 78, 557–565 (2001).

20. S.C. Cermak, K.B. Brandon, and T.A. Isbell, Synthesis and Physical Properties of Estolides from Lesquerella and Castor Fatty Esters, *Ind. Crops Prod.*, 23, 54–64 (2006).

21. S.C. Cermak and T.A. Isbell, Synthesis and Physical Properties of Cuphea-Oleic Estolides and Esters, *J. Am. Oil Chem. Soc.*, 81, 297–303 (2004).

22. T.A. Isbell and S.C. Cermak, Synthesis of Triglyceride Estolides from Lesquerella and Castor Oils, *J. Am. Oil Chem. Soc.*, 79, 1227–1233 (2002).

23. T.A. Isbell and R. Kleiman, Characterization of Estolides Produced from the Acid-Catalyzed Condensation of Oleic Acid, *J. Am. Oil Chem. Soc.*, 71, 379–383 (1994).

24. S.C. Cermak and T.A. Isbell, Physical Properties of Saturated Estolides and Their 2-Ethylhexyl Esters, *Ind. Crops Prod.*, 16, 119–127 (2002).

25. T.A. Isbell, M.R. Edgcomb, and B.A. Lowery, Physical Properties of Estolides and Their Ester Derivatives, *Ind. Crops Prod.*, 13, 11–20 (2001).

26. R.H. Purdy and C.D. Craig, Meadowfoam: New Source of Long-Chain Fatty Acids, *J. Am. Oil Chem. Soc.*, 64, 1493–1497 (1987).

27. S.M. Erhan and R. Kleiman, Meadowfoam Oil Factice and Its Performance in Natural Rubber Mixes, *Rubber World*, 203, 33–36 (1990).

28. S.M. Erhan and R. Kleiman, Vulcanized Meadowfoam Oil, *J. Am. Oil Chem. Soc.*, 67, 670–674 (1990).

29. S.M. Erhan and R. Kleiman, Factice from Oil Mixtures, *J. Am. Oil Chem. Soc.*, 70, 309–311 (1993).

30. D.A. Burg and R. Kleiman, Meadowfoam Fatty Amides: Preparation, Purification, and Use in Enrichment of 5,13-Docosadienoic Acid and 5-Eicosenoic Acid, *J. Am. Oil Chem. Soc.*, 68, 190–192 (1991).

31. H.B. Frykman, T.A. Isbell, and S.C. Cermak, 5-Hydroxy Fatty Acid Amides from Delta-Lactone and Alkyl Glucamines, *J. Surfact. Deterg.*, 3, 179–183 (2000).

32. T.A. Isbell, B.A. Plattner, and R. Kleiman, Method for the Development of δ-Lactones and Hydroxy Acids from Unsaturated Fatty Acids and Their Glycerides, U.S. Patent 5,849,935 (1998).

33. D.A. Burg and R. Kleiman, Preparation of Meadowfoam Dimer Acids and Dimer Esters, and Their Use as Lubricants, *J. Am. Oil Chem. Soc.*, 68, 600–603 (1991).

34. S.M. Erhan, R. Kleiman, and T.A. Isbell, Estolides from Meadowfoam Oil Fatty Acids and Other Monounsaturated Fatty Acids, *J. Am. Oil Chem. Soc.*, 70, 461–465 (1993).

35. T.A. Isbell, K. Kleiman, and S.M. Erhan, Characterization of Monomers Produced from Thermal High-Pressure Conversion of Meadowfoam and Oleic Acids into Estolides, *J. Am. Oil Chem. Soc.*, 69, 1177–1183 (1992).

36. R.W. Miller, F.R. Earle, I.A. Wolff, and Q. Jones, Search for New Industrial Oils. IX. Cuphea, a Versatile Source of Fatty Acids, *J. Am. Oil Chem. Soc.*, 41, 279–280 (1964).

37. Knapp, S.J., Modifying the Seed Storage Lipids of Cuphea. A Source of Medium-Chain Triglycerides, in: *Seed Oils for the Future*, S.L. MacKenzie and D.C. Taylor, eds., pp. 142–153, Champaign, IL (1993).

38. K.D. Carlson, A. Chaudhry, and M.O. Bagby, Analysis of Oil and Meal from *Lesquerella fendleri* Seed, *J. Am. Oil Chem. Soc.*, 67, 438–442 (1990).

39. K.D. Carlson, A. Chaudhry, R.E. Peterson, and M.O. Bagby, Preparative Chromatographic Isolation of Hydroxy Acids from *Lesquerella fendleri* and *L. gordonii* Seed Oils, *J. Am. Oil Chem. Soc.*, 67, 495–498 (1990).

40. K. Brahim, D.K. Stumpf, D.T. Ray, and D.A. Dierig, *Lesquerella fendleri* Seed Oil Content and Composition: Harvest Date and Plant Population Effects, *Ind. Crops Prod.*, 5, 245–252 (1996).

41. C.E. Penoyer, W. von Fischer, and E.G. Bobalek, Synthesis of Drying Oils by Thermal Splitting of Secondary Fatty Acid Esters of Castor Oil, *J. Am. Oil Chem. Soc.*, 31, 366–370 (1954).

42. D.G. Hayes and R. Kleiman, Lipase-Catalyzed Synthesis and Properties of Estolides and Their Esters, *J. Am. Oil Chem. Soc.*, 72, 1309–1316 (1995).

43. B.H. Zoleski and F.J. Gaetani, Low Foaming Railroad Diesel Engine Lubricating Oil Composition, U.S. Patent 4,428,850 (1984).

44. L.A. Nelson, C.M. Pollock, and G.J. Achatz, Secondary Alcohol Esters of Hydroxyacids and Uses Thereof, U.S. Patent 6,407,272 B1 (2002).

45. T.A. Isbell, B.A. Lowery, S.S. DeKeyser, M.L. Winchell, and S.C. Cermak, Physical Properties of Triglyceride Estolides from Lesquerella and Castor Oils, *Ind. Crops Prod.*, 23, 256–263 (2006).

46. T.A. Isbell and R. Kleiman, Mineral Acid-Catalyzed Condensation of Meadowfoam Fatty Acids into Estolides, *J. Am. Oil Chem. Soc.*, 73, 1097–1107 (1996).

47. S.C. Cermak, A.L. Skender, A.B. Deppe, and T.A. Isbell, Synthesis and Physical Properties of Tallow-Oleic Estolide 2-Ethylhexyl Esters, *J. Am. Oil Chem. Soc.*, 84, 449–456 (2007).

48. J.A. Zerkowski and D. Solaiman, Polyhydroxy Fatty Acids Derived from Sophorolipids, *J. Am. Oil Chem. Soc.*, 84, 463–471 (2007).

49. T.A. Isbell, T.P. Abbott, and J.A. Dworak, Shampoos and Conditioners Containing Estolides, U.S. Patent 6,051,214 (2000).

50. B.R. Moser, S.C. Cermak, and T.A. Isbell, Evaluation of Castor and Lesquerella oil Derivatives as Additives in Biodiesel and Ultralow Sulfur Diesel Fuels, *Energy Fuels*, 22, 1349–1352 (2008).

51. S.C. Cermak and T.A. Isbell, Synthesis and Physical Properties of Estolides and Their 2-Ethylhexyl Esters, *Ind. Crops Prod.*, 18, 183–196 (2003).

52. Firestone, D., ed., *Official and Tentative Methods of the American Oil Chemists' Society*, 4th edn., AOCS, Champaign, IL (1994).

53. M.F. Ansell and M.H. Palmer, The Lactonisation of Olefinic Acids: The Use of Sulphuric and Trifluoroacetic Acids, *J. Chem. Soc.*, 16, 2640–2644 (1963).

54. C.C. Akoh, Oxidative Stability of Fat Substitutes and Vegetable Oils by the Oxidative Stability Index Method, *J. Am. Oil Chem. Soc.*, 71, 211–216 (1994).

55. S.C. Cermak and T.A. Isbell, Improved Oxidative Stability of Estolide Esters, *Ind. Crops Prod.*, 18, 223–230 (2003).

56. H.B. Frykman and T.A. Isbell, Decolorization of Meadowfoam Estolides Using Sodium Borohydride, *J. Am. Oil Chem. Soc.*, 76, 765–767 (1999).

57. ASTM (D 445-97), *Standard Test Method for Kinematic Viscosity of Transparent and Opaque Liquids (The Calculation of Dynamic Viscosity)*, West Conshohocken, PA, (1997).

58. ASTM (D 2270-93), *Standard Practice for Calculating Viscosity Index from Kinematic Viscosity at 40 and 100°C*, West Conshohocken, PA (1993).

59. ASTM (D 97-96a), *Standard Test Method for Pour Point of Petroleum Products*, West Conshohocken, PA (1996).

60. ASTM (D 2500-99), *Standard Test Method for Cloud Point of Petroleum Products*, West Conshohocken, PA (1999).

61. W.F. Bowman and G.W. Stachowiak, Application of Sealed Capsule Differential Scanning Calorimetry. Part I: Predicting the Remaining Useful Life of Industry-Used Turbine Oils, *Lubr. Eng.*, 54, 19–24 (1998).

62. D.C. Du, S.S. Kim, J.S. Chun, C.M. Suh, and W.S. Kwon, Antioxidation Synergism Between ZnDTC and ZnDDP in Mineral Oil, *Tribol. Lett.*, 13, 21–27 (2002).

63. S. Asadaushas, J.M. Perez, and J.L. Duda, Lubrication Properties of Castor Oil–Potential Basestock for Biodegradable Lubricants, *Lubr. Eng.*, 53, 35–41 (1997).

64. S.C. Cermak, G. Biresaw, and T.A. Isbell, Comparison of a New Estolide Oxidative Stability Package, *J. Am. Oil Chem. Soc.*, 85, 879–885 (2008).

65. ASTM (D 2272-98), *Standard Test Method for Oxidation Stability of Steam Turbine Oils by Rotating Pressure Vessel*, West Conshohocken, PA (1998).

66. T.A. Isbell, T.P. Abbott, and K.D. Carlson, Oxidative Stability Index of Vegetable Oils in Binary Mixtures with Meadowform Oil, *Ind. Crops Prod.*, 9, 115–123 (1999).

67. J.L. Glancey, S. Knowlton, and E.R. Benson, Development of a High Oleic Soybean Oil-Based Hydraulic Fluid, in: *Proceedings International Off-Highway and Power Plant Congress and Exposition*, Milwaukee, WI, pp. 149–154 (1998).

68. G.J. Cochrac and S.Q.A. Rizvi, Oxidation of Lubricants and Fuels, in: *Fuels and Lubricant Handbook: Technology, Properties, Performance, and Testing*, G.E. Totten (ed.), pp 787–824, West Conshohocken, PA (2003).

69. S.Z. Erhan, B.K. Sharma, and J.M. Perez, Oxidation and Low Temperature Stability of Vegetable Oil-Based Lubricants, *Ind. Crops Prod.*, 24, 292–299 (2006).

70. ASTM (D 5800-00a), *Standard Test Method for Evaporation Loss of Lubricating Oils by the Noack Method*, West Conshohocken, PA (2000).

71. S.C. Cermak and T.A. Isbell, Pilot-Plant Distillation of Meadowfoam Fatty Acids, *Ind. Crops Prod.*, 15, 145–154 (2002).

13 Development and Tribological Behavior of Advanced Biobased Lubricants

Atanu Adhvaryu

CONTENTS

ABSTRACT

Research and development of industrial fluids and lubricants has gained significant importance due to the rising cost and uncertainty of steady petroleum-based crude supply. This paper presents a comprehensive account of biobased lubricant development and their physicochemical and tribological behavior. A variety of analytical techniques have been employed to understand molecular level complexity of triacylglycerol-based molecules, and chemical transformations achieved to create suitable industrial products. Advanced approaches like molecular modeling and statistical predictive algorithms were developed to understand biobased lubricant properties and ways to expand non-food industrial applications of seed oils.

13.1 INTRODUCTION

Vegetable oils are readily biodegradable, safe to handle, environmentally friendly, non-toxic fluids that are also readily renewable resources [1]. The triacylglycerol (TG) structure of vegetable oil, which is also amphiphilic in character, gives it an excellent potential as a candidate for use as a lubricant or functional fluid [2]. TG molecules orient themselves with the polar end at the solid surface making a close-packed monomolecular [3] surface film on the material being lubricated. In addition, the vegetable oil structure provides sites for additional functionalization, offering opportunities for improving on the existing technical properties such as thermo-oxidative stability, low-temperature stability, and lubricity. These properties make them very attractive for industrial applications that have potential for environmental contact through accidental leakage, dripping, or generation of large quantities of after-use waste materials requiring costly disposal [4].

Lubricants typically comprise a base oil that has been blended with any number of additives that enhance the ability of the base oil to withstand mechanical stresses of interacting surfaces under lubricated conditions. Most of the lubricants and many of the additives currently in daily use originate from petroleum-base stocks that may be toxic to environment, making it increasingly difficult for safe and easy disposal. During the last couple of decades, the level of public awareness of environmental issues has risen considerably and anything that does not comply with standards of biodegradability rates a big "NO" with environmentalists and government bodies. There has been an increasing demand for "green" lubricants [5] and lubricant additives in recent years due to concerns about loss of mineral oil-based lubricants to the environment and increasingly strict government regulations controlling their use.

Though most lubricants used currently originate from petroleum-base stocks, vegetable oils have seen an increased promise as biodegradable fluids over the last decade. Environmental concerns, as well as economic and performance issues, will drive the market share for these oils. Today, less than 2% of the base stocks are products of oleochemical and related industries; the primary area of their application has been as hydraulic fluids. This is consumed at approximately 5 million metric ton (MMT)/year in the U.S. market and has the highest need for biodegradable lubricants [6].

The beneficial aspects of vegetable oils as base stocks are mainly their biodegradable and non-toxic properties that are not exhibited by conventional mineral-base oils [7]. The TG structure of vegetable oil, which is also amphiphilic in character, gives it an excellent potential as a candidate for use as a lubricant or functional fluid [2].

Limitations on the use of vegetable oil in its natural form as an industrial-base fluid or as an additive relate to poor thermal/oxidation stability [8,9], poor low-temperature behavior [10,11], and other tribochemical degrading processes [12,13] that occur under severe conditions of temperature, pressure, and shear stress. To meet the increasing demands for stability during various tribochemical processes, the oil structure has to withstand extremes of temperature variations, shear degradation, and maintain excellent boundary lubricating properties through strong physical and chemical adsorption on the metal. The film-forming property of TG molecules is believed to inhibit metal-to-metal contact and progression of pits and asperities on the metal surface. Strength of the protective fluid film and extent of its adsorption on the metal surface dictate the efficiency of a lubricant's performance. It has also been observed that friction coefficient and wear rate are dependent on the adsorption energy of the lubricant [14].

The performance limitations of vegetable oil-base stocks can be overcome by genetic modification, chemical modification, processing changes, and developments in the additive technology. Soybean oil accounts for the largest and cheapest seed based oil in the U.S. market among others (i.e., corn, canola, safflower, sunflower, and their various genetically modified forms) and could have a distinct advantage over other oils if it can be modified to improve stability (oxidative and low temperature), a major step for commercialization as a base fluid. A major application area is as industrial hydraulic fluid that represents a 222 million gallon market in the United States, with potential use in waterways, farms, and forests. In applications that result in total loss of lubricant during use (e.g., 2-cycle engines, chain oils, drip oils, rail flange oils, etc.), biodegradable vegetable oil base products stand a good chance as lubricant in such situations. Similarly, vegetable base oils are suitable as metal cutting oils and fluids, due to hazardous mist formation from mineral oils during use.

Vegetable oils when subjected to low temperature environment undergo solidification through crystallization and pose a major drawback for use in industrial applications. The relatively poor low-temperature flow properties of vegetable oils arise from the presence of waxy crystals that rapidly agglomerate resulting in the solidification of the oil. Vegetable oil is a complex molecular system and, therefore, the transition from liquid to solid state does not occur at a particular temperature, but over a wide temperature range involving several polymorphic forms (α, β', β) [15–17] contributing to the wax appearance and crystallization process. This deposition of waxy materials from oil results in a rapid viscosity increase leading to poor pumpability, lubrication, and rheological behavior.

Several standard methods are currently available for determining the low temperature properties of vegetable oils, such as pour point (ASTM D-97) and cloud point (ASTM D-2500). These methods are extremely time consuming and data reproducibility between laboratories is poor. Differential scanning calorimetry (DSC) is a relatively simple and reproducible technique capable of providing a direct measurement of the ΔH (change in enthalpy) for a system undergoing physical and chemical changes during heating or cooling [18–20]. A number of studies have been carried out on the cooling behavior of different wax-bearing crudes and mineral base oils using DSC [21–26]. Reports are also available on the application of DSC for the identification of various stable polymorphs of single acid TG molecules by heating and cooling experiments [27–29]. However, the situation in an unmodified or genetically modified vegetable oil is far more complex. This is primarily due to the presence of different fatty acid (FA) moieties, their chain length, abundance, structural difference, and configuration in the TG that makes vegetable oils significantly different from other pure molecules. All the above factors individually and collectively are capable of influencing the physical properties of vegetable oils.

The use of vegetable oil derivatives as lubricant base stocks can significantly impact the utilization of locally grown seed oils for non-food applications, improving the local agro-economy. This approach would potentially lower dependence on petroleum crude, widespread use of which continues to pose a threat to environment and the eco-system. The TG structure of vegetable oil, which is also amphiphilic in character, makes it an excellent candidate as a lubricant and functional fluid [2]. For non-food industrial applications, vegetable oil possesses technical drawbacks such as poor thermo-oxidative stability, poor cold-flow properties, and poor hydrolytic stability. The ester linkage contributes to hydrolytic instability while the C=C poly-unsaturation present in the FA portion of the oil contributes to oxidative instability. The poly-unsaturated sites on the TG molecule can be easily functionalized, leading to a wide range of chemical derivatives with varied properties. Thus this system offers an excellent platform for molecular design studies. However, the challenge is to arrive at an optimal vegetable oil derivative with improved thermo-oxidation and improved low temperature stability, while maintaining the lubricity properties that are essential for any industrial application.

Synthesis of vegetable oil derivatives followed by detailed characterization of the product's physical properties is time consuming, labor intensive, and expensive. On the other hand, use of molecular models to design vegetable oil derivatives and calculation of their computationally derived descriptors could make the task easier by synthesizing only those derivatives that show promising results. This would then be followed by laboratory testing pertinent to a given application. This approach can greatly reduce the efforts to develop a new product and has been applied to design novel molecular structures for soybean oil-based fuel additives [30]. Development of quantitative structure–property relationship using only computationally derived descriptors has already been successfully applied to the prediction of water solubility [31] and vapor pressure [32]. This approach can be extended to seed oils to predict high and low temperature behavior. For example, Tan et al. [33] reported the use of molecular orbital indexes as the criterion to study the interaction between lubricant polar end groups and metal surface that was simulated with a metal cluster model. In another study, Sundaram et al. [34] demonstrated the use of design-relevant building blocks for a real industrial problem in fuel additive design. They implemented a hybrid model based on functional descriptors derived from these building blocks and evolutionary search procedures to determine optimal additive molecules that met desired performance criteria. As experimentalists probe tribological interfaces with surface analysis techniques and other laboratory tools, computational methods are brought to bear on the problem by theoreticians. Jabbarzadeh et al. [35] investigated the effects of branching on rheological properties and behavior of molecularly thin liquid films of alkane (>C5) in extreme conditions in the thin film lubrication regime. They found that dynamics of the molecules and their orientation are affected by the degree of branching. The use of experimental approaches needs a long time to find new lubricants, so the prediction of lubricant properties by computational chemistry is crucial to accelerate the designing process for lubricants. Konno et al. [36] reported evaluation of viscosity employing molecular dynamics simulation and prediction based on neural network, while Yang et al. [37] predicted viscosity, viscosity index, and viscosity-pressure coefficients of alkanes (>C5) by applying non-equilibrium molecular dynamics simulations using united atom models for intermolecular interactions. All these studies show how molecular dynamics simulation and molecular designing has allowed researchers to gain an insight into designing new lubricants at the molecular level.

13.2 EXPERIMENTAL PROCEDURES

13.2.1 Thin Film Micro-Oxidation (TFMO)

The thin film micro-oxidation (TFMO) test was developed to oxidize oils under thin film condition. The procedure has been found to simulate engine test conditions and other thermally stressed environments to which lubricants are exposed. The procedure allows quantitative estimation of evaporation loss during high temperature oxidation, oil-soluble high molecular weight oxidation product, and oil-insoluble deposit. The test is performed by oxidizing a small amount (25 μL) of oil sample (that is spread as a thin film) on a freshly polished high carbon steel catalyst surface (test coupon) under a blanket of dry air/O_2 (flow rate ~20 cm^3/min). The tests are carried out at various temperatures (175–300°C) and time periods (30–180 min) inside a bottomless glass reactor with uniform heating of the test coupon. Constant airflow ensures removal of volatile oxidation products from the reaction zone.

After the oxidation, the catalyst coupon containing the oxidized oil sample is removed from the chamber and cooled rapidly under a steady flow of dry N_2 and transferred to a desiccator for temperature equilibration. The catalyst is then weighed to determine the volatile loss (or gain) due to oxidation and soaked (~30 min) in tetrahydrofuran (THF) to dissolve the soluble portion of oxidized oil. The oil-insoluble deposit (after removing the soluble portion) is dried and weighed and reported as wt.%. The THF-soluble portion of the extract is used to obtain molecular weight distribution of the oxidized oil using gel permeation chromatography (GPC). An injection volume of 100 μL of the sample solution in THF (~1 %wt.) is used.

13.2.2 Pressurized Differential Scanning Calorimetry (PDSC) and Sub-Ambient DSC

For PDSC, 1.5–2.0 mg of sample is taken in a hermetically sealed aluminum pan with a pinhole lid for reaction of the sample with reactant gas (oxygen). The sample amount has significant effect on the shape and reproducibility of DSC exotherm. A film thickness of less than 1 mm is required to ensure proper oil–O_2 reaction [38,39]. For kinetic studies, the system is equilibrated at 35°C and heated at a specific heating rate (1°C–20°C/min). Oxygen gas (dry, 99% pure) is pressurized in the module to a constant pressure (3450 kPa) and maintained for the duration of the experiment. The inverse of peak height temperature, corresponding to maximum oxidation, from the exotherm is plotted against log heating rate (β). Using linear regression method and subsequent data computation, various kinetic parameters are obtained.

A modulated DSC is used to study the low temperature crystallization property of oil. Typically 15 mg of the oil sample is measured in an open aluminum pan and placed in the DSC module against an identical empty pan as reference. The procedure involves rapidly heating the sample to 50°C and holding isothermally for 10 min to homogenize and melt any crystalline structure present in the oil (that may act as seed to accelerate wax crystal growth during cooling). Later, the system is cooled to −100°C at a steady rate of 10°C/min. The heat flow (W/g) vs. temperature for each sample is analyzed to determine the wax appearance temperature (T_{C1}, °C) and onset temperature (T_{C2}, °C) of freezing.

13.2.3 NUCLEAR MAGNETIC RESONANCE SPECTROSCOPY

Quantitative NMR spectroscopy has proved to be a useful technique for the structural characterization of mineral [40,41] and vegetable base oils and their various genetically modified versions [42,43]. All the spectra were recorded in Fourier Transform mode on a BRUKER instrument. For ^1H measurements, deuterated chloroform ($CDCl_3$) was used to prepare the sample solution (15%, w/v) containing 1% tetramethylsilane (TMS) as an internal standard. For quantitative ^{13}C NMR measurements, sample solution was similarly prepared without any spin-lattice relaxation agent. Optimally, 2000 repetitive scans were taken for good signal-to-noise ratio. Selective ^{13}C NMR experiments (i.e., DEPT-135, HMQC, and COSY-45) were performed to identify various CH_n (n = 0–3) environments and resolve *cis*- and *trans*-protons in the TG molecule. A combination of the above procedures helped in recognizing and computing the various structural parameters of vegetable oils.

13.2.4 MOLECULAR MODELING

A molecular modeling program (SpartanPro, Wavefunction, Irvine, CA) was used to build representative TG molecules using a minimum energy calculation approach [44]. Modifications were made in the FA chains by changing the oleic or linoleic acid moiety. Different TG molecules were designed having both oleic chains, both linoleic chains, and one each of oleic and linoleic chains on the two end sides of "tuning fork" structure, while the stem side had either oleic or linoleic chain. Geometrical and energy calculations on bond angle, planar distance between hydrogens in the same and adjacent chains, molecular energy associated with a sterically hindered geometry and conformation, etc., were done on individual molecules.

13.2.5 STATISTICAL METHOD

A professional version of Minitab® 12 (MINITAB Inc., State College, PA) was used to develop various correlations between vegetable oil structures and their thermal and kinetic properties. The program analyses individual NMR-derived predictor variables (structural data) one at a time and estimates its influence on the selected vegetable oil property. The corresponding p-values were calculated and student's t-tests were done to identify the variables that imparted maximum influence. The final statistical model will include the variables based on the magnitudes of their individual and collective influence on these properties.

13.3 RESULTS AND DISCUSSION

13.3.1 Epoxidized Oil as a High Temperature Lubricant

Table 13.1 presents the physicochemical properties of soybean (SBO), high oleic soybean (HOSBO), and epoxidized soybean oil (ESBO). The viscosity of ESBO at 40°C is significantly higher than the other oils. Its higher molecular weight and more polar structure compared to SBO and HOSBO result in stronger intermolecular interactions in ESBO. Gas chromatography data show conversion of nearly all $-C=C-$unsaturation in FA chain to epoxy group. These properties will cumulatively influence the thermal and oxidative behaviors of ESBO.

Proton NMR spectrum of ESBO indicates complete conversion of $-C=C-$ to epoxy group (conversion level >90% with minor level of C16/18 FAs). The peak assignments were made using DEPT 135 and COSY 45 NMR experiments. Methine proton of $-CH_2-CH-CH_2-$ backbone at δ 5.1–5.3 ppm, methylene proton of $-CH_2-CH-CH_2-$ backbone at δ 4.0–4.4 ppm, CH_2 proton adjacent to two epoxy

TABLE 13.1
Physical Properties of Vegetable Oils

Vegetable Oil	SBO	HOSBO	ESBO
Appearance	Light yellow	Pale yellow	Colorless
Kinematic viscosity, cSt (ASTM D445) at 40°C	32.93	41.34	170.85
Kinematic viscosity, cSt (ASTM D445) at 100°C	8.08	9.02	20.41
Acid value (mg KOH/g) (AOCS, Ca 5a-40, 1997)	0.16	0.12	0.09
Peroxide value (mequiv/kg) (AOCS, Cd 8-53, 1997)	9.76	4.78	0.0
Iodine value (mg I₂/g) (AOCS, Cd 1-25, 1997)	144.8	85.9	9.11
Fatty acid composition by GC (AACC, 58-18, 1993) in %			
16: 0 Palmitic	6.0	6.0	<7.0
18: 0 Stearic	5.5	3.0	<9.0
18: 1 Oleic	22.0	85.0	—
18: 2 Linoleic	66.0	4.0	—
18: 3 Linolenic	0.5	2.0	—

groups at δ 2.8–3.0 ppm, –CH– protons of the epoxy ring at δ 3.0–3.2 ppm, α-CH$_2$ to >C=O at δ 2.2–2.4 ppm, α-CH$_2$ to epoxy group at 1.7–1.9 ppm, β-CH$_2$ to >C=O at δ 1.55–1.7 ppm, β-CH$_2$ to epoxy group at δ 1.4–1.55 ppm, saturated methylene groups at δ 1.1–1.4 ppm, and terminal –CH$_3$ groups in δ 0.8–1.0 ppm region were observed. The olefin and bisallylic protons in SBO and HOSBO have high susceptibility to thermal and oxidative degradation. The protons on these carbon atoms are highly labile and trigger radical-initiated oxidation in the presence of oxygen. The removal of these reactive protons in ESBO would suggest enhanced oxidative stability of these oils.

The data obtained from TFMO at 175°C at different time intervals are shown in Figure 13.1. The reaction was carried out on 25 μL of sample in static mode under a steady flow of air in a bottomless glass reactor. After stipulated time, the oxidized oil was washed with THF and residue left on the coupon is referred to as insoluble deposit. During oxidation process, primary oxidation products undergo further oxy-polymerization reaction in presence of O$_2$ to form oil insoluble deposits. The tendency to form such deposits prevents limits vegetable oils from high temperature lubricant applications. Figure 13.1 shows that ESBO remains fairly stable to oxidative degradation for 60 min, followed by a sharp increase in deposit formation, suggesting a rapid oxidative polymerization involving oxygen atoms. The induction time for deposit formation of ESBO is roughly twice that of HOSBO under similar oxidation temperature (175°C). The percent insoluble deposit becomes stable after 2 h of oxidation for ESBO and remains low relative to the other oils. High insoluble deposit observed in SBO and HOSBO at different time lengths compared to ESBO is mainly due to their linoleic (66%) and oleic (85%) contents, respectively, in the FA chains. SBO shows higher deposit primarily due to the presence of higher conjugated unsaturation in the FA chain relative to HOSBO.

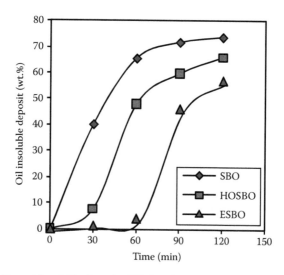

FIGURE 13.1 Percent insoluble deposit of SBO, HOSBO, and ESBO obtained from TFMO at 175°C under air flow (20 mL/min); sample size = 25 μL.

PDSC is an effective way to measure oxidative tendency of vegetable oils in an accelerated mode. At high air pressure (3450 kPa), the concentration of O_2 is in excess and at equilibrium with the sample. The peak height (T_p) and onset (T_o) temperatures of oxidation for SBO, HOSBO, and ESBO are 180.3°C, 201.1°C, and 214.9°C and 167.8°C, 185.8°C and 188.1°C, respectively.

Table 13.2 presents the effect of an antioxidant additive on the T_p and T_o of oils used in the study. An alkylated phenolic antioxidant was blended into the oil at 40°C in different concentrations (0.5–2.0 wt.%). ESBO showed maximum improvement in oxidation stability at low additive concentration (0.5 wt.%) followed by HOSBO and SBO. A plateau is reached at higher additive concentration (>1.5 wt.%) and no further significant improvement in stability was observed. Similarly, oxidation induction time (I_t) measurements under isothermal conditions are as follows: ESBO (at 195°C) shows maximum stability as compared to SBO (at 150°C) and HOSBO (at 160°C). The I_t data show that isothermal temperature for ESBO is significantly higher compared to other oils with no further improvement beyond 1.5 wt.% additive concentration. PDSC can, therefore, be used to effectively optimize the additive concentration in vegetable oil blends and for a rapid assessment of their thermal and oxidative behavior.

Vegetable oils are known to provide excellent lubricity due to their ester functionality [41]. These polar heads in the FA chains attach to metal surfaces by a process called chemisorption resulting in a monolayer film formation. The FA $-CH_2$ chains offer a sliding surface that prevents participating metals from coming in direct

TABLE 13.2
PDSC Data of Vegetable Oils at Different Antioxidant (AO) Additive Concentrations

Vegetable Oil	0.5% AO		1.0% AO		1.5% AO		2.0% AO	
	T_p (°C)	T_o (°C)	T_p (°C)	T_o (°C)	T_p (°C)	T_o (°C)	T_p (°C)	T_o (°C)
SBO	190.3	179.0	194.3	184.3	197.8	188.6	199.6	191.0
HOSBO	213.6	202.9	220.5	211.7	225.6	217.3	229.9	221.8
ESBO	251.3	243.8	259.9	252.2	262.9	254.9	264.0	256.1

Oxidation induction time (I_t) under isothermal condition

	0.5% AO (min)	1.0% AO (min)	1.5% AO (min)	2.0% AO (min)
SBO (150°C)$^\theta$	7.8	12.8	17.4	23.0
HOSBO (160°C)$^\theta$	28.7	57.3	88.9	127.1
ESBO (195°C)$^\theta$	52.2	76.8	86.4	89.8

Notes: Scanning rate = 10°C/min, constant air pressure = 3450 kPa. θ, isothermal temperature; AO, antioxidant (an alkylated phenolic compound). T_p, T_o, and I_t data are averages of three independent experiments.

contact with each other. Under boundary-lubricated conditions, bond cleavage of FA molecules takes place as a result of tribochemical reaction on the metal surface. Under such tribological conditions, it is expected that the epoxy group offers active oxygen sites that trigger polymerization reaction, forming a protective film on the surface. This protective film builds further with time, thereby reducing metal-to-metal friction considerably.

13.3.2 Low Temperature Behavior and Effect of Molecular Structure

Literature studies have shown that cooling rate can influence the shape of the DSC thermogram as well as the resulting wax appearance temperature [26]. Table 13.3 presents the wax appearance temperature (T_{C1}, °C) of SBO as a function of cooling rate. At slower cooling rates (1°C–5°C/min) the signal-to-noise ratio is low and the onset temperature is hard to estimate. The wax appearance onset temperature is relatively well defined from the base line at a faster cooling rate. An optimum rate of 10°C/min has been found to have the advantage of being fast with greater data repeatability.

The cooling scan produced two distinct exothermic peaks (Pk-1 and Pk-2) for various oils where Pk-1 was observed in the temperature range −10°C to −30°C and Pk-2 was present between −35°C and −50°C. T_{C1} and T_{C2} were defined as the temperatures at the intersection of the line tangent on Pk-1 and Pk-2, respectively, with the baseline on the "hot side" of the cooling curve. During the cooling process in seed based oils, micro-crystals start forming at temperature T_{C1} resulting from symmetrical packing of individual TG molecules in bent conformation. On further cooling, the solubilizing ability of oil is lowered and it transforms from an

TABLE 13.3
Wax Appearance Temperature of Soybean Oil (SBO) at Different Cooling Rates

Rate of Cooling (°C/min)	Wax Appearance Temperature T_{C1} (°C)
1.0	−8.3
2.0	−9.3
3.0	−10.3
5.0	−11.1
7.0	−12.1
10.0	−13.2
12.0	−14.1
15.0	−14.5

Notes: Range=+50 to −50°C, sample size: 9.9646 mg. DSC data reported are average values of duplicate experiments.

increasingly viscous fluid to a solid material, preceded by an onset temperature of freezing (T_{C2}).

Seed-based oils do not crystallize over a narrow temperature range due to the presence of various TG polymorphs in the crystal packing and the energy involved in low temperature crystallization [45]. Instead, it is a slow, continuous process in which the microcrystalline structures initially formed become macro-crystalline and rapidly change to a solid-like consistency. The unstable (α) polymorph constitutes the major proportion of the crystals formed at lower temperature. Separate work at different heating rates by other workers has shown [27] that the formation of this unstable material is almost entirely absent at slow cooling rate (0.1°C/min), where only the stable polymorph (β) was observed.

A comparison of the effect of oleic content on T_{C1} of SBO, Safflower, and Sunflower oils indicates that wax appearance temperature increases with higher oleic content. Moreover, the relative distribution of palmitic (16:0) and stearic (18:0) acids has an opposing effect from the linoleic (18:2) acid moiety on the crystallization process. Earlier work on pure triolein and trilinolein molecules has shown that the melting points of α and β forms of triolein are higher than that of the trilinolein molecule [29]. Similar variations in the freezing onset temperature (T_{C2}) were observed for the other oils studied (Table 13.4).

Figure 13.2 presents the wax appearance temperature of soybean oil and chemically modified soybean oil. In ESBO, C=C bond on the FA chain was converted to epoxy ring and T_{C1} for ESBO was −9.5°C compared to −13.61°C in SBO. High T_{C1} for epoxy oil apparently resulted from low steric hindrance of epoxy ring during

TABLE 13.4
Wax Appearance Temperature of Vegetable Oils from Cooling Curves

Vegetable Oils	T_{C1} (°C)	T_{C2} (°C)
Cotton seed oil	−10.8	−35.0
Corn oil	−25.3	−43.4
Safflower oil	−24.2	−43.7
HO-safflower oil	−26.2	−48.9
HL-safflower oil	−21.3	−39.8
Sunflower oil	−14.9	−36.7
HO-sunflower oil	−21.5	−45.8
Soybean oil	−13.2	−35.5
HO-soybean oil	−19.9	−45.4

Notes: Cooling rate = 10°C/min; range = +50 to −100°C. DSC data reported are average values of three independent experiments. Repeatability of the data ± 0.7; HO = high oleic, HL = high linoleic.

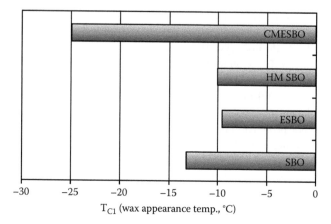

FIGURE 13.2 Wax appearance temperature (T_{Cl}, °C) of heat and chemically modified soybean oil using TMDSC; cooling rate = 10°C/min.

molecular stacking at low temperature. During this time, microcrystalline wax is formed and the oil becomes cloudy (referred to as cloud point) before rapid crystallization sets in. It has been reported that the wax appearance temperature of oils can be correlated with their cloud point [26].

A significant lowering in T_{Cl} was observed for chemically modified soybean oil (CMESBO) (−24.77°C) when compared to SBO due to the presence of additional branching in the molecule. This highly branched structure prevents individual molecules from being able to orient in a geometry that would allow them to stack in a low energy conformation to initiate crystal growth. Also, during such stacking process, the molecule requires an enormous amount of kinetic energy to overcome the steric interaction for packing encountered during the cooling process. This is often not the case and such molecules remain fairly separate until they reach a very low temperature and actual crystallization sets in.

HMSBO was prepared by controlled heat polymerization of SBO under N_2 atmosphere. During this process, the triglyceride molecules undergo polymerization involving the unsaturated double bonds. The viscosity and therefore the degree of heat-induced polymerization can be controlled by optimizing the temperature and time of the heat soaking process.

13.3.3 Analysis of Molecular Geometry

A modeling software (Spartan Pro) was used to build several representative TG molecules with different FA chains (mainly oleic and linoleic). Energy minimization approach was applied to calculate the E_{min} molecular conformations and their three-dimensional orientations in space. The reported angular and distance measurements are from model output for the entire molecule, rather than for individual atom pairs. The approach was to understand the mechanism of stacking in these molecules with different FA chains during the cooling process. It has been observed that different three-dimentional orientations were obtained for TG molecules where the two ends of the "fork" structure were 2-oleic, 2-linoleic, or 1-oleic and 1-linoleic chains. This difference is mainly in the orientation of the $-CH_2-$ carbons separated by $-C=C-$ bonds and the associated dihedral angles. In a TG molecule with two adjacent oleic acid chains in the bent fork-like structure, there is a "zigzag" in the molecule due to the presence of a single $C=C$ in the FA. The two $-CH_2-$ planes are oriented differently and make an angle in the same chain but are more or less parallel to the adjacent chain. Similarly, a molecule with an oleic and a linoleic chain constituting the double-end side of the "fork" structure would show the linoleic $-CH_2-$ plane's orientation significantly different from the oleic $-CH_2-$ planes (Figure 13.3). This is, however, not the situation if both the chains are linoleic, and adjacent, in the bent fork conformation (Figure 13.4). The presence of two $C=C$ bonds separated by a bisallylic $-CH_2-$ in each linoleic chain results in the $-CH_2-$ planes that are in staggered orientation to the adjacent chain, thus increasing the overall steric bulk when viewed along the chain length.

The steric hindrance experienced by two approaching molecules determines to a fairly large extent the condition for wax appearance. During the cooling process, the steric interaction of out-of-plane $>C=O$ group and orientation of $-CH_2-$ planes in the FA chains control the approach of the second molecule for a minimum energy conformation. The lesser the steric bulk of the FA chain, the faster and closer the

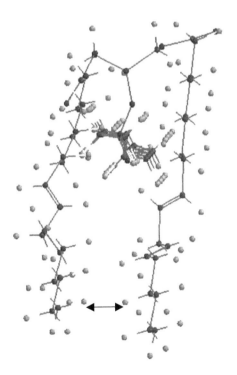

FIGURE 13.3 (See color insert.)   Minimum energy conformation of oleic-linoleic acid pairs forming the double-end side of "tuning fork" structure, with the third oleic chain forming the stem in the bent conformation.

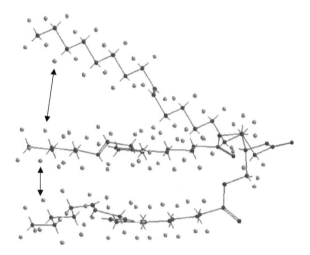

FIGURE 13.4 (See color insert.) Minimum energy conformation of 2-linoleic acid pairs forming the double-end side of "tuning fork" structure, with the third oleic chain forming the stem in the bent conformation.

approach of the second molecule, and the higher the wax appearance temperature. This is a dynamic process and each time an additional molecule is stacked on the parent body, there is a slight steric adjustment to a new low energy conformation.

13.3.4 STATISTICAL ANALYSIS OF DATA

Quantitative NMR spectroscopy (mainly ^{1}H and ^{13}C) is now an integral part of various products and process optimization studies [46–49]. The technique is nondestructive and able to generate data at the molecular level for a more detailed and accurate approach to build statistical models. This is in sharp contrast to the bulk physical property data earlier used in developing predictive models. Not surprisingly, such models fell short on various occasions, but now it is widely accepted that all the chemical and physical properties of base oils (both mineral and bio-origin) are controlled by the relative distribution of their molecular structures. Table 13.5 presents the quantitative ^{1}H NMR data on different vegetable oils used in this study. The details of experimental protocol and peak assignments are discussed in a previous publication [10].

Low temperature crystallization behavior of seed based oils could be modeled using ^{1}H NMR structural parameters. This is also supported from the molecular modeling of various TG structures by energy minimization approach in the current study. There are, however, several limitations for using a single structural parameter to explain the wax appearance and freezing characteristics of vegetable oils due to the complexity of TG structures and presence of various FA moieties. The relative extent to which these structures affect T_C can be reliably determined by adopting a logical approach of multi-component statistical analysis of the NMR data.

TABLE 13.5

Quantitative ¹H NMR Derived Structural Parameters of Vegetable Oils

Vegetable Oils	% Olefin Proton	% Bisallylic– CH_2-Proton	% Allyl– CH_2-Proton
Cotton seed oil	8.8	2.8	8.8
Corn oil	10.0	3.4	10.5
Safflower oil	11.1	4.2	10.9
HO-safflower oil	7.2	0.7	10.7
HL-safflower oil	10.9	4.4	10.9
Sunflower oil	10.1	3.8	9.6
HO-sunflower oil	6.9	0.6	10.5
Soybean oil	10.2	4.1	10.4

Notes: NMR data reported are average values of three independent experiments. The repeatability of the data is ±0.2%; HO = high oleic, HL=high linoleic.

The results from the multi-component analysis of T_{C1} yielded the following equation with coefficient of determination (R^2 adjusted) 0.93 and standard error of Y-estimate (σ_Y) 1.569.

$$T_{C1}(°C) = 91.2 - 15.8 * H_0 + 17.9 * H_b - 1.5 * H_a, \tag{13.1}$$

where

T_{C1} is wax appearance temperature in °C

H_0, H_b, and H_a are % olefinic, bisallylic, and allylic hydrogens, respectively

Analysis of the p values and student t-tests on individual predictor variables suggests allylic CH_2 has a significant influence (R^2 adjusted=0.57) while bis-allylic CH_2 the least influence on T_{C1}. However, a combined olefin H and bisallylic CH_2 has a tremendous impact on T_{C1} (R^2 adjusted=0.92), which improved further when the third component was introduced in the final equation. Among the various other structural parameters available in vegetable oils [50], it is observed that the extent and nature of unsaturation in the FA chain, in combination with its steric conformation, plays the decisive role in determining T_{C1}. Figure 13.5 presents the plot of experimental vs. calculated data for T_{C1} for a total of eight different seed based oil samples. The model (Equation 13.1) holds well for the determination of T_{C1} in the range 0°C to −30°C. Most of the available vegetable oils (both unmodified and genetically modified) would conveniently fall in this range.

Similar equation was derived for T_{C2} with the same set of predictor variables (Equation 13.2). It was observed that T_{C2} showed a better correlation as demonstrated by a lower σ_Y value of 0.669 and higher adjusted coefficient of determination

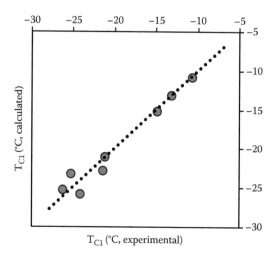

FIGURE 13.5 Correlation plot of calculated vs. experimentally observed T_{C1} (°C) data. R^2 (adjusted) = 0.93; error of Y-estimate (σ_Y) = 1.569.

$R^2 = 0.98$. The p value and student's t-test data on individual predictor variables show similar relative influence on T_{C2} but with higher percent accuracy in determination.

$$T_{C2}(°C) = 32.9 - 10.9 * H_0 + 13.6 * H_b - 1.2 * H_a, \tag{13.2}$$

where

T_{C2} is crystallization temperature in °C

H_0, H_b, and H_a are % olefinic, bisallylic, and allylic hydrogens, respectively

13.3.5 MOLECULAR MODELING AND STRUCTURE–PROPERTY RELATIONSHIPS

Molecular modeling and subsequent computation of energy in static equilibrium geometry is often helpful prior to actual synthesis which is expensive and time consuming. In most cases, a great deal of information on the physical and chemical nature of the molecule can be predicted, based on energy calculation, electron charge distribution, steric environment, and structure–property correlations. Semiempirical models are particularly useful where the equilibrium structure involves a large molecule and the molecule is treated in its global minimum energy conformation by applying semiempirical calculations. Semiempirical models follow directly from Hartree–Fock models. Here the entire molecule is treated with respect to its valence electrons [51,52].

Vegetable oils are primarily composed of TG molecules. The ester chains are derived from FAs, typically stearic ($C_{18:0}$), oleic ($C_{18:1}$), linoleic, ($C_{18:2}$), and linolenic ($C_{18:3}$) (numbers in parentheses indicate 18 carbon chain with 0, 1, 2, or 3 unsaturated sites, respectively). The TG molecule of commonly available seed oils is stable and generally maintains a liquid physical state under normal temperature and pressure.

Poly-unsaturation present in the FA chains is highly susceptible to oxidation, and complete saturation results in poor low-temperature fluidity. This inherent drawback of TG molecules limits their use in industrial applications. Chemical modification is a relatively fast alternative approach applicable to all seed oils, irrespective of their source and origin. This allows selective changes in the TG structure to obtain desired properties unattainable from regular seed and genetically modified oils. The physical and chemical properties of seed oils are largely based on the molecular structures and their relative distribution. Once a correlation between structure and property is established, a computer-generated model of a hypothetical molecule can be developed and its corresponding energy calculated. This approach holds considerable potential in designing new molecules, including those not yet synthesized.

Epoxidation of C=C in TG molecule eliminates poly-unsaturation and improves oxidation stability. It has been observed that the three-dimentional orientations of the TG molecule with different FA (oleic, linoleic and linolenic) chains, are different. This is primarily due to the orientation of $-CH_2-$ carbons separated by $-C=C-$ bonds and the associated dihedral angles. In an oleic acid chain, there is a "zigzag" due to the presence of a single C=C in the FA. The "zigzag" is due to the two $-CH_2-$ planes next to the double bond being oriented differently. The presence of two C=C bonds (as in linoleic chain) separated by bisallylic hydrogens results in $-CH_2-$ planes that are in staggered orientation, thus increasing the overall steric bulk when viewed along the chain length. The distances between allylic hydrogens (in oleic) or allylic and bis-allylic hydrogens (in linoleic) on the same FA chain are 13.73 and 11.40 Å, respectively. The linoleic chain would, therefore, have greater steric bulk compared to the oleic chain. Once epoxidized, much of the zigzag nature of the FA chain is eliminated. Greater uniformity in the epoxidized chain offers low steric hindrance when cooled and systems containing such molecules would readily undergo crystallization (pour points of soybean oil and epoxidized soy oil (ESBO) are −6°C and 3°C, respectively).

The minimum energy (E_{min}) in equilibrium conformation of epoxy structure **1** is 8.93 kcal/mol (Table 13.6). The electron density plot of the epoxy structure is shown in Figure 13.6. The molecule is thermally and oxidatively stable up to a moderate temperature range compared to a TG molecule. However, when exposed to higher temperatures over an extended period time, epoxy oils undergo rapid radical initiated polymerization. Epoxy ring opening in the presence of a catalyst (perchloric or sulfuric acid) and a suitable alcohol (ROH) would result in a hydroxyl-ether compound **2**. The equilibrium geometry E_{min} of **2** in trans-conformation is 27.48 kcal/mol, which is significantly higher than that of **1**. This increase is a result of steric interaction from the electronic charge density of adjacent oxygen lone pairs (Ö). Due to the presence of free −OH groups in **2**, individual molecules could undergo intermolecular H-bonding leading to high viscosity of the resulting fluid. Further reaction of **2** with an anhydride resulted in an ester-ether **3** with significant lowering of E_{min} in equilibrium geometry (16.42 kcal/mol). The addition of a carbonyl group at −OH displaces the electron cloud towards the more electronegative C=O group. The overall steric interaction from Ö is significantly lowered in this case. The electron potential map on **3** (Figure 13.6) illustrates that the electron-deficient H sites are more distributed in the molecule (near the $-COCH_3$ hydrogens) when compared to **2**, where

TABLE 13.6
Calculated Minimum Energy and Experimental Physical Property Data of Synthetic Oils

Synthetic Oil	E_{min} (kcal/mol)	OT[a] (°C)	OT[b] (°C)	OT[c] (°C)	Oil Insol. Deposit[d] (wt.%)	Pour Point[e] (°C)	Viscosity (at 40°C, cSt)
1, Epoxy	8.93	195.5	217.4	203.8	70	3	170
2, OH-ether	27.48	—	—	—	—	—	—
3, Ester-ether	16.42	—	—	—	—	—	—
Syn A	23.87	221.1	246.1	152.4	36	0	Viscous
Syn B	14.42	190.6	228.3	188.3	43	−6	210
Syn C	9.92	230.6	236.8	200.8	30	−9	—
Syn D	13.58	151.0	159.4	166.4	10	3	40
Syn E	16.30	159.4	198.0	179.2	44	−18	50

Note: DSC data reported are averages values of three independent experiments; repeatability of the data ±0.8; OT is oxidation onset temperature.
[a] PDSC data with pinhole hermatic aluminum pan.
[b] PDSC data with open aluminum pan.
[c] PDSC data with open steel pan.
[d] Micro-oxidation, 2 h, 200°C; average of two experiments; repeatability of the data ±1.2%.
[e] ASTM D 97.

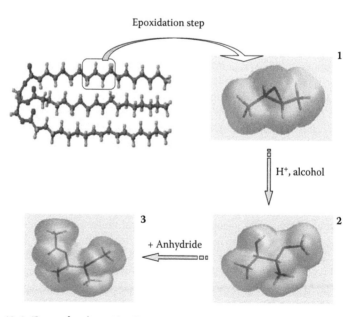

FIGURE 13.6 (See color insert.) Formation of epoxy (**1**), hydroxyl-ether (**2**), and ester-ether (**3**) compounds from triacylglycerol molecule.

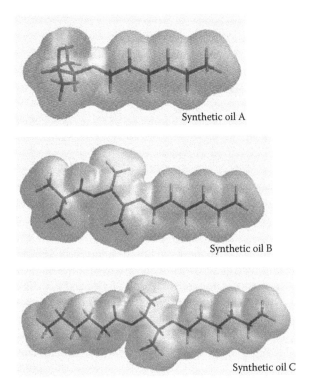

Synthetic oil A

Synthetic oil B

Synthetic oil C

FIGURE 13.7 (See color insert.)   Electron cloud distribution of synthetic fluids A, B, and C shown in Scheme 13.1.

the sole electron-deficient H was directly attached to O. This labile H on **2** is highly susceptible to de-protonation under oxidizing conditions.

In a large TG molecule, such chemical transformation can bring about a dramatic change in chemical and physical characteristics and help make a generalized prediction for chemically modified derivatives of the TG molecule. Figure 13.7 presents the optimized minimum-energy conformations of the molecules representing synthetic fluids A, B, and C. The structures depict only the branched carbon sites, and the long FA chains in the molecule have been truncated to avoid cluttering and offer clarity to the substitution pattern. Synthetic fluid A has a residual hydroxyl group that is separately functionalized with a short-branched chain (yielding Syn B) and long chain (yielding Syn C) (Scheme 13.1).

The electron charge density map indicates the presence of a labile proton in Syn A that can be easily de-protonated. The molecule experiences significant steric interaction between the oxygen lone-pair electron cloud resulting in E_{min} value of 23.87 kcal/mol (Table 13.6). The –OH group can participate in H-bonding and consequently increase the fluid viscosity. The electron cloud can be somewhat delocalized by the presence of a C=O group, and six-Hs of the two-Me groups share the partial electron-deficient charge cloud (Syn B). However, this geometry induces certain amount of instability from adjacent C=O and -Me steric interaction in the

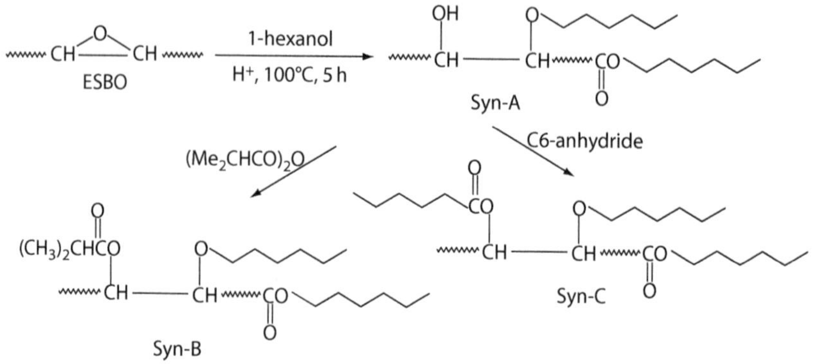

SCHEME 13.1 Reaction sequence for synthetic fluids A, B, and C.

molecule. The global E_{min} computed for this structure was 14.42 kcal/mol. In Syn C, incorporation of a long chain substitution can release the bulk strain in the molecule and also improve the low-temperature properties and viscosity characteristics of the fluid. The optimized minimum energy value for this geometry is 9.92 kcal/mol, which suggests significant improvement in the overall stability of the molecule. Scheme 13.2 presents the synthesis routes for the fluids D and E. Realistically it is a challenge to improve both oxidation and low-temperature properties using the same chemical modification. In most cases improving one property produces a negative effect on the other.

This fact is illustrated in Figure 13.8 for synthetic fluids D and E (the FA chains in the figures have been truncated for clarity). The molecules were designed to illustrate the effect of chain length and chain branching on their minimum energy values which can be helpful for prediction of their physical and chemical properties. A short chain suggests a high pour point in Syn D (Table 13.6) while a bulky and asymmetrically branched chain will sterically prevent the formation of micro-crystals during

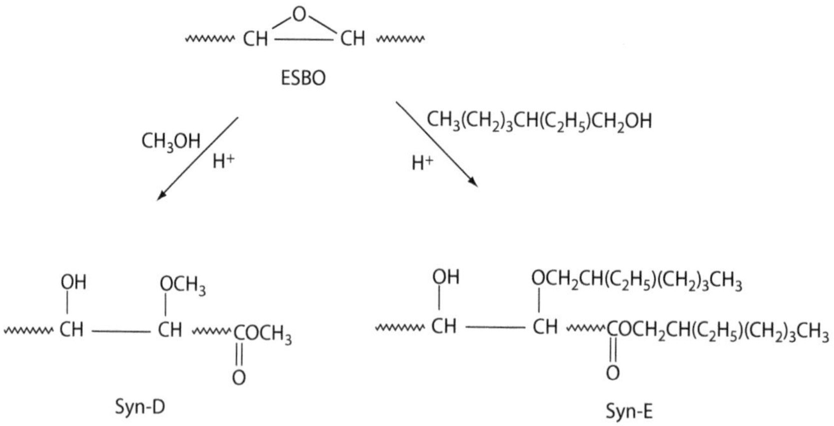

SCHEME 13.2 Reaction sequence for synthetic fluids D and E.

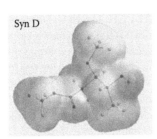
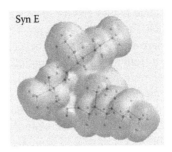

FIGURE 13.8 (See color insert.) Electron cloud distribution of synthetic fluids D and E shown in Scheme 13.2.

the cooling process in Syn E, thus delaying solidification [12]. Scheme 13.2 presents the synthesis routes for preparation of Syn D and E.

Pour point decreased with increasing chain length of the alcohol (Table 13.6). Substitution of a bulkier and branched alcohol (2-ethyl hexanol) in Syn E shows a significant lowering in the pour point (−18°C) when compared to methanol in Syn D (3°C). The equilibrium geometry E_{min} for bulkier molecule (Syn E) is 16.3 kcal/mol compared to 13.6 kcal/mol for Syn D. This factor will also contribute to lower oxidation stability for Syn E (Table 13.6).

PDSC data on the synthetic fluids using open aluminum pan, pinhole aluminum pan, and low carbon steel pans are presented in Table 13.6. Typically, test samples in pinhole aluminum pan gave a lower OT compared to an open pan. Removal of volatile organic compounds (VOCs) formed during the initial phase of oxidation in pinhole pans is largely restricted within the pan volume and close to the metal surface. Oxidation in the vapor phase is more rapid than in the liquid phase, thus shortening the OT. These VOCs generally contribute to an early onset temperature of oxidation. Further, the rate of diffusion of air into the pinhole capsule is much faster compared to high molecular weight VOCs (diffusing out of the capsule), resulting in an accumulation of volatile products of oxidation within the capsule. This is clearly not the situation in open pans where low molecular weight VOCs are continuously lost from the metal surface to the surrounding gas phase. It was also observed that the oils are more reactive on a steel surface than on aluminum, which can trigger an early OT. The increase in reactivity corresponds to a decrease in onset temperature of oxidation. Free hydroxyl groups of synthetic oils A, D, and E make strong chemical interaction with the more reactive steel surface. This interaction is as a result of iron atoms being able to offer coordination sites for polar oxygen atoms (making the vacant outer d-orbital of transition metal available to the oxygen lone pairs). This phenomenon becomes more complex with the proximity of adjacent oxygen atoms (from ether group) in these molecules, with lone electron pairs (Ö) available for intramolecular steric interaction. This situation leads to lowering of activation energy for the system and, therefore, lowers onset temperature for oxidation.

The models for Syn A, B, and C based on minimum energy calculation in their equilibrium trans-conformation, illustrate the effect of substitution on steric interaction and electronic charge density. Using the more reactive metal surface in DSC (steel pans) there is a progressive increase in the OT as the free –OH group is substituted by –COCH(CH$_3$)$_2$ in Syn B and –CO(CH$_2$)$_4$CH$_3$ in Syn C. Decrease in the free electron charge density around the oxygen atom (Ö) linked to the CO group, is partly compensated from the electron donating effect of the iso-butyl group in Syn B. The resulting electronic charge density on Ö is now less than that of the unesterified –OH group. The electron withdrawing effect of CO group is further pronounced on Ö in Syn C due to the absence of any electron-donating group adjacent to CO. The steric hindrance from Ö is similar and relatively higher in Syn D and E as compared to Syn B and C. This fact is supported from the DSC data using steel pans.

In order to qualify for most industrial applications, base fluids are required to be thermo-oxidatively stable over a broad temperature range. Much of the high temperature behavior of a lubricant is controlled by the stability of oil molecules and their interaction with atmospheric oxygen in the presence of a metal catalyst. Thin film micro-oxidation test can be conveniently designed to simulate conditions that closely resemble industrial processes. During the oxidation process, the oil molecules undergo reaction with atmospheric oxygen to form primary oxidation products, which in the presence of a metal catalyst, undergo further oxidative polymerization leading to oil insoluble deposits. The rate of deposit formation with time and temperature gives a good indication of the test fluid's stability to perform as a lubricant. It was observed that percent insoluble deposit remained fairly low for all the Syn fluids A–E at temperatures lower than 175°C, after which Syn D showed the lowest deposit at 200°C, followed by others (Table 13.6). However, the deposit levels remained fairly constant until the temperature reached 275°C. The relatively large difference in the % insoluble deposits for Syn D and E is a result of steric bulk in Synthetic fluid E. This is also reflected from their minimum energy calculations (Syn D = 13.58 kcal/mol; Syn E = 16.30 kcal/mol). The electron charge density of Ö in branched chains of Syn E does not contribute as much to the molecular instability as does steric interaction in the molecule. Therefore, localized electron charge density and steric interactions in the molecule together contribute to thermal and oxidative instability of the oil. Synthetic oils A and B have higher insoluble deposits compared to Syn C. This difference could be a result of C5 hydrocarbon chain in Syn C that does not significantly contribute to steric hindrance as it does to low temperature flow properties. Relatively low insoluble deposit of Syn C can also be explained from its low E$_{min}$ value.

13.3.6 SYNTHESIS APPROACH TO CHEMICAL MODIFICATION

Chemical modification of seed oil to improve its thermo-oxidative and low temperature stability was carried out in two stages: (i) synthesis of di-hydroxylated soybean oil from ESBO and (ii) reaction of anhydride with the di-hydroxylated product. The formation of di-OHx-soybean oil derivative (di-OHx-SBO) from ESBO via di-OH-SBO is an effective way of introducing branching on the FA chains of TG molecules. The reaction was carried out in two stages and the product had significantly improved functional properties compared to unmodified seed oils.

The removal of unsaturation in the soybean oil by converting it to epoxy-group **1**, significantly improves the thermal and oxidative stability of the oil. This was

TABLE 13.7

Thin Film Micro-Oxidation (TFMO)

Results of the Oils[a]: 175°C, 25 µL, 1 h

Test Oils	Volatile Loss (%)	Oil Insoluble Deposit (%)
SBO	12.17	65.85
1, ESBO	7.02	9.53
3, OAc-SBO	12.09	9.14
4, OBu-SBO	28.02	15.07
5, OHx-SBO	57.83	28.33

Note: SBO, soybean oil; ESBO, epoxidized soybean oil; OAc-SBO, acetoxy-SBO; OBu-SBO, butoxy-SBO; OHx-SBO, hexanoyl-SBO.

[a] Average of two independent experiments.

TABLE 13.8

Pour Point[a] and Pressurized DSC Results of the Oils at 10°C/min with Air at 1379 kPa Constant Pressure[b]

Test Oils	Oxidation Start Temperature (T_S, °C)	Oxidation Onset Temperature (T_O, °C)	Pour Point (°C)
SBO	161.3	178.2	−6
1, ESBO	177.4	203.9	0
3, OAc-SBO	135.7	165.1	−3
4, OBu-SBO	140.1	170.2	−3
5, OHx-SBO	171.9	196.6	−18

Note: SBO, soybean oil; ESBO, epoxidized soybean oil; OAc-SBO, acetoxy-SBO; OBu-SBO, butoxy-SBO; OHx-SBO, hexanoyl-SBO.

[a] ASTM D 97.

[b] T_S and T_O values are averages of three independent experiments. Standard error = ±1°C.

observed from the TFMO and PDSC data presented in Tables 13.7 and 13.8, respectively. However, the low temperature fluidity of **1** was poor and the oil was found to solidify at 0°C.

The epoxy group on the FA chain is highly susceptible to acid catalyzed ring opening in aqueous medium. Low-temperature-flow behavior of **1** can be improved by adding branching sites at the epoxy carbons. This was achieved by careful ring opening to obtain the di-hydroxy product **2** by refluxing an aqueous solution of **1** in the presence of an acid (perchloric acid, $HClO_4$). The acid has a tendency to hydrolyze the ester group, and, thereby, destroying the TG structure. Care must be taken not to exceed temperatures above 100°C to prevent molecular cleavage at the ester linkage. The resulting product **2** has high viscosity, probably through hydrogen bonding.

^{1}H NMR measurements on **1** indicate that the epoxy group is present in the δ 3.0–3.2 ppm region. The methine proton of $-CH_2-CH-CH_2-$ backbone at δ 5.1–5.3 ppm, methylene proton of $-CH_2-CH-CH_2-$ backbone at δ 4.0–4.4 ppm, CH_2 proton adjacent to two epoxy groups at δ 2.8–3.0 ppm, $-CH-$ protons of the epoxy ring at δ 3.0–3.2 ppm, α-CH_2 to >C=O at δ 2.2–2.4 ppm, α-CH_2 to epoxy group at 1.7–1.9 ppm, β-CH_2 to >C=O at δ 1.55–1.7 ppm, β-CH_2 to epoxy group at δ 1.4–1.55 ppm, saturated methylene groups δ 1.1–1.4 ppm, and terminal $-CH_3$ groups at δ 0.8–1.0 ppm region are presented. Compound **2** retains most of the characteristic peaks of compound **1** except the ones at δ 2.8–3.2 and δ 1.4–1.55 ppm region, corresponding to **H**s attached to epoxy groups and methylene groups adjacent to epoxy group, respectively.

The di-hydroxy compound **2** was then reacted with different anhydrides in equimolar ratio to obtain the corresponding products **3, 4**, and **5** (Schemes 13.3 and 13.4).

The TFMO data on these samples are presented in Table 13.7. It is observed that both volatile losses and insoluble deposits for **3, 4**, and **5** after oxidation in air at 175°C for 1 h are higher than those of **1**. Removal of carbon-carbon unsaturation through epoxidation lowered volatile loss and deposit-forming tendency in **1**. Similar trend was observed with PDSC where the addition of a branching site to the epoxy carbons resulted in a gradual increase in the start (T_S) and onset (T_O) temperatures of oxidation.

The low-temperature-flow property of vegetable oils is extremely poor and this limits their use in sub-zero temperatures. SBO has a pour point of −6°C, whereas ESBO freezes at 0°C. The excellent thermal and oxidative stability of ESBO due to removal of unsaturation, however, results in poor cold flow property, and therefore limits its use in low temperature applications. Attachment of a side chain with optimum length at the epoxy carbons improves the pour point significantly in **5** (−18°C). However, it was observed that short chains (C_2–C_4) did not make significant improvement in the cold flow behavior.

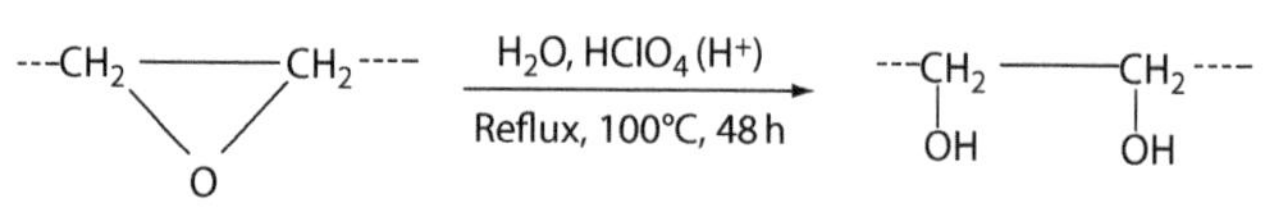

SCHEME 13.3   Reaction sequence for synthesis of di-hydroxy compound.

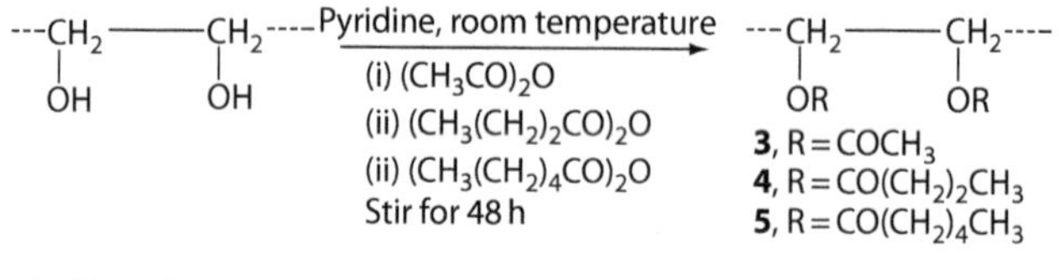

SCHEME 13.4   Reaction sequence for synthesis of **3, 4**, and **5**.

In a separate study, ESBO was reacted with 2-ethylhexanol (2-EH) in the presence of catalytic amount of sulfuric acid. Although the reaction may produce other minor products (oligomers obtained by intra- and intermolecular epoxy-epoxy reactions), epoxy ring-opening and transesterification with alcohol give the major product (Scheme 13.5).

Ring-opening reaction was confirmed from the disappearance of epoxy group signals at 1158, 845, and 822 cm^{-1} in the IR spectrum and also at 2.8–3.0 ppm in ^{1}H NMR spectrum. It was observed that transesterification followed the ring-opening reaction under the present conditions (Figure 13.9). The percent ring-opening reaction was calculated from the relative area of the epoxy ring hydrogen signals (2.8–3.0 ppm, multiplet) compared to that of the starting material. The extent of transesterified product formation was monitored by the appearance of signals for alkyl hydrogens α to the ester oxygen atom derived from the alcohol portion of the newly formed ester's α proton (3.9 ppm, doublet) and also from the disappearance of hydrogen signals of TG backbone (5.2 ppm, multiplet; 4.1–4.3 ppm, multiplet).

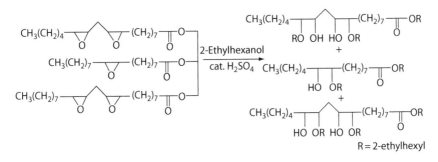

SCHEME 13.5   Reaction sequence for transesterified products.

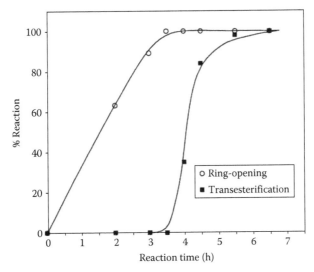

FIGURE 13.9   Plot showing percent ring-opening and transesterification reactions.

To investigate the effect of other catalysts, p-toluenesulfonic acid, Dowex 50W-X8, boron trifluoride, and a base catalyst, sodium methoxide were tested. All the reactions were carried out with 60 g of ESO and 79 g (1.3 equivalents) of 2-EH. When p-toluenesulfonic acid (2.0 g) was used as catalyst, the reaction was so slow as to result in only 82% ring opening and 0% transesterification after 22 h at 130°C. A widely used heterogeneous catalyst, Dowex 50W-8X (10 g), resulted in 100% ring-opening and 83% transesterification after 24 h at 130°C. In the reaction with sulfuric acid as catalyst, ring opening occurred prior to transesterification. When boron trifluoride (BF_3, 10 drops) was used, ring-opening reaction was completed in 2 h at 100°C while no transesterification occurred. Even after a prolonged reaction time (24 h), and higher reaction temperature (130°C), no trace of transesterified product was found. In contrast, when a base catalyst, sodium methoxide (1 %wt.) was used, transesterification was completed in 2 h at 130°C while no ring opening occurred. Even after a longer reaction time (24 h), no ring-opened product was found.

The hydroxyl groups in ring-opened/transesterifed products were further esterified with acetic anhydride or hexanoic anhydride (Scheme 13.6). Pour points, viscosities, and viscosity indices of the products (1–5) are summarized in Table 13.9, which were determined by ASTM D97, ASTM D 445, and ASTM D 2270 standard methods, respectively. The ring-opened/transesterifed product (1) obtained by reaction with catalytic amount of sulfuric acid at 120°C and 1.3 equivalents of 2-ethylhexanol has pour points −12°C, −15°C, and −21°C without pour point depressant (PPD), with 0.5% and 1% PPD, respectively. The ring-opened product obtained with

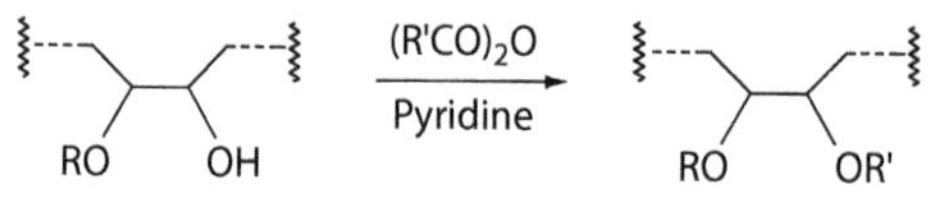

SCHEME 13.6 Reaction showing esterification with anhydride compound.

TABLE 13.9

Pour Points, Viscosities, and Viscosity Indices of Products from Reactions Shown in Scheme 13.6

Product	R′	Pour Points with (°C)			Viscosity (cSt)		Viscosity Index
		0% PPD	0.5% PPD	1% PPD	40°C	100°C	
Products obtained with 1.3 eq. of 2-EH							
1	H	−12	−15	−21	74.4	9.6	113
2	CH_3CO-	−18	−27	−30	57.2	8.7	127
3	$CH_3(CH_2)_4CO-$	−21	−33	−36	41.1	7.7	159
Products obtained with 3.0 eq. of 2-EH							
4	H	−18	−27	−30	54.4	7.89	112
5	CH_3CO-	−21	−27	−30	44.4	7.2	120

boron trifluoride showed the same pour point without PPD and with 0.5% PPD while showing a higher pour point with 1% PPD (−15°C) compared to product **1**. Viscosities of the product **1** at 40°C and 100°C were 74.4 and 9.6 cSt., respectively. The esterified product (**2**) obtained by esterification of **1** with acetic anhydride showed lower pour points, lower viscosities, and higher viscosity index than the product before esterification. The esterification product with hexanoic anhydride (**3**) displayed further lowering of pour points and viscosities and further increase of viscosity index.

To investigate the effect of reactant amount (2-EH), on physical properties, product **4** was prepared with 3.0 molar equivalents of 2-EH using the same reaction conditions as for product **1**. When compared with **1**, product **4** showed lower pour points and viscosities. It can be stated that when smaller amount of alcohol is used, inter- and intramolecular reactions between epoxy groups are predominant, resulting in high molecular weight and high viscosity products. The acetylated product **5** was prepared to compare its properties with parent product **4** as and acetylated product **2**. Product **5** has slightly lower pour point without PPD and similar pour points as product **4** when 0.5% and 1% PPD were added. Also, when compared to product **2**, product **5** shows the same pour points at higher additive concentration, lower viscosities, and comparable viscosity index.

Transesterification of TG molecules using carefully selected reagents and suitable catalysts can result in synthetic fluids with excellent physicochemical properties. It may be concluded from this study that synthetic esters with two to three branching sites bearing C_2–C_7 chains show marked improvement in low temperature stability, with high viscosity index. Such molecules are also thermally and oxidatively stable due to the elimination of C=C unsaturation from the ester chain. The scientific impact is that it demonstrates a low cost approach to develop agro-based synthetic fluids through chemical derivatization of seed oil with excellent functional properties.

13.3.7 Tribological Behavior of Chemically Modified Seed Oil

The TG structure of seed oils makes them excellent candidate for potential use as a base stock for lubricants and functional fluids [2]. This makes them very attractive for industrial applications that have potential for environmental contact through accidental leakage, dripping, or generating large quantities of after-use waste materials requiring costly disposal. A vegetable oil in its natural form has limited use as an industrial fluid due to its poor thermal/oxidation stability, poor low-temperature behavior, and tribochemically degrading processes [13,14] that occur under severe conditions of temperature, pressure, and shear stress. To meet the increasing demands for stability during various tribochemical processes, the oil molecules have to withstand extremes of temperature variations, shear degradation, and maintain excellent boundary lubricating properties by strong physical and chemical adsorption on the metal surface. TG molecules orient themselves with the polar end directed towards the metal surface making a close-packed monomolecular [3] surface film believed to inhibit metal-to-metal contact and progression of pits and asperities on the metal surface. Strength of the fluid film and extent of adsorption on the metal surface dictate the lubricant's performance, and it has also been observed that friction coefficient and wear rate are dependent on the adsorption energy of the lubricant [15].

Chemically modified soybean oil-1 (CMSBO#1) was prepared by reacting dihydroxylated soybean oil with hexanoic anhydride. This is a two-step process involving first preparation of a dihydroxylated product from ESBO. Chemically modified soybean oil-2 (CMSBO#2) is a one-step process of converting ESBO to its esterified derivative using hexanoic anhydride and BF_3 ether as catalyst. Thermally modified soybean oil (TMSBO) was prepared by heat polymerization process under inert atmosphere.

Friction coefficient is largely dependent on the concentration of vegetable oils and their derivatives in hexadecane (base oil used for the friction studies). It is clearly observed from Figure 13.10 that the coefficient of friction (COF) sharply decreases with increase in oil content and levels off at higher concentration. The rate of decrease in COF (as observed from the slope at low oil concentration) is largely influenced by oil structure and its ability to form a thin lubricating film at the point of metal contact. It was also observed that CMSBO #1 and #2 show a higher rate of decrease in COF at lower concentration (0–0.01 M) as compared to SBO and TMSBO.

Figure 13.11 presents the friction coefficient data for SBO and its derivatives at 0.07 M concentration. It is observed that SBO has the highest COF followed by HMSBO and CMSBO #1 and 2. In the boundary regime, chemically modified oils show significantly lower COF as compared to SBO and HMSBO (observed at 0.07 M concentration). It has already been established in earlier studies that polar functional groups in the TG molecule make physical and chemical interactions with the metallic surfaces under high load and sliding contact [53]. The point of attachment to the metal is through the polar group in the molecule, with the non-polar end forming a molecular layer separating the rubbing surfaces. Increasing the number of polar groups in the molecule through chemical modification at the FA

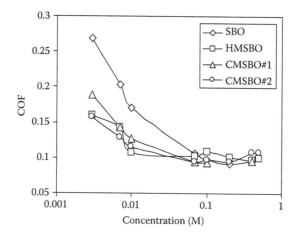

FIGURE 13.10 Coefficient of friction data of SBO and its derivatives at various additive concentrations.

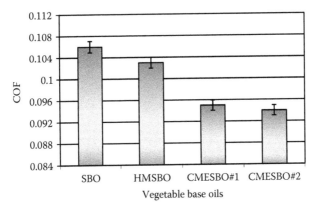

FIGURE 13.11   Coefficient of friction (at 5 rpm, 400 lb, 0.07 M conc. in hexadecane) of SBO and its derivatives.

C=C unsaturated sites improves this property. The similarity in COF for CMSBO #1 and #2 (within the % error limit) is primarily due to the same number and nature of polar groups in oil structure. The residual −OH groups in CMSBO #1 (those not converted to ester groups) contribute to polarity in the molecule but lack the hydrocarbon chain (of ester) to provide effective molecular separation between mating surfaces. One-step synthesis of CMSBO #2 does not involve intermediate hydroxylation, and the intramolecular rearrangement step converts all C=C unsaturated sites to diester functionality, providing additional polarity and non-polar hydrocarbon chains to the existing molecule. Boundary lubrication occurs when polar materials are allowed to form films by bonding chemically to the metal surfaces in contact. During metal working process, for example, significant amount of heat is generated from the primary deformation zone which accounts for nearly 60% of the total heat produced in the cutting process. This heat is generated due to metal deformation and from other friction zones (tool/chip and tool/workpiece interfaces).

13.4   CONCLUSIONS

This study presents the effect of chemical structure on the tribochemistry of seed oils for use in industrial applications. In order to qualify for broad industrial use, bio-based fluids must have both high- and low-temperature stability as well as excellent lubricating properties. These properties are inherent weaknesses in regular seed oils. Thermal and chemical modifications of oil structure can significantly influence the wear and load-carrying properties under boundary lubrication regimes. Chemical modification of TG structures has great potential in achieving a broad temperature range stability as well as excellent wear/friction characteristics.

Increasing the polar functionality in vegetable oil structure has a positive impact on wear protection resulting from stronger adsorption potential on metal surface, as well as greater lateral interaction between the ester chains. This was shown by increasing the polarity by chemical modification of soybean oil.

ACKNOWLEDGMENTS

The author wishes to thank NCAUR, USDA Peoria, and Department of Chemical Engineering, The Pennsylvania State University, for their continued support during this work.

REFERENCES

1. D.K. Salunkhe, *World Oil Seed Chemistry, Technology and Utilization, Fats and Oils Handbook*, AOCS Press, Champaign, IL, pp. 1–8, 838 (1998).
2. A. Willing, Lubricants based on renewable resources—An environmentally compatible alternative to mineral oil products, *Chemosphere*, 43, 89–98 (2001).
3. L.O. Brockway and J. Karle, Electron diffraction study of oleophilic films on copper, iron and aluminum, *J. Colloid Sci.*, 2, 277–289 (1947).
4. S.J. Randles and M. Wright, Environmentally considerate ester lubricants for automotive and engineering industries, *J. Synth. Lubr.*, 9, 145–161 (1992).
5. I. Rhee, Evaluation of environmentally acceptable hydraulic fluids, *NLGI Spokesman*, 60(5), 28 (1996).
6. R.A. Padavich and L. Honary, A market research and analysis report on the vegetable oil-based industrial lubricants, SAE Tech paper 952077, p. 13 (1995).
7. N.S. Battersby, S.E. Pack, and R.J. Watkinson, A correlation between the biodegradability of oil products in the CEC L-33-T-82 and modified Strum tests, *Chemosphere*, 24, 1998–2000.
8. R. Becker and A. Knorr, An evaluation of antioxidants for vegetable oils at elevated temperatures, *Lubr. Sci.*, 8, 95–117 (1996).
9. A. Adhvaryu, S.Z. Erhan, Z.S. Liu, and J.M. Perez, Oxidation kinetic studies of unmodified and genetically modified vegetable oils using pressurized differential scanning calorimetry and nuclear magnetic resonance spectroscopy, *Thermochim. Acta*, 364, 87–97 (2000).
10. S. Asadauskas, Depression of pour points of vegetable oils by blending with diluents used for biodegradable lubricants, *J. Am. Oil Chem. Soc.*, 76, 313–316 (1999).
11. A. Adhvaryu, S.Z. Erhan, and J.M. Perez, Wax appearance temperatures of vegetable oils determined by differential scanning calorimetry: Effect of triacylglycerol structure and its modification, *Thermochim. Acta*, 395, 191–200 (2003).
12. J.E. Brophy and W.A. Zeisman, Surface chemical phenomena in lubrication, *Annals N.Y. Academy Sci.*, 53, 836–861 (1951).
13. A. Miller and A.A. Adderson, Apparent adsorption isotherm of oleic acid from hydrocarbons as obtained from coefficient of friction measurements, *Lubr. Eng.*, 13, 553–556 (1957).
14. E.P. Kingsbury, The heat of adsorption of a boundary lubricant, *ASLE Trans.*, 3, 30–33 (1960).
15. D. Chapman, The polymorphism of glycerides, *Chem. Rev.*, 62, 433 (1962).
16. C.W. Hoerr and F.R. Paulicka, The role of x-ray diffraction in studies of the crystallography of monoacid saturated triglycerides, *JAOCS*, 45, 793 (1968).
17. J.W. Hagemann and J.A. Rothfus, Computer modeling of theoretical structures of monoacid triglyceride α-forms in various subcell arrangements, *JAOCS*, 60, 1308 (1983).
18. A.P. Bentz and B.G. Breidenbach, Evaluation of the differential scanning calorimetric method for fat solids, *JAOCS*, 46, 60 (1969).
19. Y. Miyake and K. Yokomizo, Determination of cis-and trans-18:1 fatty acid isomers in hydrogenated vegetable oils by high resolution carbon nuclear magnetic resonance, *J. Am. Oil Chem. Soc.*, 75, 801 (1998).

20. C.K. Cross, Oil Stability: A DSC alternative for the active oxygen method, *JAOCS*, 47, 229 (1970).
21. P. Claudy, J.M. Letoffe, B. Neff, and B. Damin, Diesel fuels: Determination of onset crystallization temperature, pour point and filter plugging point by differential scanning calorimetry. Correlation with standard test methods, *Fuel*, 65, 861 (1986).
22. P. Claudy, J.M. Letoffe, B. Chague, and J. Orrit, Crude oils and their distillates: Characterization by differential scanning calorimetry, *Fuel*, 67, 58 (1988).
23. F. Noel, Thermal analysis of lubricating oils, *Thermochim. Acta*, 4, 377 (1972).
24. P. Redelius, The use of DSC in predicting low temperature behavior of mineral oil products, *Thermochim. Acta*, 85, 327 (1985).
25. J.C. Hipeaux, M. Born, J.P. Durand, P. Claudy, and J.M. Letoffe, Physico-chemical characterization of base stocks and thermal analysis by differential scanning calorimetry and thermomicroscopy at low temperature, *Thermochim. Acta*, 348, 147 (2000).
26. Z. Jiang, J.M. Hutchinson, and C.T. Imrie, Measurement of the wax appearance temperatures of crude oils by temperature modulated differential scanning calorimetry, *Fuel*, 80, 367 (2001).
27. D.J. Cebula and K.W. Smith, Differential scanning calorimetry of confectionery fats. Pure triglycerides: Effects of cooling and heating rate variation, *JAOCS*, 68, 591 (1991).
28. J.W. Hagemann, W.H. Tallent, and K.E. Kolb, Differential scanning calorimetry of single acid triglycerides: Effect of chain length and unsaturation, *JAOCS*, 49, 118 (1972).
29. J.W. Hagemann and J.A. Rothfus, Polymorphism and transformation energetics of saturated monoacid triglycerides from differential scanning calorimetry and theoretical modeling, *JAOCS*, 60, 1123 (1983).
30. K.V. Camarda, P. Sunderesan, S. Siddhaye, G.J. Suppes, and J. Heppert, *An Optimization Approach to the Design of Value-Added Soybean Oil Products*, Elsevier, Amsterdam, the Netherlands, 369–374 (2001).
31. C. Liang and D.A. Gallagher, Prediction of physical and chemical properties by quantitative structure-property relationships, *Am. Lab.*, 34–40 (March 1997).
32. C. Liang and D.A. Gallagher, QSPR prediction of vapor pressure from solely theoretically-derived descriptors, *J. Chem. Inf. Comput. Sci.*, 38, 321–324 (1998).
33. Y. Tan, W. Huang, and X. Wang, Molecular orbital indexes criteria for friction modifiers in boundary lubrication, *Tribol. Int.*, 35, 381–384 (2002).
34. A. Sundaram, V. Venkatasubramanianan, and J.M. Caruthers, *Molecular Design of Fuel Additives*, Elsevier, Amsterdam, the Netherlands, 329–353 (2003).
35. A. Jabbarzadeh, J.D. Atkinson, and R.I. Tanner, The effect of branching on slip and rheological properties of lubricants in molecular dynamics simulation of Couette shear flow, *Tribol. Int.*, 35, 35–46 (2002).
36. K. Konno, D. Kamei, T. Yokosuka, S. Takami, M. Kubo, and A. Miyamoto, The development of computational chemistry approach to predict the viscosity of lubricants, *Tribol. Int.*, 36, 455–458 (2003).
37. Y. Yang, T.A. Pakkanen, and R.L. Rowley, NEMD simulations of viscosity and viscosity index for lubricant-size model molecules, *Int. J. Thermophys.*, 23, 1441–1454 (2002).
38. S. Shankwalkar and D. Placek, Oxidation kinetics of tricresyl phosphate (TSP) using differential scanning calorimetry (DSC), *Lubr. Eng.*, 50, 261 (1993).
39. A. Adhvaryu, J.M. Perez, and I.D. Singh, Application of quantitative NMR spectroscopy to oxidation kinetics of base oils using pressurized differential scanning calorimetry technique, *Energy Fuels*, 13, 493 (1999).
40. A. Adhvaryu, J.M. Perez, I.D. Singh, and O.S. Tyagi, Spectroscopic studies of oxidative degradation of base oils, *Energy Fuels*, 12, 1369–1374 (1998).
41. A. Adhvaryu, Y.K. Sharma, and I.D. Singh, Studies on the oxidative behavior of base oils and their chromatographic fractions, *Fuel*, 78, 11, 1293 (1999).

42. Y. Miyake, Y. Kazuhisa, and N. Matsuzaki, Determination of unsaturated fatty acid composition by high-resolution nuclear magnetic resonance spectroscopy, *J. Am. Oil Chem. Soc.*, 75, 1091 (1998).

43. M.M. Bergana and T.W. Lee, Structure determination of long chain polyunsaturated triacylglycerols by high-resolution ^{13}C nuclear magnetic resonance, *J. Am. Oil Chem. Soc.*, 73, 551 (1996).

44. H.H. Rosenbrock and C. Storey, *Computational Methods for Chemical Engineers*, Pergamon Press, Oxford, p. 64 (1966).

45. J.W. Hagemann and J.A. Rothfus, Effects of chain length, conformation and α-form packing arrangement on theoretical monoacid triglyceride β'-forms, *JAOCS*, 65, 4, 638 (1988).

46. A. Adhvaryu, J.M. Perez, and J.L. Duda, Quantitative NMR spectroscopy for the prediction of base oil properties, *Tribol. Trans.*, 43, 2, 245 (2000).

47. A.S. Sarpal, G.S. Kapur, S. Mukherjee, and S.K. Jain, Estimation of oxygenates in gasoline by ^{13}C NMR spectroscopy, *Energy Fuels*, 11, 662 (1997).

48. Y. Miyake, K. Yokomizo, and N. Matsuzaki, Quantitative determination of trans fatty acid content in hydrogenated edible vegetable oils by ^{13}C-nuclear magnetic resonance, *Nihon Yukagaku Kaishi*, 47, 333 (1998).

49. C.M. Yang, A.A. Grey, M.C. Archer, and W.R. Bruce, Rapid quantitation of thermal oxidation products in fats and oils by 1H-NMR spectroscopy, *Nutr. Cancer*, 30(1), 64 (1998).

50. G. Vlahov, A.D. Shaw, and D.B. Kell, Use of ^{13}C nuclear magnetic resonance distortionless enhancement by polarization transfer pulse sequence and multivariate analyses to discriminate olive oil cultivars, *JAOCS*, 76, 1223 (1999).

51. T. Clark, *A Handbook of Computational Chemistry*, Wiley, New York (1986).

52. J.J.P. Stewart, MOPAC: A semi empirical molecular orbital program, *J. Comput. Aided Mol. Des.*, 4, 1 (1990).

53. G. Biresaw, A. Adhvaryu, S.Z. Erhan, and C.J. Carriere, Friction and adsorption properties of normal and high-oleic soybean oils, *J. Am. Oil Chem. Soc.*, 79, 53–58 (2002).

14 Characterization of Surface-Active Materials Derived from Farm Products

Girma Biresaw

CONTENTS

ABSTRACT

Surface-active materials obtained by the chemical modification of plant protein isolates (lupin, barley, and oat), corn starches (dextrin, normal, high amylose, and waxy), and soybean oil (soybean oil–based polysoaps, SOPS) were investigated for their surface and interfacial properties using axisymmetric drop shape analysis. All of the materials were effective at reducing the surface tension (ST) of water and the interfacial tension (IT) at the water–hexadecane or water–soybean oil (SBO) interface. From concentration vs. equilibrium ST data, the surface energy (SE) of these materials were determined. The SE of chemically modified starches was in the range of 41–48 mJ/m². The SE of proteins varied in the range of 31–55 mJ/m² and was dependent on the crop origin. The SE of SOPS was in the range of 20–33 mJ/m² and displayed molecular weight (mol wt) dependence. The SE data were analyzed to predict interfacial energy using the Antonow, geometric mean (GM), and harmonic mean (HM) methods. Predictions of the Antonow method relative to measured values were substantially lower for SOPS–hexadecane, substantially higher for starch ester–hexadecane, and reasonably good for LPI–SBO. In the GM and HM analyses, the fraction of nonpolar SE component $\left(X_S^d\right)$ was used as a fitting parameter. The result showed that X_S^d values that predicted interfacial energies close to measured values increased in the order: SOPS < starch esters < protein isolates. This suggests that the surface polarities of the surface-active agents increased in the order: protein isolates < starch esters < SOPS.

14.1   INTRODUCTION

Agricultural products have the potential to become attractive alternative raw materials to dwindling petroleum supply for the manufacture of consumer and industrial products [1–5]. Some of the benefits of agriculture-based raw materials, compared to petroleum-based raw materials, include

1. *Renewability.* Agriculture-based raw materials can be regenerated, at least annually, as opposed to petroleum, which cannot be regenerated in a reasonable time frame.
2. *Abundance.* Whereas petroleum reserves continue to shrink, large surpluses of agricultural products are produced year after year. In the United States alone, billions of bushels of surplus crops are produced yearly. Table 14.1 shows recent projections [6] of surplus for five major crops in the United States: corn, wheat, soybean, barley, and oats. Surplus is defined as domestic production plus import, less domestic consumption plus export. According to this forecast, the combined surplus for these five crops in the United States for the 2009–2010 growing season will be nearly three billion bushels. At 40–60 lb (18–27 kg) per bushel, the total surplus of these five crops amounts to approximately 70 million metric tons per year. Such a huge surplus, accumulated over many years,

TABLE 14.1
Estimated and Projected U.S. Crop Surplus[a] (Million Bushels[b])[c]

Crop	2007/2008 Actual	2008/2009 Estimated	2009/2010 Projected, October 2009
Corn	1624	1674	1672
Wheat	306	657	864
Soybean	205	138	231
Barley	68	89	111
Oats	67	84	74

[a] Surplus = (U.S. production + U.S. import) − (U.S. consumption + U.S. export).

[b] 1 bushell = 40–60 lb (18–27 kg).

[c] Data from Ref. [6].

has caused a considerable decline in prices farmers get for their crops. Thus, successful application of agricultural products in material manufacturing not only will help the manufacturing sector transition to a more reliable and abundant raw material base but also will offer farmers reasonable returns for their crops, thereby encouraging them to continue to farm.

3. *Biodegradability.* Application of farm-based raw materials in manufacturing also presents major advantages to the environment. While petroleum-based products persist in the environment for decades after they have been used and discarded, similar biobased products readily biodegrade in a very short time into benign and harmless products. Thus, converting the manufacturing base to farm-based raw materials will have no negative impact on the quality of air, land, and water as well as on the plants and animals that depend on it.

4. *Safety.* Farm-based raw materials are also extremely safe since they are also part of the food chain for humans and animals. As a result, the application of agriculture-based raw materials for the manufacture of consumer and industrial products poses no risk to workers. In addition, the use of products manufactured from farm-based raw materials, as well as their disposal after use produces little or no risk to consumers.

Agricultural crops comprise a number of important components. Each of these components, by itself, or in combination with others, can be used as raw materials for a variety of food and nonfood applications. The three major components of seed crops are starch, oil, and proteins. Table 14.2 compares the relative quantities of these three components in five major seed crops produced in the United States. While these components provide a number of benefits (discussed above) over petroleum-based raw materials, they cannot be used as simple substitutes of

TABLE 14.2

Major Components (%) of Selected Seed Crops

Cereal	Starch	Protein	Fat/Oil	β-Glucan
Corn	80.0	9.8	4.8	0.0
Barley	78.0	9.9	1.2	3.0
Oat	66.0	17.0	7.0	4.0
Wheat	52.0	23.0	9.7	0.0
Soy	30.0	37.0	20.0	0.0

petroleum-based ingredients in materials applications. For example, application of ag-based raw materials in biomaterials development requires overcoming a number of rather difficult technical, engineering, and scientific problems. Two of the major problem areas are

1. *Overcoming inherent weaknesses.* Agriculture-based raw materials have a number of inherent weaknesses that must be overcome to make them suitable for materials applications. Examples of such weaknesses include poor oxidation stability, biostability, and hydraulic stability. These properties, while desirable for the prompt biodegradation of materials after use, are undesirable during manufacture and application of the materials.

2. *Poor or lack of understanding of the material properties of ag-based raw materials.* While petroleum-based raw materials have been investigated and developed over a period of more than a century, work on ag-based raw materials has just begun. As a result, we do not have a full understanding of their strengths and weaknesses other than those that can be easily and readily observed. Thus, there is a need to develop an in-depth understanding of their properties relevant to materials application. Of particular interest is that of relating their material properties with their chemical structures. Such an understanding will allow for the development of new approaches and strategies for countering some of the inherent weaknesses and developing ag-based ingredients with improved properties for materials application.

One of the most important properties of ingredients used in materials development is their surface activity [7–9]. Surface activity refers to the ability of materials to adsorb at interfaces and change various types of surface properties. Surface-active materials can do this because they have unique chemical structure: both polar and nonpolar components in the same molecule [7–9]. Such chemical structure allows these materials to adsorb on surfaces and also form a variety of structures in solution.

The most widely known surface property is the surface tension (ST) of liquids, which changes when surface-active materials adsorb at the liquid–gas interface.

Another equally well-known surface property is interfacial tension (IT), which is affected by the adsorption of surface-active materials at liquid–liquid interface. Adsorption of surface-active materials on solid surfaces (solid–gas interface) affects a number of important material properties, including [7] friction, adhesion, corrosion, cleaning, etc.

Surface activity is an important property in tribology. Surface-active materials play a variety of roles in lubrication [10,11]. Surface-active materials adsorb on friction surfaces and reduce boundary friction [12–16]. This has allowed widespread application of certain surface-active materials as antifriction boundary additives. Under elastohydrodynamic (EHD) conditions, adsorbed surface-active materials have been found to produce thicker lubricant film than that predicted by theoretical models [17–19]. Another major application of surface-active agents in lubrication is for solubilizing oil formulations in water to produce water-based lubricants. Various types and combinations of surface-active materials are used to formulate lubricants that can be used to generate a wide range of water-based formulations. These include clear solutions (synthetics) and emulsions with large concentrations of oil (soluble oils). Other roles of surface-active materials in lubrication include antifoaming, anticorrosion, biocides, etc. [10,11].

This chapter deals with the investigation of the surface and interfacial properties of surface-active materials derived from protein, starch, and oil components of seed crops. The modified plant proteins were isolates of barley, lupin, and oat, which were chemically modified using one or more of the following methods: acetylation; succinylation; cross-linking; cross-linking and acetylation. The modified starches were normal, high amylose and waxy starch that were esterified to various degrees using one or more of the following anydrides: acetic anydride; dodecenylsuccinic anhydride, and octenylsuccinic anhydride. Soybean oil–based polysoaps (SOPS) of varying mol wts and counterions were obtained by ring-opening polymerization of epoxidized soybean oil (ESO). All of these materials have been used in separate studies involving investigations into their properties including solubility, thermal, rhelogical, emulsification, foaming, friction, surface hydrophobicity, and ST and IT [20–25]. In this chapter, the studies specifically dealing with only surface and interfacial aspects are grouped, reviewed, and discussed from the perspective of farm-based surface-active materials. In addition, ST and IT data that were not included in previous publications are added and analyzed. Also, some of the previously published data were re-examined, reanalyzed, and used for evaluating surface properties between the various surface-active systems.

14.2   MATERIALS

14.2.1   Proteins and Chemically Modified Proteins

Detailed procedures for isolation and chemical modification of plant proteins from lupin, barley, and oat can be found elsewhere [20–22]. A brief description of the procedures for each type of crop is given below. A general flowchart for the isolation and chemical modification of plant proteins is illustrated in Figure 14.1.

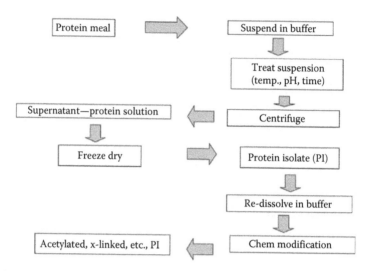

FIGURE 14.1 Flowchart for isolation and chemical modification of plant proteins.

14.2.1.1 Lupin Proteins

Lupin proteins were isolated from milled lupin meal by treating it with phosphate buffer at different pH and temperatures, as described elsewhere [20]. The protein solutions (supernatants) obtained from each temperature-pH treatment were freeze-dried and used to prepare a series of concentrations of protein solutions in phosphate buffer for use in ST and IT measurements.

14.2.1.2 Barley Protein Isolate and Derivatives

Barley protein isolate (BPI) was prepared from hexane-defatted barley powder (Honeyville Grains, Salt Lake City, UT) as described before [21]. Briefly, the powder was stirred in 0.015 N aq. NaOH (room temperature, 1 h), centrifuged (4000 g, 10 min, 10°C), the supernatant treated with 2 N aq. HCl to pH 4.5 to precipitate the BPI, which was recovered by centrifugation (10,000 g for 20 min). The BPI was further chemically modified by acetylation (Ac-BPI), cross-linking (XL-BPI), and cross-linking and acetylation (AcXL-BPI). Details of the synthesis procedures for chemical modification of BPI have been given by Mohamed et al. [21]. BPI and modified BPI samples were extracted in 0.5 M aq. NaCl solution by suspending 0.35 g of BPI in a liter of aq. NaCl solution. The suspension was stirred for 2 h, centrifuged at 300 g for 20 min, the supernatant isolated, diluted as needed with 0.5 M aq. NaCl solution, and used in ST measurements. The concentration of protein in the supernatant was determined using a LECO CHN-2000 instrument (Thun, Switzerland).

14.2.1.3 Oat Protein Isolate and Derivatives

Oat protein isolate (OPI) was prepared from hexane-defatted oats (Honeyville Grains, Salt Lake City, UT) using the procedure described above for preparation of BPI [21,22]. OPI was further used to prepare chemically modified OPI by acetylation (Ac-OPI); succinylation (Su-OPI); cross-linking (XL-OPI); cross-linking and

TABLE 14.3
Protein Isolates and Derivatives Investigated in This Work

Protein	Isolation Medium (Temp., pH)	ID
Barley protein isolate	0.5 M NaCl buffer	BPI
Acetylated BPI	0.5 M NaCl buffer	Ac-BPI
Cross-linked BPI	0.5 M NaCl buffer	XL-BPI
Acetylated and cross-linked BPI	0.5 M NaCl buffer	AcXL-BPI
Lupin protein isolate	0.01 M Phosphate buffer (amb, 4.0)	LPI-A4
Lupin protein isolate	0.01 M Phosphate buffer (amb, 6.8)	LPI-A7
Lupin protein isolate	0.01 M Phosphate buffer (amb, 8.0)	LPI-A8
Lupin protein isolate	0.01 M Phosphate buffer (100°C, 4.0)	LPI-B4
Lupin protein isolate	0.01 M Phosphate buffer (100°C, 6.8)	LPI-B7
Lupin protein isolate	0.01 M Phosphate buffer (100°C, 8.0)	LPI-B8
Oat protein isolate	0.5 M NaCl buffer	OPI
Acetylated OPI	0.5 M NaCl buffer	Ac-OPI
Succinilated OPI	0.5 M NaCl buffer	Su-OPI
Cross-linked OPI	0.5 M NaCl buffer	XL-BPI
Acetylated and cross-linked OPI	0.5 M NaCl buffer	AcXL-OPI

acetylation (AcXL-OPI). Details of the synthetic procedures have been given by Mohamed et al. [22]. OPI and modified OPI solutions in 0.5 M aq. NaCl solution used in ST measurements were prepared using the procedure described before for preparing the corresponding BPI solutions [21,22]. Protein isolates and derivatives investigated in this work are listed in Table 14.3.

14.2.2 CHEMICALLY MODIFIED STARCH

A detailed procedure for chemical modification of starch using a microwave reactor has been given before [23]. Therefore, only a brief description of the materials and procedure will be given here.

Starches used in chemical modifications were obtained from commercial sources and are listed in Table 14.4. Glacial acetic acid and/or anhydrides (acetic, >99%; octenylsuccinic, 97%; dodecenylsuccinic, 95%) used for chemical modification of starch were purchased from Aldrich Chemical Co. (Milwaukee, WI) and used as supplied.

TABLE 14.4
Some Characteristics and Source of Starches Used in This Work

Starch	Amylose (%)	Comment	Source
Normal corn	27	Pure food grade	Tate & Lyle; Decatur, IL
Maltodextrin	27	Star-dri 1	Tate & Lyle; Decatur, IL
Waxy maize	1	7350	Tate & Lyle; Decatur, IL
High amylose corn	70	Hylon 7	National Starch; Bridgewater, NJ

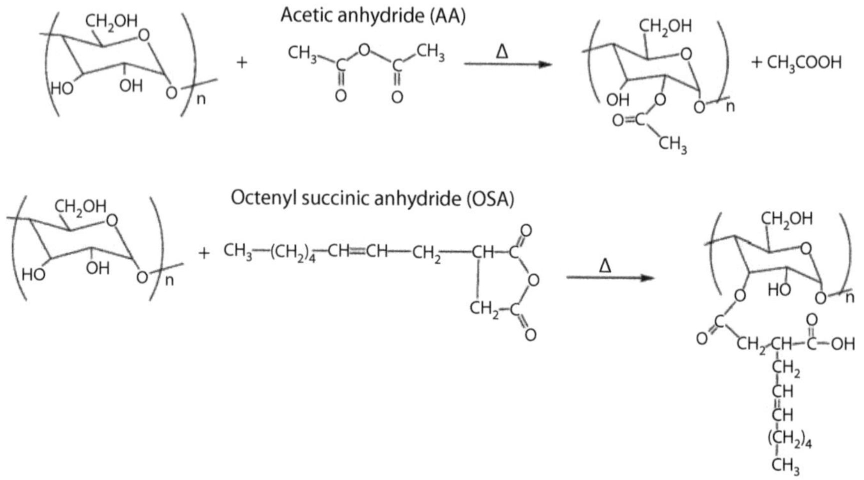

FIGURE 14.2 Esterification of starch. (From Shogren, R. and Biresaw, G., *Colloids Surf. A*, 298, 170, 2007. With permission.)

The chemical modification of the starch involves converting the free hydroxyl groups of its glucose units into esters by reacting it with fatty acids and/or anydrides [23]. A schematic depicting such a reaction is shown in Figure 14.2. The reaction was conducted using a microwave heating system (Milestone Microwave Labstation 1600, Milestone Inc., Shelton, CT). In a typical procedure [23], the appropriate quantities of starch and anydride were added to a 270 mL Teflon vessel and then stirred for 5 min with a magnetic stir bar. The vessel was then sealed, the thermocouple inserted and heated in a microwave heater from 25°C to 150°C for 3.5 min, and from 150°C–160°C for 1.5–2.5 min. The reaction mixture was then combined with 400 mL ethanol (99.5%) in a Waring blender and blended until the precipitate was broken into fine particles (about 1 min). The ethanol supernatant was poured off, and four additional ethanol extractions were performed. The starch ester product was isolated by filtration on a Buchner funnel and dried in a forced air oven overnight at 50°C and in a vacuum oven overnight at 80°C.

By varying the reaction conditions, modified waxy maize starches with varying degrees of substitution (DS) were synthesized. The DS is the percent of hydroxyl groups on the glucose units of the starch that reacted with the anhydride to form the starch ester. The chemically modified starches prepared using the microwave reactors were found to be readily soluble in water [23]. Table 14.5 shows the list of starch esters of varying DS that were obtained by chemical modification and used in this study. A series of concentrations of aqueous solutions of chemically modified starches were prepared and used in ST and IT investigations.

14.2.3 Soybean Oil–Based Polysoaps

As the detailed procedure for a two-step synthesis of soybean oil based polysoaps (SOPS) is available elsewhere [24,26], only a brief description of each of the steps will be given below.

TABLE 14.5
Chemically Modified Starches Investigated in This Work

Chemically Modified Starch	ID	Degree of Substitution (DS), %	
		Acetate	Succinate
Maltodextrin	Dextrin	0.00	
Waxy maize starch acetate	WaxAc-1	0.35	
Waxy maize starch acetate	WaxAc-2	0.70	
High amylose corn starch acetate	HiAmylAc	0.57	
Normal corn starch acetates	NormAc	0.78	
Waxy maize starch acetate/octenylsuccinate	WaxAcOcSuc2	0.36	0.046
Waxy maize starch acetate/octenylsuccinate	WaxAcOcSuc-1	0.80	0.030
Waxy maize starch acetate/dodecenylsuccinate	WaxAcDodSuc	0.31	0.022

SOPS were synthesized from epoxidized soybean oil (ESO) using a two-step synthesis procedure depicted in Figure 14.3 [24]. The first step was the ring-opening polymerization of ESO using a cationic initiator. In the second step, the polymerized ESO was hydrolyzed and converted to SOPS. ESO was obtained from Elf Atochem (Philadelphia, PA) and used as received. All other solvents and reagents used in the synthesis were obtained from the indicated commercial sources and used as supplied. A brief description of the synthesis procedure for each step is given next.

14.2.3.1 Ring-Opening Polymerization of ESO

To a solution of ESO in methylene chloride (obtained from Fisher Scientific, Fair Lawn, NJ), under nitrogen, and at 0°C (ice bath), was added dropwise the required amount of catalyst, purified and redistilled $BF_3 \cdot$ diethyl etherate (obtained from Aldrich Chemical, Milwaukee, WI). The mixture was stirred at 0°C for 3 h, the solvent removed on a rotary evaporator, the residue washed twice with hexane, and dried under vacuum at 70°C to a constant weight. The polymerized ESO product was positively identified using FTIR and NMR spectroscopies and was obtained in >99% yield. Polymerized ESO of varying molecular weights was obtained by varying the reaction temperature and/or the amount of catalyst. A detailed description of the ring-opening polymerization procedure of ESO as well as spectroscopic identification of the polymerized ESO products are given elsewhere [24,26].

14.2.3.2 Hydrolysis of Polymerized ESO

A solution of polymerized ESO in 0.4 M NaOH was refluxed for 24 h, filtered on a filter paper, and the filtrate cooled to room temperature. The resulting gel was precipitated using 1.0 M HCl, washed several times with water, and finally washed twice with 10% aqueous acetic acid. The resulting polymer was first dried overnight in an oven at 80°C, and then under vacuum at 70°C to a constant weight. The procedure gave the polycarboxylic acid product in 84% yield, which was identified using FTIR and NMR spectroscopies. Further details of the synthesis procedures for hydrolysis

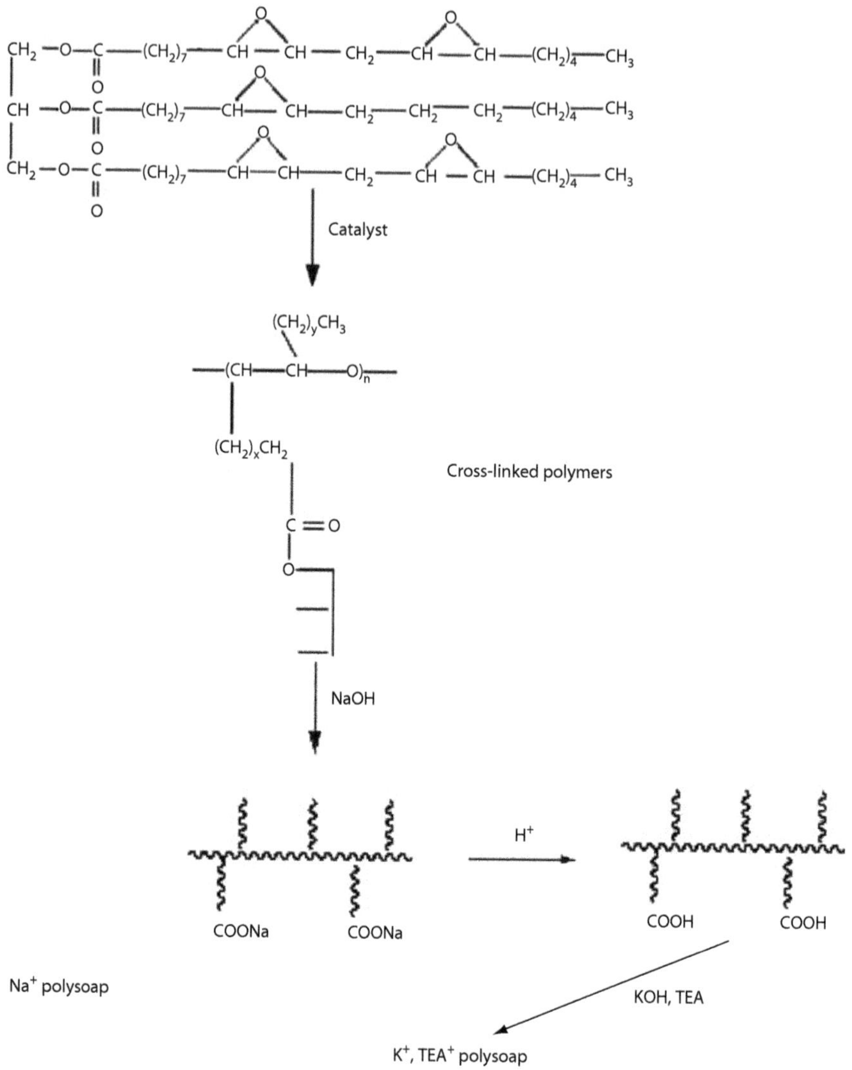

FIGURE 14.3 Ring-opening polymerization and hydrolysis of ESO.

of polymerized ESO as well as spectroscopic identification of the SOPS products are given elsewhere [24].

14.2.3.3 Saponification

SOPS with sodium and potassium counterions were prepared by neutralizing the polycarboxylic acid product with one equivalent of aq. NaOH and aq. KOH, respectively. SOPS with triethanol ammonium counterions were prepared using two equivalents of triethanol amine (TEA) obtained from Sigma (St. Louis, MO). A typical procedure for preparing 100 mL of a 1% aqueous stock solution of SOPS was as follows: a beaker containing a mixture of the required quantities of polycarboxylic

TABLE 14.6
SOPS Investigated in This Work

SOPS ID	Mol. Wt (kg/mol)	Counterion
024A	2.44	TEA+
024N	2.44	Na+
025A	2.46	TEA+
025N	2.46	Na+
026A	2.62	TEA+
026K	2.62	K+
032A	3.22	TEA+
032K	3.22	K+
032N	3.22	Na+

acid powder and aqueous base (NaOH, KOH, TEA in deionized water) was placed in a 75°C water bath and stirred with a glass rod until the powder was dissolved. The solution was transferred to a volumetric flask, and the beaker rinsed three times with 10 mL of deionized water and added to the volumetric flask. The volumetric flask was cooled to room temperature and then filled with deionized water to the 100 mL mark. These solutions were further diluted as needed and used in ST and IT investigations.

SOPS of varying mol wts and counterions prepared and investigated in this work are shown in Table 14.6.

14.2.4 OTHER MATERIALS

Deionized water used in surface and interfacial investigations was further purified to a resistivity of 18.3 megohms-cm on a Barnstead EASYpure UV/UF water purification system (EASYpure UV/UF, Model: D8611, Barnstead International, Dubuque, IA). Freshly purified water was then filtered on a 0.22 µL sterile disposable filter (MILLEX-GS 0.22 mL Filter Unit; Millipore Corp., Bedford, MA) prior to use in the preparation of various concentrations of the biobased surface-active materials.

Hexadecane (99+ % anhydrous) used in IT investigations was obtained from Aldrich Chemical (Milwaukee, WI) and used as supplied. Soybean oil (SBO) used in IT investigations was obtained from Pioneer Hi-Bred International Inc. (Des Moines, IA), and used without further purification.

14.3 METHODS

14.3.1 AXISYMMETRIC DROP SHAPE ANALYSIS

Dynamic ST and IT were measured using the axisymmetric drop shape analysis (ADSA) method [27]. In this method, ST or IT are obtained by analyzing the change in the shape of a pendant drop of a liquid suspended in air or a second liquid, respectively, as a function of time. The ADSA method is based on the Bashforth–Adams equation (Equation 14.1), which relates drop shape geometry to ST or IT as follows [8,28]:

$$\gamma = \frac{(\Delta\rho g a^2)}{H},$$ (14.1)

where

γ is the ST or IT
$\Delta\rho = (\rho_1 - \rho_2)$ is the difference in the densities of the drop and the medium
g is gravitational acceleration
a is the maximum or equatorial diameter of drop (see Figure 14.4)
H is the drop shape parameter

H is a function of the drop-shape factor S and is obtained from drop geometry as follows:

$$S = \frac{b}{a},$$ (14.2)

where b is the diameter of the drop at height a from the bottom of the pendant drop.

A schematic of pendant drop with the appropriate dimensions for use in the Bashforth–Adams equations is illustrated in Figure 14.4.

During dynamic ST or IT measurement, surface-active molecules diffuse from the bulk of the pendant drop to the surface or inter- face. This causes changes in the concentration of surface-active agents at the surface or interface as a function of time. The result will be a change in drop geometry associated with changes in the ST or IT between the drop and the medium as a function of time. A net diffusion of surface-active agents to the interface continues until an equilibrium surface or interface concentration is attained, usually after a relatively long period of time. This leads to equi- librium drop geometry and, hence, equilibrium ST or IT.

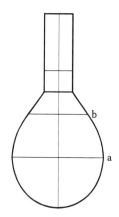

14.3.2 INSTRUMENT

Dynamic ST and IT were measured using the ADSA method on the FTA 200 automated goniometer (First Ten Ångstroms, Portsmouth, VA). The instrument comprises hardware and software that also allows for the measurement of contact angles. A schematic of the instrument hardware configured for dynamic ST and IT measurements on a pendant drop is shown in Figure 14.5.

FIGURE 14.4 Schematics of a pendant drop geometry where a is maximum or equa- torial drop diameter, and b is drop diameter at height a from the bottom of the drop. (From Mohamed, A. et al., *J. Am. Oil Chem. Soc.*, 84, 281, 2007. With permission.)

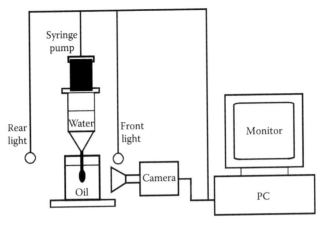

FIGURE 14.5   Schematic of an automated pendant drop goniometer. (From Mohamed, A. et al., *J. Am. Oil Chem. Soc.*, 84, 281, 2007. With permission.)

The main features of the hardware relevant to ST and IT measurements are an automated pump that can be fitted with various sizes of syringes and needles to allow for control of pendant drop formation; CCD camera (Sanyo B/W, model VCB-3512T) for automated drop image viewing, adjusting, and capturing; computer and appropriate software for setting experimental conditions (e.g., images per second), acquiring, storing, and manipulating video and numerical data; and monitor for viewing of pendant drop images and data. The instrument software (fta32 v2.0; First Ten Ångstroms) is used to carry out various tasks, including setting the experimental conditions (such as maximum drop volume, liquid pump rate, image capture triggering options, total run time, total number of images to be captured, and rate of image capture in images/s); acquiring and storing images; calculating ST or IT by analyzing each image; storing and displaying the results of each image analysis; and storing and displaying the ST or IT vs. time data for each measurement. During IT measurement using the ADSA method on the FTA 200, the image of the drop is recorded as a function of time using the high-speed CCD camera. At the end of the experiment, the image analysis software accurately measures the drop dimensions a and b from each image, and uses it to automatically calculate the ST or IT using Equations 14.1 and 14.2.

14.3.3 PROCEDURE

Farm-based surface-active materials dissolved in purified water at various concentrations were used. The STs of these solutions and their ITs with purified hexadecane or SBO were measured. All ST and IT measurements were conducted at room temperature ($23°C \pm 2°C$). Prior to conducting the ST or IT measurements, the instrument was calibrated with purified water and then checked by measuring the IT between the purified water and pure hexadecane. Typical water-hexadecane IT data obtained from such calibration measurements are illustrated in Figure 14.6. The literature value for water-hexadecane IT is 51.3 mN/m [29].

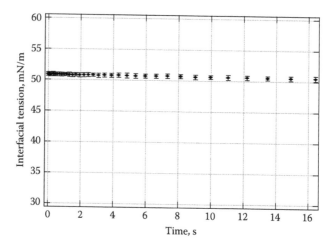

FIGURE 14.6 Dynamic IT between water and neat hexadecane (hxdcn). (From Biresaw, G. Properties of Ag-based amphiphiles, in: *Surfactants in Tribology*, G. Biresaw and K.L. Mittal, eds., pp. 259–290, CRC Press, Boca Raton, FL, 2008. With permission.)

In a typical procedure for ST measurement, a 10 mL disposable syringe (Becton Dickinson & Co., Franklin Lakes, NJ), equipped with a 17 gauge (1.499 mm OD) blunt disposable needle (KDS 17-1P, Kahnetics Dispensing Systems, Bloomington, CA), was used to generate a pendant drop of the solution in air.

For IT measurements, the disposable syringe with the blunt disposable needle was locked into place, so that the end of the needle was under the surface of the hexadecane or SBO contained in a glass cuvette (10 mm Glass Spectrophotometer Cell; Model: 22153D, A. Daigger & Company, Vernon Hills, IL). Manual trigger was then used to start the pump, so that a few drops of the aqueous solution fell to the bottom of the cuvette.

For both ST and IT measurements, the instrument is programmed to automatically deliver a specified volume of the aqueous solution at 1 µL/s to create the pendant drop, and also to automatically trigger image capture when the pump stops. The volume of aqueous solution to be automatically pumped was selected to generate the largest possible pendant drop that will not fall off before image acquisition was complete. All runs were programmed to acquire images at a rate of 0.067 s/image, with a predetermined trigger period multiplier to allow for a total of 35 images to be captured during the acquisition period. At the end of the acquisition period, each image was automatically analyzed and a plot of ST or IT vs. time was automatically displayed. The data from each run were saved as both a spreadsheet and a movie. The spreadsheet contained the time and ST or IT for each image and the movie contained each of the drop images as well as calibration information. Equilibrium ST or IT values were obtained by averaging the values at very long periods, where the ST or IT values showed little or no change with time. Repeat measurements (3–5) were conducted on each sample, and average values were used in data analysis.

14.4 DATA ANALYSIS

The data from the 3–5 repeat measurements of ST and IT were used to calculate average and standard deviations for each sample. The average values were used in further analysis. Data analysis and plotting was conducted using IgorPro version 5.0.3.0 software (WaveMetrics, Lake Oswego, OR).

14.5 RESULTS AND DISCUSSION

14.5.1 SURFACE TENSION

Typical dynamic surface tension (ST) data from ADSA measurements are illustrated in Figure 14.7. The data in Figure 14.7 are for a 20% solution of waxy maize starch acetate/dodecenylsuccinate (DS = 0.33) in purified water. As shown in Figure 14.7, the ST initially decreases sharply as a function of time. The decrease in ST slows down and, after a relatively longer time, levels off to a constant value. As mentioned before, the change in ST as a function of time is brought about by a net diffusion of surface-active molecules from the bulk of the drop to the air–water interface. Initially, the bulk and surface concentrations are far from equilibrium and net diffusion is fast. As the concentrations approach the equilibrium value, net diffusion slows considerably. At equilibrium concentration, there will be no net diffusion of surface-active molecules to the surface.

Similar ST vs. time profiles were obtained for aqueous solutions of surface-active materials based on farm-based proteins and SOPS. These are illustrated in Figures 14.8 and 14.9, respectively. The data in Figure 14.8 are for cross-linked BPI (XL-BPI) solution in 0.5 M aq. NaCl buffer. The example in Figure 14.9 is for 0.02 g/100 mL of KSOPS (potassium salt of polysoap), of molecular weight 3.21 kg/mol. In all cases,

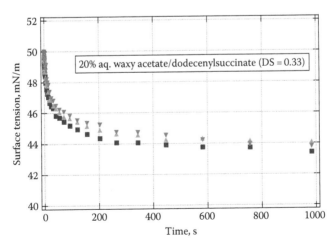

FIGURE 14.7 Multiple measurements of dynamic ST for 20% aqueous (aq.) solution of waxy maize starch acetate-dodecenylsuccinate (DS = 0.33).

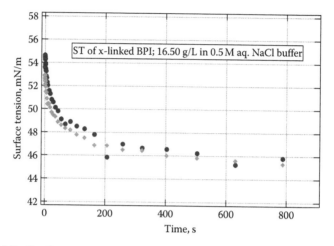

FIGURE 14.8 Duplicate measurements of dynamic ST for 16.5 g/L of cross-linked BPI solution in 0.5 M aq. NaCl buffer.

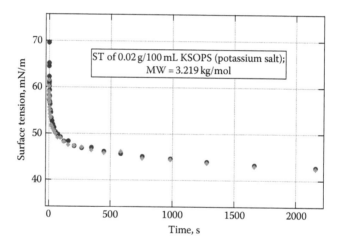

FIGURE 14.9 Duplicate measurements of dynamic ST of K⁺ SOPS (potassium salt), 0.02 g/100 mL; MW = 3.219 kg/mol.

the ST initially decreases sharply with time, and levels off to a more or less constant value at longer periods. The ST values at the longer period are more or less independent of time and correspond to the equilibrium ST of the surface-active material at the specific concentration. In this work, the equilibrium STs were obtained by averaging the last two or more ST values. The average equilibrium ST values are dependent on the chemistry and concentration of the surface-active material.

The effect of chemically modified starch concentration on ST of water is compared in Figures 14.10 and 14.11. The data in Figure 14.10 compare dextrin and starch acetates with different types of starch chemistries and various DS. Figure 14.11 compares waxy starches modified with acetate or a combination of acetate and

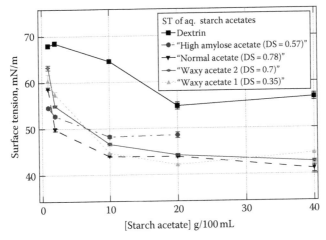

FIGURE 14.10   Effect of starch acetate concentration on ST of purified water.

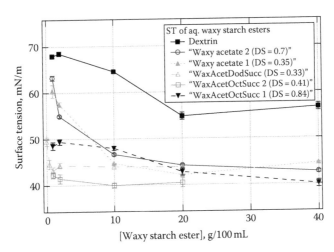

FIGURE 14.11   Effect of waxy starch ester concentration on ST of purified water.

longer chain succinates. In both cases, the ST of water decreases sharply with increasing concentration of the modified starch and levels off to a constant value at very high concentrations. As expected, dextrin was the least effective in lowering the ST of water. Among the acetates, high amylose starch was the least effective probably due to its low solubility. At high concentrations, acetates of waxy and normal starch of similar DS displayed similar STs. Also, the ST of waxy acetates at high concentration was independent of DS.

The ST of low concentrations of waxy starch esters was highly dependent on the chemistry of the ester (Figure 14.11). In general, longer chain esters were more effective at lowering the ST of water than acetates. Also, the effectiveness of the long chain waxy esters to lower the ST of water improved with reduced DS. At high concentrations (≥20 g/100 mL), the ST of all waxy esters leveled off to similar values regardless of the chemistry of the ester group or DS.

The effect of polysoap concentration on the ST of water is illustrated in Figures 14.12 and 14.13. The data in Figure 14.12 are for TEA polysoaps of varying molecular weights, and those in Figure 14.13 are for potassium polysoaps of varying molecular weights. As was the case with solutions of chemically modified starch, increasing concentrations of polysoap resulted in a sharp reduction of the ST, which leveled off to a constant value at high concentrations. The ST vs. concentration profile was found to be unaffected by differences in the mol wts of the polysoaps. However, this may be because the variations in mol wts (2.6 vs. 3.2 kg/mol) were not large enough to cause appreciable differences in ST.

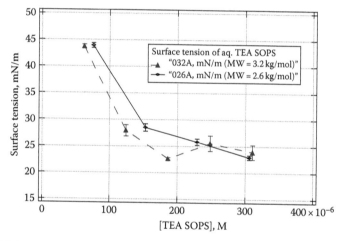

FIGURE 14.12 Effect of concentration on ST of aqueous (aq.) triethanol ammonium (TEA) SOPS.

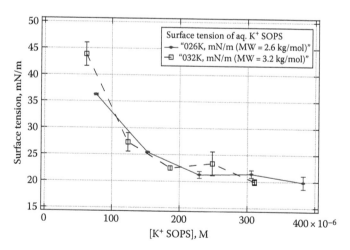

FIGURE 14.13 Effect of concentration on ST of aqueous (aq.) K⁺ SOPS.

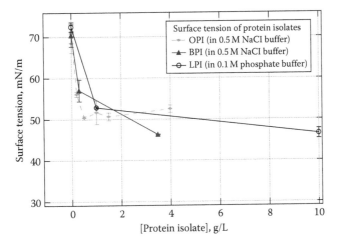

FIGURE 14.14   Effect of oat, barley, and lupin protein isolate concentration on the ST of aq. 0.5 M NaCl or 0.1 M phosphate buffers.

Figure 14.14 compares the effect of protein isolate concentration on the ST of buffer solutions. The protein isolates were from oat (OPI), barley (BPI), and lupin (LPI). OPI and BPI were investigated in 0.5 M NaCl buffer, whereas LPI was investigated in 0.1 M phosphate buffer. The LPI shown in Figure 14.14 was obtained by treating lupin meal at room temperature with phosphate buffer at pH 6.8 [20]. In all cases, the ST of the protein isolates displayed a sharp reduction with increasing concentrations. At high concentrations, however, the STs displayed very little change with increasing concentrations. This observation is similar to that discussed above for aqueous solutions of chemically modified starch and SOPS.

14.5.2   INTERFACIAL TENSION

Data from multiple measurements of dynamic IT between hexadecane and aqueous normal starch acetate (DS = 0.78) are illustrated in Figure 14.15. The data in Figure 14.15 are for starch acetate concentration of 20% (20 g/100 mL). As shown in Figure 14.15, the IT-time profile is similar to that previously described for dynamic ST measurements (Figures 14.7 through 14.9). Initially, the IT decreases sharply but levels off to a more or less constant value after a long period of time. The observed profile is consistent with expectations based on the time dependence of concentration of surface-active starch acetate at the water–hexadecane interface. Initially, the concentration of the starch acetate at the interface is low, and the IT is high, close to that of water-hexadecane (~51 mN/m). This causes a high rate of diffusion of the starch acetate from the water droplet bulk to the water–hexadecane interface, which results in a rapid drop of the IT. As time progresses, the concentration at the interface increases, the rate of diffusion slows and, with it, the rate of drop of IT. After a very long period of time, the concentration of the starch acetate at the interface attains its equilibrium value and the IT remains constant and independent of measurement time.

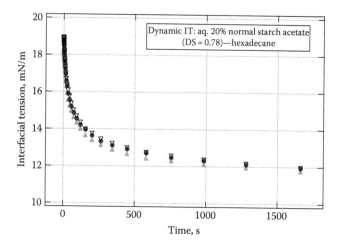

FIGURE 14.15 Multiple measurements of dynamic IT between hexadecane and 20% aqueous solution of normal corn starch acetate (DS = 0.78).

Duplicate measurements of dynamic IT between hexadecane and purified water in the presence of TEA⁺ SOPS are illustrated in Figure 14.16. The data in Figure 14.16 are for SOPS of molecular weight 3.2 kg/mol at a concentration of 310 μM. The data in Figure 14.16 display all the expected features of dynamic IT described above: initial sharp reduction of IT with time; no change of IT with time at very long times; and slow reduction of IT with time at intermediate periods.

The effect of LPI on the dynamic IT between SBO and water with 0.01 M phosphate buffer is illustrated in Figure 14.17. The LPI used to generate the data in Figure 14.17 was isolated from lupin meal with aq. 0.01 M phosphate buffer, at pH 4.0 and ambient temperature. As can be seen in Figure 14.17, the dynamic IT data display the

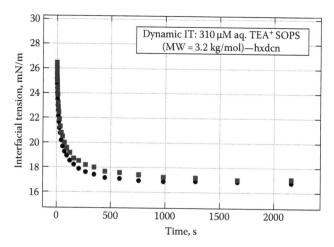

FIGURE 14.16 Multiple measurements of dynamic IT between hexadecane (hxdcn) and aq. triethanol ammonium (TEA⁺) SOPS (MW = 3.2 kg/mol, 310 μM).

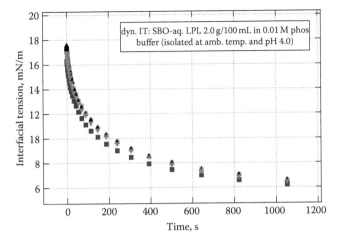

FIGURE 14.17 Multiple measurements of dynamic IT between soybean oil (SBO) and aq. LPI (2.0 g/100 mL in 0.01 M phosphate (phos) buffer; isolated at pH 4.0 and ambient temperature (amb. temp.)).

familiar time dependence of IT observed with the other farm-based surface-active materials.

The dynamic IT data similar to those shown in Figures 14.15 through 14.17 were used to calculate the equilibrium ITs between solutions of farm-based surface-active materials in water or aqueous buffer, and hexadecane or SBO. This was done by averaging the last two or three data points, where the IT displays little or no change with time.

Figure 14.18 shows the effect of starch acetate concentration on water–hexadecane equilibrium IT. The data in Figure 14.18 compare the equilibrium IT between hexadecane and aqueous acetate solutions of normal (DS = 0.78), high amylose (DS = 0.57),

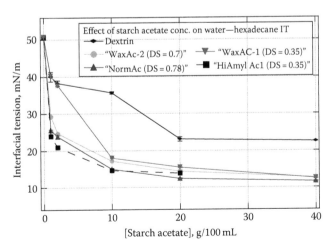

FIGURE 14.18 Effect of starch acetate concentration on equilibrium IT between water and hexadecane.

and waxy (DS=0.35, 0.7) starches. Also included in the comparison in Figure 14.18 is the equilibrium IT between hexadecane and aqueous dextrin. As shown in Figure 14.18, solubilization of acetates or dextrin in water results in a sharp decrease of the equilibrium IT between hexadecane and water. The equilibrium IT continues to decrease with a further increase of concentrations of the acetates or dextrin. At very high concentrations of acetates or dextrin, the equilibrium IT levels off to a more or less constant value. Figure 14.18 also shows that dextrin was the least effective at lowering water–hexadecane equilibrium IT. At low concentrations (<10 g/100 mL), the effectiveness of waxy acetate at lowering equilibrium IT improved with increasing DS. At high concentrations (>10 g/100 mL), all acetates, regardless of starch chemistry or DS, displayed a similar equilibrium IT value of 15±2 mN/m.

Figure 14.19 compares equilibrium IT between hexadecane and water with solubilized waxy esters. The waxy esters compared in Figure 14.19 are acetates (DS=0.35, 0.7), octenyl succinates (DS=0.41, 0.84), and dodecenyl succinates (DS=0.33). Also shown in Figure 14.19 is the equilibrium IT of hexadecane–aqueous dextrin. As shown in Figure 14.19, the equilibrium IT data vs. starch ester concentration displayed a similar profile as that described above for starch acetate solutions. Thus, the equilibrium IT decreased sharply with the addition of starch ester in water; decreased further with an increasing concentration of starch ester in water; and leveled off to a constant value at a very high concentration of starch ester in water. Close examination of Figure 14.19 shows that the long chain esters, regardless of the DS, were the most effective at reducing the equilibrium IT between hexadecane and water. Thus, the equilibrium IT between hexadecane and aqueous waxy esters of octenyl and dodecenyl succinates reached its minimum constant value at about 2 g/100 mL of concentration, whereas the corresponding value for aqueous waxy acetates and dextrin was 20 g/100 mL. In addition, long chain waxy ester resulted in a lower minimum equilibrium IT between water and hexadecane than waxy acetates.

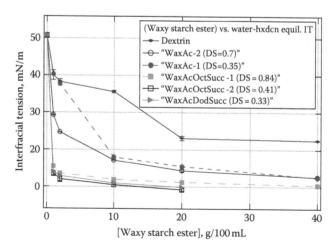

FIGURE 14.19 Effect of waxy starch ester concentration on equilibrium IT between water and hexadecane.

The effect of TEA⁺ SOPS and Na⁺ SOPS concentrations in water on water–hexadecane equilibrium IT is illustrated in Figures 14.20 and 14.21, respectively. The data in Figures 14.20 and 14.21 compare SOPS with molecular weights of 2.439, 2.461, and 3.219 kg/mol. The data in Figures 14.20 and 14.21 display the familiar equilibrium IT vs. concentration profile. Thus, as the concentration of SOPS in water increases, the equilibrium IT decreases sharply, and levels off to a constant value. Close examination of Figures 14.20 and 14.21 shows that the effectiveness of TEA⁺ SOPS and Na⁺ SOPS in reducing the water–hexadecane equilibrium IT was highly dependent on the molecular weight of SOPS. Thus, the equilibrium IT between water and hexadecane, at all concentrations and for both counterions, decreased with decreasing molecular weight of SOPS.

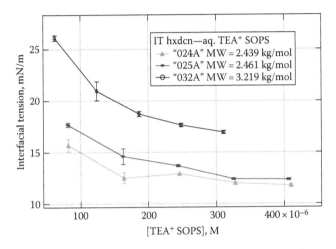

FIGURE 14.20 Effect of triethanol ammonium (TEA⁺) SOPS concentration on water-hexadecane (hxdcn) equilibrium IT.

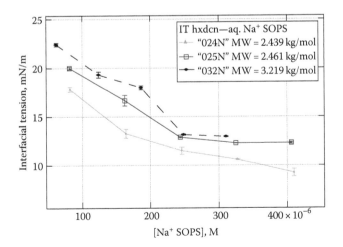

FIGURE 14.21 Effect of Na⁺ SOPS concentration on water–hexadecane (hxdcn) equilibrium IT.

14.5.3 SURFACE ENERGY

The minimum equilibrium ST for the various farm-based surface-active materials were obtained from ST vs. concentration data such as those shown in Figures 14.10 through 14.14. Two methods were used to obtain these values. In most cases, the average and standard deviation of minimum equilibrium ST values were calculated from the data at a high concentration, which displays little or no change with increased concentration (e.g., OPI in Figure 14.14). In a few cases, there were insufficient data at the high concentrations (e.g., BPI in Figure 14.14), and the lowest available ST data were used. Tables 14.7 through 14.9 summarize the calculated average and standard

TABLE 14.7
Minimum Equilibrium ST of Aqueous Solutions of Chemically Modified Starch

Chemically Modified Starch (DS)	Minimum Equilibrium ST (mN/m)
Dextrin (0.00)	55.7 ± 1.3
WaxAc-1 (0.35)	43.8 ± 1.4
WaxAc-2 (0.70)	43.4 ± 1.1
HiAmylAc (0.57)	48.4 ± 0.2
NormAc (0.78)	43.0 ± 1.6
WaxAcOcSuc2 (0.36)	40.6 ± 0.8
WaxAcOcSuc-1 (0.80)	41.4 ± 1.9
WaxAcDodSuc (0.31)	44.0 ± 0.5

TABLE 14.8
Minimum Equilibrium ST of Aqueous Solutions of SOPS

SOPS ID	Molecular Weight (kg/mol)	Minimum Equilibrium ST (mN/m)
024A	2.44	26.4 ± 0.6
024K	2.44	26.7 ± 0.2
024N	2.44	32.9 ± 0.6
025A	2.46	21.2 ± 0.6
025K	2.46	28.6 ± 0.5
025N	2.46	26.4 ± 1.2
026A	2.6	22.9 ± 0.4
026K	2.6	19.9 ± 0.6
032A	3.2	23.9 ± 1.4
032K	3.2	19.9 ± 1.1
032N	3.2	21.6 ± 0.5

TABLE 14.9

Minimum Equilibrium ST of Aqueous Solutions of Protein Isolates and Their Derivatives

Protein ID	Minimum Equilibrium ST (mN/m)
BPI	46.3±0.1
Ac-BPI	43.1±0.8
XL-BPI	46.7±1.1
AcXL-BPI	44.2±1.6
LPI-A4	40.2±2.7
LPI-A7	38.8±0.5
LPI-A8	42.0±0.3
LPI-B4	39.9±0.6
LPI-B7	35.1±2.1
LPI-B8	31.0±1.4
OPI	51.3±1.0
Ac-OPI	44.7±0.3
Su-OPI	48.3±1.0
XL-OPI	54.8±2.7
AcXL-OPI	45.0±0.9

deviation of the minimum equilibrium STs for the starch esters, SOPS, and protein isolate and derivatives, respectively, discussed in this work.

As explained before, the minimum equilibrium ST is obtained from data at high concentrations where the ST values are independent of concentration. At the high concentrations, the surface-active materials had attained full coverage of the surface of the water or buffer solutions. As a result, the minimum equilibrium ST values in Tables 14.7 through 14.9 correspond to the surface energies of the farm-based surface-active materials fully occupying the surface. This is clearly illustrated in Table 14.7, which displays the minimum equilibrium ST values for the various chemically modified starch esters. Literature values for the surface energy (SE) of the different types of starch vary in the range of 36–59 mJ/m^2 [30–33], and the values shown in Table 14.7 are clearly within this range. Amazingly, the various ester chemical modifications did not result in appreciable changes in the surface energies of the various types of starches.

Table 14.8 compares the surface energies of SOPS of varying molecular weight and counterions. It appears that the SE of SOPS shows a mild decrease with increasing molecular weight. Thus, SOPS with molecular weight above 2.5 kg/mol display SE values around 20 mJ/m^2 whereas those with molecular weight below 2.5 kg/mol display SE values around 25 mJ/m^2.

Table 14.9 compares the surface energies of protein isolates and derivatives, which varied in the range of 31–55 mJ/m^2. LPI displayed the lowest SE whereas OPI and derivatives displayed the highest. The SE of protein isolates and derivatives increased in the order: LPI-B < LPI-A < BPI and derivatives < OPI and derivatives.

The SE of LPI-B displayed a pH dependence, which increased in the order: pH 8 < pH 6.8 < pH 4. The surface energies of chemically modified BPI and OPI displayed dependence on the type of chemical modification. Thus, acetylation resulted in significant reductions of the surface energies of BPI and OPI, whereas cross-linking had no effect at all. The surface energies of oat and barley proteins increased in the order: Ac ~ AcXL < Su < XL ~ OPI or BPI (Table 14.9).

14.5.4 INTERFACIAL ENERGY

Following the same approach discussed above, the minimum equilibrium IT between hexadecane or SBO and aqueous solutions of farm-based surface-active materials was determined from the respective IT vs. concentration data similar to those shown in Figures 14.18 through 14.21. As an example, the minimum equilibrium IT from measurements on the hexadecane-aqueous starch ester solutions is illustrated in Table 14.10. The values in Table 14.10 are the equilibrium ITs at full coverage of the hexadecane–water interface. As a result, they represent the interfacial energy between the starch ester and hexadecane. The same method was employed to determine the interfacial energies of the following systems: hexadecane and SOPS; SBO and protein isolates; and SBO and protein isolate derivatives. Representative interfacial energy data for these systems are illustrated in Table 14.11.

The interfacial energy between the farm-based surface-active materials and the oils, γ_{SO}, can be estimated from the corresponding surface energies of the surface-active materials (γ_S) and of the oils (γ_O). This can be accomplished using a variety of methods, some of which require knowledge of the SE components for each material [34,35]. The SE of materials comprises polar (γ_S^p) and dispersion (γ_S^d) components, which are related as follows.

$$\gamma_S = \gamma_S^d + \gamma_S^p \tag{14.3}$$

TABLE 14.10
Minimum Equilibrium IT between Hexadecane and Aqueous Solutions of Chemically Modified Starch

Aqueous Starch	IT (mN/m)
Dextrin (DS = 0.00)	22.7 ± 0.4
WaxAc-1 (DS = 0.35)	12.4 ± 0.2
WaxAc-2 (DS = 0.7)	13.5 ± 1.2
NormAc (DS = 0.78)	12.0 ± 0.6
HiAmylAc (DS = 0.57)	14.2 ± 0.6
WaxAcOctSuc-1 (DS = 0.84)	10.8 ± 0.6
WaxAcOctSuc-2 (DS = 0.41)	9.8 ± 0.9
WaxAcDodSuc (DS = 0.33)	10.4 ± 0.8

TABLE 14.11
Measured IT and Predicted Interfacial Energies at Aqueous SOPS–Hexadecane, Aqueous Starch Ester–Hexadecane, and Aqueous LPI–Soybean Oil Interfaces

	Interfacial Tension (mN/m)									
			Calculated[a]							
			$X_S^d = 0.95$		$X_S^d = 0.8$		$X_S^d = 0.7$		$X_S^d = 0.6$	
Aqueous Solution ID	Measured	Ant	GM	HM	GM	HM	GM	HM	GM	HM
Protein–SBO[b]										
LPI-A4	6.6±0.2	5.1	2.1	2.1	8.1	8.2	12.4	12.8	17.1	18.1
LPI-A7	4.4±0.1	3.7	2.0	2.0	7.9	8.0	12.1	12.7	16.7	17.9
LPI-A8	3.8±0.1	6.9	2.2	2.4	8.4	8.4	12.8	13.1	17.6	18.4
Starch (DS)–hxdcn[c]										
WaxAcOct-1 (0.84)	10.8±0.6	13.9	3.1	4.2	8.5	8.8	12.4	12.5	16.6	16.7
NormAc (0.78)	12±0.6	15.5	3.5	4.8	9.0	9.4	13.0	13.0	17.2	17.3
WaxAcDod (0.31)	10.4±0.8	16.5	3.7	5.2	9.3	9.7	13.3	13.4	17.6	17.6
WaxAc-2 (0.7)	13.5±1.2	15.9	3.6	4.9	9.1	9.5	13.1	13.2	17.4	17.4
HiAmylAc (0.57)	14.2±0.6	20.9	4.8	7.1	10.6	11.6	14.9	15.2	19.4	19.4
SOPS–hxdcn[c]										
026K	11.9±0.1	7.6	1.8	2.6	5.6	7.1	8.3	10.4	11.2	14.1
026A	14.2±0.1	4.6	1.5	1.8	5.5	6.4	8.4	9.9	11.5	13.8
032N	13.0±0.0	5.9	1.6	2.1	5.5	6.7	8.3	10.1	11.3	13.9
032K	12.7±0.2	7.6	1.8	2.6	5.6	7.1	8.3	10.4	11.2	14.1
032A	16.9±0.2	3.6	1.4	1.7	5.5	6.3	8.5	9.8	11.7	13.7
024A	11.8±0.1	1.1	1.4	1.4	5.7	6.1	8.8	9.7	12.2	13.7
024K	8.3±0.6	0.8	1.4	1.4	5.7	6.1	8.9	9.7	12.2	13.7
024N	9.3±0.4	5.4	1.8	1.9	6.6	6.6	10.1	10.3	13.8	14.4
025A	12.3±0.1	6.3	1.6	2.2	5.5	6.7	8.3	10.1	11.3	13.9
025K	10.1±0.1	1.1	1.4	1.4	5.9	6.1	9.2	9.8	12.7	13.8
025N	12.3±0.1	1.1	1.4	1.4	5.7	6.1	8.8	9.7	12.2	13.7

[a] Calculation methods: Ant, Antonow, Equation 14.4; HM, harmonic mean method, Equation 14.8; GM, geometric mean method, Equation 14.9.

[b] ST of soybean oil, 35.1 mN/m (Figure 14.21).

[c] ST of hexadecane, 27.5 mN/m [38].

The simplest method for estimating interfacial energy between two materials from the SE of the individual materials is that proposed by Antonow [36,37], and is given as follows:

$$\gamma_{so} = |\gamma_s - \gamma_o|,$$

(14.4)

where

γ_s, γ_o are the surface energies of the two substances, here surface-active material and oil (liquid), respectively

γ_{so} is the interfacial energy

The Antonow method is simple because it does not require knowledge of the polar and dispersion SE components of the materials.

Two other methods that require knowledge of the SE polar and dispersion components are the GM and HM methods [34]. In the HM method, the relationship between IT and ST parameters is as follows:

$$\gamma_{so} = \gamma_s + \gamma_o - \frac{4\gamma_s^d\gamma_o^d}{\gamma_s^d + \gamma_o^d} - \frac{4\gamma_s^p\gamma_o^p}{\gamma_s^p + \gamma_o^p}$$

(14.5)

For the GM method, the corresponding relationship is given by

$$\gamma_{so} = \gamma_s + \gamma_o - 2\left(\gamma_o^d\gamma_s^d\right)^{1/2} - 2\left(\gamma_o^p\gamma_s^p\right)^{1/2}.$$

(14.6)

For some mixtures, the SE of the oil components could be purely dispersive, i.e., γ_o^p in Equations 14.5 and 14.6) is zero. This will eliminate the last term in Equations 14.5 and 14.6. Further simplification of Equations 14.5 and 14.6 can be achieved by replacing γ_o^d by γ_o and by defining γ_s^d as follows:

$$\gamma_s^d = X_s^d\gamma_s,$$

(14.7)

where $X_s^d = \gamma_s^d/\gamma_s$ is the fraction of nonpolar component of γ_s and has values between 0 and 1 [34].

The simplified HM and GM equations for predicting interfacial energy will be as follows:

$$\gamma_{so} = \gamma_s + \gamma_o - \frac{4X_s^d\gamma_s\gamma_o}{X_s^d\gamma_s + \gamma_o}$$

(14.8)

$$\gamma_{so} = \gamma_s + \gamma_o - 2\left(X_s^d\gamma_s\gamma_o\right)^{1/2}$$

(14.9)

Hexadecane used in the current work has purely dispersive SE [35]. Also, based on the fact that esters of long chain fatty acids have purely dispersive SE [35], it is safe to assume that SBO also has purely dispersive SE. Thus, Equations 14.8 and 14.9 can be used to estimate the interfacial energies between the farm-based surface-active materials and hexadecane or SBO.

In applying Equations 14.4, 14.8, and 14.9 to predict the interfacial energies, we used for γ_s the SE values for the farm-based surface-active materials which correspond to the minimum equilibrium ST values, shown in Tables 14.7 through 14.9. For ST of hexadecane, γ_0 in Equations 14.4, 14.8, and 14.9, we used the literature value of 27.0 mN/m [38], which we verified. For SBO, γ_0 was obtained from the dynamic ST data generated in this work, which is illustrated in Figure 14.22. The average ST value from three repeat measurements (total of 51 points) was 35.1 ± 0.3 mN/m. This value is within the range of 31.7–35.4 mN/m reported by others [39,40], and was used as the γ_0 value for SBO.

Application of Equations 14.8 and 14.9 also requires values for X_S^d. The X_S^d values for the various farm-based surface-active materials are unknown. Thus, we used X_S^d as a fitting parameter in Equations 14.8 and 14.9 with values varying between 0 and 1.

Table 14.11 compares γ_{so} values calculated using Antonow's (Equation 14.4), HM (Equation 14.8) and GM (Equation 14.9) methods for SOPS–hexadecane, starch ester–hexadecane, and LPI–SBO systems. The calculated values from the GM and HM methods were obtained for X_S^d values that varied in the 0.60–0.95 range. Also shown in Table 14.11 are the measured IT values along with the corresponding standard deviations. Table 14.11 reveals that the predictions of the Antonow method relative to measured IT values varied depending on the source of the farm-based surface-active materials. For aqueous SOPS–hexadecane systems, the Antonow method underpredicted the interfacial energy but overpredicted for the aqueous starch ester–hexadecane system. On the other hand, for the LPI–SBO system, the Antonow method gave a reasonable prediction relative to the measured IT values.

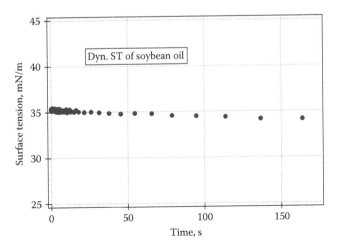

FIGURE 14.22 Dynamic ST (dyn ST) of soybean oil (SBO) used in this work.

Close examination of the predictions of the GM and HM methods reveals a number of interesting features. In general, interfacial energy predicted with these methods increased with decreasing X_S^d. In addition, predicted SE values from the GM method were generally equal or slightly larger than those from the HM method.

Comparison of the measured vs. predicted surface energies by the GM and HM methods display an interesting dependence on X_S^d. Of particular interest is the value of X_S^d that predicts interfacial energy comparable to the measured values using the GM or HM method. Table 14.11 also shows that this X_S^d value is dependent on the source of the farm-based surface-active material. This value for aq. LPI–SBO is in the 0.80–0.95 range. The corresponding values for aqueous starch ester–hexadecane and aqueous SOPS–hexadecane are 0.7–0.8, and 0.6–0.7, respectively. These values suggest that, among the three types of farm-based surface-active materials investigated here, SOPS have the most polar surface, LPI the least polar surface, and starch esters have polarity that is intermediate between the two. The reason for this trend of polarity between these farm-based surface-active materials is not clear. It is of great interest if these conclusions, based on the predictions of the HM and GM methods, can be verified by other methods. Our group is looking into independent methods of verifying this polarity trend.

14.6   SUMMARY AND CONCLUSION

Farm-based raw materials provide a number of benefits as potential replacements for petroleum in a number of consumer and industrial products. These include renewability, abundance, biodegradability, and safety. Widespread application of agriculture-based raw materials, especially those from surplus crops, will increase the demand for crops, improve the income of farmers, generate "green" jobs, reduce reliance on imported oil, and strengthen the rural and national economy.

However, farm-based raw materials have a number of inherent weaknesses and cannot be used as direct substitutes to petroleum-based products currently in the market. Overcoming these weaknesses requires a sustained research and development effort to understand and improve their properties. Once this is accomplished, cost-effective manufacturing methods need to be developed and applied. Successful introduction of biobased products into the market requires that they are competitive against petroleum-based products currently in the market, both in performance and cost.

Surface activity is one of the critical properties of ingredients used in materials application. ST/energy and IT/energy are two of the surface properties used to characterize ingredients used in materials. These properties affect a number of critical material performance properties, such as compatibility, emulsification, adsorption, wetting, adhesion, etc.

Starch, protein, and oil are major ingredients of commercial farm crops, such as soybean and corn. In this work, surface-active materials were obtained by chemical modification of proteins, starches, and oils from major crops and their surface and interfacial properties investigated. The proteins investigated were LPI; BPIs and derivatives (acetylated, cross-linked, acetylated and cross-linked); OPI and derivatives (acetylated, succinilated, cross-linked, acetylated and cross-linked). The starches were chemically modified by reaction of their free hydoxyl groups with acetyl, dodecenylsuccinyl, and octenylsuccinyl anhydrides, and combinations. The chemically modified starches were normal, high amylose, and waxy corn starches. The degree of substitution (esterification) of the free alcohols of various starches was varied, and unmodified dextrin was used as the control. SOPS of varying mol wts were synthesized by ring-opening polymerization of ESO. A series of concentrations of the farm-based surface-active materials in purified water or buffer solutions were prepared. The dynamic STs of the solutions were measured using the ADSA method, and the data used to determine the equilibrium ST at each concentration. All of the farm-based materials investigated here were found to be effective at reducing the ST of water or buffer solutions. The dynamic ST data for all solutions displayed a sharp decrease in ST with time, which leveled off to a constant equilibrium value at relatively long periods of time. For all farm-based surface-active materials investigated, the equilibrium ST decreased with increasing concentration and leveled off to a constant value, corresponding to the SE of the surface-active material.

The data showed that the SE of the various starch esters was in the $41-48\,\mathrm{mJ/m^2}$ range, which is within range of the reported values for starch. The SE of SOPS displayed mol wt dependence. Thus, SOPS with molecular weight above $2.5\,\mathrm{kg/mol}$ displayed SE values in the $20-24\,\mathrm{mJ/m^2}$ range, whereas most SOPS with molecular weight below $2.5\,\mathrm{kg/mol}$ display SE values in the range of $26-33\,\mathrm{mJ/m^2}$. Protein derivatives displayed a wide range of SE ($31-55\,\mathrm{mJ/m^2}$) that was dependent on the crop source of the protein. Thus, SE was lowest for LPI, highest for OPI and derivatives, with BPI and derivatives intermediate between the two.

ADSA was also used to measure the dynamic IT between aqueous solutions of farm-based surface-active materials and oils as follows: hexadecane and aqueous solutions of SOPS; hexadecane and starch esters; SBO and aqueous protein isolates; SBO and protein isolate derivatives. The dynamic IT data were used to determine the equilibrium IT between the oil and each concentration of the surface-active material. For all farm-based surface-active materials, the equilibrium IT decreased with increasing concentrations of the surface-active materials in the purified water or buffer, and leveled off at a constant minimum value at very high concentrations. The minimum value corresponds to the interfacial energy between the surface-active material and the oil (hexadecane or SBO).

The measured STs of hexadecane, SBO, and the farm-based surface-active materials were analyzed and used to estimate the interfacial energies between the oils (SBO or hexadecane) and the surface-active materials. The analysis was conducted using the Antonoff, GM, and HM methods. The predictions of the Antonoff method were compared to those measured using ADSA. The results showed that, relative to the measured values, the interfacial energy predicted using the Antonoff method was substantially low for SOPS–hexadecane, substantially high for starch ester–hexadecane, and reasonably good for LPI–SBO.

In the analysis of SE data using the GM and HM methods, the fraction of non-polar SE component for the farm-based surface-active materials, X_S^d, was used as a fitting parameter for predicting the interfacial energy. The value of X_S^d varies between 0 and 1. The results showed that for the same X_s^d value, predicted SE from the GM method was generally equal to, or slightly larger than, that from the HM method. The results also showed that X_s^d values that predict interfacial energies close to the measured values using the GM and HM methods increased in the order: SOPS < starch esters < protein isolates. This result suggests that the polarity of the surfaces of these farm-based surface-active materials increase in the order: protein isolates < starch esters < SOPS. The reason for this trend of polarity is not obvious. Our group is looking into independent methods of verifying this polarity trend.

ACKNOWLEDGMENTS

The author expresses profound gratitude to the following colleagues and collaborators whose work has been reviewed and cited in this chapter: Dr. Sevim Z. Erhan, Dr. Mila P. Hojilla-Evangelista, Dr. Zengshe (Kevin) Liu, Dr. Abdellatif Mohamed, Dr. Steven C. Peterson, Dr. Patricia Rayas-Duarte, Dr. David J. Sessa, Dr. Randal L. Shogren, and Dr. Jingyuan (James) Xu. The author is also profoundly grateful to Megan Goers, Natalie Lafranzo, and Xiaozhou Cao for their help with the ST and IT measurements.

Names are necessary to report factually on available data; however, the USDA neither guarantees nor warrants the standard of the product, and the use of the name by the USDA implies no approval of the product to the exclusion of others that may also be available.

REFERENCES

1. D. L. Kaplan (Ed.), *Biopolymers from Renewable Resources*, Springer, Berlin, Germany (1998).
2. A. L. Mohanty, M. Misra, and G. Hinrichsen, Biofibres, biodegradable polymers and biocomposites: An overview, *Macromol. Mater. Eng.* 276/277, 1–24 (2000).
3. N. J. Fox and G. W. Stachowiak, Vegetable oil-based lubricants—A review of oxidation, *Tribol. Int.* 40, 1035–1046 (2007).
4. G. F. Fanta, H. M. Muijs, K. Eskins, F. C. Felker, and S. M. Erhan, Starch-containing lubricant for oil field application, U.S. Patent 6,461,999 B1 (2002).
5. M. P. Schneider, Plant-oil-based lubricants and hydraulic fluids, *J. Sci. Food Agric.* 86, 1769–1780 (2006).
6. Annon WASDE-475; http://usda.mannlib.cornell.edu/ (October 9, 2009).
7. K. L. Mittal (Ed.), *Contact Angle, Wettability and Adhesion*, VSP, Utrecht, the Netherlands (1993).
8. P. C. Hiemenz, *Principles of Colloid and Surface Chemistry*, Marcel Dekker, New York (1986).
9. M. L. Free, Introduction to surfactants, in: *Surfactants in Tribology*, G. Biresaw and K. L. Mittal (Eds.), pp. 3–10, CRC Press, Boca Raton, FL (2008).
10. J. C. Childers, The chemistry of metalworking fluids, in: *Metalworking Fluids*, J. P. Byers (Ed.), pp. 165–189, Marcel Dekker, New York (1994).
11. G. Biresaw, Surfactants in lubrication, in: *Lubricant Additives: Chemistry and Applications*, L. R. Rudnick (Ed.), 2nd edn., pp. 399–419, Taylor & Francis, Boca Raton, FL (2009).
12. S. Jahanmir and M. Beltzer, An adsorption model for friction in boundary lubrication, *ASLE Trans.* 29, 423–430 (1986).

13. M. Beltzer and S. Jahanmir, Effect of additive molecular structure on friction, *Lubr. Sci.* 1, 3–26 (1988).
14. G. Biresaw, A. Adharyu, S. Z. Erhan, and C. J. Carriere, Friction and adsorption properties of normal and high oleic soybean oils, *J. Am. Oil Chem. Soc.* 79, 53–58 (2002).
15. A. Adhvaryu, G. Biresaw, B. K. Sharma, and S. Z. Erhan, Friction behavior of some seed oils: Biobased lubricant applications, *Ind. Eng. Chem. Res.* 45, 3735–3740 (2006).
16. T. L. Kurth, J. A. Byars, S. C. Cermak, B. K. Sharma, and G. Biresaw, Non-linear adsorption modeling of fatty esters and oleic estolides via boundary lubrication coefficient of friction measurements, *Wear* 262, 536–544 (2007).
17. G. Guangteng and H. A. Spikes, Fractionation of liquid lubricants at solid surfaces, *Wear* 200, 336–345 (1996).
18. G. Biresaw and G. Bantchev, Effect of chemical structure on film-forming properties of seed oils, *J. Synth. Lubr.* 25, 159–183 (2008).
19. G. Bantchev and G. Biresaw, Elastohydrodynamic study of vegetable oil— Polyalphaolefin blends in the low film thickness regime, submitted to *Tribol. Int.*, August 2009.
20. A. A. Mohamed, S. C. Peterson, M. P. Hojilla-Evangelista, D. J. Sessa, P. Rayas-Duarte, and G. Biresaw, Effect of heat treatment and pH on the thermal, surface, and rheological properties of *Lupinus albus* protein, *J. Am. Oil Chem. Soc.* 82, 135–140 (2005).
21. A. Mohamed, M. P. Hojilla-Evangelista, S. C. Peterson, and G. Biresaw, Barley protein isolate: Thermal, functional, rheological and surface properties, *J. Am. Oil Chem. Soc.* 84, 281–288 (2007).
22. A. Mohamed, G. Biresaw, J. Xu, M. P. Hojilla-Evangelista, and P. Rayas-Duarte, Oats protein isolate: Thermal, rheological, surface and functional properties, *Food Res. Int.* 42, 107–114 (2009).
23. R. Shogren and G. Biresaw, Surface properties of water soluble starch, starch acetates and starch acetates/alkenylsuccinates, *Colloids Surf. A* 298, 170–176 (2007).
24. G. Biresaw, Z. S. Liu, and S. Z. Erhan, Investigation of the surface properties of polymeric soaps obtained by ring opening polymerization of epoxidized soybean oil, *J. Appl. Polym. Sci.* 108, 1976–1985 (2008).
25. G. Biresaw and R. Shogren, Friction properties of chemically modified starch, *J. Synth. Lubr.* 25, 17–30 (2008).
26. Z.S Liu and S. Z. Erhan, U.S. patent application pending (file number: 007,7298; filed in 2007).
27. Y. Rotenberg, L. Boruvka, and A. W. Neumann, Determination of surface tension and contact angle from the shapes of axisymmetric fluid interfaces, *J. Colloid Interface Sci.* 93, 169–183 (1983).
28. A. P. Adamson and A. W. Gast, *Physical Chemistry of Surfaces*, Wiley, New York (1997).
29. C. J. van Oss, M. K. Chaudhury, and R. J. Good, Monopolar surfaces, *Adv. Colloid Interface Sci.* 28, 35–64 (1987).
30. J. W. Lawton, Biodegradable coatings for thermoplastic starch, in: *Cereals: Novel Uses and Processes*, G. M. Campbell, C. Webb, and S. L. McKee (Eds.), pp. 43–47, Plenum Press, New York (1997).
31. J. W. Lawton, Surface energy of extruded and jet cooked starch, *Starch* 47, 62–67 (1995).
32. I. Krycer, D. G. Pope, and J. A. Hersey, An evaluation of tablet binding agents. Part I. Solution binders, *Powder Technol.* 34, 39–51 (1983).
33. G. Biresaw and C. J. Carriere, Correlation between mechanical adhesion and interfacial properties of starch/biodegradable polyester blends, *J. Polym. Sci. Polym. Phys.* 39, 920–923 (2001).
34. S. Wu, *Polymer Interface and Adhesion*, Marcel Dekker, New York (1982).

35. C. J. van Oss, *Interfacial Forces in Aqueous Media*, Marcel Dekker, New York (1994).
36. G. Antonoff, On the validity of Antonoff's rule, *J. Phys. Chem.* 46, 497–499 (1942).
37. G. N. Antonow, Sur La Tension Superficielle Des Solutins Dans La Zone Critique, *J. Chim. Phys.* 5, 364 (1907).
38. J. J. Jasper, The surface tension of pure liquid compounds, *J. Phys. Chem. Ref. Data* 1, 841–1010 (1972).
39. R. Subramanian, S. Ichikawad, M. Nakajima, T. Kimurac, and T. Maekawac, Characterization of phospholipid reverse micelles in relation to membrane processing of vegetable oils, *Eur. J. Lipid Sci. Technol.* 103, 93–97 (2001).
40. W. M. Formo, Physical properties of fats and oils, in: *Bailey's Industrial Oil and Fat Products*, D. Swern (Ed.), Vol. 1, 4th edn., pp.177–232, Wiley, New York (1979).

15 Chemically Modified Fatty Acid Methyl Esters: Their Potential Use as Lubrication Fluids and Surfactants

Brajendra K. Sharma, Kenneth M. Doll, and Sevim Z. Erhan

CONTENTS

ABSTRACT

Recent developments in the synthesis and characterization of lubrication fluids and surfactants from methyl oleate are reviewed. The synthesis of materials using an epoxidation route is discussed. In this route, epoxidation of fatty acid methyl esters improves their oxidation stability while maintaining their good lubricity characteristics. Further reaction of the epoxide to form ester structures enables an even more versatile fluid, which has the stability advantages of the epoxides, but the flow characteristics of the olefins. Finally, a simple glyceride ester with an epoxidized fatty chain is shown to be an oil-in-water emulsifier.

15.1 INTRODUCTION

Vegetable oils have been used for a variety of applications since prehistoric times, where the first lubricants were natural in origin. However, the nineteenth century brought competition, first from mineral oils [1], then from synthetic hydrocarbons, which have a history in lubrication dating back to 1877 [2]. By 1997, the use of biobased lubricants was down to 8 million gallons out of a market of 2.66 billion gallons. However, recently there has been a reversal of this trend, and annual biobased lubricant sales have doubled in a market where overall lubricant sales have been relatively stable. Methods to utilize highly unsaturated and oxidatively unstable, semidrying vegetable oils, such as soybean oil (SBO) [3–8], have received a great deal of attention from our research group [9–21] as well as from the many other groups [22–29].

Another approach, which is gaining in popularity, is the use of alkyl esters instead of the complete oil triacylglyceride structure (Figure 15.1). Transesterification is carried out on hundreds of million gallons of vegetable oils annually, yielding methyl esters, which are commonly sold as biodiesel [30]. However, due to market fluctuations, biodiesel production profitability is marginal, where the current selling price of competing petroleum-derived diesel is in the $2–3.00 per gallon range. Because the average lubricant sells at greater than $7.00 per gallon, the opportunity to synthesize alkyl ester-based lubricants looks attractive.

The oxidative stability of many alkyl esters has been studied closely [31] and is in many cases actually worse than that of SBO. This means that these fluids will also require chemical modifications before they are suitable for lubricant use.

Because the rate of oxidation of an alkyl ester is roughly proportional to the number of doubly-allylic hydrogen atoms in the structure [3], chemical modifications, which remove the double-bond structure can most readily improve this property. The simple act of hydrogenation of the double bonds increases the stability of SBO [32–34], or biodiesel [35], but the resultant materials do not have the flow characteristics necessary for lubrication use.

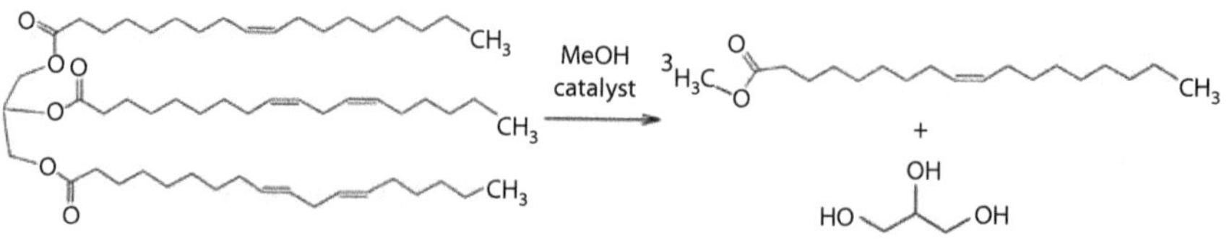

FIGURE 15.1   The transesterification reaction of a triacylglyceride with methanol to produce fatty acid methyl esters and glycerol.

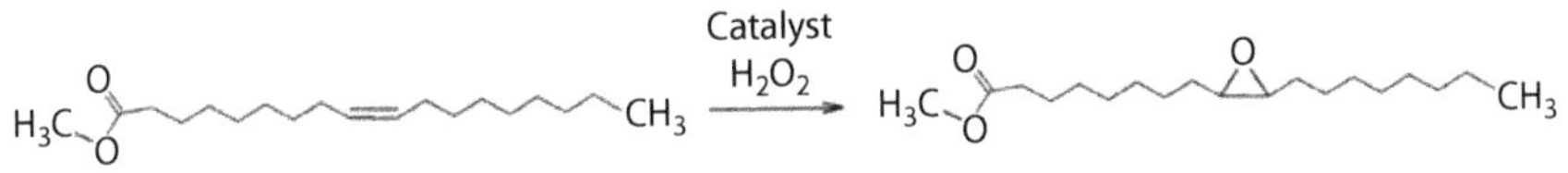

FIGURE 15.2 The epoxidation of methyl oleate with hydrogen peroxide. The catalyst can be enzymatic, heterogeneous, or homogeneous. Formic acid is also a commonly used catalyst, and was used in this work.

A more chemically interesting route is through epoxidation (Figure 15.2), which has been performed on natural oils as far back as 1945 [36,37]. This well-studied reaction [38–48] has been shown to make polymer precursors and plasticizers [45,49–52], but surprisingly, no systematic lubrication studies of epoxidized methyl oleate (EMO), epoxidized methyl linoleate (EMLO), and epoxidized methyl linolenate (EMLEN) were done prior to our work [53–55].

Additionally, although the epoxidized materials are very useful on their own, the addition of a ring-opening step to the synthesis has opened up an even broader range of materials for study [56–58]. This ring-opening reaction has also been studied [59,60] commonly leading to poly-hydroxy [45], ether [14,61,62], or ester compounds [56–58,63,64]. These compounds, especially the esters, show properties that are highly desirable in lubrication fluids.

Fatty acid methyl esters have also proven to be useful in surfactant synthesis (Figure 15.3). Their large lipophilic fatty chains can be combined with an appropriate hydrophile. The simplest method for doing this is a simple mixing of fatty acids with base to form alkali metal soaps, the earliest surfactant [65], and a common

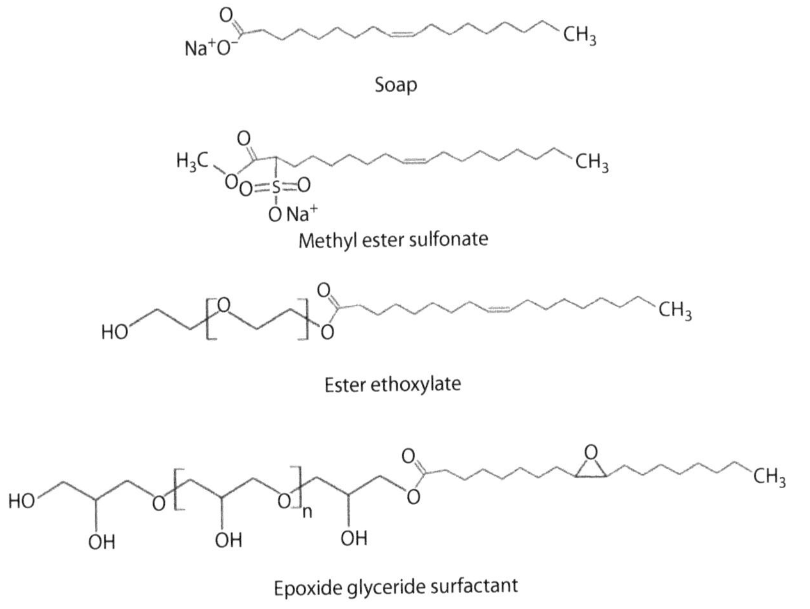

FIGURE 15.3 The structures of common monomeric surfactants and the EMO-based surfactant (Epoxide Glyceride Surfactant) studied in this work.

nuisance to biodiesel producers [66]. Nonionic surfactants can be made by the addition of ethylene oxide, a sugar group, or both, to the ester end of the fatty acid methyl ester. Using this technology, surfactants can be made with a wide range of hydrophile–lipophile balance (HLB) values making them desirable in the fast-growing personal care surfactant market.

A more recent type of fatty acid-based methyl ester surfactants are methyl ester sulfonates (MES), which are produced from the sulfonation of highly saturated methyl esters, and have a sulfate group near the ester group [67–69]. One technical difficulty which has hindered the development of this material is that while the MES are good surfactants, their easily formed di-anionic salts are not. Recent developments and understanding of this complex speciation have finally allowed the development of a modern MES industry [70–75].

Less common is the use of epoxidized oleochemicals to build surfactants. We have made a nonionic surfactant from EMO and glycerol, resulting in a highly promising biobased material [76–78]. This material is a good emulsifier and, more importantly, retains the epoxide group, opening up the possibility of further chemical modification of the hydrophobic group.

Chemical modification of fatty acid methyl esters overcomes some of their limitations, and increases their potential for use in the lubrication and surfactant industries. Although many different reaction chemistries are possible, the following examples show how the use of an epoxidation route can be especially versatile.

15.2 SYNTHESIS OF CHEMICALLY MODIFIED FATTY ACID METHYL ESTERS

An effective lubricant has a few specific tasks. First, it must be able to reach, and lubricate the surface that is being protected. Second, it must reduce friction. Third, it must not decompose too rapidly under the conditions employed. In order to quantify the ability of a lubricant to meet the first criterion, we measure viscosity, pour point, and cloud point. A measurement of the second criterion is the coefficient of friction (CoF), which can be determined by the ball-on-disk configuration of the friction and wear test machine. The final criterion can be best studied by pressure differential scanning calorimetry (DSC) and thin-film micro-oxidation (TFMO).

In general, saturated fatty acid esters have good lubricity and stability properties, but they suffer from poor flow characteristics. Unsaturated fatty acid esters, on the other hand, have good lubricity and flow characteristics but are oxidatively unstable. Our first order of attack on this problem is to remove the unsaturation through epoxidation (Figure 15.2). Our next step involves a ring-opening of the epoxy materials (Figure 15.4), to give products with an ester function at the 9th or the 10th position of the fatty chain.

15.2.1 MATERIALS

Expeller pressed, refined, bleached, degummed SBO was obtained from Nexsoy, a company located in Springfield, IL. Methyl oleate and methyl linoleate of greater than 99% purity were purchased from Nu Chek Prep, a company headquartered

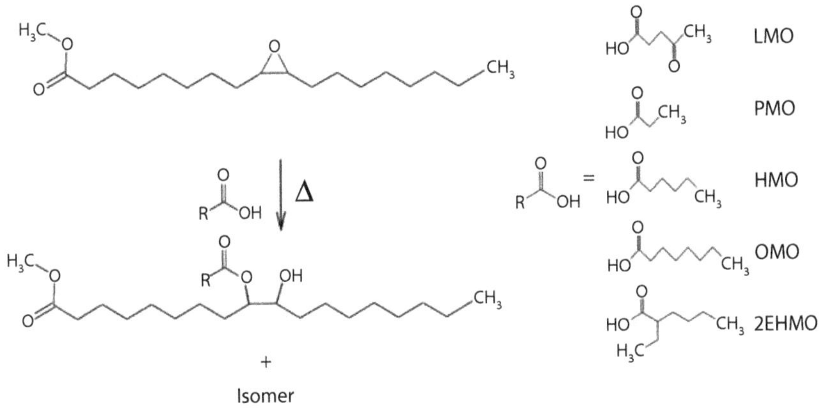

FIGURE 15.4 The ring-opening reaction of EMO to form the 9-ester-10-hydroxy derivative of methyl stearate. The 10-ester-9-hydroxy isomers also form. The material synthesized from EMO and propanoic acid is abbreviated PMO; hexanoic acid, HMO; octanoic acid, OMO; 2-ethyl hexanoic, 2EHMO; and levulic acid, LMO. Note that levulinic acid contains a ketone structure.

in Elsyian, MN. All other reagents were purchased from Sigma-Aldrich, or Fisher Scientific. They were reagent grade or better and they were used as received. Surfactants used for comparison in emulsification experiments were obtained from commercial manufacturers as free samples and used as received. The commercial surfactants were: Pluronic® L43, from BASF and Caprol® MPGO from Abitec Corporation, Janesville, WI.

15.2.2 SYNTHESIS OF EPOXIDIZED MATERIALS

The materials were epoxidized using hydrogen peroxide with formic acid as a catalyst (Figure 15.2). Progress of this Swern epoxidation [36,37,79] must be followed directly in order to avoid ring opening or other side reactions of the desired product. Typically, the methyl ester is placed in a round-bottomed flask equipped with an overhead stirrer and about 0.25 equivalents of formic acid are slowly added forming a layered mixture. The reactor is cooled in an ice bath and a molar excess of 30% hydrogen peroxide solution is added over about 5 min while monitoring the temperature of the solution. The peroxide must be added slowly enough such that the temperature of the solution remains near room temperature or below. Reaction time varies with the type of olefin, stirring rate, peroxide concentration, and temperature. The progress of the reaction must be monitored by taking aliquots of the solution and analyzing by gas chromatography. The product is purified by addition of hexane, discarding the aqueous/formic acid layer, then extracting the hexane solution with a saturated sodium bicarbonate solution repeatedly until the pH of the hexane solution became slightly basic. The hexane solution is then subjected to final washing with water and drying via rotary evaporation and anhydrous sodium sulfate to yield the epoxidized products. More detailed description of this synthesis is available elsewhere [80].

15.2.3 RING-OPENING REACTION OF EPOXIDIZED MATERIALS

The ring-opening reactions of epoxides (Figure 15.4) typically requires a fivefold or greater excess of the carboxylic acid, which functions as both the reactant and solvent, although the use of butanol as a solvent also gives high quality product. Carboxylic acids, from three to eight carbons, were added to a round-bottomed flask equipped with an overhead stirrer. Next, the epoxidized material was added and the reaction mixture was heated to 100°C using an oil bath or heating mantle. As with the epoxidation, these reactions were periodically monitored by injection of aliquots into the gas chromatograph. Typically, the reaction was complete in 7 h. The products of shorter chain carboxylic acids were purified by washing with water and then drying under vacuum. Products made from longer chain carboxylic acids were not washed due to the lower water solubility of the starting materials. The residual acid was removed by rotary evaporation at 40°C followed by Kugel–Rohr evaporation at 70°C. Detailed syntheses and spectra of these products are available elsewhere [56].

15.2.4 SYNTHESIS OF EMO GLYCERIDE ESTER SURFACTANT

Our syntheses started with the synthesis of sodium polyglyceride, followed by trans-esterification of EMO to form a surfactant product (Figure 15.3). The sodium polyglyceride was made by stirring glycerol with NaOH pellets while heating in an oil bath at 140°C for 2 h yielding a gelled product with a viscosity of ~23,000 ± 3,000 mPa/s. Using the catalyst already in the system, the epoxidized materials were added in various molar ratios, and the reactions were run for 10–16 h at 70°C.

15.2.5 STRUCTURAL CHARACTERIZATION

Detailed FTIR and NMR spectra of these products are described in detail elsewhere [54,55,57,81]. Generally, these reactions were monitored by removal of aliquots of solution, which were then analyzed by gas chromatography. These kinetic investigations are also described elsewhere [56].

15.3 PHYSICOCHEMICAL AND PERFORMANCE PROPERTIES

15.3.1 FRICTION AND WEAR INVESTIGATION

The primary and obvious purpose of a lubricant is to reduce friction and wear. The CoF was measured by the traditional ball-on-disk that is commonly used for boundary lubrication testing. It uses a ball in contact with a stationary cylindrical disk, which moves around the disk at a specified speed. Under appropriate conditions, this geometry allows a steady state to be reached, where the surface contact area has increased to the point that harder steel ball cannot create more macroscopic wear or deformation to the softer disk.

In this work, we used a Falex multispecimen friction measurement apparatus. The friction-testing specimen, balls (52,100 steel; 12.7 mm diameter; 64–66 Rc hardness; extreme polish) and disks (1018 steel; 25.4 mm OD; 15–25 Rc hardness; 0.36–0.46 μm roughness), were thoroughly cleaned by consecutive sonication in fresh methylene chloride and in hexane before use. Ball-on-disk experiments were carried out under low speed of 6.22 mm/s (5 rpm) and high load of 181.44 kg at 25°C where both CoF and torque were measured. Similar experiments using a four-ball configuration with a set speed of 1200 rpm and a normal load of 40 kg were also carried out. In these experiments, the diameters of the wear scars on the balls were measured using a digital optical microscope. These tests are described in greater detail elsewhere [53,54].

We tested the lubricants as 0.01 or 0.02 M solutions in hexadecane. The data clearly shows (Figure 15.5) that all of the oleochemicals reduce friction considerably compared to the control (pure hexadecane). These are also comparable to unmodified methyl oleate, indicating that, although we have not significantly improved the lubricity of this already high-performing compound, we have not destroyed that property either.

Using this methodology, CoF values measured for several different concentrations of oleochemicals in hexadecane solution were fitted to an equation derived from the Langmuir model [82] or the Temkin model if an extra variable was used. In either form, it gives the effective adsorption energy of the lubricant. This treatment has also been used in the past [53,83–86], and it yields an adsorption energy, which is useful for comparison of materials. Testing of the olefinic and epoxidized oleochemicals: MO, ML, MLEN, EMO, EMLO, and EMLEN (Table 15.1) yielded different trends.

Within the group of olefinic methyl esters, a clear trend of adsorption energy vs. the amount of unsaturation can be seen. As the lubricant is changed from a

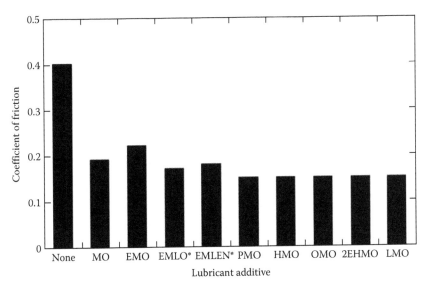

FIGURE 15.5 The measured CoF of 0.01 M solutions in hexadecane of methyl oleate (MO), epoxidized methyl oleate (EMO), epoxidized methyl linoleate (EMLO), epoxidized methyl linolenate (EMLEN), and the compounds from Figure 15.4, in hexadecane solution. *EMLO and EMLEN measurements utilized 0.02 M solutions.

TABLE 15.1
Adsorption Energies of Oleochemicals Calculated by Fitting Temkin Model to CoF Data

Oil or Ester[a]	Adsorption Energy (kJ mol⁻¹)
Oils	
SBO	−15.1
ESO	−13.8
Methyl esters	
MO	−9.6
ML	−10
MLEN	−10.5
EMO	−11.3
EMLO	−9.2
EMLEN	−9.2

[a] SBO, soybean oil; ESO, epoxidized soybean oil; MO, methyl oleate; ML, methyl linoleate; MLEN, methyl linolenate; EMO, epoxidized methyl oleate; EMLO, epoxidized methyl linoleate; EMLEN, epoxidized methyl linolenate.

singly unsaturated compound to one that is triply unsaturated, the adsorption energy becomes stronger. However, the opposite trend is observed as the lubricant is changed from one to three epoxides. The less epoxidized EMO displays the strongest adsorption among the epoxy methyl esters. As for the case of epoxides vs. olefins, the epoxy species adsorbs more strongly in two of the three cases.

This data treatment is useful, but it has some drawbacks. The equation for the adsorption requires mathematical fitting, and data need to be collected at various concentrations over a wide range to accurately calculate the surface coverage (θ). An easier method involves a simple examination of the wear scars on the balls and disk, which can be used as an indicator of lubricant performance. Using any of the lubricating conditions, wear scars of consistent width are carved onto the disk and leaving a scar along one axis of the ball, which will closely match. However, measuring the scar perpendicular to this dimension (termed the y-axis), gives a scar width that indicates lubricant performance and correlates well with calculated adsorption energies. All of the epoxidized materials as well as the olefinic materials studied here can be correlated using this treatment, although the two families of lubricants cannot be accurately fitted to the same trend line. The new method, while being less quantitative than the mathematical models, allows a simple and rapid qualitative comparison for evaluating materials.

A high magnification scanning electron microscope (SEM) was also used to analyze the worn disk surface and provided information on the extent of the wear protection by the lubricants. SEM pictures of disk scars obtained using MO as a lubricant (Figure 15.6) were compared to disk scar where EMO was used as a lubricant (Figure 15.7). The EMO samples show a reduction in dislodged fragments, suggesting the possibility that the structure of EMO results in better protective film formation than that of MO. This also correlates with the stronger adsorption energy of EMO. However, multiple repetitions and control experiments show that the effect was not always observed in a predictable and quantifiable way.

The four-ball tribological investigations tell the same story as the ball-on-disk experiments. The observed wear scars using only hexadecane or poly alpha olefin (PAO) are large, but addition of only 5 wt% of chemically modified fatty acid methyl ester lubrication fluid was able to reduce the wear scar diameter by a significant

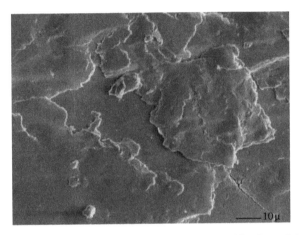

FIGURE 15.6   Scanning electron micrograph at, 1000× magnification, of the disk surface after a ball-on-disk test with 0.02 M methyl oleate in hexadecane.

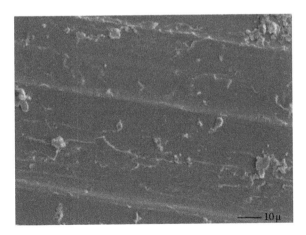

FIGURE 15.7   Scanning electron micrograph at, 1000× magnification, of the disk surface after a ball-on-disk test with 0.02 M EMO in hexadecane.

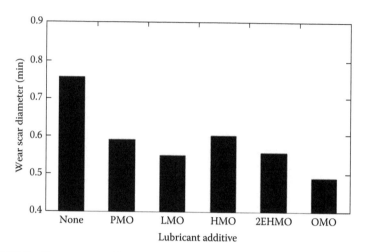

FIGURE 15.8 The wear scar diameter observed in a four-ball experiment with 5 wt% concentration of chemically modified fatty acid methyl esters.

margin (Figure 15.8). In the PAO system, the wear scar diameter was lowered by more than 0.1 mm in all cases, compared to PAO alone (Figure 15.8). Even more impressive is that these materials display similar reduction in wear in hexadecane solution (not shown). However, the effect of the addition of the chemically modified esters to SBO was much less evident since SBO itself is a good lubricant in its natural state. The overall trend is that as the chain length of the ester branch increases, the esters become more effective. One possible reason is increased viscosity due to long branches. The improved tribological properties of ring-opened products may also be attributed to the two extra polar functional groups apart from the ester group of fatty acid ester. Oxygen moieties, like the hydroxyl and ester functionalities at 9th and 10th positions on the fatty acid, help the compounds adsorb onto the substrate surface and reduce friction, especially under excessive load. This mechanism has been demonstrated for the epoxy moieties [17] and could be functioning in the hydroxyl ester compounds as well. As the friction process starts, the active groups of these molecules undergo transformation at the contact zone and develop a stable tribochemical film to protect further wear. The branching structure may also increase the strength of this film, thereby reducing friction and wear. Because of these phenomena, these compounds demonstrate excellent antifriction and antiwear properties on both the ball-on-disk and four-ball tests [84,86]. Overall, there can be no doubt that these chemically modified natural esters can significantly reduce friction and wear.

15.3.2 Viscosity and Low-Temperature Flow Properties

The dynamic viscosity at 25°C was measured on a Brookfield (Middleboro, MA), DV-III programmable Rheometer controlled by Rheocalc 2.4 software using CP-40 spindle and programmed to vary the sheer rate from 0.5 to 10.0 rpm. The kinematic viscosity was measured using Cannon-Fenske calibrated viscometers (Cannon

Instrument Co., State College, PA) in a Cannon temperature bath (CT-1000) at 40°C and 100°C, as per ASTM D445-95 method. Low-temperature flow properties were studied by determining pour points and cloud points as per ASTM D-5949 and ASTM D-5773 methods, respectively, using the phase technology analyzer, Model PSA-70S.

Viscosity index is a measurement which compares the viscosity of a sample at two different temperatures, and is considered a good indicator of the temperature dependence of lubricant viscosity. Lubricants with indices over 100 are often considered suitable for many applications. Unmodified SBO has an impressively high viscosity index that is over 200, while EMO and EMLO also have viscosity index over 125. The epoxy ring-opened products have viscosity index in the range of 80–100, while EMLEN has a very low viscosity index (Table 15.2).

Along with its superior wear-reducing properties, methyl oleate also has very good cold flow characteristics. However, its viscosity is too low for many applications. The kinematic viscosity of methyl oleate at 100°C is below the 2.0 mm² s⁻¹ threshold, which is specified in the viscosity index standard, so its index cannot even be correctly defined. Through the use of chemical modification [54,55,57], we have synthesized materials which all have a significantly increased viscosity. Each of the first two epoxidations, as in EMO and EMLO, give materials with a significantly increased viscosity. However, the third epoxidation, as in EMLEN, gives a large viscosity increase, resulting in a material with a low-viscosity index. Higher viscosity of EMLEN can be attributed to an increase in the interaction of the lone electron

TABLE 15.2
Viscosity and Viscosity Indices of the Epoxidized Compounds and Ring-Opened Esters

Oil[a]	Dynamic Viscosity 25°C (mPa s)	Kinematic Viscosity 40°C (mm² s⁻¹)	Kinematic Viscosity 100°C (mm² s⁻¹)	Viscosity Index
MO[b]	NA	4.5	1.6	NA
EMO	12.4	8	2.5	151
EMLO	24.8	14.3	3.5	125
EMLEN	1488.5	308	19.3	63
PMO	38.2	21.1	4.0	84
HMO	48.5	NA	NA	NA
OMO	46.5	26.9	4.8	101
2EHMO	52.5	27.7	4.7	82
LMO	132.6	57.9	7.5	89

[a] MO, methyl oleate; EMO, epoxidized methyl oleate; EMLO, epoxidized methyl linoleate; EMLEN, epoxidized methyl linolenate; PMO, propionic ester of methyl hydroxy-oleate; HMO, hexanoic ester of methyl hydroxy-oleate; OMO, octanoic ester of methyl hydroxy-oleate; 2EHMO, 2-ethyl hexyl ester of methyl hydroxy-oleate; LMO, levulinic ester of methyl hydroxy-oleate.

[b] MO values are from various literature sources.

pairs on the extra oxygen groups in the epoxy compounds. There is also a strong possibility that there is some viscosity-increasing polymer present in the EMLEN sample. The ring-opened products all have viscosities in the range of SBO, ~50 mPa s at 50°C. This is a good value adjusting the viscosity of a variety of lubricant blends. Overall, from a viscosity standpoint, either epoxidation or epoxidation with ring opening give suitable material.

The cold flow properties tell a different story (Table 15.3). Unmodified methyl oleate has good cold temperature flow characteristics, which get better in methyl linoleate (pour point −48°C) and methyl linolenate (pour point −60°C). Unfortunately, upon epoxidation, the materials generally display very unsatisfactory pour point and cloud point properties. EMLEN displays a much lower cloud point as expected from the molecular structure where it can be seen that the larger number of epoxide groups will cause considerable kinking in the chain inhibiting crystallization. The pour points of the epoxy oleochemicals indicate that they will flow at temperatures above 0°C, and therefore are suitable for work at room temperature. However, these pour points would limit use of these materials in outdoor winter conditions. The ring-opening reaction allows for the synthesis of products, which not only regain the methyl oleate pour point values, but actually surpass them by using longer chain carboxylic acids [54,55,57].

TABLE 15.3

Pour Points and Cloud Points of Epoxidized and Ring-Opened Ester Compounds Synthesized in This Work

Oil[a]	Pour Point (°C)	Cloud Point (°C)
MO	−21	−15
EMO	0	4.1
EMLO	−1.5	2.6
EMLEN	−7.5	−36.6[b]
PMO	−15	−9
HMO	−18	−11
OMO	−33	−31
2EHMO	−24	−19
LMO	−5	−5

[a] MO, methyl oleate; EMO, epoxidized methyl oleate; EMLO, epoxidized methyl linoleate; EMLEN, epoxidized methyl linolenate; PMO, propionic ester of methyl hydroxy-oleate; HMO, hexanoic ester of methyl hydroxy-oleate; OMO, octanoic ester of methyl hydroxy-oleate; 2EHMO, 2-ethyl hexyl ester of methyl hydroxy-oleate; LMO, levulinic ester of methyl hydroxy-oleate.

[b] High viscosity may have affected the pour point and cloud point readings of the instrument.

This may be because branching on the fatty acid chains can disrupt the tendency of fatty acid chains to form macrocrystalline structures through uniform stacking at low temperatures. This explains the success of the approach of using the ring-opening reaction of epoxides to improve the low-temperature flow behavior of fatty acid esters by attaching ester branching at the double-bond sites of the fatty acid chains.

15.3.3 OXIDATIVE STABILITY

DSC is a useful techniques in evaluating the effect of temperature on properties of materials. These changes are represented as exothermic or endothermic peaks as a function of temperature. In general, chemical decomposition and oxidation cause exothermic peaks, while physical properties such as melting and boiling cause endothermic peaks. Pressurized DSC (PDSC) is an effective way of measuring the oxidative stability of lubricant base oils, vegetable oils, and oleochemicals, in an accelerated mode [87]. At high air pressure, the concentration of oxygen is in excess and at equilibrium with the sample. Thus, any inconsistency due to the access rate of oxygen and egress rate of volatile degradation products is effectively eliminated. PDSC was used to measure the temperature at which each material starts to show oxidation. This temperature is called the onset temperature (OT), and it is an effective measure of a lubricant's oxidative stability. It is obtained by extrapolating the tangent drawn on the steepest slope of reaction exotherm to the baseline. A high OT suggests a high oxidation stability of the oil.

Oxidative stability of oils was measured using PDSC 2910 thermal analyzer from TA Instruments, which was calibrated using the melting point of indium metal (156.6°C) before use. The measurements were performed under pressurized dry air (1379 kPa). The OT and signal maximum temperature (SMT) of oxidation were calculated from the exotherm in each case giving a good measure of the oxidation properties of the materials.

Another method of oxidation stability evaluation is the commonly called Rancimat® test, which was used by the European Committee for Standardization, method EN14112 "Determination of oxidative stability (accelerated oxidation test)," of 2003. The method measures the amount of time, the induction period (IP), that it takes for the oil to oxidize. To measure this, a Metrohm USA, Inc (Riverview, FL) model 743 Rancimat instrument is used with an air flow rate (10 L/h) of air through 3 ± 0.01 g of sample and the block temperature set to 110°C with a correction factor (ΔT) of 1.5°C. The glass conductivity measuring vessel contains 50 ± 0.1 mL of distilled water. Conductivity was measured in a cell which trapped volatiles from the oxidation. The IP was mathematically determined as the inflection point of a computer-generated plot of conductivity (μS/cm) of distilled water vs. time. Methyl oleate is relatively stable to oxidation showing an oxidative stability test result of 2.8 h [31]. However, the polyunsaturated methyl linoleate and methyl linolenate are very susceptible, giving test results of only 0.94 and ~0.0 h, respectively [17]. Different studies on the oxidation of SBO and SBO-based products have shown the same trend [3,6,7,15,88,89]. More unsaturation always leads to faster oxidation. The PDSC method [90,91] also gives the same trend, where oxidation of the three materials starts at 177°C, 139°C, and 117°C, respectively (Table 15.4). Epoxidation is a

TABLE 15.4
Oxidative Stability of Compounds in This Work:
PDSC OT, TFMO Volatilization Percentage, and
Tetrahydrofuran Insoluble Deposit Formation

Oil[a]	OT (°C)	Volatile Loss 200°C (wt%)	Deposit Formation 200°C (wt%)
MO	177	NA	NA
Methyl linoleate	139	NA	NA
Methyl linolenate	117	NA	NA
EMO	189	87	5
EMLO	180	82	5
EMLEN	131	40	12
PMO	175	80	<2
OMO	160	55	<2
2EHMO	166	70	<2
LMO	162	40	<2[b]

[a] MO, methyl oleate; EMO, epoxidized methyl oleate; EMLO, epoxidized methyl linoleate; EMLEN, epoxidized methyl linolenate; PMO, propionic ester of methyl hydroxy-oleate; HMO, hexanoic ester of methyl hydroxy-oleate; OMO, octanoic ester of methyl hydroxy-oleate; 2EHMO, 2-ethyl hexyl ester of methyl hydroxy-oleate; LMO, levulinic ester of methyl hydroxy-oleate.

[b] LMO did form significant deposits at 225°C.

simple chemical modification, which can improve this shortcoming of oleochemically based lubricants [54]. Epoxidation increases these values by 12–41°C increasing the range of applications, especially for the less stable materials. The ring-opened products, while not as stable as the epoxides, still show considerably greater stability than that of methyl linoleate or methyl linolenate. Among the ring-opened products, PMO shows the highest OT, followed by 2EHMO, LMO, and OMO. These data show that oxidative stability decreases with an increase in the chain length of the ester side chain. One of the possible reasons may be that the longer side chains have more easily accessible sites for oxidation leaving them more susceptible to cleavage. Similar results were reported in some of the earlier studies also on chemically modified vegetable oils [10] and synthetic esters [92] where short chain acids were found to be more stable than longer chain acids.

Oxidation tests were also conducted using the TFMO test method. This method is designed to study the volatility and deposit-forming tendencies of the modified oil samples. In most lubrication applications, the lubricant functions as a thin film so the TFMO test is considered the test of choice to simulate the actual conditions [19]. The TFMO test has also been shown to correlate well with high-temperature bearing tests and other tests where evaporation can occur. TFMO can also correlate well with both the PDSC method [54,90] and the time-consuming rotary bomb oxidation test (RBOT). During the oxidation process, some small primary oxidation products are measured as volatile loss, while others, in the presence of excess oxygen, undergo oxy-polymerization to form insoluble deposits. Unsaturated oleochemicals have a higher tendency to form such deposits, which limits their use in high-temperature lubricants. Using the volatile loss and insoluble deposit data obtained from TFMO tests, the amount of oil left for lubrication can be predicted, giving an indication of useful lubricant lifetime.

TFMO was performed by oxidizing a small amount of oil (25 μL) as a thin film on a freshly polished high carbon steel surface with a steady flow (20 cm³/min) of dry air. The tests were done at various temperatures between 150°C and 225°C for 120 min. Both volatile loss and tetrahydrofuran-insoluble deposits were measured.

In this study, TFMO showed slightly different results from PDSC. The volatile loss decreases with increasing epoxy rings in the molecule, possibly due to increasing molecular weight and viscosity. At 200°C, it can be seen that the volatile loss of EMLEN is less than half of that of the closest competitor. In general, the volatile loss of the epoxides is significantly larger, except for the value for EMLEN. However, this apparently good value for EMLEN was actually caused by the increased formation of deposits. TFMO deposit results show no significant oxidative degradation occurring up to 175°C for all epoxy samples [54]. This OT for deposit formation is very similar to the PDSC OT. The percent insoluble deposit for EMO and EMLO became stable after 200°C with EMO remaining lower compared to the other oils. A sharp increase in the deposit formation of EMLEN after175°C was observed, which suggests a rapid breakdown of the epoxy group leading to oxidative polymerization through oxygen bonding. There is no appreciable change in deposit formation even after 175°C for EMO and EMLO. Larger insoluble deposits were observed in EMLEN compared to the other samples, which may be due to higher epoxy content. The presence of more epoxy rings in the EMLEN are attractive sites for reaction with primary oxidation products, which results in more polymerization leading to more insoluble deposits and less volatile product formation. In other words, EMLEN still undergoes decomposition, but to a nonvolatile material.

Based on TFMO results, it appears that the longer chain esters, OMO, 2EHMO, are the most oxidatively stable. With an increase in temperature, the volatile loss also increases. This indicates that breakdown of ring-opened products, as a result of oxidative degradation, increases with temperature. At 150°C, volatile loss for ring-opened products is less than half that of EMO. Negligible amounts of insoluble deposits were observed up to 200°C during oxidation process of the ring-opened products, indicating stability against polymerization. Up to 200°C, the percent insoluble deposit for EMO is greater than for the ring-opened products. However, above 200°C, there was an increase in the deposit formation (4–23%) for the ring-opened products. This suggests that above 200°C, oxidative polymerization is occurring, perhaps through the generation of reactive oxygen containing radicals. From the volatile loss data, the longer side chains have more sites for reactive cleavage and they also have more sites for possible polymerization reactions. When the side chains were chemically different, as in the case for LMO, this extra reactivity was especially evident. Because LMO's carbonyl functionality allows the formation of different polymer types, it displays a sharp increase in the deposit formation (24%).

15.3.4 SURFACTANTS FROM EPOXIDIZED METHYL OLEATE AND GLYCEROL

Materials utilizing an epoxidized natural oil have been synthesized in the past [93,94], but the methodology has not been pushed to its limit. Because there is at least the potential for a large surplus of glycerol from the biodiesel industry [95], the synthesis of a glycerol-based surfactant family is of interest.

The surfactant systems were investigated using dynamic surface tension measurements with a Sita T-60 bubble pressure tensiometer using Sita online V2.1 software. Static surface tension and critical micelle concentration were measured on a Krüss K10T digital tensiometer. The duNouy ring method of surface tension measurement was used. Emulsion droplet size was measured by a Malvern Mastersizer/E instrument equipped with a small volume sample delivery unit. Data were analyzed and processed using Malvern Mastersizer software. Drop size is reported as volume mean diameter (VMD), where 50% of the volume of the dispersed liquid is contained in drops of the VMD or smaller.

Using very simple reaction methodology, we were able to synthesize a variety of surfactant materials (Figure 15.3) from EMO and glycerol [76]. These materials could reduce the surface tension of water to ~34 mN m^{-1} and had hydrophiles of 2–7 glyceride units as determined by NMR spectroscopy. Their HLB values ranged from 7 to 13, as calculated by the Griffin formula [96]. This shows that

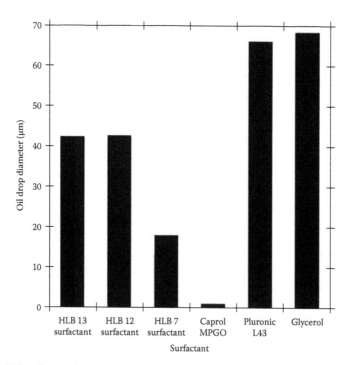

FIGURE 15.9 Comparison of measured droplet diameters of oil in water emulsions prepared with surfactants synthesized in this work, Pluronic® L43, and Caprol® MPGO, and glycerol, which has no surfactant capability. In general, smaller drop size is an indication of a superior emulsifier.

these materials have properties in the range applicable for use as oil-in-water emulsifiers. Tests on these materials showed they could stabilize a 2% SBO in water emulsion yielding a drop size (Figure 15.9) as small as 18 μm. This is good compared to a nonsurfactant, glycerol, 68 μm oil drops, and even an ethylene oxide/propylene oxide-based surfactant, which gave 66 μm diameter oil drops. However, a commercial emulsifier specifically designed for this type of emulsion had a much better result, 0.8 μm diameter drops. Considering the ease of this synthesis, this is a promising result, but much more work needs to be done to improve the emulsifying power of these systems.

15.4 CONCLUSION

Epoxidation has proven itself to be a valuable synthetic method for the production of value-added products from naturally derived esters. This chapter has reviewed some of the recent work which has produced valuable lubrication fluids and surfactants. Epoxidation of SBO is also a commercially viable reaction, which is commonly practiced in the synthesis of polymer plasticizers. Using this readily available commercial material, the esters which we used in this work can be easily attained [97]. Additionally, a commercially available ester, Vikoflex 4050, has also been shown to have some of the excellent lubricity similar to that of EMO, EMLO, and EMLEN [53].

Perhaps the most valuable lesson here is that natural oils are versatile. A good example not shown here, is the ability to synthesize an acetal (ketal) using the epoxy fatty acid methyl ester. This has the potential to be used as a valuable lubricant [98], or as an acid-cleavable extraction surfactant [99]. Just one more example of the bright future of biobased products.

ACKNOWLEDGMENTS

We would like to acknowledge Donna I. Thomas and Richard H. Henz for assistance in the work presented here.

The use of trade, firm, or corporation names in this publication is for the information and convenience of the reader. Such use does not constitute an official endorsement or approval by the United States Department of Agriculture or the Agricultural Research Service of any product or service to the exclusion of others that may be suitable.

REFERENCES

1. A. R. Lansdown, *Lubrication and Lubrication Science*. ASME Press, New York (2004).
2. C. Barrett, Perception Vs. Reality, *Tribol. Lubr. Technol.* 63, 28–35 (2007).
3. J. P. Cosgrove, D. F. Church, and W. A. Pryor, The Kinetics of the Autoxidation of Polyunsaturated Fatty Acids, *Lipids* 22, 299–304 (1987).
4. J. W. Harris, Relative Rates of Grease Oxidation in a Penn State Microoxidation Apparatus on Glass and on Steel Sample Pans, *NLGI Spokesman* 65, 18 (2002).

5. N. Canter, Developing a New Type of Antioxidant, *Tribol. Lubr. Technol.* 61, 10–12 (2005).
6. A. R. Coscione and W. E. Artz, Vegetable Oil Stability at Elevated Temperatures in the Presence of Ferric Stearate and Ferrous Octanoate, *J. Agric. Food Chem.* 53, 2088–2094 (2005).
7. B. K. Sharma, A. Adhvaryu, J. M. Perez, and S. Z. Erhan, Influence of Composition on Thermo-Oxidative and Tribochemical Behavior, *J. Agric. Food Chem.* 53, 2961–2968 (2005).
8. W. E. Artz, P. C. Osidacz, and A. R. Coscione, Acceleration of the Thermoxidation of Oil by Heme Iron, *J. Am. Oil Chem. Soc.* 82, 579–584 (2005).
9. S. Z. Erhan, A. Adhvaryu, and B. K. Sharma, Chemically Functionalized Vegetable Oils. In: *Synthetics, Mineral Oils, and Biobased Lubricants Chemistry and Technology*, L. R. Rudnick (ed.), pp. 14–30, CRC Press, Boca Raton, FL (2005).
10. B. K. Sharma, A. Adhvaryu, Z. Liu, and S. Z. Erhan, Chemical Modification of Vegetable Oils for Lubricant Applications, *J. Am. Oil Chem. Soc.* 83, 129–136 (2006).
11. S. Z. Erhan, A. Adhvaryu, and Z. Liu, Chemically Modified Vegetable Oil-Based Industrial Fluid, U.S. Patent 6583302 (2003).
12. H.-S. Hwang, A. Adhvaryu, and S. Z. Erhan, Preparation and Properties of Lubricant Basestocks from Epoxidized Soybean Oil and 2-Ethylhexanol, *J. Am. Oil Chem. Soc.* 80, 811–815 (2003).
13. H.-S. Hwang and S. Z. Erhan, Modification of Epoxidized Soybean Oil for Lubricant Formulations with Improved Oxidative Stability and Low Pour Point, *J. Am. Oil Chem. Soc.* 78, 1179–1184 (2001).
14. H.-S. Hwang and S. Z. Erhan, Lubricant Base Stocks from Modified Soybean Oil. In: *Biobased Industrial Fluids and Lubricants*, S. Z. Erhan and J. M. Perez (eds.), pp. 20–34, AOCS Press, Champaign, IL (2002).
15. S. Z. Erhan, B. K. Sharma, and J. M. Perez, Oxidation and Low Temperature Stability of Vegetable Oil-Based Lubricants, *Ind. Crops Prod.* 24, 292–299 (2006).
16. B. K. Sharma, A. Adhvaryu, S. K. Sahoo, A. J. Stipanovic, and S. Z. Erhan, Influence of Chemical Structures on Low-Temperature Rheology, Oxidative Stability, and Physical Properties of Group II and III Base Oils, *Energy Fuels* 18, 952–959 (2004).
17. A. Adhvaryu and S. Z. Erhan, Epoxidized Soybean Oil as a Potential Source of High-Temperature Lubricants, *Ind. Crops Prod.* 15, 247–254 (2002).
18. B. K. Sharma, A. Adhvaryu, J. M. Perez, and S. Z. Erhan, Biobased Grease with Improved Oxidation Performance for Industrial Application, *J. Agric. Food Chem.* 54, 7594–7599 (2006).
19. A. Adhvaryu, B. K. Sharma, H. S. Hwang, S. Z. Erhan, and J. M. Perez, Development of Biobased Synthetic Fluids: Application of Molecular Modeling to Structure–Physical Property Relationship, *Ind. Eng. Chem. Res.* 45, 928–933 (2006).
20. B. K. Sharma, J. M. Perez, and S. Z. Erhan, Soybean Oil-Based Lubricants: A Search for Synergistic Antioxidants, *Energy Fuels* 21, 2408–2414 (2007).
21. B. K. Sharma, Z. Liu, A. Adhvaryu, and S. Z. Erhan, One-Pot Synthesis of Chemically Modified Vegetable Oils, *J. Agric. Food Chem.* 56, 3049–3056 (2008).
22. E. R. Booser, Lubrication and Lubricants. In: *Lasers to Mass Spectrometry, Kirk-Othmer Encyclopedia of Chemical Technology*, J. I. Kroschwitz and M. Howe-Grant (eds.), pp. 463–517, John Wiley & Sons, New York (1995).
23. R. J. Sturwold and F. O. Barrett, Water Soluble Triglyceride Compositions and Method for Their Preparation, U.S. Patent 3970569 (1976).
24. G. Hillion and D. Proriol, Synthesis of a High-Grade Lubricant from Sunflower Oil Methyl Esters, *OCL—Oleagineux Corps Gras Lipides* 10, 370–372 (2003).
25. E. Durak and F. Karaosmanoglu, Using of Cottonseed Oil as an Environmentally Accepted Lubricant Additive, *Energy Sources* 26, 611–625 (2004).
26. I. Gawrilow, Vegetable Oil Usage in Lubricants, *Inform* 15, 702–705 (2004).

27. A. P. Pratap, A. S. Kadam, and D. N. Bhowmick, Modified Oils and Fats as Biolubricants, *Inform* 16, 282–285 (2005).
28. W. B. Wan Nik, F. N. Ani, H. H. Masjuki, and S. G. Eng Giap, Rheology of Bio-Edible Oils According to Several Rheological Models and Its Potential as Hydraulic Fluid, *Ind. Crops Prod.* 22, 249–255 (2005).
29. M. A. Farajzadeh, M. Ebrahimi, A. Ranji, E. Feyz, V. Bejani, and A. A. Matin, Hplc and Gc Methods for Determination of Lubricants and Their Evaluation in Analysis of Real Samples of Polyethylene, *Microchim. Acta* 153, 73–78 (2006).
30. G. Knothe, Biodiesel and Renewable Diesel, *Inform* 19, 149–152 (2008).
31. G. Knothe, Designer Biodiesel: Optimizing Fatty Ester Composition to Improve Fuel Properties, *Energy Fuels* 22, 1358–1364 (2008).
32. I. Karabulut, S. Yaprak, and M. Kayahan, Determination of Changes in Some Physical and Chemical Properties of Soybean Oil During Hydrogenation, *Food Chem.* 81, 453–456 (2003).
33. G. R. List, W. C. Byrdwell, K. R. Steidley, R. O. Adlof, and W. E. Neff, Triacylglycerol Structure and Composition of Hydrogenated Soybean Oil Margarine and Shortening Basestocks, *J. Agric. Food Chem.* 53, 4692–4695 (2005).
34. K. Mondal and S. B. Lalvani, Mediator-Assisted Electrochemical Hydrogenation of Soybean Oil, *Chem. Eng. Sci.* 58, 2643–2656 (2003).
35. B. R. Moser, M. J. Haas, J. K. Winkler, M. A. Jackson, S. Z. Erhan, and G. R. List, Evaluation of Partially Hydrogenated Methyl Esters of Soybean Oil as Biodiesel, *Eur. J. Lipid Sci. Technol.* 109, 17–24 (2007).
36. T. W. Findley, D. Swern, and J. T. Scanlan, Epoxidation of Unsaturated Fatty Materials with Peracetic Acid in Glacial Acetic Acid Solution, *J. Am. Chem. Soc.* 67, 412–414 (1945).
37. W. R. Schmits and J. G. Wallace, Epoxidation of Methyl Oleate with Hydrogen Peroxide, *J. Am. Oil Chem. Soc.* 31, 363–365 (1954).
38. P. A. Z. Suarez, M. S. C. Pereira, K. M. Doll, B. K. Sharma, and S. Z. Erhan, Epoxidation of Methyl Oleate Using Heterogeneous Catalyst, *Ind. Eng. Chem. Res.* 48, 3268–3270 (2009).
39. M. Guidotti, N. Ravasio, R. Psaro, E. Gianotti, S. Coluccia, and L. Marchese, Epoxidation of Unsaturated Fames Obtained from Vegetable Source Over Ti(IV)-Grafted Silica Catalysts: A Comparison between Ordered and Non-Ordered Mesoporous Materials, *J. Mol. Catal. A* 250, 218–225 (2006).
40. S. A. Grabovskiy, N. N. Kabal'nova, C. Chatgilialoglu, and C. Ferreri, Epoxidation of Polyunsaturated Fatty Acid Double Bonds by Dioxirane Reagent: Regioselectivity and Lipid Supramolecular Organization, *Helv. Chim. Acta* 89, 2243–2253 (2006).
41. G. L. Crocco, W. F. Shum, J. G. Zajacek, and H. S. J. Kesling, Epoxidation Process, U.S. Patent 5166372 (1992).
42. K. D. Carlson, R. Kleiman, and M. O. Bagby, Epoxidation of Lesquerella and Limnanthes (Meadowfoam) Oils, *J. Am. Oil Chem. Soc.* 71, 175–182 (1994).
43. A. Campanella, M. A. Baltanás, M. C. Capel-Sánchez, J. M. Campos-Martín, and J. L. G. Fierro, Soybean Oil Epoxidation with Hydrogen Peroxide Using an Amorphous Ti/SiO₂ Catalyst, *Green Chem.* 6, 330–334 (2004).
44. S. Sinadinovic-Fiser, M. Jankovic, and Z. S. Petrovic, Kinetics of In Situ Epoxidation of Soybean Oil in Bulk Catalyzed by Ion Exchange Resin, *J. Am. Oil Chem. Soc.* 78, 725–731 (2001).
45. Z. S. Petrovic, A. Zlatanic, C. C. Lava, and S. Sinadinovic-Fiser, Epoxidation of Soybean Oil in Toluene with Peroxoacetic and Peroxoformic Acids—Kinetics and Side Reactions, *Eur. J. Lipid Sci. Technol.* 104, 293–299 (2002).
46. C. Orellana-Coca, D. Adlercreutz, M. M. Andersson, B. Mattiasson, and R. Hatti-Kaul, Analysis of Fatty Acid Epoxidation by High Performance Liquid Chromatography Coupled with Evaporative Light Scattering Detection and Mass Spectrometry, *Chem. Phys. Lipids* 135, 189–199 (2005).

47. J. A. Nowak, T. A. Zillner, I. Mullin, and L. Patrick, Thin-Film Epoxidation of an Unsaturated Oil or Alkyl Fatty Acid Ester, U.S. Patent 6734315 (2004).
48. G. J. Piazza and T. A. Foglia, One-Pot Synthesis of Fatty Acid Epoxides from Triacylglycerols Using Enzymes Present in Oat Seeds, *J. Am. Oil Chem. Soc.* 83, 1021–1025 (2006).
49. Z. Liu, S. Z. Erhan, and J. Xu, Preparation, Characterization and Mechanical Properties of Epoxidized Soybean Oil/Clay Nanocomposites, *Polymer* 46, 10119–10127 (2005).
50. R. L. Shogren, Z. Petrovic, Z. Liu, and S. Z. Erhan, Biodegradation Behavior of Some Vegetable Oil-Based Polymers, *J. Polym. Environ.* 12, 173–178 (2004).
51. C. K. Hong and R. F. Wool, Development of a Biobased Composite Material from Soybean Oil and Keratin Fibers, *J. Appl. Polym. Sci.* 95, 1524–1538 (2005).
52. J. La Scala and R. P. Wool, Effect of Fa Composition on Epoxidation Kinetics of Tag, *J. Am. Oil Chem. Soc.* 79, 373–378 (2002).
53. T. L. Kurth, B. K. Sharma, K. M. Doll, and S. Z. Erhan, Adsorption Behavior of Epoxidized Fatty Esters via Boundary Lubrication Coefficient of Friction Measurements, *Chem. Eng. Commun.* 194, 1065–1077 (2007).
54. B. K. Sharma, K. M. Doll, and S. Z. Erhan, Oxidation, Friction Reducing, and Low Temperature Properties of Epoxy Fatty Acid Methyl Esters, *Green Chem.* 9, 469–474 (2007).
55. K. M. Doll, B. K. Sharma, and S. Z. Erhan, Friction Reducing Properties and Stability of Epoxidized Oleochemicals, *CLEAN—Soil Air Water* 36, 700–705 (2008).
56. K. M. Doll, B. K. Sharma, and S. Z. Erhan, Synthesis of Branched Methyl Hydroxy Stearates Including an Ester from Biobased Levulinic Acid, *Ind. Eng. Chem. Res.* 46, 3513–3519 (2007).
57. B. K. Sharma, K. M. Doll, and S. Z. Erhan, Ester Hydroxy Derivatives of Methyl Oleate: Tribological, Oxidation and Low Temperature Properties, *Bioresour. Technol.* 99, 7333–7340 (2008).
58. S. Z. Erhan, K. M. Doll, and B. K. Sharma, Method of Making Fatty Acid Ester Derivatives, U.S. Patent 20080154053 (2008).
59. A. Campanella and M. A. Baltanás, Degradation of the Oxirane Ring of Epoxidized Vegetable Oils in Liquid-Liquid Systems: II. Reactivity with Solvated Acetic and Peracetic Acids, *Latin Am. Appl. Res.* 35, 211–216 (2005).
60. A. Campanella and M. A. Baltanas, Degradation of the Oxirane Ring of Epoxidized Vegetable Oils in Liquid-Liquid Heterogeneous Reaction Systems, *Chem. Eng. J.* 118, 141–152 (2006).
61. B. R. Moser and S. Z. Erhan, Synthesis and Evaluation of a Series of α-Hydroxyethers Derived from Isopropyl Oleate, *J. Am. Oil Chem. Soc.* 83, 959–963 (2006).
62. B. R. Moser and S. Z. Erhan, Preparation and Evaluation of a Series of α-Hydroxyethers from 9,10-Epoxystearates, *Eur. J. Lipid Sci. Technol.* 109, 206–213 (2007).
63. B. R. Moser, B. K. Sharma, K. M. Doll, and S. Z. Erhan, Diesters from Oleic Acid: Synthesis, Low Temperature Properties, and Oxidation Stability, *J. Am. Oil Chem. Soc.* 84, 675–680 (2007).
64. X. Pages-Xatart-Pares, C. Bonnet, and O. Morin, Synthesis of New Derivative from Vegetable Oil Methyl Esters via Epoxidation and Oxirane Opening. In: *Recent Developments in the Synthesis of Fatty Acid Derivatives*, G. Knothe and J. T. P. Derksen, (eds.), AOCS Press, Champaign, IL (1999).
65. L. Spitz, The History of Soaps and Detergents. In: *Sodeopec, Soaps, Detergents, Oleochemicals, and Personal Care Products*, L. Spitz (ed.), pp. 1–72, AOCS Press, Champaign, IL (2004).
66. R. Tesser, M. Di Serio, E. Santacesaria, M. Guida, and M. Nastasi, Kinetics of Oleic Acid Esterification with Methanol in the Presence of Triglycerides, *Ind. Eng. Chem. Res.* 44, 7978–7982 (2005).

67. L. Cohen and F. Trujillo, Synthesis, Characterization, Surface Properties of Sulfoxylated Methyl Esters, *J. Surfact. Deterg.* 1, 335–341 (1998).
68. L. Cohen and F. Trujillo, Performance of Sulfoxylated Fatty Acid Methyl Esters, *J. Surfact. Deterg.* 2, 363–365 (1999).
69. N. C. Foster, Manufacture of Methyl Ester Sulfonates and Other Derivatives. In: *Sodeopec, Soaps, Detergents, Oleochemicals, and Personal Care Products*, L. Spitz (ed.), pp. 261–287, AOCS Press, Champaign, IL (2004).
70. D. G. Krzysik and J. M. Utschig, Liquid Cleanser Compositions, U.S. Patent 2005239669 (2005).
71. S. Shmad, Mes Overview, *Inform* 17, 13 (2006).
72. L. Cohen, F. Soto, C. Pratesi, and L. Faccetti, Sulfoxidation of Fatty Acid Methyl Esters: Conversion and Selectivity, *J. Surfact. Deterg.* 9, 47–50 (2006).
73. J. L. Berna, Raw Materials for Sulfonation and Sulfation: Production, Characteristics, and Uses. In: *Sodeopec, Soaps, Detergents, Oleochemicals, and Personal Care Products*, L. Spitz (ed.), pp. 238–260, AOCS Press, Champaign, IL (2004).
74. M. A. Sibila, M. C. Garrido, J. A. Perales, and J. M. Quiroga, Ecotoxicity and Biodegradability of an Alkyl Ethoxysulphate Surfactant in Coastal Waters, *Sci. Total Environ.* 394, 265–274 (2008).
75. D. W. Roberts, L. Giusti, and A. Forcella, Chemistry of Methyl Ester Sulfonates, *Inform* 19, 2–9 (2008).
76. K. M. Doll and S. Z. Erhan, Synthesis and Performance of Surfactants Based on Epoxidized Methyl Oleate and Glycerol, *J. Surf. Deterg.* 9, 377–383 (2006).
77. K. M. Doll, B. R. Moser, B. K. Sharma, and S. Z. Erhan, Current Uses of Vegetable Oil in the Surfactant, Fuel, and Lubrication Industries, *Chem. Oggi/Chem. Today* 24, 41–44 (2006).
78. K. M. Doll and S. Z. Erhan, Polyol and Amino Acid-Based Biosurfactants, Builders, and Hydrogels. In: *Biobased Surfactants and Detergents Synthesis, Properties, and Applications*, D. G. Hayes, K. Dai, D. K. Y. Solaiman, and R. D. Ashby (eds.), pp. 425–448, AOCS Press, Urbana, IL (2009).
79. S. P. Bunker and R. P. Wool, Synthesis and Characterization of Monomers and Polymers for Adhesives from Methyl Oleate, *J. Polym. Sci. Part A: Polym. Chem.* 40, 451–458 (2002).
80. K. M. Doll and S. Z. Erhan, Synthesis of Carbonated Fatty Methyl Esters Using Supercritical Carbon Dioxide, *J. Agric. Food Chem.* 53, 9608–9614 (2005).
81. K. M. Doll, B. R. Moser, and S. Z. Erhan, Surface Tension Studies of Alkyl Esters and Epoxidized Alkyl Esters Relevant to Oleochemically Based Fuel Additives, *Energy Fuels* 21, 3044–3048 (2007).
82. I. Langmuir, The Constitution and Fundamental Properties of Solids and Liquids. II. Liquids, *J. Am. Chem. Soc.* 39, 1848–1906 (1917).
83. T. L. Kurth, J. A. Byars, S. C. Cermak, B. K. Sharma, and G. Biresaw, Non-Linear Adsorption Modeling of Fatty Esters and Oleic Estolide Esters via Boundary Lubrication Coefficient of Friction Measurements, *Wear* 262, 536–544 (2007).
84. T. L. Kurth, G. Biresaw, and A. Adhvaryu, Cooperative Adsorption Behavior of Fatty Acid Methyl Esters from Hexadecane via Coefficient of Friction Measurements, *J. Am. Oil Chem. Soc.* 82, 293–299 (2005).
85. A. Adhvaryu, G. Biresaw, B. K. Sharma, and S. Z. Erhan, Friction Behavior of Some Seed Oils: Biobased Lubricant Applications, *Ind. Eng. Chem. Res.* 45, 3735–3740 (2006).
86. G. Biresaw, A. Adhvaryu, and S. Z. Erhan, Friction Properties of Vegetable Oils, *J. Am. Oil Chem. Soc.* 80, 697–704 (2003).
87. B. K. Sharma and A. J. Stipanovic, Development of a New Oxidation Stability Test Method for Lubricating Oils Using High-Pressure Differential Scanning Calorimetry, *Thermochim. Acta* 402, 1–18 (2003).

88. A. S. Colakoglu, Oxidation Kinetics of Soybean Oil in the Presence of Monoolein, Stearic Acid and Iron, *Food Chem.* 101, 724–728 (2007).
89. Y. Watanabe, E. Ishido, X. Fang, S. Adachi, and R. Matsuno, Oxidation Kinetics of Linoleic Acid in the Presence of Saturated Acyl L-Ascorbate, *J. Am. Oil Chem. Soc.* 82, 389–392 (2005).
90. R. O. Dunn, Effect of Antioxidants on the Oxidative Stability of Methyl Soyate (Biodiesel), *Fuel Process. Technol.* 86, 1071–1085 (2005).
91. R. O. Dunn, Antioxidants for Improving Storage Stability of Biodiesel, *Biofuel Bioprod. Biorefin.* 2, 304–318 (2008).
92. S. J. Randals, Esters. In: *Synthetic Lubricants and High Performance Functional Fluids*, L. R. Rudnick and R. L. Shubkin (eds.), pp. 63–102, Marcel Dekker, New York (1999).
93. S. Warwel, M. Rüsch Gen. Klaas, H. Schier, F. Brüse, and B. Wiege, Surfactants from Glucamines and Epoxy Fatty Acid Esters, *Eur. J. Lipid Sci. Technol.* 103, 645–654 (2001).
94. G. Biresaw, Z. S. Liu, and S. Z. Erhan, Investigation of the Surface Properties of Polymeric Soaps Obtained by Ring-Opening Polymerization of Epoxidized Soybean Oil, *J. Appl. Polym. Sci.* 108, 1976–1985 (2008).
95. J. B. Rattay, Glycerine-How Sweet It Is, *Inform* 17, 285 (2006).
96. W. C. Griffin, Calculation of Hlb Values of Non-Ionic Surfactants, *J. Soc. Cosmet. Chem.* 5, 249–256 (1954).
97. R. A. Holser, Transesterification of Epoxidized Soybean Oil to Prepare Epoxy Methyl Esters, *Ind. Crops Prod.* 27, 130–132 (2008).
98. J. Filley, New Lubricants from Vegetable Oil: Cyclic Acetals of Methyl 9,10-Dihydroxystearate, *Bioresour. Technol.* 96, 551–555 (2005).
99. M. Iyer, D. G. Hayes, and J. M. Harris, Synthesis of Ph-Degradable Nonionic Surfactants and Their Applications in Microemulsions, *Langmuir* 17, 6816–6821 (2001).

16 Vaporization and Carbonization Tendency of Vegetable Oils as a Function of Chemical Composition: Morphology of Carbon Deposits on Steel Surfaces at Elevated Temperature

Leslie R. Rudnick, Mohamed A. Abdellatif,
Ömer Gül, and Girma Biresaw

CONTENTS

ABSTRACT

The objective of this study was to investigate the volatility of a series of vegetable oils and to relate the results to the vegetable oil fatty acid profile and
deposit forming tendency. Since the amount of maximum deposit is related
to what remains to carbonize, volatility is a contributing factor in the overall result. Volatility was measured using thermogravimetric analysis (TGA).
In the thermal treatment experiments, the vegetable oils were thermally and
oxidatively treated at high temperatures (250°C). The temperature regime of
these studies was in the range of crown head and upper piston temperatures in
a conventional and high performance spark ignition (SI) engine. Differences
in deposit forming tendencies of the oils studied were significant. These large
differences in deposit-forming tendencies under the same conditions can be
attributed to the differences in the chemical composition of the vegetable
oils. The amount of carbonaceous deposit left on a stainless steel (SS) 304
strip was found to be directly proportional to the amount of palmitic and
total saturated acids present in the vegetable oil. Carbon deposit was found
to be independent of the amount of stearic acid present in the vegetable oil.
Oxidative stability was found to be dependent on the total unsaturated fatty
acid amount. Different deposit morphologies or surface coverages were
observed from vegetable oils having different fatty acid compositions. In general, after thermo-oxidative treatment the SS 304 surface was covered more
uniformly with vegetable oils having higher oleic acid content vegetable oils,
whereas a layered surface coverage was seen with vegetable oil containing
higher palmitic acid content. A sponge-like deposit morphology was observed
with coconut oil, a vegetable oil that is solid at room temperature having 90%
total saturates.

16.1 INTRODUCTION

The reality of today's lubricant world is that many of the environmental issues that
have been discussed for decades are now being regulated into reality. The design features of low toxicity, biodegradability, low ash, and low volatility are being considered
during the design phase of formulation rather than as an afterthought in the process.
In Europe, the implementation of REACH (Registration, Evaluation, Authorization,
and Restriction of Chemical substances) now requires the registration of lubricants
with a greater degree of detail concerning the above factors. REACH was originally
conceived as a means of encouraging changes in manufacturing practices so as to

"ensure a high level of protection of human health and the environment as well as the free movement of substances, on their own, in preparations and in articles, while enhancing competitiveness and innovation" [1].

Commercially available lubricants can be considered to fall into three main categories: mineral oils, synthetics, and biobased oils [2]. Mineral oils can be considered as all of those oils derived from extraction or distillation of petroleum at the refinery. Synthetic fluids are those lubricants that are formed from the chemical reactions of smaller monomers or chemical entities by specific reactions. Biobased materials are those that come from natural sources and can be considered renewable.

Mineral oils and vegetable oils as well as most synthetic lubricants are mixtures of molecules and are not pure compounds that have sharp boiling or melting points. Only a few pure and generally expensive synthetics are composed of one or a few components. For most synthetics, for example, there is some benefit in having a mixture of compound molecules. By having a mixture of structures present in a lubricant, there is a reduction in the pour point based on the fact that structures with different shapes are less prone to orienting in such a way as to crystallize and becoming slow to pour. For example, decene-based polyalphaolefins (PAOs) are discrete mixtures of C30, C40, C50 molecules, but there are many isomers within each group. The pour points are much lower than they would be if only one isomer was present.

For a single component lubricant, the volatility is a property that is fundamental to its molecular structure. Most lubricants are mixtures of a few to thousands of components. In general, the volatility of a mixture is related to the molar concentration of the most volatile components under the conditions of the thermal treatment.

Volatility is also affected by the components of mixtures present in mineral oils, vegetable oils, and synthetics. Vegetable oils are a complex mixture of different triglycerides. The triglycerides differ from each other in the composition and structure of the fatty acid residues, that is, the fatty acids from which the triglycerides are made. The fatty acid residues can have structural variations in chain lengths, degrees of unsaturation, and relative concentrations within a triglyceride as well as among triglycerides in the same vegetable oil. The mix of triglycerides in vegetable oils also varies depending on the source of the crop from which it is extracted. Thus, vegetable oil from a specific crop variety (e.g., corn) will have its own characteristic composition of triglyceride mixture that is distinguishable from other crops [3]. Vegetable oils are generally less volatile than mineral oils and synthetics. This is because most of the structures of most of the components present in vegetable oil mixtures are triacylglycerides with approximately similar molecular weight and are generally higher than those of comparable synthetic and petroleum-based oils.

Most oils used in engine and industrial lubricants are mineral oils derived from petroleum sources. Synthetic lubricants offer significant advantages over mineral oils in several applications; however, cost and biodegradability are disadvantages of many synthetics. Lately, interest has shifted to the replacement of mineral oils with vegetable oils in many applications [4–6] because mineral oils have limited biodegradability and, therefore, have a greater tendency to persist in the environment [6]. Vegetable oils are both biodegradable and renewable and can replace mineral oils in many industrial applications where temperatures are moderate.

Vegetable oils have shown potential as biodegradable lubricants in applications that include engine oils, hydraulic fluids, and transmission oils [7–12]. Rudnick [13] and Erhan and Asadauskas [14] have reported on lubricant base stocks based on vegetable oils. Environmentally acceptable hydraulic fluids, based on vegetable oil-base fluids, have been reported by several researchers [15–17].

Studies have shown that in normal equipment operation, including in bearings, gears, and piston rings, temperatures rarely go above 150°C [18,19]. Automotive design aimed at fuel economy, however, has resulted in higher engine temperatures for passenger car engine oils and for high-performance diesel engine oils. This is partly due to smaller engines with smaller oil sumps aimed at reducing weight and aerodynamic drag. There are shifts to exhaust gas recirculation (EGR) and other modes of operation where engine temperatures are more severe resulting in enhanced oil oxidation. Thus, thermal and oxidative stabilities of lubricants need to be addressed to accommodate these changes.

Normal sump temperatures are generally below 150°C. For example, engine sump temperatures under normal loads are of the order of 100°C–120°C for conventional automotive systems, and about 90°C for high-performance diesel engines. These temperatures are increasing with increases in performance requirements and the constant desire to lower vehicle weight by designing smaller, hotter running engines.

Based on reported temperatures in combustion engines, lubricants experience temperatures much higher than 150°C [18]. Lubricants, even if for only short residence times, experience temperatures through the autoxidation range (below 260°C) and into the intermediate range (260–480°C). Evaporation of lower boiling components begins at much lower temperatures [20].

These repetitive excursions into the combustion zone decrease the useful life of oil and result in carbonaceous deposits that cause wear and reduce performance. These may only represent transient temperatures for a small portion of the oil; however, it should be noted that at the molecular level, these higher temperatures are sufficient to break covalent bonds, initiate free-radical reactions, and permit reactions to occur that may have very low rates at lower temperatures.

However, vegetable oils are known to have less oxidative stability than mineral oils and synthetic base fluids. This is due to the fact that they contain unsaturated fatty acids having allylic hydrogens that are oxidatively unstable compared to fully saturated hydrocarbons. Saturated vegetable oils, while oxidatively more stable, exhibit poor low-temperature properties relative to the higher branched hydrocarbons. This results in some vegetable oils, notably palm and coconut oils, being cloudy or solid at room temperature, which makes them, on first inspection, poor candidates for lubricants.

From previous studies, it has been shown that the amount of residue deposited depends on the metal substrate and also on the chemical structure of the oil [20]. Thermal stability refers to the resistance of the oil to decomposition at elevated temperatures to form deleterious solid deposits. Oxidative stability refers to the resistance to degradation of the oil in the presence of air or oxygen.

The objective of this study was to investigate the volatility of a series of vegetable oils and to relate the results to the vegetable oil fatty acid profile and deposit forming tendency. Since the amount of maximum deposit is related to what remains to carbonize, volatility is a contributing factor in the overall result.

This chapter describes the volatility of biobased lubricants derived from vegetable sources using thermogravimetric analysis (TGA) methods. TGA was conducted in air at 200 psi (1 psi = 6.89 kPa). This study also examines the effects of thermal and oxidative stressing on the stability of the vegetable oil lubricants. The data gathered in this study are related to results reported earlier [21] on the same fluids where these oils were thermally and oxidatively stressed. By modifying a technique originally developed for thermal stressing of fuels [22] the oils in this study were thermally stressed and the amounts of carbon residue deposited on a stainless steel metal coupon (SS 304) were measured.

16.2 EXPERIMENTAL

16.2.1 Materials

The following seed oils were investigated: refined, bleached, and deodorized (RBD) soybean oil; refined canola oil; refined corn oil; refined cottonseed oil; refined sunflower oil; refined palm oil; and refined coconut oil. (Note: Palm and coconut oils were not fully refined.) The refined seed oils were obtained from commercial sources and used as supplied. The viscosities and suppliers of the seed oils used in this study are listed in Table 16.1. The samples of seed oils used in this study were taken from the same batch used in our previous study [21], so that a direct comparison with the

TABLE 16.1
Kinematic Viscosity and Suppliers of Seed Oils Used in This Study

Oil	Kin Viscosity at 40°C (cSt)[a]	Supplier
Soybean, refined	31	Avatar Co., University Park, IL
Canola, refined	34	Welch, Holmes and Clark, Co. Newark, NJ
Corn, refined	32	Welch, Holmes and Clark, Co. Newark, NJ
Cottonseed, refined	40[b]	Welch, Holmes and Clark, Co. Newark, NJ
Sunflower, refined	30	Welch, Holmes and Clark, Co. Newark, NJ
Palm, fully refined	(solid at RT)	Fuji Vegetable Oil, Inc. Savannah, GA
Coconut, fully refined	28	Fuji Vegetable Oil, Inc. Savannah, GA
Rapeseed, refined	40	Welch, Holmes and Clark, Co. Newark, NJ
Sesame, refined	35	Welch, Holmes and Clark, Co. Newark, NJ

Source: Gül, Ö. and Rudnick, L.R., Vegetable oil fatty acid composition and carbonization tendency on steel surfaces at elevated temperature: Morphology of carbonaceous deposits, In: *Surfactants in Tribology*, G. Biresaw and K.L. Mittal, eds., pp. 291–307, CRC Press, Boca Raton, FL, 2008. With permission.

[a] Data for unrefined seed oils from Ref. [3].
[b] Data for unrefined cottonseed oil from Ref. [29].

earlier results could be made. In addition to the refined seed oils, polyalpha olefin (PAO) (Durasyn 166, ~31 cSt at 40°C, BP–Amoco, Naperville, IL) and solvent-refined highly paraffinic ISO VG 32 mineral oil (150 N oil, 30 cSt at 40°C, American Refining Group, Bradford, PA) were also investigated. The Noack reference oil (RL 172/5), was obtained from Ametek Petrolab (Albany, NY) and has a Noack evaporation loss of 14.5%.

Stainless steel strips used in thermal stress experiments were obtained from Goodfellows (Devon, PA). The SS specifications are as follows: AISI 304—foil, Fe/Cr18/Ni10, Item No. FE220323, typical analysis: Cr 17–20%, Mn <2%, Ni 8–11%, C <800 ppm, Fe balance.

The gas used for reactor purging before the thermal treatment experiments was ultrahigh purity (UHP) argon. The gas used in the LECO carbon analyzer was UHP oxygen and the gas used for the pressure differential scanning calorimetry (PDSC) experiments was UHP air.

16.2.2 METHODS

16.2.2.1 Volatility Analysis

Volatility of the oil samples was measured using the American Society for Testing and Materials (ASTM) Standard D6375–99a (TGA Noack Method), on a TA Instruments 2050 TGA (TA Instruments, New Castle, DE). The first step in this procedure involves measuring the Noack reference time, that is, the time required for the Noack standard sample to achieve the specified evaporation loss as specified by Ametek Petrolab who provided the standard. The Noack reference oil used here (RL 172/5) had a listed Noack evaporation loss of 14.5%.

To determine the time needed for the sample to lose 14.5%, the standard sample (~20 mg) was held at 50°C for 2 min to equilibrate and was then heated to 249°C in air atmosphere at 30°C/min, and held isothermal for 20 min. The resulting weight versus time thermogram was used to determine the time needed to attain evaporation loss of (14.5%) as specified by the manufacturer, where the time was found to be 8.56 min. This is the time used to determine the weight loss of our samples, which directly relates to Noack volatility.

The Noack volatility method used to determine the volatility of the oil samples was carried out in a two-step procedure, where the fist step is considered screening of the sample and the second is the actual step used to determine its volatility behavior. First, the thermogram of the oil sample was generated by heating oil samples (20–25 mg) in platinum TGA pan (TGA 2050, TA Instruments, New Castle, DE) in an air atmosphere up to 50°C and allowing it to equilibrate for 2 min followed by heating at 30°C/min up to 251°C, are then holding it for 20 min. A second test, on the same sample, was conducted with a fresh sample, and the TGA thermograms (in duplicate runs) were examined to determine the volatility of the sample, which is the percentage weight loss at 8.56 min of the sample heated at 251°C, the Noack reference time. The data from the duplicate measurements were used to determine the average volatility and standard deviation for each sample. An overlay of thermograms of the Noack standard oil along with high- and low-volatile samples from this study is illustrated in Figure 16.1.

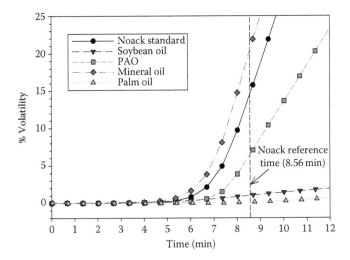

FIGURE 16.1 Noack volatility of oils with varying chemical structures but similar viscosities (30–35 cSt at 40°C): mineral oil, polyalphaolefin, soybean, and palm oils.

16.2.2.2 Decomposition Kinetics

Decomposition kinetics analysis was conducted on a TA Instruments 2050 TGA (TA Instruments, New Castle, DE). Samples (~20 mg) were heated in air atmosphere from ambient temperature to 800°C, at three different heating rates: 10, 15, and 20°C/min. The peak degradation temperature plotted against the percentage weight loss for each heating rate was analyzed according to ASTM Standard E1641 method, which is part of the TA Specialty Library software provided with the instrument. In this method, activation energy is calculated using the Flynn–Walls equation [23]:

$$\log \beta \cong 0.457 \left(-\frac{Ea}{RT} \right) + \left[\log \left(\frac{AEa}{R} \right) - \log F(b) - 2.315 \right], \qquad (16.1)$$

where
β is the heating rate (K/min)
T is the absolute temperature (K)
R is the gas constant (8.314×10^{-3} kJ/mol/K)
b is the conversion (weight %)
Ea is the activation energy (kJ/mol)
A is the pre-exponential factor

According to this equation, Ea can be obtained from the slope of the plot of log β versus 1000/T (K) for each percent conversion per minute. Activation energy values were calculated for degradation (conversion) ranging from 10% to 50%, in 5% increments. A typical example of data from TGA decomposition experiment displaying the effect of temperature on sample weight (weight %) and derivative of weight

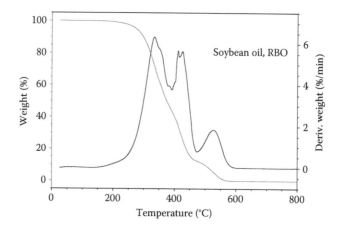

FIGURE 16.2   Weight (%) and derivative of weight (%/min) from typical TGA decomposition kinetic experiment on RBD soybean oil.

(weight % per min) is illustrated in Figure 16.2. The data in Figure 16.2 are for RBD soybean oil and were obtained at a heating rate of 10°C/min.

16.2.2.3   Thermal Treatment

A detailed description of the flow reactor system used in these studies has been given earlier [22]. In these earlier studies [22] on the thermal and oxidative stability of fresh jet fuel, we used a procedure whereby fuel was pumped once through the reactor during the experiment. In the present work, the study was performed in the absence of oxygen, using 100 mL vegetable oil that was recirculated continuously through the reactor for the duration of the experiment. This allows the oil to come in contact with the metal strip many times during the test. The reactor was allowed to run for 5 h, and subsequently, the metal strip was removed and analyzed. This method represents a regime where the oil is repetitively thermally treated during the experiment, as opposed to the tests on fuels that receive only one thermal treatment.

The vegetable oils were tested on SS 304 strips. The metal strips, 15 cm × 3 mm × 0.6 mm, were cut and rinsed in acetone. Thermal treatment was carried out at 250°C oil temperature (approximately 260°C wall temperature) in the presence of SS 304 strip. The oil was subjected to heating for 5 h at a 4 mL/min oil flow rate. The metal strip was placed in a 26.5 cm and 0.635 cm outside diameter (OD) glass-lined stainless steel tube reactor. Throughout the experiment, the reactor outlet temperature and oil flow rate were maintained at 250°C and 4 mL/min, respectively, for 5 h of circulating oil. The SS preheating section was 2 mm inner diameter (3.175 mm o.d.) and 61 cm in length. The oil residence time in this preheating zone was 22 s at a liquid oil flow rate of 4 mL/min. The oil residence time in the reactor (4 mm i.d., 6.35 mm o.d. and 31.75 cm length) was 59 s at the same oil flow rate. At the end of the reaction period (5 h), the strips were cooled under an argon flow in the reactor.

16.2.2.4 Temperature Programmed Oxidation

In the above thermal treatment experiments, deposits were collected on a SS 304 strip. After removing the SS 304 strip from the reactor, it was first carefully rinsed with acetone to remove residual vegetable oil without removing the deposits.

The total amount of carbon deposits on the metal strips was determined using a LECO RC-412 Multiphase Carbon Analyzer (LECO Corporation, St. Joseph, MI). Conventionally, this instrument has been used to measure the amount of deposit on metal surfaces [22]. In the carbon analyzer, carbon in the deposit is oxidized to carbon dioxide by reaction with ultrahigh pure O_2 in a furnace over a CuO catalyst bed. The CO_2 product is quantitatively analyzed as a function of furnace temperature using a calibrated infrared (IR) detector. In this study, the deposit on the metal surface was characterized by temperature-programmed oxidation (TPO). In this procedure, the metal coupon with the deposit was heated at a rate of 30°C/min in flowing O_2 (750 mL/min) to a maximum temperature of 900°C and held for 6 min at the final temperature. Only the deposits collected on the metal strips were measured. The application of the technique has been previously reported [22]. To investigate the morphology of the carbon deposit, 1 cm long pieces were cut from the center of a 15 cm-long Inconel strip and examined using an ISI-DS 130 dual-stage scanning electron microscope (SEM).

16.2.2.5 Pressure Differential Scanning Calorimetry

The oxidative stability of the vegetable oils was determined using PDSC. The baseline and temperature calibration of the instrument was performed using indium, which was used for the standard calibration temperature (156.6°C).

A small amount of oil sample (−1 mg) was weighed into an aluminum flat bottom sample pan. The uncovered pan containing the sample was placed on the left platform (sample platform) in the PDSC cell and an empty reference pan was placed on the right platform. The cell was capped and all gas valves except the outlet valve were closed. The air cylinder regulator was then adjusted to the desired pressure. The inlet valve was slowly opened and air allowed to purge the cell. The air pressure in the cell was then allowed to reach a constant value of 13.6 atm. In the constant pressure operational mode, the outlet valve was opened and adjusted to provide an air flow of 20 mL/min.

The oxidation onset temperature (temperature ramp) was measured at 13.6 atmosphere pressure and 10°C/min heating rate. The sample temperature was programmed to increase at a rate of 10°C/min from 100°C to 320°C. After the test was completed, the air inlet valve on the cell was closed and the pressure was released by opening the pressure release valve.

16.3 RESULTS AND DISCUSSION

The thermal and oxidative stabilities of vegetable oils depend on the composition of the oil, the temperature, the nature of metals, etc. Vegetable oils contain triglycerides, which are naturally synthesized from glycerol and various types of fatty acids such as stearic, palmitic, oleic, linoleic and linolenic acids. Various combinations of these fatty acids are found in the triglycerides of the vegetable oils from various crops. The compositions of the fatty acids in the triglycerides of the vegetable oils investigated in this study are listed in Table 16.2. In Table 16.2, the first seven vegetable oils are liquid at room temperature, whereas the last two (palm and coconut oils) are solid at room temperature.

The chemical structures of some of these fatty acids are shown in Table 16.3. These fatty acids contain 16 (palmitic acid) or 18 carbon atoms with different amounts of unsaturation. Oleic acid (18:1), linoleic acid (18:2), and linolenic acid (18:3) have one, two, and three carbon–carbon double bonds, respectively. Stearic acid has 18 carbon atoms and does not contain any carbon–carbon double bonds. The conventional designation of this is 18:0. Another fatty acid found in most vegetable oils is palmitic acid that contains 16 carbon atoms. Palmitic acid also contains no carbon–carbon double bonds and is designated (16:0).

The vegetable oils investigated in this study contained triglycerides primarily comprising palmitic, oleic, and linoleic acid components. They also contain lower concentrations of other fatty acid components. Palm and coconut oils are cloudy or solid at room temperature because their triglycerides comprise high concentrations of saturated fatty acids (50.0% and 91.2% total saturated fatty acids, respectively). This makes these two vegetable oils poor candidates for low temperature lubricant applications (Table 16.2). On the other hand, the other vegetable oils in this study had 25% or lower total saturates, or conversely, 75% or higher total unsaturated fatty acid compounds in their triglycerides. High concentrations of unsaturated components impart a fluid character to vegetable oils.

Vegetable oils are known to have less oxidative stability than mineral oils and synthetic base fluids [21]. This is directly related to the fact that their triglycerides contained unsaturated fatty acids. Chemically, it is the allylic hydrogens that make vegetable oils oxidatively unstable compared to fully saturated hydrocarbons. Saturated fatty acids, while being oxidatively more stable, exhibit poor low-temperature properties relative to more branched hydrocarbons.

The carbon deposit from oxidative decomposition of vegetable oils on the SS 304 strips, expressed as micrograms of carbon deposit per square centimeter ($\mu g/cm^2$) of strips, is summarized in Table 16.4 and Figure 16.3. A wide variation in the amount of carbon deposits from the different vegetable oils was observed. The vegetable oils studied displayed carbon deposits that varied from 2.0 to 86.3 $\mu g/cm^2$. Sunflower oil gave the least amount of carbon deposit, while cottonseed oil gave the highest amount of carbon deposit.

This large difference in deposit-forming tendency under the same stress conditions can be attributed to the differences in the fatty acid chemical compositions of the triglycerides in the vegetable oils. In this study, the relationship between chemical composition and carbon deposit was investigated. The chemical composition of the lubricant is an important predictor of the formation of deposits. Exposure of vegetable oils to high temperatures triggers free radical reactions that lead to deposition of carbonaceous solids on metal surfaces.

To assess the effect of chemical composition on carbon deposit, the relationship between chemical group concentration and the amount of carbonaceous deposit was analyzed. A plot of carbonaceous deposit versus concentration of palmitic

TABLE 16.2

Fatty Acid Composition of Triglycerides in Vegetable Oils Studied in This Study, (%)[a]

Oil (Refined)	Caprylic 8:0	Capric 10:0	Lauric 12:0	Myristic 14:0	Palmitic 16:0	Stearic 18:0	Total Saturates	Oleic 18:1	Linoleic 18:2	Linolenic 18:3	Gadoleic 20:1	Erucic 22:1	Total Unsaturates
Sunflower					7.0	4.5	11.5	18.7	67.3	0.8			86.8
Sesame					10.7	3.5	14.2	40.0	45.9				85.9
Rapeseed					5.5	1.4	6.9	23.6	14.6	5.6	7.5	41.4	92.7
Canola					10.0	2.9	12.9	58.4	20.4	8.3			87.1
Corn					12.1	1.7	13.8	29.2	57.0				86.2
Soybean					10.6	4.0	14.6	23.2	53.7	7.6			84.5
Cottonseed				0.9	21.9	2.2	25.0	19.3	55.7				75.0
Palm				2.4	44.4	3.4	50.2	36.8	10.6				47.4
Coconut	8.1	6.1	48.0	16.3	9.9	2.3	90.6	7.9	1.5				9.4

a Data from Ref. [3].

TABLE 16.3
Some Structures of Fatty Acids in Vegetable Oil Triglycerides

Fatty Acid	Structure	Number of Carbon Atoms: Unsaturation
Palmitic acid		16:0
Stearic acid		18:0
Oleic acid		18:1
Linoleic acid (omega-6)		18:2
Alpha linolenic acid (omega-3)		18:3

Source: Gül, Ö. and Rudnick, L.R., Vegetable oil fatty acid composition and carbonization tendency on steel surfaces at elevated temperature: Morphology of carbonaceous deposits, In: *Surfactants in Tribology*, G. Biresaw and K.L. Mittal, eds.,
pp. 291–307, CRC Press, Boca Raton, FL, 2008. With permission.

acid component in the triglycerides of the vegetable oils shows that the amount of carbon deposit increases as the palmitic acid concentration increases (Figure 16.4) ($R^2=0.87$). However, similar analysis showed no such correlation between carbon deposit and percent stearic acid (Figure 16.5).

When the amount of carbon deposit is plotted against the percent total saturates, a similar trend to that of palmitic acid was observed, that is, as the total percent saturates increases, the carbon deposit amount increases (Figure 16.6) ($R^2=0.73$). On the other hand, an inverse correlation was observed when amount of carbon deposit was plotted as a function of total percent unsaturates (Figure 16.7) ($R^2=0.73$).

The oxidation temperatures for vegetable oils obtained using PDSC are plotted in Figure 16.8. Oxidation temperature can be used to characterize the oxidative stability of the oils. The oxidation temperatures of the vegetable oils investigated in this study varied in the range of 148.3°C–210.3°C. Among the oils investigated, soybean oil had the lowest, while coconut oil had the highest oxidation temperatures.

In Figure 16.9, vegetable oil oxidation temperatures are plotted against the amount of total unsaturates. The data show an inverse relationship, that is, the oxidation temperature decreases as the total amount of unsaturates increases (Figure 16.9) ($R^2=0.69$). The mechanism of oxidation of vegetable oils is consistent with that of hydrocarbon oxidation. The allylic hydrogens of unsaturated fatty acids are the most susceptible sites for initiation of oxidation.

TABLE 16.4
Thermal Oxidation Temperature and Carbon Deposit Amounts of Biobased Lubricant Oils

Oil	Oxidation Temperature as Determined by Pressure DSC (°C)	Carbon Deposit (μg/cm^2)
Soybean, refined	148.3	38.5
Canola, refined	173.6	5.3
Corn, refined	180.4	31.2
Cottonseed, refined	166.9	86.3
Sunflower, refined	165.3	2.0
Palm, fully refined	191.7	31.3
Coconut, fully refined	210.3	43.2
Rapeseed, refined	170.6	5.5
Sesame, refined	180.9	3.1

Source: Gül, Ö. and Rudnick, L.R., Vegetable oil fatty acid composition and carbonization tendency on steel surfaces at elevated temperature: Morphology of carbonaceous deposits, In: *Surfactants in Tribology*, G. Biresaw and K.L. Mittal, eds., pp. 291–307, CRC Press, Boca Raton, FL, 2008. With permission.

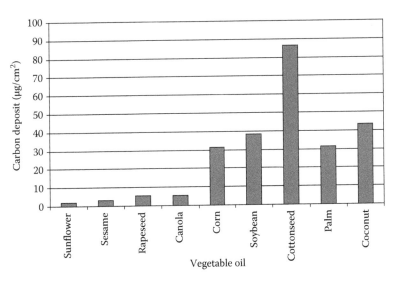

FIGURE 16.3 Carbon deposit amounts from vegetable oils investigated in this study. (From Gül, Ö. and Rudnick, L.R., Vegetable oil fatty acid composition and carbonization tendency on steel surfaces at elevated temperature: Morphology of carbonaceous deposits, In: *Surfactants in Tribology*, G. Biresaw and K.L. Mittal, eds., pp. 291–307, CRC Press, Boca Raton, FL, 2008. With permission.)

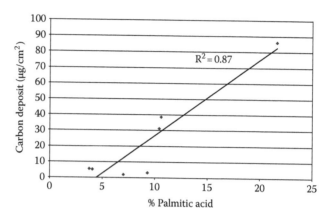

FIGURE 16.4 Effect of palmitic acid content (% w/w) of triglycerides on carbon deposition from vegetable oils. (From Gül, Ö. and Rudnick, L.R., Vegetable oil fatty acid composition and carbonization tendency on steel surfaces at elevated temperature: Morphology of carbonaceous deposits, In: *Surfactants in Tribology*, G. Biresaw and K.L. Mittal, eds., pp. 291–307, CRC Press, Boca Raton, FL, 2008. With permission.)

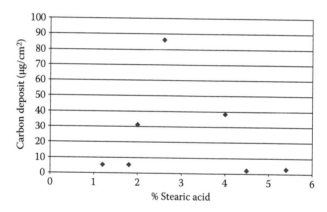

FIGURE 16.5 Effect of stearic acid content (% w/w) of triglycerides on carbon deposition from vegetable oils. (From Gül, Ö. and Rudnick, L.R., Vegetable oil fatty acid composition and carbonization tendency on steel surfaces at elevated temperature: Morphology of carbonaceous deposits, In: *Surfactants in Tribology*, G. Biresaw and K.L. Mittal, eds., pp. 291–307, CRC Press, Boca Raton, FL, 2008. With permission.)

16.3.1 TEMPERATURE-PROGRAMMED OXIDATION AND SCANNING ELECTRON MICROSCOPY EVALUATION OF CARBON DEPOSITS

In this study, the solid carbon deposit morphology observed was different for different vegetable oils depending on their chemical composition. The stressing temperature is also a significant parameter having a major influence not only on the rate of oxidation but also on the type of carbonaceous deposit [24–26].

TPO profiles show peaks as a function of temperature (Figure 16.10). The peaks observed in these plots are related to the nature of the deposits. Peak intensities are

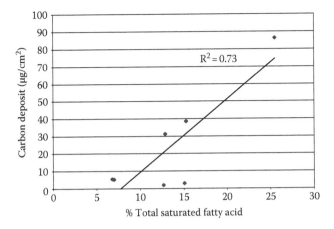

FIGURE 16.6 Effect of total saturated fatty acid content (% w/w) of triglycerides on carbon deposition from vegetable oils. (From Gül, Ö. and Rudnick, L.R., Vegetable oil fatty acid composition and carbonization tendency on steel surfaces at elevated temperature: Morphology of carbonaceous deposits, In: *Surfactants in Tribology*, G. Biresaw and K.L. Mittal, eds., pp. 291–307, CRC Press, Boca Raton, FL, 2008. With permission.)

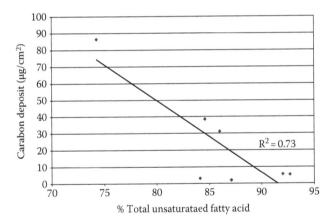

FIGURE 16.7 Effect of total unsaturated fatty acid content (% w/w) of triglycerides on carbon deposition from vegetable oils. (From Gül, Ö. and Rudnick, L.R., Vegetable oil fatty acid composition and carbonization tendency on steel surfaces at elevated temperature: Morphology of carbonaceous deposits, In: *Surfactants in Tribology*, G. Biresaw and K.L. Mittal, eds., pp. 291–307, CRC Press, Boca Raton, FL, 2008. With permission.)

directly proportional to the amount of the carbon deposit. Observations from the TPO and SEM images of the deposits are discussed below.

16.3.2 TPO Profile Evaluation

Figure 16.10 shows the TPO profiles of the carbon deposits from thermal stressing of the different vegetable oils on SS 304 at 250°C and 4 mL/min flow rate for 5 h. Most

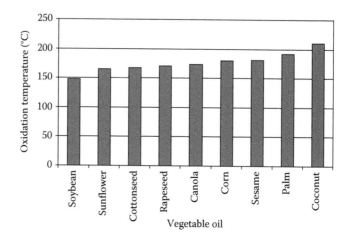

FIGURE 16.8 Oxidation temperature of vegetable oils determined using PDSC (°C). (From Gül, Ö. and Rudnick, L.R., Vegetable oil fatty acid composition and carbonization tendency on steel surfaces at elevated temperature: Morphology of carbonaceous deposits, In: *Surfactants in Tribology*, G. Biresaw and K.L. Mittal, eds., pp. 291–307, CRC Press, Boca Raton, FL, 2008. With permission.)

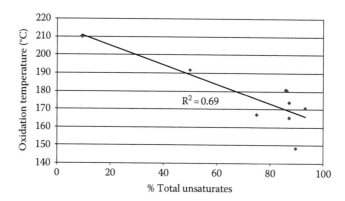

FIGURE 16.9 Effect of total unsaturated fatty acid content (% w/w) of triglycerides on oxidation temperature of vegetable oil determined by PDSC (°C). (From Gül, Ö. and Rudnick, L.R., Vegetable oil fatty acid composition and carbonization tendency on steel surfaces at elevated temperature: Morphology of carbonaceous deposits, In: *Surfactants in Tribology*, G. Biresaw and K.L. Mittal, eds., pp. 291–307, CRC Press, Boca Raton, FL, 2008. With permission.)

of the vegetable oils gave two major peaks: one between 300°C and 350°C (amorphous carbon) and the second between 450°C and 500°C (more ordered amorphous carbon). Each vegetable oil gave a different TPO profile with different relative carbon signals.

TPO profiles in the low temperature region (between 100°C and 200°C) correspond to a less ordered (more reactive and possibly hydrogen-rich chemisorbed carbonaceous substance) deposit that may result from secondary deposition

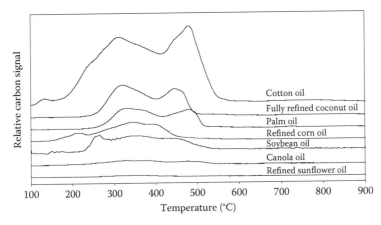

FIGURE 16.10 TPO profiles of carbon deposits from thermal stressing of vegetable oils on stainless steel 304 at 250°C, 5 h, 4 mL/min flow rate. (From Gül, Ö. and Rudnick, L.R., Vegetable oil fatty acid composition and carbonization tendency on steel surfaces at elevated temperature: Morphology of carbonaceous deposits, In: *Surfactants in Tribology*, G. Biresaw and K.L. Mittal, eds., pp. 291–307, CRC Press, Boca Raton, FL, 2008. With permission.)

processes promoted by the presence of incipient carbon [22]. Secondary deposition refers to thermally driven deposit growth on incipient carbon deposit that leads to the thickening and coating of the deposits by pyrolytic carbon formation [27]. Reactive deposits burn off at the lower temperatures, while less-reactive deposits burn off at higher temperatures.

High-temperature peaks (600°C) in the TPO profiles indicate the presence of relatively low-reactive carbon deposits. Highly ordered structures in the deposits, having pre-graphitic or graphitic order, would reduce oxidation reactivity compared to amorphous carbon with no apparent structural order [22,28]. The TPO profiles of the vegetable oils examined in this study display no deposits of this type.

Among the seed oils investigated, cottonseed oil, which has the highest amount of palmitic acid component and the lowest amount of oleic acid component in its triglycerides, gave the highest carbon signal. Sunflower oil, which has a lower amount of palmitic acid component and the highest amount of linoleic acid component in its triglycerides, gave the lowest carbon signal. Cottonseed oil gave the dominant peaks at 300°C and 500°C and gave a relatively high-carbon signal at higher TPO temperature (500°C).

The TPO profile data of the seed oils in Figure 16.10 are arranged so that the lower temperature peak intensity (~300°C) to the higher temperature peak intensity (~500°C) ratio of the oils decreases from bottom (TPO profile of sunflower) to top (TPO profile of cottonseed oil). Higher burn off peak intensity implies that a more ordered carbon deposit is formed.

16.3.3 SEM Image Evaluation

Some vegetable oils (e.g., soybean and canola oils) showed that the SS 304 strip surface covered more uniformly although their unsaturated fatty acid content levels (e.g., oleic and linoleic acid) are quite different (Figures 16.11 and 16.12, respectively). Corn oil, a vegetable oil with a high linoleic acid content, gave a uniformly covered surface and some agglomeration on the covered surface (Figure 16.13).

The SEM image from the thermo-oxidative treatment of the cottonseed oil sample shows layered surface coverage. Of the oils studied (liquid at room temperature), cottonseed oil has the highest palmitic acid content. A defect structure of layered coverage from the thermo-oxidative stressing of cottonseed oil can be seen in Figure 16.14.

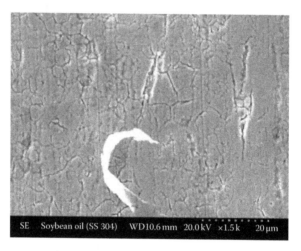

FIGURE 16.11   SEM image of carbon deposit from soybean oil. (From Gül, Ö. and Rudnick, L.R., Vegetable oil fatty acid composition and carbonization tendency on steel surfaces at elevated temperature: Morphology of carbonaceous deposits, In: *Surfactants in Tribology*, G. Biresaw and K.L. Mittal, eds., pp. 291–307, CRC Press, Boca Raton, FL, 2008. With permission.)

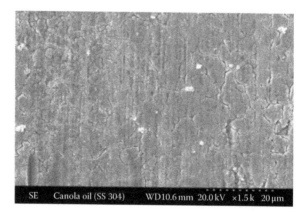

FIGURE 16.12   SEM image of carbon deposit from canola oil. (From Gül, Ö. and Rudnick, L.R., Vegetable oil fatty acid composition and carbonization tendency on steel surfaces at elevated temperature: Morphology of carbonaceous deposits, In: *Surfactants in Tribology*, G. Biresaw and K.L. Mittal, eds., pp. 291–307, CRC Press, Boca Raton, FL, 2008. With permission.)

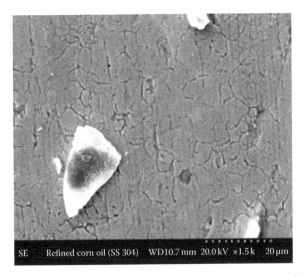

FIGURE 16.13 SEM image of carbon deposit from corn oil. (From Gül, Ö. and Rudnick, L.R., Vegetable oil fatty acid composition and carbonization tendency on steel surfaces at elevated temperature: Morphology of carbonaceous deposits, In: *Surfactants in Tribology*, G. Biresaw and K.L. Mittal, eds., pp. 291–307, CRC Press, Boca Raton, FL, 2008. With permission.)

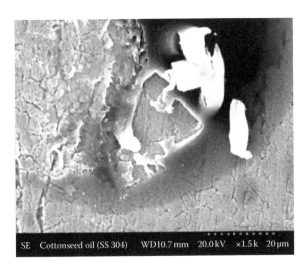

FIGURE 16.14 SEM image of carbon deposit from cottonseed oil. (From Gül, Ö. and Rudnick, L.R., Vegetable oil fatty acid composition and carbonization tendency on steel surfaces at elevated temperature: Morphology of carbonaceous deposits, In: *Surfactants in Tribology*, G. Biresaw and K.L. Mittal, eds., pp. 291–307, CRC Press, Boca Raton, FL, 2008. With permission.)

The two vegetable oils that are solid at room temperature (palm and coconut oils) produced deposits with different surface coverages and different microstructures. Palm oil covered the surface more uniformly (Figure 16.15), while coconut oil produced a sponge-like surface coverage (Figure 16.16). Fatty acid compositions of these two vegetable oils are quite different (Table 16.2). Palm oil has higher palmitic,

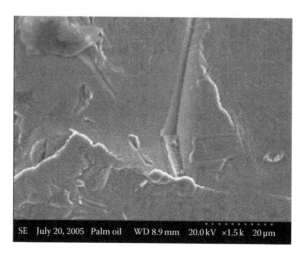

FIGURE 16.15 SEM image of carbon deposit from palm oil. (From Gül, Ö. and Rudnick, L.R., Vegetable oil fatty acid composition and carbonization tendency on steel surfaces at elevated temperature: Morphology of carbonaceous deposits, In: *Surfactants in Tribology*, G. Biresaw and K.L. Mittal, eds., pp. 291–307, CRC Press, Boca Raton, FL, 2008. With permission.)

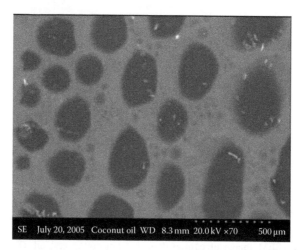

FIGURE 16.16 SEM image of carbon deposit from coconut oil. (From Gül, Ö. and Rudnick, L.R., Vegetable oil fatty acid composition and carbonization tendency on steel surfaces at elevated temperature: Morphology of carbonaceous deposits, In: *Surfactants in Tribology*, G. Biresaw and K.L. Mittal, eds., pp. 291–307, CRC Press, Boca Raton, FL, 2008. With permission.)

oleic, and linoleic acid contents, but coconut oil has higher total saturates (sum of all saturated fatty acid components).

Hydrocarbon composition and stressing conditions both affect the type and the amount of deposit. Different optical textures were observed in samples of deposits formed with different vegetable oils. For example, the deposits formed from cottonseed and coconut oils are obviously quite different (Figures 16.14 and 16.16).

16.3.4 VOLATILIZATION

The Noack volatilization of the purified seed oils determined using TGA method (ASTM D6375–99a) is shown in Figure 16.1, where the volatility time of the control (8.5 min) was used to determine the volatility of the samples. Figure 16.1 shows one sample, mineral oil, with volatility higher than the control, and three samples, PAO, palm oil, and soybean oil, lower than the control. The control represents an average volatility for Noack volatility and, therefore, samples are expected to have volatilities higher and lower than this reference. Chemical structure and refining process can affect volatility. For example, the temperature where a refinery cut is made between two viscosities of oil can affect Noack volatility. Table 16.5 compares the volatility of the purified seed oils with those of PAO and mineral oil of similar viscosities, where refined palm oil exhibited the least weight loss after heating for 8.5 min at 251°C at 30°C/min. Figure 16.17 compares the volatility of the seed oils.

TABLE 16.5
TGA Noack Volatility of Purified Vegetable PAO and Mineral Oils[a]

Description	Kin Visc at 40°C, cSt[b]	Volatility, %
Noack control		14.5
Refined rapeseed oil	40	0.36
Refined palm oil[c]		0.13
Refined corn oil	32	0.21
Refined sesame oil	35	0.25
Refined sunflower oil	30	0.30
Refined canola oil	34	0.22
Fully refined coconut oil	28	0.30
Refined cottonseed oil	40	0.30
Soybean oil, RBD	31	1.01
Polyalphaolefin	30[d]	6.41
Mineral oil (highly paraffinic)	30[e]	19.89

[a] Values are averages of duplicate measurements.
[b] Data from Table 16.1.
[c] Solid at room temperature.
[d] Data from Ref. [30].
[e] Data from Ref. [31].

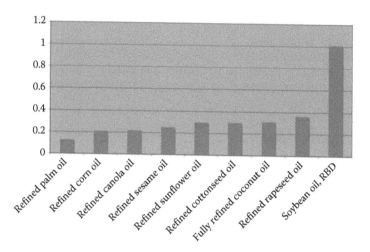

FIGURE 16.17   TGA Noack volatility of purified vegetable oils.

16.3.5 DEGRADATION KINETICS

The degradation activation energy (Ea) of the samples was calculated from the TGA data, where samples were heat degraded. The linear form of Flynn–Walls equation (Equation 16.1) [23] was used for the calculation, where the slope of the line was the Ea. The equation was based on heating samples at different heating rates and the percent conversion of the material, in the form of weight loss, was determined and recorded after 10% of the material had degraded and continued at a 10% increment. The calculated Ea can be used to determine whether the oil sample was degraded in

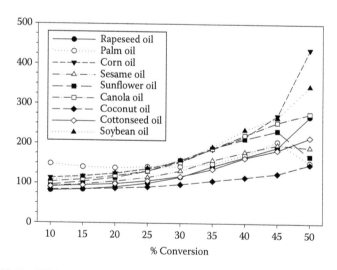

FIGURE 16.18   TGA decomposition kinetics of refined vegetable oils.

one- or multistep process. Ea was plotted against percent conversion, where straight line indicates one-step degradation process. Figure 16.18 compares calculated Ea values of the refined seed oils. As shown in Figure 16.18, the Ea of coconut oil indicates that it was the only oil that displayed a one-step degradation mechanism, indicating a homogeneous structure. Total unsaturation levels of sunflower oil and corn oil are 86.8% and 86.2%, respectively. This means unsaturation is not an effect. The unsaturated fatty acid types (e.g., oleic and linoleic) are quite different. Corn oil and sunflower oil exhibit a multi-step degradation mechanism, that signifies a diverse fatty acid composition, possibly due to unsaturated fatty acid types (e.g., oleic and linoleic). Overall, most oils showed a minimum of two-step degradation mechanism possibly due to the mixed triglyceride composition, that is, the presence of saturates and unsaturates in addition to diverse chain length. The higher Ea reported for palm oil is in agreement with its lower volatility value relative to the other purified seed oils, as shown in Figure 16.17 and Table 16.5.

16.4   CONCLUSION

In this study, the relationships between the amount of carbon deposit or morphology and the chemical composition of the vegetable oil were sought. The differences in carbon forming tendencies for the oils studied were significant. These large differences in deposit-forming tendency under the same stressing conditions can be attributed to the differences in the chemical compositions of the vegetable oils. In this study, it was found that a vegetable oil containing higher amount of palmitic acid or total saturates gave a higher amount of carbon deposits at 250°C. The variation in the fatty acid composition was also noticed in the volatility values as well as in the degradation kinetics of these oils. Except for the mineral oil, the PAO and vegetable oils investigated here displayed volatility properties far below the Noack standard sample. The vegetable oils displayed a multistep heat degradation kinetics, indicating diverse fatty acid composition. Oxidation temperatures of the vegetable oils investigated here were found to be dependent on the amount of total unsaturated fatty acids. Vegetable oils with different fatty acid contents showed different deposit morphologies or surface coverages. The SS 304 strip surface covered more uniformly although their unsaturated fatty acid content levels are quite different. Corn oil, a vegetable oil with a high linoleic acid content, gave a uniformly covered surface and some agglomeration on the covered surface (Figure 16.13). Layered surface coverage was observed with vegetable oils that are liquid at room temperature that contain higher palmitic acid content. A sponge-like deposit morphology was observed with coconut oil, which is solid at room temperature and contains 90% total saturates.

ACKNOWLEDGMENTS

Special thanks to Welch, Holmes and Clarke Co., Avatar Co., BP Amoco and Fuji Vegetable Oil, Inc., for providing the materials used in this study. We thank Quinta Nwanosike and Motunrayo Kemiki for their efforts in these studies. We are grateful to Jason Adkins for help with the TGA measurements.

REFERENCES

1. Annon. Corrigendum to Regulation (EC) No 1907/2006 of the European Parliament and of the Council of 18 December 2006 concerning the Registration, Evaluation, Authorisation and Restriction of Chemicals (REACH), establishing a European Chemicals Agency, amending Directive 1999/45/EC and repealing Council Regulation (EEC) No 793/93 and Commission Regulation (EC) No 1488/94 as well as Council Directive 76/769/EEC and Commission Directives 91/155/EEC, 93/67/EEC, 93/105/ EC and 2000/21/EC. *Official Journal of the European Union, English Edition*, L136, 50, 1–280 (May 29, 2007).
2. L. R. Rudnick (ed.), *Synthetics, Mineral Oils and Bio-Based Lubricants: Chemistry and Applications*, CRC Press, Boca Raton, FL (2006).
3. S. S. Lawate, K. Lal, and C. Huang, Vegetable oils—Structure and performance, In: *Tribology Data Handbook*, E. R Booser (ed.), pp. 103–116, CRC Press, Boca Raton, FL (1997).
4. L. R. Rudnick and W. J. Bartz, Comparison of synthetic, mineral oil, and bio-based lubricant fluids, In: *Synthetics, Mineral Oils and Bio-Based Lubricants: Chemistry and Applications*, L. R. Rudnick (ed.), pp. 331–349, CRC Press, Boca Raton, FL (2006).
5. D. Whitby, Lubricant base stock potential of chemically modified vegetable oils, *Lipid Technol.*, 16, 125–129 (2004).
6. L. R. Rudnick and S. Z. Erhan, Natural oils as lubricants, In: *Synthetics, Mineral Oils and Bio-Based Lubricants: Chemistry and Applications*, L. R. Rudnick (ed.), pp. 353–360, CRC Press, Boca Raton, FL (2006).
7. A. Permsuwan, D. J. Picken, K. D. R. Seare, and M. F. Fox, Vegetable oil-based lubricants—A review of oxidation, *Int. J. Ambient Energy*, 17, 157–161 (1996).
8. L. A. T. Honary, An investigation of the use of soybean oil in hydraulic systems, *Bioresour. Technol.*, 56, 41–47 (1996).
9. J. Z. Adamczewska and D. Wilson, Development of ecologically responsive lubricants, *J. Synthetic Lubr.*, 14, 129–142 (1997).
10. A. Arnsek and J. Vizintin, Lubrication properties of rapeseed-based oils, *J. Synthetic Lubr.*, 16, 281–296 (1999).
11. A. Arnsek and J. Vizintin, Scuffing and load capacity of rapeseed-based oils, *Lubr. Eng.*, 55(8), 11–18 (1999).
12. A. Arnsek and J. Vizintin, Pitting resistance of rapeseed-based oils, *Lubr. Eng.*, 57, 17–21 (2001).
13. L. R. Rudnick, A comparison of synthetic and vegetable oil esters for use in environmentally friendly fluids, In: *Bio-Based Industrial Fluids and Lubricants*, S. Z. Erhan and J. M. Perez (eds.), pp. 46–58, AOCS Press, Champaign, IL (2002).
14. S. Z. Erhan and S. Asadauskas, Lubricant basestocks from vegetable oils, *Ind. Crops Products*, 11, 277–282 (2000).
15. I. Rhee, Evaluation of environmentally acceptable hydraulic fluids, *NLGI Spokesman*, 60, 28–35 (1996).
16. R. P. S. Bisht, G. A. Sivasankaran, and V. K. Bhatia, Vegetable oils as lubricants and additives, *J. Sci. Ind. Res.*, 48, 174–180 (1989).
17. S. Lavate, Environmentally friendly hydraulic fluids, In: *Bio-Based Industrial Fluids and Lubricants*, S. Z. Erhan and J. M. Perez (eds)., pp. 35–46, AOCS Press, Champaign, IL (2002).
18. L. R. Rudnick, R. P. Buchanan, and F. Medina, Evaluation of oxidation-mediated volatility of hydrocarbon lubricant base fluids, *J. Synthetic Lubr.*, 23, 11–26 (2006).
19. M.M. Sukirno, Oxidative degradation of mineral oil under trobcontact and ineffectiveness of inhibitors, In: *Proceedings of ASME/STLE Tribology Conference*, Toronto, Ontario, Canada (1998).

20. H. Kopsch, Thermal methods of petroleum analysis, *Thermochem Acta*, 2, 229–230 (1997).
21. Ö. Gül and L. R. Rudnick, Vegetable oil fatty acid composition and carbonization tendency on steel surfaces at elevated temperature: Morphology of carbonaceous deposits, In: *Surfactants in Tribology*, G. Biresaw and K. L. Mittal (eds.), pp. 291–307, CRC Press, Boca Raton, FL (2008).
22. Ö. Gül, L. R. Rudnick, and H. H. Schobert, Effect of the reaction temperature and fuel treatment on the deposit formation of jet fuels, *Energy Fuels*, 20, 2478–2485 (2006).
23. J. H. Flynn and L. A. Wall, Direct method of determination of activation energy from thermogravimetric data, *Polym. Lett.*, 4, 323–328, (1966).
24. D. L. Trimm, Fundamental aspects of the formation and gasification of coke, In: *Pyrolysis—Theory and Industrial Practice*, L. F. Albright, B. L. Crynes, and W. H. Corcoran (eds.), pp. 203–232, Academic Press, New York (1983).
25. L. F. Albright and J. C. Marek, Mechanistic model for formation of coke in pyrolysis units producing ethylene, *Ind. Eng. Chem. Res.*, 27, 755–759 (1988).
26. R. T. K. Baker, D. J. C. Yates, and J. A. Dumesic, Filamentous carbon formation over iron surfaces, In: *Coke Formation on Metal Surfaces*, L.F. Albright and R.T. Baker (eds.), ACS Symposium Series 202, pp. 1–21, American Chemical Society, Washington, DC (1982).
27. J. Li and S. Eser, Carbonaceous deposit formation on metal surfaces from thermally stressed dodecane, In: *Carbon '95, Extended Abstracts, 22nd Biennial Conference on Carbon*, p. 314, American Carbon Society, San Diego, CA (1995).
28. O. Altin and S. Eser, Analysis of solid deposits from thermal stressing of a JP-8 fuel surfaces in a flow reactor, *Ind. Eng. Chem. Res.*, 40, 596–603 (2001).
29. G. Biresaw and G. Bantchev, Elastohydrodynamic (EHD) traction properties of seed oils, *Tribology Transaction*, 53(4), 573–583 (2010).
30. G. Biresaw and G. Bantchev, Effect of chemical structure on film-forming properties of seed oils, *J. Synthetic Lubrication*, 25(4), 159–183 (2008).
31. S. J. Asadauskas, G. Biresaw, T. G. McClure, Effects of chlorinated paraffin and ZDDP concentrations on boundary lubrication properties of mineral and soybean oils, *Tribol. Lett.*, 37(2), 111–121 (2010).

Part IV

General Topics and Applications

17 Surface Chemistry at the Tribological Interface

Wilfred T. Tysoe

CONTENTS

ABSTRACT

The film growth kinetics and the nature of the surface film found for clean metal samples measured either in ultrahigh vacuum or with a microbalance are used to model the tribological chemistry occurring under extreme pressure conditions for a model lubricant consisting of methylene chloride dissolved in poly α-olefin (PAO). In particular, the variation in seizure load with additive concentration measured in a pin and V-block tribometer can be calculated as a function of additive concentration.

It is also found that a monolayer of film covering the surface can reduce friction to its minimum value and that the shear strength of this film depends on the contact pressure. The friction coefficient increases as the film thickness increases due to the increased contact area between the rough tribopin and the surface, and a final regime is detected when the film thickness exceeds the maximum peak-to-valley distance of the asperities where the load is entirely supported by the film.

17.1 INTRODUCTION

In general, lubricants are rather complex mixtures of components, each of which is added to provide a particular function, for example, to improve the viscosity, prevent the growth of bacteria, or inhibit the formation of foams. This chapter focuses on a class of lubricant additives that react at the tribological interface to form some type of surface film. Depending on the particular application of the lubricant, this film may be required to reduce friction or minimize wear, or both. However, these additives react with the surface to form a film of average thickness X where the resulting film thickness depends on the nature of the additive, the surface, and the reaction conditions. In fact, one of the most widely used antiwear additives, zinc dialkyl dithiophosphate, was originally used as an antioxidant, but it was also found to reduce wear by reacting with the surface to form a film. In the following, we refer to these as "tribofilms," and they are formed by a chemical reaction with the contacting surfaces that are moving relative to each other. The resulting thickness of the film arises from a balance between the rate that it is formed by reaction with the surface and the rate at which it is removed (the wear rate) and is therefore inherently a kinetic process. Accurately modeling the wear rate is a complex issue, but is most simply discussed in terms of a simple Archard wear model [1] where the wear (the amount of material removed) per unit sliding length is proportional to the applied load and inversely proportional to the shear strength or hardness of the materials that constitute the contacting interface. In principle, modeling the film-formation kinetics should be relatively straightforward since chemical kinetic processes are rather well understood [2].

Reaction rates follow a law that depends on the reactant concentrations raised to some power, the reaction orders, and a rate constant. The reaction orders depend on the reaction pathway, and the rate constant is a function of the entropy and enthalpy changes between the transition state and the reactants and is proportional to $\exp(-\Delta G_a/kT))$, where ΔG_a is the Gibbs free energy of activation, T is the temperature, and k the Boltzmann constant. It is implicitly assumed in deriving this equation that the system is in thermal equilibrium. If this were indeed the case, understanding the tribochemical reactions between the additive and the surface would require a knowledge of ΔG_a for a particular additive + surface combination. In practice, the situation is rather more complex. First, in many engineering applications, the surface is rather ill-defined, being covered by, for example, surface oxides or carbonaceous deposits. Second, real surfaces are generally rough. This means that the surface–surface contact does not occur over the whole of the interface but over a relatively small fraction of the surface [3]. One important implication of this effect is that, during sliding, energy is not dissipated uniformly over the whole surface but sporadically at the contacting "asperities" that occur on a relatively small proportion of the surface. Thus, as these asperities ride over each other as two surfaces rub, a large amount of energy is dissipated in a local region for a relatively short amount of time, given roughly by the diameter of the asperity contacts divided by the velocity. Since temperature can be thought of as an energy density, the energy dissipated in the small region of the asperity causes a rapid increase in energy density, often denoted the "flash temperature" [4]. However, the hot contacting asperities are not necessarily in thermal equilibrium with their surroundings since they are likely not to have had sufficient time to thermalize, so that these "flash temperatures" are, in general, not really temperatures in a thermodynamic sense.

A further complication is that reaction selectivities can change substantially with heating rate. If we imagine two reactions from a common starting reactant A, that can produce either product B or C, then if the rate constants of these reactions have Gibbs free energies of activation that have very different entropic and enthalpic contributions, i.e., that have different Arrhenius pre-exponential factors and activation energies, merely varying the heating rate in a kinetic process can drastically change the selectivity from one product to another. There are numerous examples in the literature, but an excellent example for surface reactions is provided by a ruthenium surface covered by a mixture of CO and atomic oxygen [5]. At low heating rates (around 10 K/s), the preferred reaction pathway is the desorption of CO to leave oxygen on the surface. However, if the surface is heated much more rapidly using a laser pulse (at 10^6 K/s), the reaction changes completely so that CO now reacts with adsorbed oxygen to form carbon dioxide. This implies that the surface chemistry measured at constant temperature may differ completely from that at tribological interfaces where the kinetic conditions might be completely different.

It is not surprising, therefore, that it has often been felt that tribological interfaces have peculiar properties and that "tribochemistry" was thought to be different from other surface chemistry. This arises since the tribological interface may not be well defined due to the presence of contaminants, and because energy is dissipated at contacting asperities, which may not be at thermal equilibrium and the rapid heating rates may alter the reaction selectivity. These are extremely complex issues to fully address, so that, as a basis for fully understanding these complexities, regimes must first be sought in which the chemistry at the tribological interface can be fully understood in terms of the surface chemistry measured under conditions of thermodynamic equilibrium. This will allow us to test the proposition that the chemistry occurring at a solid surface can be used to understand the chemistry of lubricant additives at the tribological interface. In order to explore this issue, we classify the tribological regimes in terms of the power P dissipated at the interface. If the applied load is L, the sliding velocity V, and the friction coefficient μ, then $P = LV\mu$. That is, the total power dissipated and, therefore, the severity of the contact conditions will increase with applied load and velocity. This will cause an increase in the temperature of the surface region due to Joule (frictional) heating where the average surface temperature increase is proportional to P. This temperature rise occurs at contacting asperities leading to issues described above. Clearly, one could minimize these problems for a tribological interface if the temperature rise were extremely low, potentially resulting in a surface that remains close to thermodynamic equilibrium. In this case, the loads and sliding velocities would have to be sufficiently small that the surface is not strongly perturbed. Certainly, measurements made using the atomic force microscope (AFM) are likely to fulfill this condition. In engineering applications, brushes in electric motors are designed to operate with low friction coefficients and low applied loads, although in this case shear at the interface may induce surface reactions.

Another region where thermodynamic equilibrium might be found is under very severe tribological conditions. In this regime, if the asperity–asperity contacts occur sufficiently rapidly that they start to overlap, the "flashes" will start to coalesce so that the temperature fluctuations become lower. A point may thus be reached at which the resulting fluctuations are sufficiently small that the interface becomes close to thermal equilibrium once again. It is in this region that the correspondence between surface chemistry and tribochemistry should be sought. This "extreme pressure" (EP) regime, where the contact conditions are severe, is interesting in its own right since the lubricant additives (used for applications such as machining and wire drawing) are generally not environmentally benign so that a conceptual understanding of how they operate will guide the design of more environmentally benign alternatives.

It thus appears that very mild and very severe tribological interfaces will be those for which the surface chemistry determined under conditions of thermal equilibrium should apply. In the intermediate region, it is likely that the situation will be complicated by the existence of "flash temperatures" at contacting asperities. It is anticipated that an understanding of the limiting conditions will eventually enable theses regimes also to be fully understood.

This chapter thus addresses two central questions:

1. Can the tribochemistry of so-called EP additives be understood in terms of their surface chemistry measured under conditions of thermal equilibrium?
2. What are the tribological properties of the resulting film?

These questions are addressed by measuring the surface chemistry of the reactants under ultrahigh vacuum (UHV) conditions, which enables clean, well-characterized surfaces to be obtained. In addition, the high wear rates that occur under EP conditions will ensure that the realistic tribological interface is free from contaminants.

17.2   EXPERIMENTAL SECTION

The surface chemistry was followed, as described elsewhere, under UHV conditions by using molecular beam method [6,7], where the nature of the surface was analyzed using Auger electron spectroscopy or x-ray photoelectron spectroscopy. Experiments were carried out at intermediate pressures using a microbalance, in which mass changes of an iron foil due to film formation were continually monitored. In this case, the resulting samples were analyzed ex situ by a number of techniques including Raman and Mössbauer spectroscopy [8,9]. The friction of well-characterized samples was measured using a UHV tribometer that has been described in detail elsewhere [10,11]. In this case, a thin film could also be evaporated directly onto the surface, where the film deposition rate was measured using a quartz crystal microbalance. Films could also be reactively grown onto metal surfaces in an isolatable cell attached to the vacuum chamber that houses a UHV-compatible tribometer so that films grown on metal surfaces can be transferred directly into the UHV tribometer without intervening exposure to air.

17.3　RESULTS AND DISCUSSION

17.3.1　SURFACE CHEMISTRY OF MODEL CHLORINATED HYDROCARBON ADDITIVES ON IRON SURFACES

As noted above, the first part of this chapter focuses on testing the postulate that the surface chemistry of the materials at the tribological interface (under conditions of thermodynamic equilibrium) can be used to understand the tribochemistry under EP conditions. EP lubricants used with steel generally consist of a base oil that includes chlorine-, sulfur- or phosphorus-containing additives. Accordingly, in the following, we examine the surface chemistry of simple chlorinated hydrocarbons on iron and use the results, in particular the film growth kinetics, to model the tribological properties.

17.3.2　SURFACE CHEMISTRY OF TRIBOLOGICAL ADDITIVES

In order to establish a connection between the surface chemistry of model lubricant additives and their operation under tribological conditions, it is first necessary to understand their surface chemistry. This is initially carried out under UHV conditions, where the background pressure is $\sim 10^{-10}$ Torr. This enables clean surfaces to be obtained and the nature of the reactively formed film to be interrogated without intervening exposure to air. Subsequently, the film growth kinetics is examined under higher pressures of the model additives, using a microbalance in which the film growth rate is monitored by the time dependence of the change in the mass of the sample. These experiments are illustrated in the following using a simple model chlorinated hydrocarbon lubricant additive, methylene chloride (CH_2Cl_2), although similar experiments have been performed for a number of other model additives [12,13].

In order to explore the formation of reactive (tribo) films at elevated temperatures, experiments were carried out using d.c. molecular beams of methylene chloride impinging onto a clean iron surface [6,7]. In this case, a beam of methylene chloride effusing from a capillary source was incident on the surface, and the reaction products were detected using a mass spectrometer placed in line of sight of the sample. The resulting reflected fluxes of methylene chloride (o, 49 amu) and hydrogen (•, 2 amu) are plotted versus sample temperature in Figure 17.1. This reveals that the intensity of the reflected methylene chloride signal is relatively constant up to a temperature of ~700 K but then decreases at higher temperatures. This indicates that there is no continuous reaction of methylene chloride with the surface below ~700 K, and the decrease in signal due to reflected methylene chloride becomes larger at higher temperatures indicating that the surface reaction is temperature dependent. This is accompanied by an increase in the hydrogen signal, and the increase in hydrogen yield exactly follows the decrease in methylene chloride signal, and no other reaction products are found. An analysis of the reactively formed film indicates that it comprises ferrous chloride. This indicates that the surface chemical reaction proceeds as

$$Fe + CH_2Cl_2 \rightarrow FeCl_2 + C + H_2$$

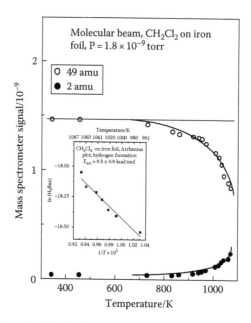

FIGURE 17.1 Results of molecular beam experiments in which a beam of methylene chloride is incident on clean iron foil in UHV, monitoring 49 amu (○) due to methylene chloride and 2 amu (●) due to reactively formed hydrogen. Inset: An Arrhenius plot of ln(rate) versus $1/T$ for the molecular beam data. (From Lara, J. and Tysoe, W.T., *Langmuir*, 14, 307, 1998. With permission.)

Similar simple surface reaction pathways are found for other chlorinated hydrocarbon additives [12,13]. In these cases, the amount of carbon incorporated into the ferrous chloride film merely reflects the stoichiometry of the reactant.

Since the decrease in methylene chloride signal is proportional to the reaction rate, its temperature dependence is plotted in Arrhenius form (ln(rate) versus $1/T$) and the result is shown as an inset to Figure 17.1. The slope of this linear plot yields an activation energy for the reaction of 9.5 ± 0.9 kcal/mol. Thus, the capability of both detecting the reaction products and analyzing the nature of the resulting reactively formed film enables the chemistry of this (relatively simple) reaction to be established.

In order to determine whether a similar reaction occurs at higher pressures, the film growth rate was measured using a microbalance, which continually monitors mass changes of an iron foil immersed in methylene chloride vapor [8,9]. The thickness of the resulting film is proportional to the change in sample mass. In this case, however, it is not possible to measure the gas-phase reaction products as was done for the UHV experiments above. Shown in Figure 17.2a is a plot of the film thickness as a function of time for various reaction temperatures. In all cases, there is an initial rapid growth of a film, which then decreases to zero at longer times. That is, the film growth kinetics is self-limiting. This is clearly an important property of a lubricant additive since continuous reaction with the surface would lead to an eventual complete corrosion of the sample. Again, an activation energy can be

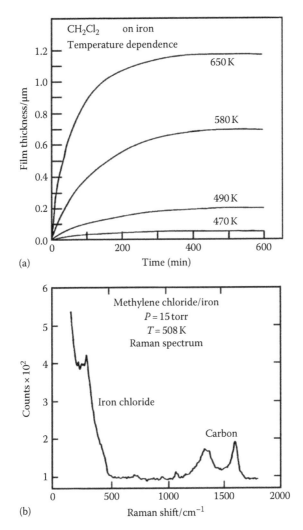

FIGURE 17.2 (a) Plots of film thickness as a function of time for the growth of a film from methylene chloride on iron at a methylene chloride pressure of 15 Torr as a function of reaction temperature, where the temperatures are marked adjacent to the corresponding curves. (b) The Raman spectrum of a film grown from methylene chloride at a pressure of 15 Torr and a sample temperature of 508 K. (From Huezo, L.A. et al., *Langmuir*, 10, 3571, 1994; Kotvis, P.V. et al., *Langmuir*, 9, 467, 1993. With permission.)

obtained by measuring the initial film growth rate from the slope of these curves, and constructing an Arrhenius plot yields an activation energy of 9.8 kcal/mol, in excellent agreement with the value measured above in UHV. Several techniques have been used to analyze the resulting films [8,9]. In particular, Raman spectroscopy (Figure 17.2b) not only reveals the presence of ferrous chloride but also shows that the resulting carbon is incorporated into the film in the form of small carbonaceous particles.

Similar results are obtained for other small, chlorinated hydrocarbons. In addition, it is found that some carbon diffuses into the bulk of the iron substrate, in particular when carbon tetrachloride is used as an additive, and the resulting carbide can also form a tribofilm at high temperatures [14–16].

17.3.3 RELATIONSHIP BETWEEN SURFACE CHEMISTRY AND CHEMISTRY AT TRIBOLOGICAL INTERFACE

The goal is to test the postulate that the surface chemistry of lubricant additives found under conditions of thermodynamic equilibrium (as shown in Section 17.3.2) can be used to model the chemistry under tribological conditions. As noted above, it is anticipated that this can be done for very mild contacts, where the surface is not strongly perturbed and also under conditions where the contact conditions are very severe, under so-called EP conditions.

Since it is difficult to monitor the thickness of the tribological films in situ, tribological experiments were carried out to measure the conditions under which the films failed [14]. Experiments were performed using a pin and V-block configuration in which a pin rotates (at 290 rpm) between two V-shaped blocks. The pin and V-block are immersed in a lubricant consisting of a model additive (in this case, methylene chloride) dissolved in a poly α-olefin (PAO) to mimic the base oil. Initial experiments were carried out with the apparatus enclosed in a nitrogen-purged glove box to eliminate the effect of dissolved gases (oxygen and water vapor), but identical results were obtained with the system in air. In this experiment, the load applied to the V-blocks is increased and the torque needed to maintain a constant angular velocity is recorded. The slope of the torque versus applied load can be used to yield a friction coefficient, resulting in values of ~0.08 for methylene chloride dissolved in PAO. However, a load is eventually reached at which the friction coefficient increases drastically and this load is denoted the seizure load. Note that material is worn away from the surface during the experiment, so that any surface contaminants that might have initially been present are removed, leaving bare metal exposed to the lubricant. The measured values of seizure load are remarkably reproducible and vary with additive concentration and the resulting plots of seizure load versus additive concentration are shown in Figure 17.3. The seizure load initially increases with additive concentration but reaches a plateau where the further addition of chlorinated hydrocarbon results in no increase in seizure load. This behavior is typical of many lubricant additives.

It is argued that seizure occurs when the tribofilm has been completely removed from the surface. At loads below the seizure load, a film is present on the surface, although the thickness cannot be measured in situ. However, the film thickness at any point during the experiment is assumed to result from a balance between the rate of film formation and removal. It is further postulated that, under these EP conditions, the film growth kinetics and the nature of the resulting surface film measured in Section 17.3.2 are identical to those occurring at the tribological interface. The challenge then is to measure the temperature at the interface between the pin and V-block. Briefly, this is accomplished by measuring the wear rate of a known material as a function of applied load in the pin and V-block apparatus [17]. A simple

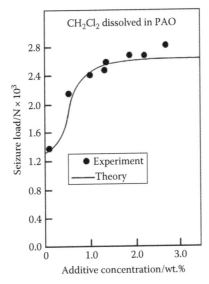

FIGURE 17.3 Plot of seizure load versus additive concentration when using methylene chloride as an additive to PAO (●). Shown as a solid line is a fit to these data using a kinetic model described in the text. (From Kotvis, P.V. et al., *Langmuir*, 9, 467, 1993. With permission.)

Archard wear law [1] suggests that the wear rate is proportional to the normal load and inversely proportional to the hardness of the materials that comprise the contact. The hardness of a material depends on temperature and, in particular, will decrease asymptotically to zero as the material melts: a liquid has zero hardness [18]. This is manifest by a nonlinearity in the rate of wear as a function of load, culminating in an asymptotically large wear rate as the interface temperature reaches the melting point of the tribocontact. This essentially provides an internal calibration of the temperature at the interface. Measuring the asymptote in the wear rate as a function of load for various materials, with different melting points, allows the interfacial temperature to be calibrated [17]. It is found that the interfacial temperature increase (above the ambient) varies linearly with the applied load and thus allows the temperature at a particular load in the pin and V-block experiment to be calculated. In this experiment, the wear, which is relatively large, is measured from the width of the wear scar on the surface of the V-block. Note that being able to reliably measure a well-defined thermodynamic property, a melting point, suggests that the surface temperature is indeed a well-behaved thermodynamic quantity under EP conditions.

The temperature calibration allows the film growth rate to be calculated (using the above measured kinetic data) at a particular applied load (and therefore an interfacial temperature). The corresponding wear rate can easily be measured as a function of applied load from the width of the wear scar as a function of time. Since film growth and wear occur simultaneously, the rate of change in film thickness X (dX/dt, due to both film growth and wear) is calculated as a function of applied load in the pin and V-block apparatus. This, in turn, is used to calculate the total film thickness X during the experiment, and the seizure load is taken to be that at which the film thickness X decreases to zero. The resulting calculated values of seizure load are plotted in Figure 17.3 as a solid line and are clearly in excellent agreement with the experimental data [19,20].

Thus, during the tribological experiment, the film thickness initially increases as the load and the resulting interfacial temperature increase. However, at higher loads, the wear rate increases more rapidly than the film growth rate, ultimately resulting in complete removal of the film, causing the system to seize. Both experiment and theory show that, in the case of methylene chloride, when the chlorine concentration exceeds ~2 wt.%, increasing the additive concentration results in no further increase in seizure load. In this case, the interfacial temperature in this plateau region (when the seizure load is ~2800 N) is ~960 K, the melting point of the $FeCl_2$ tribofilm. Thus, increasing the additive concentration, although it initially results in the formation of a thick $FeCl_2$ film, does not result in an increase in seizure load since the film is very rapidly worn from the surface when the interfacial temperature reaches the melting point of $FeCl_2$, leading to the plateau in the plot of seizure load versus additive concentration, as shown in Figure 17.3. Thus, interfacial temperatures measured for the plateaux of plots such as that shown in Figure 17.3 correspond to the melting temperatures of the materials that comprise the tribofilm and can thus be used to provide an indication of the nature of the tribofilm.

Finally, the agreement between the experimental and theoretical curves shown in Figure 17.3 indicates that, at least in the EP regime, the surface chemistry and kinetics measured either in UHV or using the microbalance can be used to model the tribochemistry under EP conditions. As noted above, under less severe conditions, where energy is dissipated at localized asperities, this correspondence may not occur and reaction rates and products found at the tribological interface may differ from those measured under conditions of true thermodynamic equilibrium.

17.3.4 TRIBOLOGICAL PROPERTIES OF THIN, LUBRICOUS FILMS

The model outlined above indicates that ferrous chloride reactively formed on the surface when using chlorinated hydrocarbons as additives to a base oil provides a low-friction film. This raises the questions of how the presence of such a film modifies the friction, and what film thickness is required to reduce the fiction to its minimum value and how does friction vary with film thickness. In the following, the friction of well-defined films on clean metal substrates is measured in UHV [10,11]. This strategy ensures that films are deposited on clean substrates and that they are not modified by interaction with an ambient atmosphere. Films can either be formed reactively from gas-phase reactants in a reaction cell attached to the UHV chamber that houses the tribometer or evaporated onto the surface in situ where the film thickness can be measured directly using a quartz crystal microbalance housed in the UHV chamber. Figure 17.4 shows a plot of the friction coefficient of a film of $FeCl_2$ grown on a clean iron substrate by reaction with carbon tetrachloride [21]. The initial clean surface has a friction coefficient of ~2 but decreases to a limiting value of ~0.08 as a film is formed on the surface. Similar results are obtained when the ferrous chloride film is evaporated directly onto the iron, indicating that it is indeed the formation of the ferrous chloride film that reduces friction [21]. In particular, the

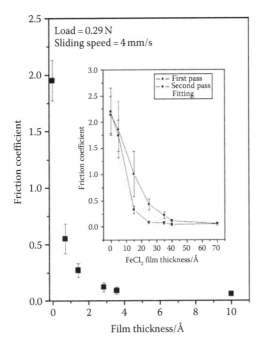

FIGURE 17.4   Plot of friction coefficient versus film thickness for a ferrous chloride film grown by reaction of methylene chloride with iron. (From Gao, F. et al., *Wear*, 256, 1005, 2004. With permission.)

minimum friction coefficient measured in the UHV tribometer (~0.08) is in good agreement with that found in the pin and V-block apparatus described above. This indicates that frictional measurements of thin films grown under UHV conditions are close to those found under more realistic conditions.

In order to explore the behavior of such thin films in greater detail, the frictional behavior of a series of structurally simpler cubic alkali halides was studied. In this case, a series of films with the same cubic structures, but with widely varying mechanical properties, can be compared on the same substrate [10]. In addition, alkali halides are sufficiently stable that they will not react with the bare metal substrate. A typical plot of friction coefficient versus film thickness for potassium chloride on iron is shown in Figure 17.5 [10,11,22]. This displays the same behavior as found for ferrous chloride (Figure 17.4) and is typical of the behavior of thin films on metal substrates. Thus, the friction coefficient decreases from the initially high value typical of sliding on the metal substrate, reducing to a constant value when more than ~4 nm of KCl has been deposited. Simultaneous measurements of the contact resistance between the pin and substrate show a rapid increase in contact resistance at the point at which the friction coefficient attains it minimum value. This implies that there are no longer any uncovered portions of the metal surface at this film thickness.

The growth kinetics of alkali halide has been explored on metal surfaces, and it is found that due to the relative immobility of the alkali halide on the surface, the

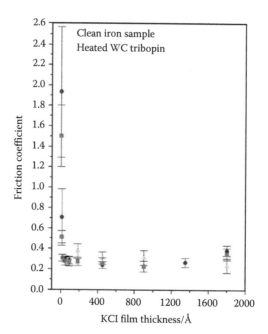

FIGURE 17.5 Plot of friction coefficient versus film thickness for potassium chloride evaporated onto iron where the different symbols represent the results of different experiments for different films. (From Gao, F. et al., *Tribol. Lett.*, 14, 99, 2003. With permission.)

growth of second and subsequent layers of the alkali halide commences before the first layer is completely covered. That is, alkali halide adsorbs at the point that it impinges, so that it adsorbs on the bare surface when it impinges on the bare surface, but can also adsorb onto portions of the surface that are already covered. This effect has also been imaged directly using atomic force microscopy, where the presence of two and three layers of the alkali halide is found while there are still bare portions of the surface present [23]. Subsequent alkali halide appears to nucleate as small, square crystallites. In addition, frictional measurements indicate that the friction of the bare metal is high, while the covered portion exhibits low friction, which is not strongly dependent on how many layers are present.

This suggests a model for friction reduction in which the friction coefficient is proportional to the relative proportion of the surface that is bare (having a high friction coefficient $\mu_0(Fe)$ ~2) and that which is covered having a lower friction coefficient $\mu_0(KCl)$ ~0.3. In this case, if the relative coverage of KCl is given by Θ_{KCl}, then the friction coefficient μ is given by

$$\mu = \mu_0(Fe)(1 - \Theta_{KCl}) + \mu_0(KCl)\Theta_{KCl}.$$

Figure 17.6 shows the data from Figure 17.5 replotted as $\mu - \mu_0(KCl)$ versus film thickness. The resulting predicted value of Θ_{KCl}, assuming that the model described above is correct, is plotted as a solid line on this curve. However, the proportion of the bare iron surface (equal to $1 - \Theta_{KCl}$) can be measured directly in UHV by titrating

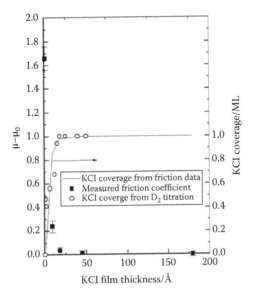

FIGURE 17.6 Replot of $\mu-\mu_0$, where μ_0 is the limiting friction coefficient, as a function of film thickness (■) and the resulting calculated coverage of the KCl film that covers the metal surface is plotted as a solid line. Shown also is the KCl covered measured by titrating the bare metal surface using deuterium (●). (From Gao, F. et al., *Tribol. Lett.*, 14, 99, 2003. With permission.)

it with deuterium [24] since deuterium adsorbs selectively on the metal surface, but not onto KCl. Deuterium desorbs from the surface when it is heated and is detected using a mass spectrometer. Deuterium is used to titrate the surface since there are no masses at 4 amu (corresponding to D_2) in the background of the UHV chamber that might interfere with the measurement. The values of Θ_{KCl} measured in this way are plotted in Figure 17.6 (o) where the agreement between the prediction of the model and the experimental measurements is good. This indicates that the assumptions used in deriving the above equation are correct, and a complete coverage of the surface by the tribofilm is required to reduce the friction to its minimum value (in this case ~0.3).

While the friction coefficient is a useful parameter, it is a function of the contact area between the tribopin and the substrate and thus depends both on the shear properties of the interface and on the contact area. Measurement of the friction coefficient using different applied loads reveals that the friction coefficient obeys Amontons' law and suggests that the contact area is proportional to the applied load. This implies that the load is supported by plastic deformation at asperity contacts on the surface. In this case, therefore, the real contact area is simply proportional to L/H_S, where L is the normal load and H_S is the hardness of the substrate. Since the lateral sliding force is given by the contact area multiplied by the shear strength of the contacting interface, plotting the friction coefficient versus $1/H_S$ for the same film deposited onto substrates with different values of H_S should yield a straight line. Such a plot for a film of KCl deposited onto various substrates, measured in UHV, is shown in Figure 17.7. This indeed results in a good straight line, within experimental

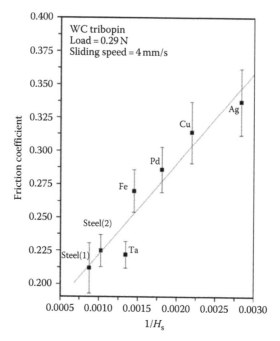

FIGURE 17.7 Plot of limiting friction coefficient of potassium chloride on various substrates versus $1/H_s$, where H_s is the hardness of the substrate material. (From Gao, F. et al., *Tribol. Lett.*, 31, 99, 2008. With permission.)

error. However, the line does not pass through zero as $1/H_s$ tends to zero. This indicates that the shear strength S of the contact depends on the contact pressure P as

$$S = S_0 + \alpha P,$$

where S_0 and α are constants. A more detailed analysis of this sliding behavior is given in Ref. [25] and suggests that S_0 lies between 40 and 60 MPa and that $\alpha = 0.14 \pm 0.02$.

Such pressure-dependent shear strengths have been measured previously and can be interpreted in terms of an atomistic sliding model where S_0 corresponds to the energy for atoms to surmount the sliding potential at the surface and the pressure dependence accounts for the work expended against the normal load as the atoms move normal to the surface when surmounting this potential [26].

These results indicate that a tribofilm that completely covers the surface with at least a monolayer of material is required to reduce the friction coefficient to its minimum value and that the sliding interface of the resulting tribofilm has a pressure-dependent shear strength. These observations are in accord with the kinetic model proposed above, where seizure is proposed to occur when the film has been completely removed. However, the model also indicates that a thicker film can be formed on the surface by reaction with the chlorinated hydrocarbon additive at intermediate loads. This, therefore, raises the question of what effect does the presence of much thicker films have on the tribological properties. The results of experiments to measure the friction coefficients of thicker films are shown in Figure 17.8, which plots the friction coefficient versus film thickness for films of KCl, NaCl, and KI of up to ~1 μm in thickness deposited onto iron [27]. For clarity, the initial rapid decrease seen, for example in Figures 17.4 and 17.5, has been removed. These results reveal that the friction coefficient does increase with increasing film thickness. The lines plotted through the data reveal that the change occurs in two regimes, one above

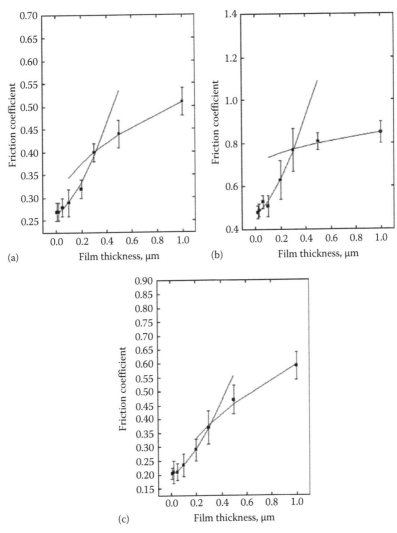

FIGURE 17.8 Experimental results for the friction coefficient versus film thickness for thicker films once the initial reduction seen in Figure 17.5 has been completed. This initial reduction has been removed from the data for clarity and results are shown for (a) potassium chloride, (b) sodium chloride, and (c) potassium iodide. (From Gao, F. et al., *Tribol. Lett.*, 15, 327, 2003. With permission.)

and another below a film thickness ~0.3 μm, irrespective of the nature of the film material. This suggests that this transition is not due to a film material property but to a geometrical property of the interface. AFM measurements of the topography of the (rougher) tribopin reveal that the maximum peak-to-valley distance of the tribopin is also ~0.3 μm. This suggests that, below this maximum value, an increase in film thickness results in more of the halide film contacting the asperities, thereby leading to a higher total contact area and thus a larger lateral force required to shear the contact. Above this critical film thickness, the load is borne entirely by the film and the asperity tips do not contact the metal substrate below. The latter situation has been investigated theoretically and suggests that, in this case, the friction coefficient should vary as $\sqrt{t}$ where t is the film thickness [28] and indeed this behavior is observed experimentally.

The behavior of films of intermediate thickness ($t < 0.3$ μm) has been analyzed theoretically by describing the contact using a modified Greenwood–Williamson model [29]. This assumes that a series of independent asperities with an exponential height distribution contact the surface. Depending on the height of the asperity, it can either contact the metal substrate below or the thick halide film on the substrate. The asperities are assumed to be supported by plastic deformation in both the metal substrate and the halide film. Balancing this with the applied load enables the contact area with both the substrate and the film to be calculated as a function of film thickness. These values depend on the asperity height distribution, which is measured directly from an AFM image of the surface of the tribopin. Thus, the contact areas (with the film and substrate) are determined by independently measured parameters; the hardness of the film and substrate, the film thickness, and the asperity height distribution. As noted above, the interfacial shear strength depends on the contact pressure. This suggests that, since the pressure in a plastic contact equals the hardness of the material, the shear strength of asperities that contact the substrate (where the contact pressure equals the hardness of iron) is different from those that contact the film (where the contact pressure equals the hardness of the film material). This allows the contributions to the total lateral force from shearing contacts with both the substrate and the film to be calculated to yield a friction coefficient as a function of film thickness. The resulting prediction of this model is plotted as a solid line along with the experimental data for KCl on iron (■) (Figure 17.9) [27]. Note that all of the parameters in this model are measured independently in separate experiments. The agreement between theory and experiment is very good, although there is some difference at larger film thicknesses.

Thus, the frictional behavior of the film depends on thickness and can be separated into three regimes. In the initial regime, friction is reduced by the proportion of the surface that is covered by at least a monolayer of the film and reaches its minimum value once the surface is completely covered. A second regime occurs when the film is thinner than the total peak-to-valley distance of the asperities in the contact, where the effective contact area increases with film thickness. A final regime is identified when the film becomes thicker than the total peak-to-valley distance of the asperities, where the load is completely supported by the film itself.

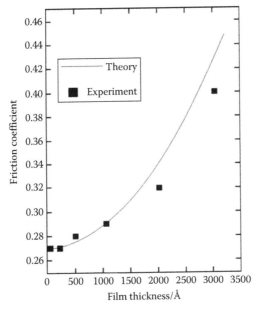

FIGURE 17.9 Theoretical plot of friction coefficient versus film thickness for thick films of KCl on iron (■) compared with the results of theory (solid line). (From Gao, F. et al., *Tribol. Lett.*, 15, 327, 2003. With permission.)

17.4 CONCLUSIONS

It has been demonstrated that the film growth kinetics and the nature of the surface film measured on clean metal samples either in UHV or with a microbalance, as a function of temperature, can successfully be used to model the tribological chemistry occurring under EP conditions. In particular, the variation in seizure load with additive concentration can be modeled for a lubricant consisting of methylene chloride dissolved in PAO. It is postulated that this occurs because of the large energy dissipation when asperities pass over each other, which is so rapid that these "flash temperatures" coalesce to once again yield a surface that is close to being in thermodynamic equilibrium. It is suggested that very low-load contacts, where the interface is not strongly perturbed, would also fulfill this condition. It is likely, however, that, in intermediate regimes, the lack of thermodynamic equilibrium and the presence of rapid, local heating rates could lead to surface properties of the tribological interface that differ from those measured for a surface that is in thermal equilibrium.

It is also found that a monolayer of film covering the surface can reduce friction to its minimum value and that the shear strength of this film depends on the contact pressure. The friction coefficient increases as the film thickness increases due to the increased contact area between the rough tribopin and the surface, and a final regime is detected when the film thickness exceeds the maximum peak-to-valley distance of the asperities where the load is entirely supported by the film.

REFERENCES

1. J.F. Archard, Contact and Rubbing of Flat Surfaces, *J. App. Phys.*, 24, 981–988 (1953).
2. K.J. Laidler, *Chemical Kinetics*. Harper and Row, New York (1987).
3. F.P. Bowden and D. Tabor, *The Friction and Lubrication of Solids*. Clarendon Press, Oxford (1950).
4. H. Blok, Theoretical Study of Temperature Rise at Surfaces of Actual Contact under Oiliness Lubricating Conditions, *Proc. Inst. Mech. Eng.*, 2, 222–235 (1937).
5. M. Bonn, S. Funk, Ch. Hess, D.N. Denzler, C. Stampfl, M. Scheffler, M. Wolf, and G. Ertl, Phonon versus Electron-Mediated Desorption and Oxidation of CO on Ru(0001), *Science*, 285, 1042–1045 (1999).
6. J. Lara and W.T. Tysoe, Interaction of Effusive Beams of Methylene Chloride and Chloroform with Clean Iron: Tribochemical Reactions Explored in Ultrahigh Vacuum, *Langmuir*, 14, 307–312 (1998).
7. M. Kaltchev, G. Celichowski, J. Lara, and W.T. Tysoe, A Molecular-Beam Study of the Tribological Chemistry of Carbon Tetrachloride on Oxygen-Covered Iron, *Tribol. Lett.*, 9, 161–165 (2000).
8. L.A. Huezo, C. Soto, C. Crumer, and W.T. Tysoe, Growth Kinetics and Structure of Films Formed by the Thermal Decomposition of Methylene Chloride on Iron, *Langmuir*, 10, 3571–3576 (1994).
9. P.V. Kotvis, L.A. Huezo, and W.T. Tysoe, Surface Chemistry of Methylene Chloride on Iron—A Model for Chlorinated Hydrocarbon Lubricant Additives, *Langmuir*, 9, 467–474 (1993).
10. F. Gao, P.V. Kotvis, and W.T. Tysoe, The Friction, Mobility and Transfer of Tribological Films: Potassium Chloride and Ferrous Chloride on Iron, *Wear*, 256, 1005–1017 (2004).
11. G. Wu, F. Gao, M. Kaltchev, J. Gutow, J. Mowlem, W.C. Schramm, P.V. Kotvis, and W.T. Tysoe, An Investigation of the Tribological Properties of Thin KCl Films on Iron in Ultrahigh Vacuum: Modeling the Extreme-Pressure Lubricating Interface, *Wear*, 252, 595–606 (2002).
12. P.V. Kotvis, W.S. Millman, L. Huezo, and W.T. Tysoe, The Surface Decomposition and Extreme-Pressure Tribological Properties of Highly Chlorinated Methanes and Ethanes on Ferrous Surfaces, *Wear*, 147, 401–409 (1991).
13. J. Lara, P.V. Kotvis, and W.T. Tysoe, The Surface Chemistry of Chlorinated Hydrocarbon Extreme-Pressure Lubricant Additives, *Tribol. Lett.*, 3, 303–310 (1997).
14. J. Lara, H. Molero, A. Ramirez-Cuesta, and W.T. Tysoe, Structure and Growth Kinetics of Films Formed by the Thermal Decomposition of CCl_4 on Iron Surfaces, *Langmuir*, 12, 2488–2494 (1996).
15. W.T. Tysoe, K. Surerus, J. Lara, T.J. Blunt, and P.V. Kotvis, The Surface Chemistry of Chloroform as an Extreme-Pressure Lubricant Additive at High Concentrations, *Tribol. Lett.*, 1, 39–46 (1995).
16. P.V. Kotvis, J. Lara, K. Surerus, and W.T. Tysoe, The Nature of the Lubricating Films Formed by Carbon Tetrachloride Under Conditions of Extreme Pressure, *Wear*, 201, 10–14 (1996).
17. T.J. Blunt, P.V. Kotvis, and W.T. Tysoe, Determination of Interfacial Temperature Under Extreme Pressure Conditions, *Tribol. Lett.*, 2, 221–230 (1996).
18. H. Ernst and M.E. Merchant, Surface Friction between Metals: A Basic Factor in the Metal Cutting Process, In: *Proceedings of Special Summer Conference Friction and Surface Finish*, MIT Rep. Vol. 15, pp. 76–101, MIT Press, Cambridge, MA (1940).
19. T.J. Blunt, P.V. Kotvis, and W.T. Tysoe, Surface Chemistry of Chlorinated Hydrocarbon Lubricant Additives—Part I: Extreme-Pressure Tribology, *Tribol. Trans.*, 41, 117–123 (1998).

20. T.J. Blunt, P.V. Kotvis, and W.T. Tysoe, Surface Chemistry of Chlorinated Hydrocarbon Lubricant Additives—Part I: Modeling the Tribological Interface, *Tribol. Trans.*, 41, 129–139 (1998).
21. F. Gao, O. Furlong, P.V. Kotvis, and W.T. Tysoe, Tribological Properties of Films Formed by the Reaction of Carbon Tetrachloride with Iron, *Tribol. Lett.*, 20, 171–176 (2005).
22. F. Gao, P.V. Kotvis, and W.T. Tysoe, The Frictional Properties of Thin Inorganic Halide Films on Iron Measured in Ultrahigh Vacuum, *Tribol. Lett.*, 15, 327–332 (2003).
23. T. Filleter, W. Paul, and R. Bennewitz, Atomic Structure and Friction of Ultrathin Films of KBr on Cu(100), *Phys. Rev. B.*, 77, 035430 (2008).
24. F. Gao, G. Wu, D. Stacchiola, M. Kaltchev, P.V. Kotvis, and W.T. Tysoe, The Tribological Properties of Monolayer KCl Films on Iron in Ultrahigh Vacuum: Modeling the Extreme-Pressure Lubricating Interface, *Tribol. Lett.*, 14, 99–104 (2003).
25. F. Gao, O. Furlong, P.V. Kotvis, and W.T. Tysoe, Pressure Dependence of Shear Strengths of Thin Films on Metal Surfaces Measured in Ultrahigh Vacuum, *Tribol. Lett.*, 31, 99–106 (2008).
26. G. He and M.O. Robbins, Simulations of Kinetic Friction Due to Adsorbed Surface Layers, *Tribol. Lett.*, 10, 7–14 (2001).
27. F. Gao, P.V. Kotvis, and W.T. Tysoe, The Frictional Behavior of Thin Halide Films on Iron, *Tribol. Trans.*, 47, 208–217 (2004).
28. E.F. Finken, A Theory for the Effects of Film Thickness and Normal Load in the Friction of Thin Films, *J. Lubr. Technol.*, 91, 551–556 (1969).
29. J.A. Greenwood and J.B.P. Williamson, Contact of Nominally Flat Surfaces, *Proc. R. Soc. Lond. A.*, 295, 300–319 (1966).

18 The Role of Surface Science in Magnetic Recording Tribology

Tom Karis

CONTENTS

ABSTRACT

Surface science is at the forefront as the limits of magnetic recording are expanded. This chapter presents case studies derived from experiences in magnetic recording disk research, development, and manufacturing. Surfactants are intermingled with product design, reliability, and processing. The most well-known surfactant in magnetic recording is the molecularly thin film of lubricant that resides on the disk overcoat. Both the lubricant and the overcoat are approaching only a few atomic layers in thickness. The nanometer-scale gap between the recording head and the disk must remain free of chemical and particulate contamination. Issues of contamination and lubricant transfer to the slider continue to challenge reliability. Large-scale, low-cost manufacturing of magnetic recording disks with sub-nanometer roughness is enabled by precise control of surface chemistry. Corrosion layers must be removed from disk substrates by washing in an aqueous solution and drying before the magnetic layers are deposited by sputtering. Since the magnetic recording layers are only about 100 nm thick, even minute amounts of surface contamination must be completely removed prior to their deposition. Any remaining particles, and particles deposited during sputtering, form asperities, which are removed by an abrasive tape polishing. To maintain the increasing magnetic recording data areal density, future magnetic recording disks may store data bits on patterns of discrete features etched into the disk topography. Fabrication of patterned media advances the state of the art in high-speed nanoimprint lithography (NIL). Surface chemical challenges currently in the forefront of patterned media technology are reviewed.

18.1   INTRODUCTION

We expect hard disk drive technology to provide ever-increasing data storage capacity and improved reliability at a lower cost for each new product. Continued advancement relies on the successful application of surface science in the drive components, manufacturing processes, and component integration.

Because the films of significance in magnetic recording are so thin, the tribological performance is dominated by surfaces. This chapter presents case studies based on experiences in product development and lessons learned in remediation of challenges encountered during introduction of new technologies for magnetic recording disks. These examples highlight the role of surfactants not only in enabling reliable operation but also in contributing to unexpected problems that are regularly surmounted by the industry.

First and foremost in magnetic recording is the ability of the magnetic recording disk and the magnetic recording head to store and retrieve magnetically encoded information. One measure of this ability is the soft error rate. The perpendicular magnetic recording disk structure comprises magnetic layers, a cap layer, and a protective overcoat (Figure 18.1). The magnetic layers are designed to obtain the desired soft error rate.

The liquid lubricant that is topically applied to the overcoat on all magnetic recording disks is an amphiphilic surfactant [1]. Polar groups attached to the nonpolar perfluoropolyether (PFPE) main chain tether the lubricant to the overcoat. Recently, the industry trend has been to increase the number of polar hydroxyl groups on each lubricant molecule. The evolution of the PFPE lubricant structure since the advent of thin-film disks to the present time is reviewed in Section 18.2. Adamson's potential distortion model [2] is applied to characterize the effect of lubricant hydroxyl group chemisorption on the water adsorption isotherm and contact angle in Section 18.3.1.

Hygroscopic cloud condensation nuclei are ubiquitous in the ambient air environment. If cloud condensation nuclei are adsorbed on lubricated magnetic recording

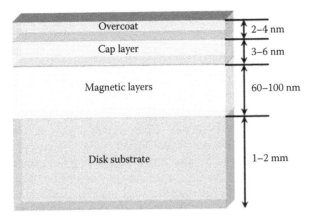

FIGURE 18.1   A schematic illustration of the typical film stack for perpendicular magnetic recording media on a disk substrate. Glass substrates may include an adhesion layer; AlMg substrates include an NiP layer adjacent to the magnetic layer.

disks by exposure to inadequately filtered air, they adsorb atmospheric moisture and become overcoated with amphiphilic disk lubricant to form liquid nanodroplets [3]. The equilibrium nanodroplet height exceeds the slider flying height. Nanodroplets are impacted by the slider resulting in lubricant transfer and smears, which causes tribological instability. The thermodynamics of liquid nanodroplet equilibration is included in Section 18.3.2.

The finished disk is susceptible to the effects of vapor phase contaminants, which preferentially adsorb at sites of asperity removal or on pinholes formed by overcoat abrasion during polishing. Silica formation at the slider disk interface by tribochemical oxidation of reactive adsorbates derived from poly(dimethylsiloxane) [4] is reviewed in Section 18.3.3.

The combined thickness of the layers on the disk substrate is only about 100 nm. Often, the lowest soft error rate is obtained with magnetic layer materials that are susceptible to corrosion. Corrosion is prevented by several nanometer of a carbon overcoat on the magnetic layers. It is then critical to avoid pinholes through the overcoat, which form a galvanic cell with moisture [5]. An interesting case of rod-like corrosion crystals that were observed during perpendicular recording media development is described in Section 18.4.1. The amphiphilic disk lubricant was unintentionally acting as a soft template for crystallization of cobalt oxalate needles [6] in the presence of moisture. A means to inhibit corrosion through in situ adsorption of vapor phase corrosion inhibitors (VCIs) [7] on the disk and recording heads is presented in Section 18.4.2.

Another way to decrease the soft error rate is by reducing the space between the magnetic recording head and the disk magnetic layers. There is an increased likelihood for development of liquid lubricant thickness nonuniformity and transfer to the slider [8] as the slider flies closer to the disk. Lubricant thickness nonuniformity and transfer to the slider leads to slider instability. The effort to maintain a stable interface at reduced flying height is driving the transition of disk lubricants from mobile liquids to highly chemisorbed monolayers [9].

The slider is also brought closer to the disk magnetic layers by decreasing the overcoat thickness. Novel overcoats are being investigated as an alternative to the highly chemisorbed liquid lubricant monolayer [10]. One important property that must be provided by the novel overcoat is decreased adhesion stress (force per unit area) in slider disk asperity contacts. Low adhesion stress is currently provided by the liquid lubricant. Adhesion stress in the sub-microsecond time scale asperity contacts at the slider disk interface is attributed to dispersion force arising primarily from interactions in the ultraviolet and visible regions of the electromagnetic spectrum [11]. Briefly, the adhesion stress is calculated as the disjoining pressure between the slider and the disk in the limit of zero air gap using the nonretarded Lifshitz theory. The adhesion stress derived in this manner was found to be nearly the same as the tensile strength of the interface between the two materials calculated from their dispersive surface energies [12]. A case study on how to calculate the dispersion spectrum that is needed to obtain the theoretical adhesion stress between a novel fluorocarbon overcoat and the slider is presented in Section 18.4.3. The adhesion stress calculated for the fluorocarbon overcoat is compared with that of the present-day liquid lubricant and carbon overcoat.

The slider can only fly as close to the disk as is permitted by the height of the tallest asperities on the disk surface. Smoothness of the disk surface after magnetic layers, overcoat deposition, and lubrication begins with the disk substrate (Figure 18.1). Both glass and aluminum–magnesium (AlMg) substrates are washed to remove particles just prior to the deposition step. AlMg substrates are coated with a layer of nickel–phosphorus alloy (NiP). The nickel (Ni) carbonate corrosion product that forms on NiP-plated AlMg alloy disk substrates is removed when the substrates are washed in an aqueous solution [13]. After washing, the Ni carbonate corrosion layer rapidly reforms on the NiP-plated disk substrates during storage in air containing moisture and CO_2. The mechanism of the carbonate corrosion is reviewed, and the use of a controlled storage environment to limit the corrosion after washing the substrate is illustrated in Section 18.5.1.

Once the magnetic layers and overcoat have been deposited on the washed substrate and lubricated, abrasive tape polishing is performed to reduce the height of surface asperities below the few nanometer air gap, or flying height, between the disk and the magnetic recording slider [14]. Details of the polishing process are critical to the reliability of the disk products, because the disks must be sufficiently smoothed without creating pinholes through the overcoat. A method for using friction force versus load measurements is shown to provide insight into the mechanism of the abrasive tape polishing process for magnetic recording disks. An investigation of the friction force between the tape and disk during the polishing sweep pass is presented in Section 18.5.2.

Finally, surface science promises to enable a technology that may be needed to continue the trend of increasing data storage density. Magnetic domains are coplanar on present-day perpendicular magnetic recording media. When the magnetic bits are packed even closer together, one way to maintain thermal stability of the domains is by magnetically isolating them from each other through spatial separation [15]. Spatial separation between tracks of bits is accomplished, for example, by placing the domains on discrete tracks separated by a 50 nm-deep groove in between (discrete track media, DTM). Spatial isolation along a track is further accomplished by placing each domain on a pedestal or mesa where each is separated from its neighbors by a depth of 50 nm (bit patterned media, BPM). Tremendous challenges and opportunities exist for surface science, that will allow manufacturing technologies to mass produce magnetic recording disks patterned with these features through step and flash nanoimprint lithography (NIL). A surfactant monolayer of adhesion promoter provides adhesion between cured photoresist and the disk on which the pattern is being etched. A surfactant monolayer of release agent provides low adhesion to enable separation of the cured photoresist on the disk from the quartz pattern template. For example, one failure mode is the plugging of template holes by the resist caused by resist fracture due to high adhesion during separation. These focus areas in patterned media technology are covered in Section 18.6.

18.2 DISK LUBRICANTS

The most widely used disk lubricants are PFPEs with the Z-type backbone chain. These are random copolymers with the linear backbone chain structure

$$X-[(O\ CF_2)_m-(OCF_2CF_2)_n-(OCF_2CF_2CF_2)_p-(OCF_2CF_2CF_2CF_2)_q]_{x_0}-OX,$$

where
   X is the end group
   Z is the backbone chain with degree of polymerization x_0 and molecular weight of 2000–4000 Da

A variety of end groups are available to optimize the lubricants chemical and tribological properties [16]. The end groups for some of the commercially available lubricants are shown in Table 18.1. The adsorption energy of end groups (other than $-CF3$) on the carbon overcoat surface is higher than that of the backbone chain [17,18]. The X1-P-type cyclic phosphazene end group on A20H [19,20] is sterically large in comparison to the lubricant polymer chain [21], and the phosphazene end group molecular weight of about 1000 Da is a significant contribution to the molecular weight of the lubricant [22]. Aside from its high molecular weight and viscosity, phosphazene has a strong UV absorption relative to Ztetraol. The UV absorption spectra of X1-P and Ztetraol 2000 are shown in Figure 18.2. The strong UV absorption of the phosphazene relative to the PFPE suggests that the dispersion spectrum of the phosphazene, which governs the dispersion interaction of the interface materials, is significantly different from that of the PFPE. The dispersion spectrum of X1-P could be estimated from the static dielectric constant, the visible index of refraction, and the UV absorption spectrum for the phosphazene (see Section 18.4.3). Reduction of the interfacial adhesion may contribute to the good tribological performance when cyclic phosphazene is combined with or attached to PFPE disk lubricant [23].

   PFPE magnetic recording disk lubricants have evolved to comprise a multiplicity of hydroxyl end groups to increase their chemisorption to the overcoat (Table 18.2).

TABLE 18.1

Molecular Structure for PFPE End Groups on Z-Type PFPE Disk Lubricants

Notation	Structure
Z	$-CF_3$
Zdol	$-CF_2CH_2OH$
Ztetraol	$-CF_2CH_2OCH_2\overset{\overset{\displaystyle OH}{\displaystyle |}}{C}HCH_2OH$
Zdiac	$-CF_2COOH$
Zdeal	$-CF_2COOCH_3$
Zdol TX	$-CF_2CH_2(OCH_2CH_2)_{1.5}OH$

AM-3001 $-CF_2CH_2OCH_2$

A20H* $-CF_2CH_2O$

ZDPA $-CF_2CH_2N\overset{\diagup CH_2CH_2CH_3}{\diagdown CH_2CH_2CH_3}$

Source: Adapted from Rudnick, L.R. (ed.), *Lubricant Additives Chemistry and Applications*, 2nd edn., CRC Press, Taylor & Francis Group, Boca Raton, FL, 2009.

*A20H has one zdol end group.

Two Zdol chains are linked with epichlorohydrin to form a Zdol multidentate (ZDMD) with one hydroxyl group at each end and one in the middle [24]. Two Ztetraol chains are linked with a fluorinated di-epoxide to form a Ztetraol multidentate (ZTMD) with two hydroxyl groups at each end and four near the middle [9]. A novel three-arm star multidentate PFPE with one hydroxyl group at the end of each arm was derived by direct fluorination of the hydrocarbon ester (LTA-30) [25]. More recently, a PFPE lubricant with four hydroxyl groups was similarly prepared (QA-40) [26]. The chemisorbed fractions of these lubricants are illustrated in Figure 18.3. The chemisorbed fraction is the ratio of the lubricant film thickness that remains on the overcoat after rinsing with solvent to the original film thickness. It is also possible to increase the chemisorbed fraction without changing the lubricant molecular structure or composition by "UV bonding" [27]. An intermediate chemisorbed fraction between Zdol and Ztetraol is provided by UV bonding a mixture of A20H and Zdol (A20H/Zdol UV in Figure 18.3). An advanced tribological

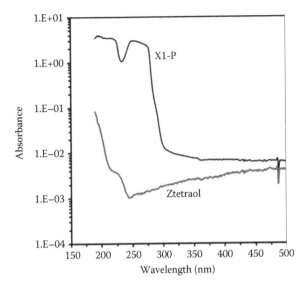

FIGURE 18.2 UV spectra of cyclic phosphazene X1-P and Solvay Solexis Ztetraol 2000.

overcoat for magnetic recording disks may not require topical lubrication by PFPE (see Section 18.4.3).

18.3 CONTAMINATION

18.3.1 WATER ADSORPTION

Disk drives experience wide ranges of temperature and humidity during storage and operation. In the following, it is shown how the potential distortion model of Adamson [2] is used to model the film thickness at saturation from the water adsorption isotherms and the water contact angle on thin-film magnetic recording disks [28]. For example, typical water adsorption isotherms are shown in Figure 18.4a.

The relationship between the equilibrium adsorbed water film thickness x at a given relative humidity %RH is given by

$$-\frac{\Delta G^0}{kT} = \ln\left(\frac{100}{\%\text{RH}}\right) = \ln\left(\frac{P^0}{P}\right) = \varepsilon_0 e^{-ax} - \beta e^{-bx} \quad (18.1)$$

where
  ΔG^0 is the adsorption free energy
  k is the Boltzmann constant
  T is absolute temperature

The relative humidity (RH) is replaced by P/P^0, where P is the partial pressure of water and P^0 is the saturation vapor pressure. The first term on the right side of Equation 18.1 is the attractive potential arising from dispersion interactions between

TABLE 18.2
Multidentate Disk Lubricant Structures Derived from Z-Type PFPEs

Notation | **Structure**

ZDMD

$HOCH_2CF_2-Z-OCF_2CH_2OCH_2CHCH_2OCH_2CF_2-Z-OCF_2CH_2OH$
$\quad$ (with $-OH$ on the central CH)

ZTMD

$HOCH_2CHCH_2OCH_2CF_2-Z-OCF_2CH_2OCH_2CHCH_2OCH_2CHCF_2CF_2CF_2CHCH_2OCH_2CHCH_2OCH_2CF_2-Z-OCF_2CH_2OCH_2CHCH_2OH$
$\quad$ (with $-OH$ on the indicated CH centers)

LTA-30

$HOCH_2CF_2-[OCF_2CF_2]_y-[OCF_2CF_2]-OCF$ bearing:
- $CF_2O-[CF_2CF_2O]_x-CF_2CH_2OH$
- $CF_2O-[CF_2CF_2O]_z-CF_2CH_2OH$

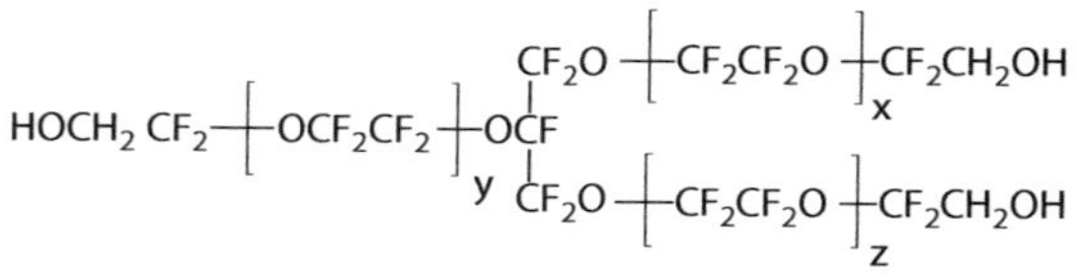

QA-40

$HOCH_2CF_2-[OCF_2CF_2]_y-O-CF$ linked through CF_2-O-CF_2 to CF bearing:
- $CF_2-O-[CF_2CF_2O]_v-CF_2CH_2OH$
- $CF_2-O-[CF_2CF_2O]_z-CF_2CH_2OH$

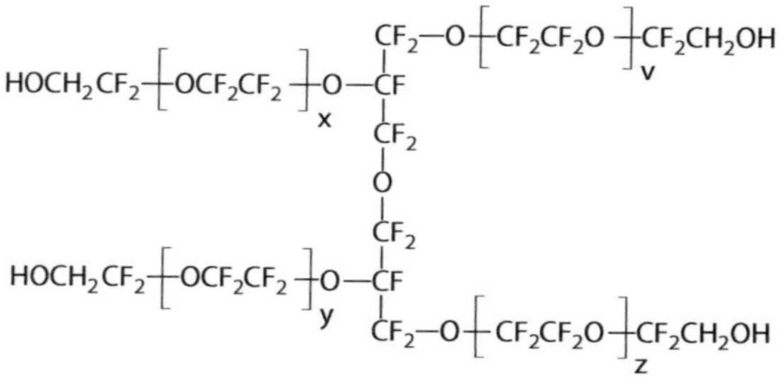

The ZDMD is derived from Zdol 1000 or Zdol 2000. The ZTMD is derived from Ztetraol 1000. The LTA-30 has a molecular weight of 3000 Da, and the QA-40 has a molecular weight of 4000 Da. Z is the PFPE chain defined in the text. Units of molecular weight should be Da as 3000 Da and 4000 Da.

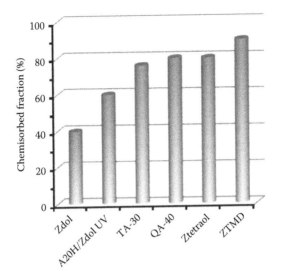

FIGURE 18.3   Chemisorbed fractions of magnetic recording disk lubricants.

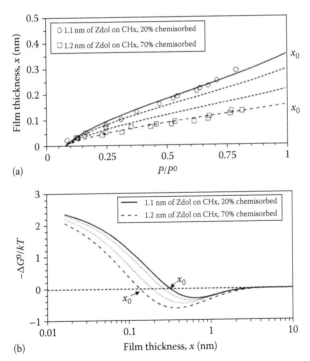

FIGURE 18.4   (a) Water adsorption isotherms and (b) adsorption free energy. The dotted curves show the effect of increasing the dispersion force decay length from its value at 20% chemisorbed Zdol. Symbols show the experimental data. The horizontal dashed line in (b) is the zero point of the adsorption free energy, and x_0 is the film thickness at saturation.

the adsorbate and the solid surface. For the first few molecular diameters, the dispersion interaction energy is approximated by $\varepsilon_0 e^{-ax}$, where ε_0 is a dispersion energy coefficient and $1/a$ is a dispersion energy decay length. The second term on the right side of Equation 18.1 is the entropic energy arising from nonrandom orientation of the water molecules in the first few molecular layers. The structural (entropic) contribution to the adsorption free energy is approximated by βe^{-bx}, where β is a structure energy coefficient and $1/b$ is a structure decay length.

The water contact angle is related to the water adsorption isotherm through the spreading coefficient. The spreading coefficient for a liquid on a solid in the presence of its saturated vapor is

$$S_{LSV} = \gamma_W \left[\cos(\theta) - 1 \right] = -\left(\frac{1}{V} \right) \int_{x_0}^{\infty} \Delta G^0 dx \qquad (18.2)$$

where $V \equiv d^3$ is the volume occupied by an adsorbate (water) molecule. The diameter of the water molecule $d \approx 0.3$ nm was derived from the data for water adsorption on polyethylene [29], and γ_W is the surface tension of water. The physical arrangement of the water molecules at the edge of a droplet with a contact angle $\theta > 90°$ is depicted in Figure 18.5. The contact angle is derived by integrating the adsorption free energy over the region in the droplet above the film thickness at saturation, $x > x_0$ in Figure 18.4b, according to the integral in Equation 18.2. In terms of the potential distortion model, Equation 18.1 parameters, and the film thickness at saturation $(P/P^0 = 1)$ is

$$x_0 = \left(\frac{1}{a-b} \right) \ln \left(\frac{\varepsilon_0}{\beta} \right) \qquad (18.3)$$

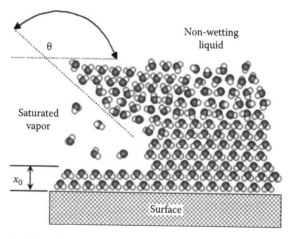

FIGURE 18.5 A schematic representation of the water molecules at the edge of a droplet on a non-wetting surface.

TABLE 18.3

Potential Distortion Model Parameters and Calculated Values from Measured Water Adsorption Isotherm and Contact Angle for Zdol 2000 on Hydrogenated Carbon Overcoated Disks

Chemisorbed Fraction (%)	ε_0	1/a (nm)	β	1/b (nm)	ΔS_{LSV} (mN/m)	x_0 (nm)	Water Contact Angle (°) Calculated	Water Contact Angle (°) Measured
20	3.7	0.208	1.1	0.735	−48.6	0.353	70.6	70.6
70	3.7	0.111	1.1	0.885	−96.9	0.154	109.4	109.5

At the edge of the droplet, the film thickness is x_0, because the film on the solid is in equilibrium with its saturated vapor.

The thermodynamic model can be used to probe the sensitivity of the water contact angle and adsorption isotherm to change in the energy parameters ε_0, a, β, and b. The parameters fitted to the experimental adsorption isotherm data and measured contact angles for two different chemisorbed fractions of Zdol 2000 are listed in Table 18.3. The model fit was able to reproduce both the isotherm and the contact angle with the same values of ε_0 and β for both 20% and 70% chemisorbed. The difference in the adsorption isotherms could mostly be accounted for by changes in the dispersion energy decay length $1/a$, and the contact angle was adjusted by small changes in the structure decay length $1/b$, where a and b are the coefficients of x in Equation 18.1. The effect of changing $1/a$ from its value at 20% chemisorbed to the value at 70% chemisorbed is shown by the dotted curves in Figure 18.4. The main effect of increasing the chemisorbed fraction was to decrease the dispersion energy decay length. This can also account for the increase in the water contact angle with chemisorption through the effect of the decay length on the water contact angle, as shown in Figure 18.6. It could be that the chemisorbed lubricant more effectively screens the dispersion interaction between the water and the overcoat. Alternatively, it is possible that non-chemisorbed lubricant is displaced by water in the droplet, exposing more of the water to the unlubricated carbon overcoat. Lubricant interaction with nanoscopic water droplets suggests that very small droplets become overcoated with the amphiphilic lubricant, as described in the following section.

18.3.2 NANODROPLETS

Ambient aerosol particle deposition was observed during laboratory testing of disks following storage of the disks in closed nonconductive polycarbonate cassettes. The aerosol particles were shown to form liquid nanodroplets on lubricated magnetic recording disks [3]. The nanodroplets produce lubricant transfer from the disk to the magnetic recording slider.

Two different types of cassettes are used in magnetic disk manufacturing: unfilled nonconductive and carbon fiber-filled conductive cassettes. The deposition rates for the two types of cassette are compared in Figure 18.7. When the disks were transferred from a conductive to a nonconductive cassette, the deposition rate was

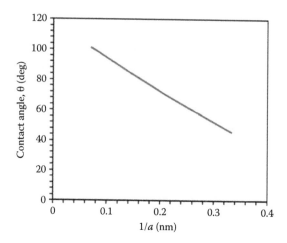

FIGURE 18.6 Calculated water contact angle showing the effect of the dispersion energy decay length. (The fixed parameters are in the first row of Table 18.3.)

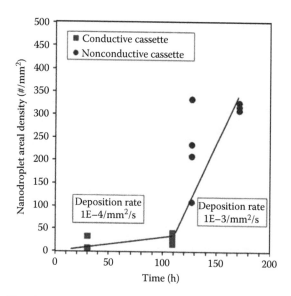

FIGURE 18.7 Nanodroplet areal density during ambient particle deposition on disks in a non-conductive cassette versus a conductive cassette exposed to air in a non-clean room environment to accelerate the deposition rate. (Measurements done with the assistance of Wendt, R. and Wang, R.-H., Hitachi Global Storage Technologies, San Jose Research Center, San Jose, CA.)

increased by an order of magnitude. The electrostatic voltage measured on disks in the conductive cassette was between 0 and 10 V. The electrostatic voltage measured on disks in the nonconductive cassette was 2000–3000 V. The electrostatic voltage and deposition rate were similarly decreased by a benchtop air ionizer blowing over a nonconductive cassette.

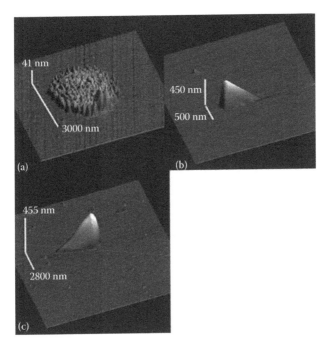

FIGURE 18.8   Tapping mode AFM images of liquid nanodroplets: (a) after brief exposure to condensing humidity on an unlubricated carbon overcoat, (b) on a lubricated overcoat with ambient RH of approximately 50%, and (c) on a lubricated overcoat after brief exposure to condensing humidity. (Measurements courtesy of Best, M., Hitachi Global Storage Technologies, San Jose Research Center, San Jose, CA.)

Water rapidly evaporates from the nanodroplet on an unlubricated carbon overcoat. Figure 18.8a shows the dried residue of a nanodroplet on an unlubricated overcoat after exposure to brief condensation. The nanodroplet does not evaporate on a Zdol or Ztetraol lubricated carbon overcoat. Figure 18.8b shows a typical nanodroplet on a lubricated overcoat with exposure only to air at 50% RH. The nanodroplet grows larger upon brief exposure to condensing humidity. Figure 18.8c shows the larger nanodroplet typically observed after brief exposure to condensing humidity on a lubricated overcoat. The nanodroplets are deformed in the scan direction by lateral force from the imaging tip.

A thermodynamic model was developed and used to estimate the size of the initial aerosol particle [3]. The nanodroplet contact angle was calculated from the tapping mode atomic force microscope (AFM) images of 10 nanodroplets on Zdol and Ztetraol lubricated disks exposed to ambient air of approximately 50% RH. The contact angle was derived from the central profile perpendicular to the scan direction and a spherical cap model

$$\theta = \arctan\left[\frac{4hd}{d^2 - 4h^2}\right] \qquad (18.4)$$

where

    θ is the contact angle
    h is the droplet height
    d is the base diameter

By analogy with the analysis employed for a liquid aerosol droplet [30], the surface free energy of a liquid nanodroplet on the disk surface is the sum of the surface energy in the spherical cap and in the surface energy in the droplet base in contact with the disk. With A_{LV} and γ_{LV} as the surface area and surface tension of the liquid–vapor interface (droplet cap), and A_{SL} and γ_{SL} as the surface area and surface tension of the solid–liquid interface (droplet base), the nanodroplet surface energy is $A_{LV}\gamma_{LV}+A_{SL}\gamma_{SL}$.

The chemical potential of the liquid in the nanodroplet equilibrates with that in the surrounding air, as determined by the ambient RH. For an aqueous solution, the vaporization free energy change is $-nRT\ln(f/f^{\circ})$, where n is the number of moles of water in the droplet, R is the gas constant, T is absolute temperature, f is the fugacity of water vapor in the ambient air, and f° is the fugacity of liquid water in the droplet. The energy balance for the nanodroplet at equilibrium is then

$$\Delta G = -\left(\frac{\pi d^2 h}{6}\right)\left(\frac{\rho RT}{M}\right)\ln\left\{\frac{y}{a_w}\right\}+\pi dh\gamma_{LV}+\frac{\pi d^2}{4}\gamma_{SL}=0 \qquad (18.5)$$

where

    h and d are the droplet height and base diameter, respectively, as in Equation 18.4
    ρ is the density
    M is the molecular weight of water

The fugacity ratio was replaced by the ratio of the relative humidity to the activity coefficient of water in the droplet, y/a_w, where $0 < y = RH\%/100 < 1$. The water activity coefficient is determined by the composition of the solutes and concentration inside the nanodroplet. (a_w is the ratio of the vapor pressure of the solution to the vapor pressure of pure water.)

The surface energy terms in the nanodroplet free energy balance are estimated as follows. The lubricated disk surface energy was $\gamma_{SV} \approx 16.8\,\text{mJ/m}^2$ from the pure water contact angle of 92° on the lubricated disk with the Girifalco–Good–Fowkes–Young equation [30].

The interface between a Ztetraol saturated water droplet and the Ztetraol lubricated disk carbon overcoat is assumed to consist of two layers of lubricant molecules arranged in an oily bilayer with the hydrophilic hydroxyl groups in the nanodroplet. This arrangement is expected to provide a very low surface energy for droplet/disk interface, so that $\gamma_{SL} \approx 0$ is used in the nucleus size calculation. The energy balance in Equation 18.5 then reduces to

$$\ln\left\{\frac{y}{a_w}\right\} \approx \left(\frac{M}{\rho RT}\right)\frac{6\gamma_{LV}}{d} \qquad (18.6)$$

Baking in a dry oven at 90°C after equilibration with ambient air at 50% RH did not change the size of the nanodroplets. Vaporization of the water in the nanodroplets could be suppressed by the presence of an amphiphilic Ztetraol surfactant layer on the spherical cap surface. The decrease in the surface tension of a water droplet by a Ztetraol 2000 surfactant layer was measured [3]. Ztetraol 2000 decreased the surface tension of water from 72 to 22.4 mJ/m². The film of Ztetraol overcoating the nano-droplets suppresses evaporation by effectively increasing the vaporization activation energy barrier (see Refs. [31,32]).

The surface energy and the relative degree of saturation y/a_w were calculated from the AFM images of the nanodroplets. Water vapor condensation continues increasing the size of the nanodroplets and diluting the solute until $y/a_w \approx 1.01$.

The size of the initial dry aerosol particle forming the nanodroplet was estimated from values of for typical inorganic aerosol particle constituents at saturated concentration and 20°C from Refs. [33–37]. A form of the Redlich-Kister expansion for the excess Gibbs free energy [38] was employed to calculate the mole fraction of the solute from a_w at 40% RH according to $\ln(a_w) = Bx^2$, where B is a coefficient determined from the literature data, and x is the mole fraction of the solute. A histogram showing the diameter of the initial "dry" nucleus that deposited on the disk to form the measured nanodroplet, calculated from the nanodroplet volume and mole fraction for each of the model salts, is plotted in Figure 18.9. The average initial dry nucleus diameter is 110 nm, with a standard deviation of 43 nm. In the ambient environment, the aerosol particle is larger due to absorption of moisture to equilibrate with the ambient RH.

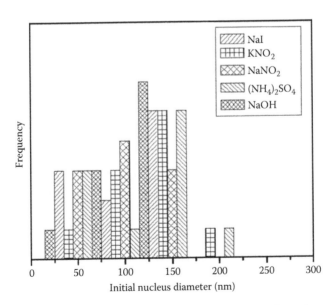

FIGURE 18.9 Histogram of the initial nucleus size calculated from the nanodroplets, assuming that the nuclei were comprised of the salts listed in the legend.

18.3.3 SILOXANE TRIBOCHEMISTRY

Sometimes, during the accelerated life testing of a new disk drive product at elevated temperature, there is a subtle but steady degradation in the magnetic performance with time. The signal loss is found to be caused by thin deposits of silicon oxide (SiOx), which increase the slider flying height (spacing between the read/write head and the disk). The origin of the SiOx is inevitably traced to a source of silicone oil, which is also referred to as poly(dimethylsiloxane) or PDMS. The source of the PDMS is usually the release liner of pressure-sensitive adhesives. The pathways for conversion of PDMS release liner residue to SiOx on the slider are shown schematically in Figure 18.10. Low molecular weight PDMS particles from the release liner or the outgassing of oligomers by depolymerization [39,40] enter the air stream in the drive enclosure as siloxanes. The siloxanes adsorb on the disk [4].

Once on the disk overcoat, the siloxanes form nanosize droplets up to 100nm high, which exceeds the flying height separation between the trailing edge of the magnetic recording slider and the disk. Due to the high relative velocity between the slider and the disk, there is a high shear impact between the droplet and the slider. The flash temperature can be as high as 450°C in an asperity contact [41]. Siloxane droplets on the disk are chemically oxidized into SiOx. It was initially proposed that acidic products from degraded disk lubricants played a key role in the conversion of PDMS to SiOx [42,43]. Subsequently, it was shown that SiOx also forms during pin-on-disk sliding contact on unlubricated magnetic recording disks, which shows that degraded lubricant is not needed. A cyclic phosphazine X1-P disk lubricant prevents SiOx formation. The adsorbed siloxanes spread on the X1-P lubricated disk overcoat rather than forming droplets with height, as they do on PFPE lubricated disks [4].

It is also likely that the siloxane droplets form on pre-existing defect sites, which act as nucleation sites. An experimental set of disks was prepared to study the effect of multiple polishing passes on the siloxane adsorption sensitivity of perpendicular magnetic recording disks. Each polishing pass sweeps an abrasive tape across the disk loaded with a weighted pad so that the tape is held in contact with the disk

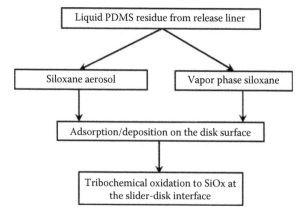

FIGURE 18.10 Schematic diagram for the pathway from PDMS release liner residue to SiOx on the slider.

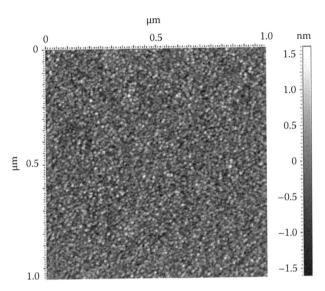

FIGURE 18.11   AFM image of a typical perpendicular magnetic recording disk surface. $1 \times 1 \, \mu m^2$ scan area. The texture shows the magnetic grain roughness. (From Joseph, J. and Dai, Q., Hitachi Global Storage Technologies, San Jose Development and Research, San Jose, CA.)

surface (see Section 18.5.2). The polishing step removes asperities that protrude from the disk surface above the slider flying height. A typical tapping mode AFM image on a $1 \times 1 \, \mu m^2$ area of the disk surface is shown in Figure 18.11. The polishing makes only subtle changes to the disk topography, which is not obvious in the AFM images.

The amount of siloxane adsorbed on disks exposed to a PDMS source and extracted for measurement is shown in Figure 18.12a. The AFM peak roughness is shown Figure 18.12b, and the pinhole count measured following exposure to HCl vapor [44] is shown in Figure 18.12c. With an increasing number of polishing passes, the siloxane adsorption amount and the pinhole count increased and the peak roughness decreased. This combination of effects is attributed to the pinhole sites acting as nucleation sites for siloxane adsorption. The siloxane is not uniformly distributed over the PFPE lubricated disk surface, because it is not detectable by reflection Fourier transform infrared (FTIR) spectroscopy or ellipsometry, which is sensitive to less than 0.01 nm of adsorbates evenly distributed on the surface over areas of several mm^2.

The most likely mechanism by which the PDMS vapor transfers to the slider from contaminated internal components is shown in Figure 18.13. The siloxane is converted to SiOx by thermal oxidation in high shear and intermittent asperity contacts. This process continues until the SiOx film deposit is thick enough to cause a detectable increase in the slider flying height, as measured by magnetic signal loss.

Siloxane contamination is prevented by monitoring the handling and composition of all disk drive internal components. For extra precaution, some of the drive

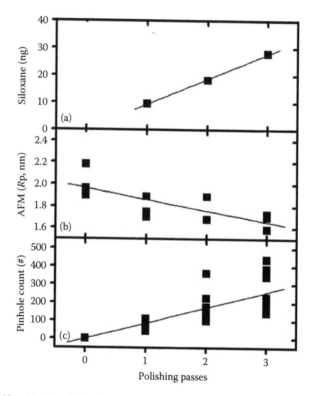

FIGURE 18.12 (a) Adsorbed siloxane amount extracted from disks (From Ohtani, T., Hitachi Production Engineering Research Laboratory, Japan.), (b) AFM peak roughness R_p (From Joseph, J. and Dai, Q., Hitachi Global Storage Technologies San Jose Development and Research Center, San Jose, CA.), and (c) pinhole count following exposure to HCl vapor (From Tu, H.-B., Hitachi Global Storage Technologies, San Jose Development, San Jose, CA.) as a function of disk polishing passes.

components are vacuum baked to extract siloxanes as well as hydrocarbons and other volatile organic compounds.

18.4 CORROSION PROTECTION

18.4.1 COBALT OXALATE CORROSION

Cobalt oxalate needles were observed on early versions of perpendicular magnetic recording media [6]. Corrosion protection, which is normally provided by the carbon overcoat, is limited by the presence of pores in the overcoat [45]. Surface roughness, overcoat thickness, and the type of overcoat deposition process (e.g., ion assisted versus sputtered) determine the overcoat porosity [44]. Pores are also formed by asperity removal or abrasion during the polishing process. Atmospheric moisture adsorbs or condenses in the pores during exposure to elevated RH. The overcoat forms the cathode for oxygen reduction along the pore walls [46–48]. Cobalt is the most active metal in the recording layers [49,50] and oxidizes to cobalt hydroxide. Corrosion

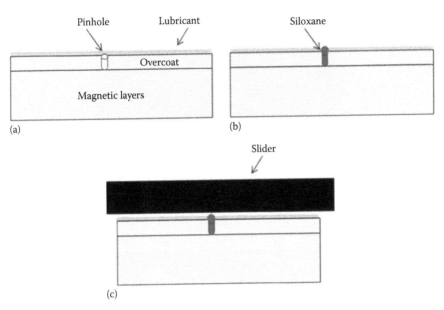

FIGURE 18.13 (a) Lubricated overcoat with a pinhole, (b) siloxane adsorbed at the pinhole, and (c) tribochemical oxidation of the adsorbed siloxane to silica.

products form mounds, which protrude from the disk surface. Corrosion thus leads to contamination and wear of the slider–disk interface [51].

An experimental cobalt alloy metallic layer with a porous carbon overcoat was designed to study corrosion. The porous overcoat is made by depositing the carbon directly on the magnetic oxide, without a cap layer (Figure 18.1). Nonporous overcoats are obtained with a cap layer (e.g., CoPtCr) deposited between the magnetic oxide and the overcoat. The disks are lubricated with 1 nm of Ztetraol 2000. The disks are exposed to elevated temperature and humidity at 65°C/100% RH (non-condensing) for 48 h (RH/T exposure). For the non-condensing RH/T exposure, the cassette of disks is sealed in a foil bag with a damp cloth and placed in the oven. When the time is up, the bag is removed from the oven and opened. Under these conditions, no liquid water droplets form on the disks.

Striking needle-like crystals and nodules were observed to form on the disks by darkfield microscopy (Figure 18.14). An AFM image of the needles and nodules is shown in Figure 18.15. The needles are rectangular shaped, 6–8 μm long, 300–600 nm wide, and 30–40 nm high. In addition, smaller nodules were observed around the needles.

The composition of the needles was determined by microanalysis, including infrared (IR) microscopy, time-of-flight secondary ion mass spectroscopy imaging, and x-ray photoelectron spectroscopy. Micro-Raman spectroscopy of the needles and nodules revealed that their spectrum matched the spectrum of cobalt oxalate.

It was determined that the cobalt hydroxide corrosion product reacts with oxalic acid to form cobalt oxalate [6]. Oxalic acid is ubiquitous in the form of atmospheric particles, and sublimes with a fairly high vapor pressure (4.7 mPa at 25°C) [52,53].

(a)   (b)

FIGURE 18.14 Darkfield micrographs of cobalt oxalate nodules (a) and needles (b) after RH/T exposure with a porous overcoat on perpendicular magnetic recording media.

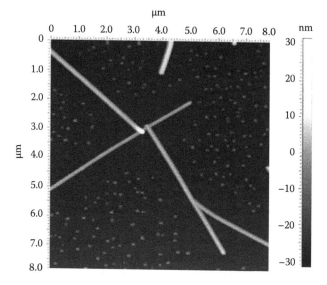

FIGURE 18.15 AFM images of cobalt oxalate nodules and needles after RH/T exposure with a porous overcoat on perpendicular magnetic recording media.

On lubricated media, the lubricant backbone chain acts as a soft template [54] for crystallization of the cobalt oxalate into microscopic needles.

18.4.2 VAPOR PHASE CORROSION INHIBITORS

One important consideration in the design of new disk drive products is corrosion of the magnetic recording head and the disk magnetic layers. An investigation was performed to determine if corrosion protection could be provided by a VCI in the disk drive. Copper corrosion is inhibited by adsorbed benzotriazoles (see Fang et al. [55]). Corrosion protection on nonferrous metals is provided through d–π^* bonding [56]. The adsorbed film thickness was measured with ellipsometry. Figure 18.16 shows the adsorbed film thickness on a silicon wafer exposed to a

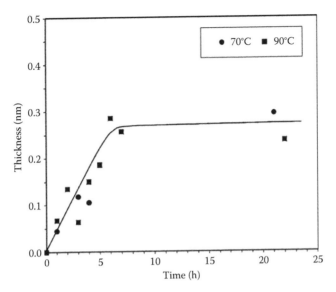

FIGURE 18.16 Thickness of VCI adsorbed on a silicon wafer, measured by ellipsometry.

commercial VCI in a $130 \, cm^3$ sealed jar at 70°C and 90°C. At the saturation vapor pressure of the VCI, an equilibrium film thickness is obtained after about 7 h. A simplified schematic drawing of how adsorbed benzotriazoles can protect nonferrous metals from corrosion by blocking access of adsorbed water to the metal surface is shown in Figure 18.17.

Subsequently, further tests were performed to optimize the inhibitor formulation for protection of metal films, including Cu, Co, $Co_{40}Fe_{60}$, $Co_{90}Fe_{10}$, IrMn, $Ni_{80}Fe_{20}$, NeFeCr, NiMn, and PtMn. Since the VCI exhibited the ability to inhibit corrosion of the magnetic layer materials, additional tests demonstrated corrosion inhibition of uncoated magnetic recording heads. However, tests done with VCIs built into disk drives exhibited (nonrecoverable) hard errors after extended thermal cycling. The errors were caused by nano-crystals of VCI, which form when the air inside the disk drive is saturated with VCI vapor. The nano-crystals caused the errors by increasing the separation between the magnetic recording head and the disk magnetic layers.

In order to avoid the formation of crystals, the VCI could only be used for corrosion protection in the disk drive if it was effective at vapor phase concentrations below the saturation concentration. Subsaturation tests could not easily be performed in a closed system because the VCI vapor dissolves in water and greatly diminishes the vapor pressure, which allows corrosion of the metal film samples. A flow-through test apparatus was set up to expose the metallic films to subsaturation concentration of VCI at nearly 100% RH. A schematic of the flow-through subsaturation VCI corrosion test apparatus is shown in Figure 18.18. The VCI was melted and poured into the sublimation flow cell and allowed to solidify. The sublimation flow cell contained a labyrinth of channels over which air flowed to equilibrate with the VCI at the set temperature. The samples were initially pretreated overnight with

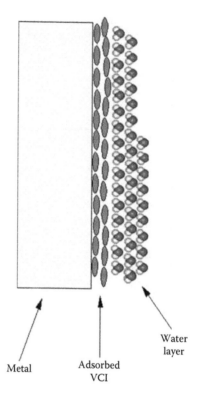

Water
layer

Metal   Adsorbed
               VCI

FIGURE 18.17 A simplified schematic model illustrating how benzotriazole-based VCI protects nonferrous metals from corrosion by preventing access of adsorbed water to the metal surface.

VCI-saturated dry air to establish the protective layer of adsorbed VCI. The VCI-saturated air was then mixed with clean humid air to obtain the subsaturation VCI concentration for the test. The concentration/ flow rate characteristics of the sublimation flow cell and the test results are shown in Figure 18.19. The VCI protected Cu from corrosion for 1 week at degrees of saturation 0.5 and 0.25.

A method was devised to maintain the vapor pressure below saturation within the disk drive enclosure using a source and drain arrangement [7,57]. Equation 18.7 gives the degree of saturation as a function of time t in volume V:

$$\frac{C(t)}{C_0} = \left[\frac{CD_s}{CD_s + CD_d}\right]\left[1 - \exp\left(-\frac{CD_s + CD_d}{V}\right)t\right], \qquad (18.7)$$

where CD_s and CD_d are the diffusive conductances of the source and drain connecting them to the disk drive enclosure. The VCI source and sink mechanism was incorporated in disk drives, and they were subjected to temperature humidity cycling tests for 10 weeks. This method for VCI application inhibited corrosion in disk drives without VCI crystallization at the slider disk interface.

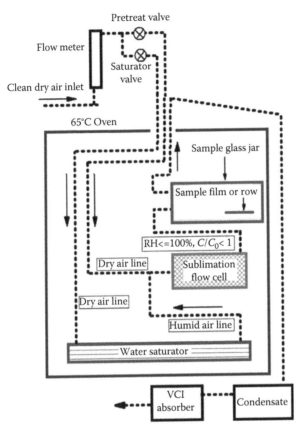

FIGURE 18.18 A schematic of the flow-through subsaturation VCI corrosion test apparatus. The sample was a metallic film on a 1 in. diameter Si or glass wafer or an uncut slider row
containing a series of non-overcoated magnetic recording heads.

18.4.3 NOVEL OVERCOATS

The functional layers for recording on thin-film magnetic recording disks comprise
metallic or metal oxide films. The body of the magnetic recording slider is a TiC/
Al_2O_3 ceramic. Early investigators in magnetic recording tribology discovered severe
sticking and wear between the metallic layers of the disk and the ceramic slider [58].
When smooth, clean ceramic and disk magnetic layer metal surfaces are brought
into contact, the adhesion exceeds the tensile strength of the metal [59]. A 30 nm
thick carbon film overcoating the metallic layers on the disk increased the wear
resistance in contact with a ceramic slider by a factor of six [60]. An excellent review
on the physical properties of thin carbon films is given by Tsai and Bogy [61]. The
slider–disk interface tribology was subsequently further improved by incorporating
a 4 nm thick carbon overcoat on the air bearing surface of the magnetic recording
slider [62]. Presently, all magnetic recording disks and sliders are overcoated with a
thin film of carbon. Non-carbon overcoats have been tested as a means to improve

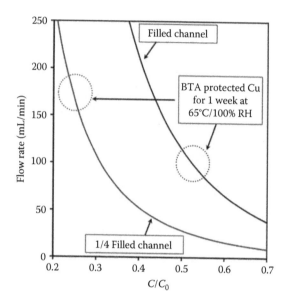

FIGURE 18.19 The flow rate through the sublimation flow cell and the degree of VCI saturation C/Co at the sample for benzotriazole. The dotted circles show the levels at which corrosion tests were performed and the VCI protected the Cu films from corrosion.

the corrosion resistance of thinner overcoats, for example, silicon nitride [63] and titanium silicon nitride [64]. However, the application of silicon nitride films as the outer overcoat on magnetic recording surfaces has been limited by hydrolytic instability [65] and tribochemistry [66].

Carbon overcoated magnetic recording disks are dip coated with a solution of fluorohydrocarbon lubricant (Section 18.2). As mentioned, a current goal in the magnetic recording industry is for the lubricant layer thickness to be approximately the monolayer thickness of the polymeric lubricant chain, and for the lubricant chain to be made shorter and stiffer and more fully chemisorbed to the disk overcoat to decrease the head–disk spacing and to prevent lubricant transfer from the disk to the slider [9]. The durability of the slider–disk interface generally decreased with decreasing lubricant thickness and increasing chemisorbed fraction [67]. Given the trend of thinner overcoats, decreasing thickness, and increased chemisorption, it seems that the next step should be to incorporate the lubricious properties imparted by the fluorohydrocarbon lubricant directly into the carbon overcoat by vacuum deposition of a fluorohydrocarbon overcoat. The feasibility of this approach has been demonstrated [10,68,69].

The novel overcoat system must provide low friction and wear. Since magnetic recording surfaces are smooth, the contact area can be large. In this case, dispersion force interaction is the main contributor to the adhesion component of the friction force on the slider. The adhesion force applies a torque about the pivot point of the slider, which is balanced by the lateral (friction) force of the suspension on the pivot point [12]. Thus, one requirement in the design of a novel overcoat system is to minimize the dispersion force between the disk and slider.

The interaction free energy per unit area between two material surfaces separated by a planar gap is [70,71]

$$E = \frac{kT}{2\pi} {\sum_{n=0}^{\infty}}' \int_0^\infty \rho \, d\rho \ln G(\xi_n),$$ (18.8)

where
  k is the Boltzmann constant
  T is absolute temperature
  ρ is the surface mode wave vector
  $\xi_n = n(2\pi kT/\hbar)$
  $\hbar$ is the reduced Planck constant

When the spacing gap is L, the force per unit area across the gap is [72]

$$\Pi = -\frac{\partial E}{\partial L} = -\frac{kT}{2\pi} {\sum_{n=0}^{\infty}}' \int_0^\infty \rho \, d\rho \frac{1}{G(\xi_n)} \frac{\partial G(\xi_n)}{\partial L}.$$ (18.9)

The prime on the summation in Equations 18.8 and 18.9 indicates that the $n=0$ term is given half weight. Retardation is not included because the film thickness is less than 10 nm [71], and only the upper layers are considered in the following. For an air gap of thickness L between two materials indicated by subscripts a, l, and c, $G(\xi_n) = 1 - \Delta_{la}\Delta_{ca}e^{-2\rho L}$. Δ_{la} and Δ_{ca} are functions of the dispersion spectrum for each material pair and they will be defined later.

The adhesion stress, σ_{adh}, is defined as the limiting case of Equation 18.9 as $L \to 0$. For dispersive adhesion between materials l and c

$$\sigma_{adh} = -\Pi(L \to 0) = \frac{kT}{3\pi} \rho_c^3 {\sum_{n=0}^{\infty}}' \frac{\Delta_{la}\Delta_{ca}}{1 - \Delta_{la}\Delta_{ca}}.$$ (18.10)

The surface energy of material l is [73]

$$\gamma_l^d = -\frac{kT}{8\pi} \rho_c^2 {\sum_{n=0}^{\infty}}' \ln\left(1 - \Delta_{la}^2\right),$$ (18.11)

where ρ_c is the cutoff wavelength. The ρ_c is determined from the dispersive surface energy measured for material l [12].

The well-known Hamaker constant is approximated by [73]

$$A_{lac} = \frac{3kT}{2} {\sum_{n=0}^{\infty}}' \Delta_{la}\Delta_{ca}.$$ (18.12)

Before calculating the surface energy and adhesion stress, it is necessary to define the quantities Δ_{la} and Δ_{ca}. The definition arises from a solution of Maxwell's equations and summing the allowed normal mode frequencies [71]

$$\Delta_{la} = \frac{\varepsilon_l(i\xi_n)-1}{\varepsilon_l(i\xi_n)+1} \tag{18.13}$$

and similarly for Δ_{ca}, with $i = \sqrt{-1}$, $\varepsilon_l(i\xi_n)$ is the dispersion spectrum of the material l, and $\varepsilon_c(i\xi_n)$ is the dispersion spectrum of material c. The dispersion spectrum of air is taken to be 1.

The dispersion force between the disk and slider is determined by the dispersion spectrum of the materials in the layers comprising the magnetic recording interface. Application of the Lifshitz theory to calculate the dispersion force for the magnetic recording interface layers has recently been shown by Dagastine et al. [72,74]. The calculation requires the film thickness and the dispersion spectrum $\varepsilon_j(i\xi)$ for each material j [75]:

$$\varepsilon(i\xi) = \varepsilon_0\varepsilon_r(i\xi) = \varepsilon_0\left(1 + \frac{2}{\pi}\int_0^\infty \frac{x\varepsilon''(x)}{x^2+\xi^2}dx\right) \tag{18.14}$$

The material subscript j is not shown, $i\xi$ is the complex frequency, and $\varepsilon''(x)$ is the electromagnetic absorption spectrum.

The biggest challenge to calculating the dispersion force for a novel overcoat system is to obtain the electromagnetic absorption spectrum [73,76]. The absorption spectrum can, in principle, be derived from dielectric, ultraviolet/visible (UV/Vis), FTIR, and ellipsometric spectroscopy. Alternatively, for ceramic materials it is more convenient to employ vacuum UV or electron energy loss spectroscopy [70]. In the following, the dispersion spectrum is approximately calculated from the dielectric and FTIR spectra and the refractive index at a single wavelength for the novel fluorocarbon overcoat.

The starting point is the low-frequency limit to obtain the static dielectric constant $\varepsilon_r(0)$. The limiting low-frequency absorption is provided by the dielectric spectra measured at elevated temperatures with temperature-frequency superposition. The dielectric spectroscopy data from Ref. [69] were frequency temperature shifted relative to reference temperature $T_o = 22°C$ to obtain the master curves shown in Figure 18.20. The shift factors are listed in Table 18.4. The dielectric loss factor in Figure 18.20 shows that the film had a significant conduction term. The conduction contribution to the loss is $\sigma/(\omega a_{T_o}\varepsilon_o)$, where $\omega = 2\pi f$, σ is the conductivity, and ε_o is the permittivity of free space. The fit to the conduction contribution at low frequency $f < 1$ (1/s) is shown by the line in Figure 18.20. The conduction contribution was subtracted from the loss to obtain the dielectric absorption spectrum. The dielectric absorption was fitted by a Debye model with three terms

$$\sum_{1}^{3} \frac{\varepsilon_i \tau_i \omega a_{T_o}}{1 + \left(\tau_i \omega \, a_{T_o}\right)^2} \tag{18.15}$$

The curve fit to the Debye model is shown by the smooth curve in Figure 18.20. The dielectric relaxation strengths and relaxation times from the nonlinear regression curve fit are listed in Table 18.5. When the Debye model of Equation 18.15 is substituted into Equation 18.14 with $\xi = 0$, then the static dielectric constant is

$$\varepsilon_r(0) = 1 + \sum \varepsilon_i = 14.4.$$

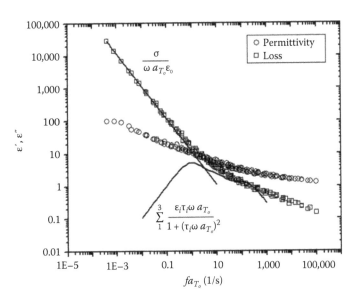

FIGURE 18.20 Master curves for the dielectric spectrum (loss ε'' and permittivity ε') of the fluorocarbon overcoat from spectra measured between 20°C and 200°C.

TABLE 18.4
Frequency Temperature Shift Factor for Fluorocarbon Overcoat Dielectric Spectrum Relative to Reference Temperature $T_o = 22°C$

Temperature (°C)	$\log_{10}\left(a_{T_o}\right)$	Temperature (°C)	$\log_{10}\left(a_{T_o}\right)$
22	0	140	−2.2
40	−0.35	160	−2.5
60	−0.6	180	−2.8
80	−1.1	200	−3.1
100	−1.4	220	−3.4
120	−1.8		

TABLE 18.5
Dielectric and Absorption Properties of Fluorocarbon Overcoat

Property	Symbol	Units	Value
Conductivity	σ	pS/m	634
Dielectric relaxation strengths	ε_i	—	9.63, 2.76, 2.04
Dielectric relaxation times	τ_i	s	1.08, 0.106, 0.0077
Static relative permittivity	$\varepsilon_r(0)$	—	14.4
Refractive index [68] (632.8 nm)	n	—	1.49
IR coefficient	C_{IR}	—	12.18
IR characteristic frequency	ω_{IR}	rad/s	2.6×10^{14}
UV coefficient	C_{UV}	—	1.22
UV characteristic frequency	ω_{UV}	rad/s	1.7×10^{16}

The summation in the Lifshitz theory, Equation 18.8, samples the dispersion spectrum at integer multiples of the complex frequency:

$$\xi_n = (2\pi k T / \hbar) n = (2.5 \times 10^{14} \text{ rad/s}) \, n \text{ at } 25°C \tag{18.16}$$

Therefore, the lowest frequency that is sampled by $n=1$ is in the IR region of the spectrum. The next 400 terms span the visible and extend into the UV region of the absorption spectrum. The dispersion spectrum is approximated by a two-frequency Ninham–Parsegian construction [71,73,76]:

$$\varepsilon_r(i\xi) = 1 + \frac{C_{IR}}{1 + (\xi/\omega_{IR})^2} + \frac{C_{UV}}{1 + (\xi/\omega_{UV})^2} \tag{18.17}$$

Coefficients C_{IR} and C_{UV} and frequencies ω_{IR} and ω_{UV} represent the contributions of the IR and UV regions, respectively. Across the transition from the low frequency to the IR region, $C_{IR} = \varepsilon_r(0) - n^2$. From Ref. [69], there are IR absorption peaks at 1800 and 1000 cm^{-1}, so that $\omega_{IR} \approx 2.6 \times 10^{14}$ rad/s. The level shift across the transition from the UV to the visible region of the spectrum gives $C_{UV} = n^2 - 1$. There are no UV data available for the fluorocarbon overcoat, so the characteristic UV frequency is estimated to be in between the values for PTFE and fluorinated ethylene propylene (FEP) in Ref. [73], Table 18.2, $\omega_{UV} \approx 1.7 \times 10^{16}$ rad/s. Values for the coefficients in Equation 18.17 are given in Table 18.5. Note that the free charge response [77] is not included, so the dispersion spectrum of the fluorocarbon overcoat may be underestimated in the limit of low n.

The dispersion spectrum for the fluorocarbon overcoat, along with two types of carbon overcoats and PFPE lubricant, is shown in Figure 18.21. The Ninham–Parsegian coefficients for the amorphous hydrogenated carbon (a:CH) and PFPE are given in Ref. [11]. The coefficients for the diamond-like carbon (DLC) are derived

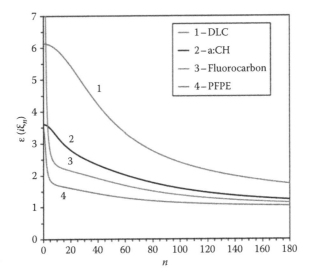

FIGURE 18.21 Dispersion spectra for DLC, a:CH, the novel fluorocarbon overcoat, and PFPE disk lubricant. The horizontal axis n is the index subscript in the summation of the Lifshitz theory, Equation 18.8. Except for the fluorocarbon, the dispersion spectra were reproduced from a curve fit to Figure 2 in Refs. [72,74].

from a nonlinear regression fit to the curve from Figure 2 in Ref. [72]. The regression fit provides two terms both in the UV region of the spectrum (Table 18.6). The visible refractive index estimated from $\sum C_{UV} = n^2 - 1$ was close to the literature value of 2.419 for diamond [78]. The static relative permittivity estimated from $\varepsilon_r(0) = 1 + \sum C_{UV}$ was within the range of 5.5–10 from the literature for diamond [79]. It is reassuring that back-calculating these properties from the dispersion spectrum and the Ninham–Parsegian construction provides reasonable values for DLC.

TABLE 18.6
Ninham–Parsegian Coefficients for DLC Derived from a Curve Fit to Figure 18.2

Property	Symbol	Units	Value
Static relative permittivity (calculated)	$\varepsilon_r(0)$	—	6.14
Refractive index (calculated)	n	—	2.48
UV coefficient	C_{UV}	—	4.352
UV characteristic frequency	ω_{UV}	rad/s	1.141×10^{16}
UV coefficient	C_{UV}	—	0.785
UV characteristic frequency	ω_{UV}	rad/s	5.590×10^{16}

Source: White, L.R. et al., *J. Appl. Phys.*, 97, 104503, 2005.

The dispersion spectrum provides the surface energy of the overcoat in air, the Hamaker constant for trilayer sytems of materials in which the middle layer is air, and the adhesion stress between material pairs (which contributes to the tensile strength of asperity junctions in sliding) [11]. The surface energies are listed in Table 18.7. The fluorocarbon surface energy lies between that of the PFPE liquid lubricant and the a:CH. The Hamaker constants calculated from the dispersion spectra are listed in Table 18.8. The adhesion stress for pairs of materials is listed in Table 18.9. The unlubricated carbon overcoats have the highest adhesion stress, which is probably why they do not make good sliding pairs alone. The adhesion stress between the carbon and PFPE lubricant is the next lowest in the series. This accounts for the tendency of lubricant to transfer to the slider unless it is chemisorbed onto the disk overcoat. The surface energy fluorocarbon/fluorocarbon material pair is slightly less than that of the a:CH/fluorocarbon material pair, suggesting that it would be best but not necessary

TABLE 18.7
Surface Energy in Air for Fluorocarbon Overcoat, Conventional Carbon Overcoats, and PFPE Disk Lubricant Calculated from the Dispersion Spectrum

Material	Surface Energy (mJ/m²)
DLC	127.8
a:CH	42.0
Fluorocarbon	27.2
PFPE	13.0

The cutoff wavelength for PFPE was 0.221 and the cutoff wavelength for a:CH carbon overcoat of 0.278 nm [11] was used for the fluorocarbon.

TABLE 18.8
Hamaker Constants for Materials across Air Calculated from Dispersion Spectra

Material	Hamaker Constant (zJ)
DLC/air/DLC	311.0
a:CH/air/a:CH	111.0
a:CH/air/fluorocarbon	87.4
Fluorocarbon/air/fluorocarbon	70.6
a:CH/air/PFPE	48
PFPE/air/PFPE	23
a:CH/PFPE/air	−27.9

TABLE 18.9
Adhesion Stress in Air for Pairs of Fluorocarbon Overcoats, Conventional Carbon Overcoats, and PFPE Disk Lubricant Calculated from Dispersion Spectrum

Materials	Adhesion Stress (MPa)	State
DLC/DLC	3002	Solid/solid
a:CH/a:CH	886	Solid/solid
DLC/PFPE	571	Solid/liquid
a:CH/PFPE	356	Solid/liquid
a:CH/fluorocarbon	346	Solid/solid
Fluorocarbon/fluorocarbon	309	Solid/solid
PFPE/PFPE	165	Liquid/liquid

to coat both the disk and the slider with the fluorocarbon overcoat. The lowest adhesion stress is between the PFPE liquid lubricant and PFPE, which accounts for the low friction and wear in the present-day magnetic recording slider–disk interface. While the adhesion stress is higher for the novel fluorocarbon overcoat than that for the PFPE liquid lubricant, it may be sufficiently lower than that of the unlubricated carbon overcoats to provide a good tribological interface with no liquid lubricant transfer to the slider [80].

18.5   DISK PROCESSING

18.5.1   SUBSTRATE CORROSION

Substrates for desktop disk drives (95 mm diameter) and most server disk drives are made from an AlMg alloy plated with NiP. A thin layer of Ni carbonates forms on the NiP during exposure to ambient air in storage and shipping. Incoming substrates are washed in a commercial disk substrate washer. Substrate washing involves sonication in a mildly acidic solution and scrubbing with brushes. Substrate washing completely removes the carbonate from the NiP without leaving any residue detectable by x-ray photoelectron spectroscopy (XPS) or FTIR spectroscopy. Carbonate decreases the orientation ratio of longitudinal magnetic recording media [13], which degrades the fidelity of the magnetic recording layers. Rapid regrowth of the adventitious carbonate layer is a form of contamination, which may also occur in the several hours' time between substrate washing and magnetic layer deposition.

Carbonate regrowth after washing NiP-plated AlMg substrates was tested on untextured substrates that are used for some types of perpendicular recording media. Figure 18.22 shows the composition of the carbonate layer on untextured perpendicular magnetic recording disk substrates. The composition of the carbonate layer on untextured perpendicular magnetic recording disk substrates is the same as that on textured longitudinal magnetic recording disk substrates.

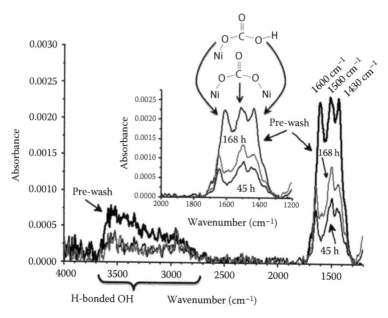

FIGURE 18.22 FTIR spectra and peak assignments showing the Ni carbonate and bicarbonate formation on untextured NiP-plated AlMg substrates for perpendicular magnetic media. Substrates were stored in a sealed polyethylene bag in ambient air after washing.

Initially, atmospheric moisture and carbon dioxide combine to form carbonic acid, as in the reaction shown in Equation 18.18.

$$H_2O + CO_2 \longrightarrow H_2CO_3 \qquad (18.18)$$

The carbonic acid, H_2CO_3, then reacts with Ni to form monodentate Ni carbonate, $NiHCO_3$, as in the reaction shown in Equation 18.19, which gives rise to the IR absorption peaks near 1430 and near 1600 cm^{-1} in Figure 18.22.

$$H_2CO_3 + Ni^+ \longrightarrow NiHCO_3 + H^+ \qquad (18.19)$$

Subsequently, the $NiHCO_3$ transforms to bidentate Ni carbonate, Ni_2CO_3, according to the reaction in Equation 18.20, which gives rise to the IR absorption peak near 1520 cm^{-1}.

$$NiHCO_3 + Ni^+ \longrightarrow Ni_2CO_3 + H^+ \qquad (18.20)$$

The reaction rate appears to be zero order in CO_2 because the reaction rate was unchanged with an order of magnitude reduction in the CO_2 concentration.

Carbonate formation and removal by washing apparently degrades the (longitudinal magnetic recording) orientation ratio through a subtle alteration of the surface topography. For example, in one case the rms roughness of the freshly textured substrate was 0.23 nm, 0.29 nm after one wash, and 0.34 nm after a second wash, with carbonate allowed to grow between texturing and washing. The differences in the surface topography were detected as subtle changes in the power spectral density of the surface images, but the exact nature of the differences could not be reproducibly quantified. Significant alteration in the surface topography by carbonic acid corrosion and removal by washing are not surprising because the length scale of the Ni carbonate molecular structure is 0.3 nm, which is comparable to the rms roughness of the substrate.

In addition to altering the disk topography, the carbonate can be decomposed into carbon and gaseous products by energetic ions during sputtering of the magnetic recording layers. Some of the carbon remains on the NiP surface at the interface with the first deposited metallic layer. An example of the carbon contamination at the NiP interface of a longitudinal media film stack sputtered on an unwashed NiP-plated AlMg substrate is shown in Figure 18.23. This level of carbon contamination significantly decreased the orientation ratio.

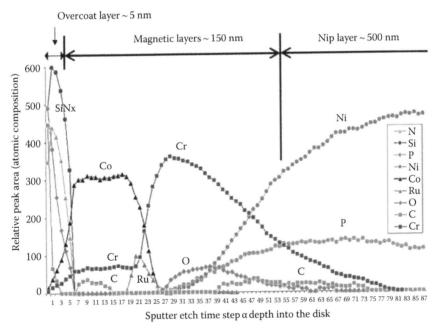

FIGURE 18.23 XPS depth profile of a longitudinal recording media sputtered on an unwashed substrate that contained a significant amount of Ni carbonate corrosion. Carbon contamination from the carbonate is present in the interface between the magnetic layers and the NiP and also in the magnetic layers. Depth scale is approximate and is not proportional to the sputter etch time step. (Courtesy of Guo, X.-C., Hitachi Global Storage Technologies, San Jose Research Center, San Jose, CA.)

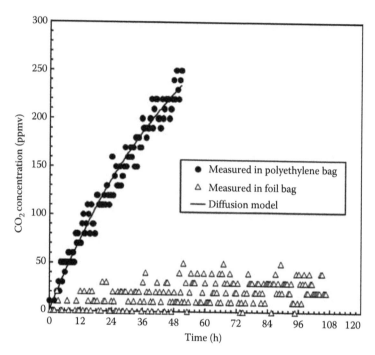

FIGURE 18.24 CO_2 concentration inside a sealed bag as a function of time for the thin polyethylene bag and a foil bag (see Table 18.10). Bags were sealed and exposed to room air containing 400–450 ppmv CO_2 (ppmv is parts per million by volume). (Measurements were done with the assistance of Brown, C., at the Hitachi Global Storage Technologies, San Jose Research Center, San Jose, CA.)

Since the carbonate formation requires CO_2 and water, its formation could, in principle, be limited by controlling either one or both reactants. Packaging methods were investigated to determine if it was possible to prevent the carbonate contamination by shipping cassettes of freshly washed substrates in bags filled with dry air or nitrogen and by storing them in dry chambers between wash and sputter.

The RH inside the bag holding a cassette of disks cannot be easily maintained at a low level because of water absorption into the polycarbonate cassettes from ambient air during storage in manufacturing. The equilibrium water absorption in the cassette material is about 0.15 wt% at 20°C and 50% RH. Even after initially filling it with dry air, the RH inside the bag containing the cassette reaches about 40% at equilibrium.

It is more practical to limit the CO_2 in the package. CO_2 permeation through a sealed bag was measured by sealing a CO_2 recorder inside the bag, initially purged with nitrogen. The CO_2 permeation rates through a polyethylene and a foil bag are shown in Figure 18.24.

The model for diffusion through the polyethylene bag is derived with a mass balance and Fick's First Law as follows:

$$\frac{dC(t)}{dt} = k\left[C_\infty - C(t)\right], \tag{18.21}$$

where

$$k = \frac{DA}{lV} \tag{18.22}$$

Combined with initial and final conditions, $C(t)$ is the CO_2 concentration inside the package at time t, C_∞ is the CO_2 concentration in the surrounding air, D is the diffusion coefficient through the polyethylene bag material, A is the polyethylene bag surface area, l is the polyethylene bag film thickness, and V is the gas volume inside the polyethylene bag. Equations 18.21 and 18.22 then provide

$$C(t) = C_\infty + \left\{C(0) - C_\infty\right\}\exp\left[-kt\right] \tag{18.23}$$

The diffusion properties for polyethylene bags with two different film thicknesses derived from these measurements are listed in Table 18.10. The diffusion coefficients are close to those reported by Michaels and Bixler [81] for a degree of crystallinity of 0.9–0.95. In order to avoid forming more than 10% of a carbonate monolayer, cassettes of freshly washed NiP-plated substrates can be vacuum sealed into polyethylene bags with a pump-down pressure of at least 99.898 kPa or more, backfilled with nitrogen to 71 kPa; the sealed polyethylene bags must be encapsulated into foil bags within less than 10 min.

It is also possible to inhibit carbonate formation on substrates in the manufacturing environment after wash by storing them in transfer carts made of glass and metal that are purged with clean dry air. Cassettes of NiP-plated AlMg substrates were placed in a transfer cart that was especially equipped with a clean dry air regulator and partially sealed to provide a low humidity within minutes after closing the door. The RH inside the cart after closing the door is shown in Figure 18.25a. The RH decreased below 5% within approximately 5 min after closing the door. The water contact angles measured on the substrates for the dry and ambient storage conditions are shown in Figure 18.25b. The contact angle provides a sensitive measurement for the presence of carbonate on the substrate [13], and there was little or no carbonate

---

TABLE 18.10

Specifications and Diffusion Properties for Two Different Polyethylene Bags Used to Store Cassettes of Magnetic Recording Disk Substrates

Bag Type	Bag Film Thickness, l (μm)	Bag Volume, V (m³)	Bag Surface Area, A (m²)	Rate Constant, k (1/s)	CO_2 Diffusion Coefficient, D (m²/s)
Thin	100	2.8×10^{-3}	0.17	4.6×10^{-6}	7.5×10^{-12}
Thick	200	6.5×10^{-3}	0.25	1.8×10^{-6}	9.1×10^{-12}

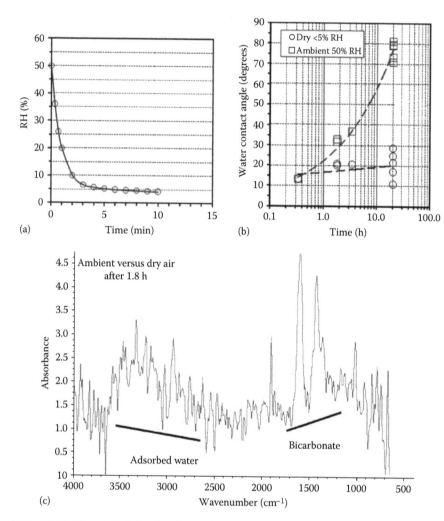

FIGURE 18.25 (a) Relative humidity after closing the door of the dry storage cart, (b) water contact angle on the substrates comparing dry and ambient storage conditions versus time after wash, and (c) FTIR absorption spectrum of the substrate stored in ambient air. (Measurements with the assistance of Villarreal, T., at the Hitachi Global Storage Technologies, San Jose Development Laboratory, San Jose, CA.)

formed on the substrates stored in the dry cart, even after nearly 10 h. The FTIR spectra were also measured on these substrates to detect the bicarbonate/carbonate formation. Figure 18.25c shows the FTIR absorption spectrum of the substrate stored in ambient air. The background spectrum was the single-beam spectrum of the substrate stored in dry air. Both spectra were measured 1.8 h after wash. The absorption peaks show that, after only 1.8 h, the surface of the freshly washed substrate stored in ambient air contained significantly more adsorbed water and bicarbonate than the substrate stored in dry air.

18.5.2   FINAL POLISHING

To satisfy the demand for increasing magnetic recording data storage density, it is necessary to provide smoother magnetic recording disk surfaces. The surface must be smooth enough to avoid asperity contacts with the read/write head sensor element. Disk asperities are removed in the final stages of the thin-film magnetic recording disk manufacturing process. After the magnetic layers and carbon overcoat are deposited in a vacuum chamber, the disks are allowed to cool in air. Next, the disks are dip coated [82] with about 1 nm of lubricant (see Section 18.2). Finally, the disks are polished with a mild abrasive tape to remove asperities [14], typically above 5 nm for advanced disk products. The disk surface roughness is less than 0.5 nm, while peak-to-valley roughness of the polishing tape is about 5 μm. The polishing tape is usually traversed across the disk surface from the inner to the outer diameter. Typically, the polishing is done by a single pass of the tape across the spinning disk surface. The disk rotation rate can be adjusted to maintain a constant linear velocity during the polishing pass. The tape is not advanced during the polishing pass. At the end of the polishing pass, the tape is advanced, so that a fresh abrasive tape surface is used for each new disk. Polishing must avoid damaging or scratching the 3–4 nm thick carbon overcoat that protects the magnetic layers from corrosion [83] while machining down the asperities.

This section describes measurements of the friction force between the polishing tape and the thin-film recording disk. A laboratory benchtop friction tester was set up to automatically load the tape with a polishing pad and sweep it radially outward across the disk in a close approximation to the automated polishing tools in disk manufacturing. The applied load and friction force are simultaneously recorded to measure the friction coefficient. A friction and wear model is developed for relating changes in the friction coefficient to the load-dependent and adhesion-dependent terms [84].

A series of measurements were done on the laboratory benchtop friction tester in which the same tape and disk were subjected to eight polishing passes with about 1 min delay in between each pass (no tape advance between passes). The friction force measured during each pass is shown in Figure 18.26a. The horizontal axis shows the elapsed time during the polishing pass. With time, the friction force gradually increased to an asymptotic level during each polishing pass. The asymptotic level increased with the number of polishing passes. The friction traces form a family of curves that can be superimposed on one another by shifting along the time axis to overlap the asymptotic region, as shown in Figure 18.26b.

The sliding distance s is the product of the disk linear velocity and the elapsed time during each polishing pass. The increase in friction with s is associated with wear of the high points of the tape surface topography. Micrographs of a polishing tape surface after an extended period of sliding are shown in Figure 18.27. The high points of the tape surface topography are abraded while sliding on the disk surface, increasing the contact area between the tape and the disk. These results show that it is not possible to measure the initial state of the tape by plotting the friction force as a function of time because the tape surface asperity contact area begins to change as

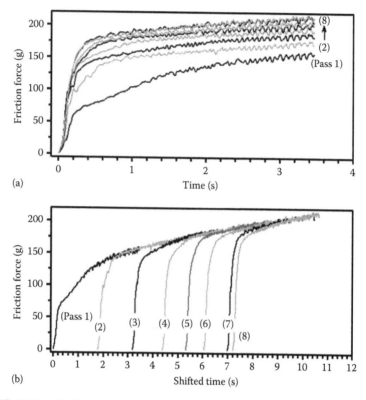

(a)

(b)

FIGURE 18.26 (a) Friction force versus time during multiple polishing passes across the disk with the same tape and disk and (b) friction traces from (a) shifted along the time axis to overlap one another after reaching steady state. Normal load 100 g, linear velocity 3.45 m/s, 7 mm diameter Poron® foam pad.

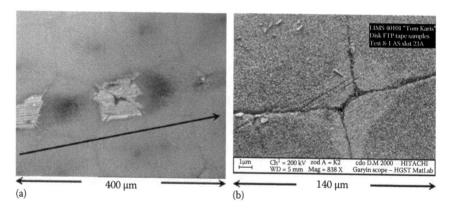

(a)   (b)

FIGURE 18.27 (See color insert.)   Micrographs of disk polishing tape after an extended period of sliding on a disk. (a) Optical micrograph of worn region. The arrow shows the sliding direction. (b) SEM micrograph of worn region. (Courtesy of Stone, G., Hitachi Global Storage Technologies, San Jose Materials Analysis Laboratory, San Jose, CA.)

soon as there is contact with the disk. The following model is developed to account for the effect of the change in contact area of the tape on the friction force.

The tape friction force is discussed on the basis of Amontons' law with adhesion [84]:

$$F = L\mu + \tau_0 A, \tag{18.24}$$

where
  F is the friction force
  L is the normal load
  μ is the friction coefficient
  τ_0 is the shear stress due to adhesion
  A is the contact area

The change in the contact area with sliding distance is estimated with Archards' wear law, as follows. The wear volume

$$V = \frac{kLs}{H}, \tag{18.25}$$

where
  k is a wear coefficient
  s is the sliding distance traveled by the tape on the disk
  H is the composite hardness

In this case, the tape hardness dominates the composite hardness because the tape hardness is much less than that of the disk. The tape nanoindentation hardness measured with a Berkovich indenter on a CSM ultrananoindentation hardness tester (UNHT) was approximately 65–125 MPa with 200–800 nm indentation depths, respectively, and a 1 mN maximum load. The tape binder predominantly undergoes plastic flow because the remnant indentation depth is almost 90% of the maximum indentation depth. The disk nanoindentation hardness is typically 5–10 GPa.

The contact area is approximately related to the sliding distance through $A = A_0 + \delta A$ and assuming that $\delta A \propto \delta(V^{2/3})$, where δA represents an incremental increase in the initial contact area A_0. Equation 18.24 is then approximately rewritten as

$$\frac{F}{L} \approx \mu + \frac{\tau_0 A_0}{L} + \tau_0 \left(\frac{k}{H}\right)^{2/3} \frac{1}{L^{1/3}} s^{2/3}. \tag{18.26}$$

Thus, a plot of F/L versus $s^{2/3}$ should be linear, with intercept

$$\mu + \frac{\tau_0 A_0}{L} \tag{18.27}$$

and slope

$$\tau_0 \left(\frac{k}{H} \right)^{2/3} \frac{1}{L^{1/3}}. \tag{18.28}$$

The experimental slope and intercept can be measured as a function of L to determine μ, $\tau_0 A_0$, and $\tau_0(k/H)^{2/3}$.

A set of measurements were done on the benchtop friction tester to explore the load dependence embodied in Equation 18.26. The disks were typical carbon overcoated 95 mm diameter desktop disks lubricated with 1 nm of 20% chemisorbed A20H/Zdol. The polishing pad was a round, 7 mm diameter, 2.5 mm thick Poron (Rogers Corp., open cell microcellular polyurethane foam with Shore D hardness 17). The polishing tape was an alumina abrasive in a binder on a Mylar® backing and was 9.525 mm wide (Figure 18.27). Fresh sections of tape and a fresh disk were used for each polishing pass.

The tape is loaded onto the spinning disk by pressing down on the Mylar backing with the foam polishing pad. The pad load is provided by an air cylinder. The normal load and the friction force are measured simultaneously by data acquisition with National Instruments LabView™. The data acquisition is started as the tape is being loaded onto the innermost radius of the disk surface. While the pad is loading, the tape holder assembly is begun traversing radially outward across the disk surface at 2 mm/s. When the outer edge of the loaded portion of the tape reaches the outer diameter of the disk, the data acquisition is stopped and the tape is unloaded from the disk. For each test at nominal load L, the disk is rotated at a constant linear velocity of 3.45 m/s. The disk rotation rate is adjusted to maintain constant linear velocity while the tape holder assembly translates across the disk. The linear velocity is $v = \omega \times r$, where the disk rotation rate is ω and the radial position of the inner edge of the polishing tape is r. The sliding distance is $s = v \times t$, where v is the linear velocity and t is the elapsed time during the polishing pass (sliding time).

Four measurements of F and L versus t were done at each load, and F/L versus $s^{2/3}$ was averaged to obtain the average F/L traces at each load, as shown in Figure 18.28a.

Notice that although the quantity F/L in Equation 18.26 should increase linearly with $s^{2/3}$ at each load, there is an initial transient. The linear regression was limited to the steady-state portion of the friction trace above $s^{2/3} = 1.5 \, m^{2/3}$ or $s = 1.84 \, m$. For example, the linear fits to the 60 and 140 g friction traces are shown in Figure 18.28a.

As mentioned previously, further insight into the friction mechanism can be derived from the slope and intercept of the linear regression fit to Equation 18.26. The intercept of the linear fit to F/L versus $s^{2/3}$ is given by Equation 18.27, and the slope is given by Equation 18.28. The experimental intercept and slope of F/L versus $s^{2/3}$ curves are plotted as a function of the nominal load L in Figure 18.28b.

The load dependence of the intercept $\mu + \tau_0(A_0/L)$ from Equation 18.27 provides information about the frictional properties of the tape before any wear has taken place, i.e., limit $F/L \big|_{s \to 0}$. It includes the friction coefficient μ and junction shear

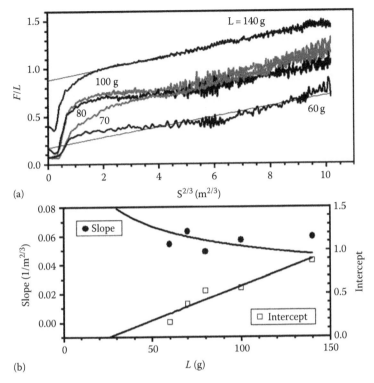

FIGURE 18.28   (a) Ratio of polishing tape friction force on the disk to the load as a function of sliding distance to the 2/3 power, F/L versus $s^{2/3}$ at several values of the nominal load, (b) slope and intercept of the line fitted to the steady state portion of the curves in (a) as a function of normal load.

stress τ_0, which are independent of L. The initial contact area A_0 is expected to change with L. For elastic Hertzian contacts, $A_0 \propto L^{2/3}$, and for elastic random rough surfaces, $A_0 \propto L$. From the lower curve in Figure 18.28b, the quantity $\mu + \tau_0(A_0/L)$ versus L increases linearly with L. The linear increase of $\mu + \tau_0(A_0/L)$ with L implies that $A_0 \propto L^2$. The strong dependence of A_0 on L is unexpected.

The zero crossing of $\mu + \tau_0(A_0/L)$ versus L is considered next. Since friction is a dissipative process, it is necessary that $\mu > 0$; however, the plot of $\mu + \tau_0(A_0/L)$ versus L has a negative intercept and crosses the load axis at 26 g. This result can be accounted for by the presence of mixed air lubrication in the tape/disk interface. An air bearing between the surface of the tape and disk provides at least 26 g of lift force on the tape. The presence of mixed lubrication could also account for $A_0 \propto L^2$. Finite element modeling of the airflow at the inlet and the measurements of the linear velocity dependence of $\mu + \tau_0(A_0/L)$ are planned to further investigate the mechanism for this surprising behavior of $\mu + \tau_0(A_0/L)$ versus L.

The load dependence of the slope $\tau_0(k/H)^{2/3}(1/L^{1/3})$ from Equation 18.28 provides information about the change in the frictional properties of the tape due to wear. The quantity $\tau_0(k/H)^{2/3}(1/L^{1/3})$, is expected to be proportional to $1/L^{1/3}$, which is shown by the smooth curve fit to the slope data points in the upper curve of Figure 18.28b. The data are consistent with the friction and wear model, but further tests should be done at more different values of L to improve the significance of the correlation.

While the tape friction study provides insight into the polishing tape/disk tribology, the key role of the polishing process is to remove from one to several dozen or more asperities from the disk surface during the polishing pass. This is clearly done by the high points of the tape surface topography, which are decorated by scuffing and debris in Figure 18.27. Thinning of the disk carbon overcoat on the disk magnetic grain roughness peaks is often collateral damage, which requires an optimization between asperity removal and corrosion due to pinhole generation.

It is also possible that a hard particle can be compressed between the tape and disk to form a scratch. Recessed areas in the polishing tape surface topography form "chip pockets," which collect the removed particles from the disk surface. Particulate debris can be seen as the dark accumulation behind the worn cusps in Figure 18.27a. A scratch is formed when a particle is trapped between the tape and the disk and the pressure applied to the particle by the tape exceeds the yield stress for plastic flow of the disk surface layers. Typical disk polishing scratches are several nm deep and several hundred nm across. These are the result of plastic deformation and cause magnetic defects through slippage of the magnetic grains [85].

18.6   PATTERNED MEDIA

Surface chemistry is essential to the future of magnetic recording systems that extend the storage capacity through patterned media up to and beyond terabits per square inch. One limit to recording density is the thermal stability or demagnetization of adjacent bits on a planar geometry [15]. In patterned media, bits are spatially isolated by placing each bit along a discrete circumferential track (discrete track media, DTM) or each bit on a separate island (bit patterned media, BPM). The "tracks" in DTM have a pitch and depth of approximately 50 nm each. The technological challenge for BPM is to manufacture a disk containing these islands, which are 10–25 nm in diameter and spaced apart similarly, which is 10^{12} islands on a 2.5 in diameter disk. The islands must lie along circumferential tracks, with little or no random variability in their circumferential separation and submicrometer radial run-out over the surface of the disk. Random variability along the track contributes timing error, or jitter, to the read and write signals. Radial run-out causes the recording transducer to move off the center of the data track due to the finite bandwidth of the track following servomechanism.

The lithographic approach is common to fabrication of both DTM and BPM. Formation of the nanoscale features is done by replication from a template (Figure 18.29). The process, referred to as NIL [86] or step and flash imprint lithography (SFIL) [87,88], must be capable of mass-producing millions of magnetic recording disks per year. Each disk is typically 65–95 mm in diameter, and few or no defects on a disk are acceptable.

A rigid quartz wafer is typically embossed with the DTM or BPM topography pattern, including servo and alignment marks. It is necessary to treat the quartz wafer with a chemical release layer. Unsaturated hydrocarbon photoresist typically

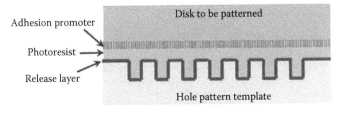

FIGURE 18.29 Schematic of the patterned medium template replication process showing the surface layers (adhesion promoter and release layer) and bulk materials (disk and template).

comprising an acrylic acid ester and containing a free-radical photoinitiator must completely wet the topography pattern on the quartz. Without a release layer, the photoresist is inseparable from the quartz following crosslinking by UV exposure through the quartz. The release layer forms an adhesion barrier between the quartz and the hydrocarbon photoresist. In order to avoid cohesive fracture in the resist upon separation, a surfactant can be included in the photoresist to act as an internal release agent [89]. The surfactant migrates to the interface between the photoresist and the quartz surface with the pattern.

Although the quartz/cured photoresist interface must have little or no adhesion, the resist must strongly adhere to the disk or substrate that is to be patterned by etching. The adhesion to the disk is accomplished by an adhesion promoter. For example, an acryloxyalkyl (trialkoxy silane) containing an unsaturated alkyl chain is applied as an adhesion promoter. The silane covalently bonds to the solid surface, while the unsaturated alkyl group diffuses into the photoresist. When the photoresist is cured, the unsaturated bond participates in the free radical polymerization to form a covalent bonding between the crosslinked matrix of the photoresist and the solid surface.

The patterned media process for disk manufacturing is still under development. Current research needs to include improved release of the photoresist from the quartz topography pattern. Adhesion force distorts the nanometer-scale pattern replica topography by inducing plastic flow and internal stress. Chemically attached release layers applied to the quartz pattern template surface wear off over a few thousand replication cycles, requiring costly and time-consuming maintenance for cleaning and re-coating. Higher adhesion leads to cohesive failure of the cured photoresist at the interface. There is a need to improve the surface cleaning process for recycling of the nanometer-scale topography patterns on the precious quartz pattern templates.

A further challenge to be overcome is in the arena of BPM. Beyond an areal data storage density of a terabit per square inch, the length scale of the quartz pattern template surface topography features becomes less than the resolution of e-beam lithography. The spatial resolution of e-beam lithography is limited by beam jitter and the minimum spot size. Higher areal data storage density can potentially be achieved through the use of directed block copolymer self-assembly. A chemical contrast pattern is generated by the e-beam in a self-assembled monolayer [90] or a molecularly thin polymer brush layer [91–93] on a substrate. The chemical contrast is formed when the polymer composition is locally altered by the energetic electrons.

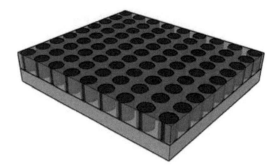

FIGURE 18.30 Schematic drawing of a phase-separated block copolymer forming cylindrical domains over a chemically patterned substrate.

The pattern guides the self-assembly of the block copolymer when it is deposited and annealed on the pattern. The block copolymer phase-separates into cylindrical domains, which are schematically illustrated in Figure 18.30. The cylindrical domains provide a selective etch barrier when one of the constituent blocks of the copolymer etches at a rate that is significantly different from that of the other block. The self-assembled pattern is thereby transferred into the quartz topography pattern needed to manufacture DTM or BPM by plasma or ion beam etching.

However, the e-beam patterning tool used to generate the chemical contrast pattern is still inadequate to provide the precision and bit areal density needed to achieve the required magnetic recording bit areal density. It has been recently shown that block copolymer self-assembly is able to correct e-beam pattern jitter and to multiply the areal bit density by a factor of four [94]. This method could provide domain densities up to 10^{12} bits (one terabit) per square inch.

18.7 CONCLUSION

This chapter focused on lubrication, contamination, corrosion protection, disk processing, and patterned media. Historically, these have been key areas for magnetic recording research and development, and will continue to be so. Reviewing the history of these areas provides insight into their continuing evolution. Contamination is now well understood. Cases of contamination that do occur are traced back to their root cause using micro-analytical methods, and the source of the contamination is eliminated. Corrosion protection is an ongoing area of development in which the overcoat is being made thinner and must remain free of porosity.

Disk lubricants have evolved from mobile fluorohydrocarbon polyethers that replenish asperity contacts to fully chemisorbed materials in order to avoid lubricant transfer from the disk to the slider. It is possible that by further understanding the dispersion force interactions between the disk and slider surface layers, overcoats with low adhesion may eliminate the need for topically applied disk lubricant.

Patterned media needs improved resist release materials. When a photoresist, which is initially a low viscosity liquid, is cured into a crosslinked solid on the surface of the solid pattern, all of the surface area is in contact. Even a small amount of adhesion necessitates a large force for separation of the resist from the hole pattern template. Sacrificial surface layers are not an option because they obscure the nanometer-scale features of the pattern template. Compatible material pairs that have inherently low interfacial adhesion are desirable. Alternatively, new surfactants are needed to provide intervening surface layers for low adhesion between the cured photoresist and the pattern template.

REFERENCES

1. T. E. Karis, B. Marchon, M. D. Carter, P. R. Fitzpatrick, and J. P. Oberhauser, Humidity effects in magnetic recording, *IEEE Trans. Magn.*, 41, 593–598 (2005).
2. A. W. Adamson, An adsorption model for contact angle and spreading, *J. Colloid Interface Sci.*, 27, 180–187 (1968).
3. T. E. Karis and U. V. Nayak, Liquid nanodroplets on thin film magnetic recording disks, *Tribol. Trans.*, 47, 103–110 (2004).
4. X.-C. Guo, V. Raman, T. E. Karis, and Y. Z. Yao, Flyability failures due to siloxanes at the head-disk interface revisited, *IEEE Trans. Magn.*, 43, 2223–2225 (2007).
5. J.-F. Ying, T. Anoikin, and C. Martner, Evolution of the corrosion process on thin-film media, *J. Vac. Sci. Technol. A.*, 18, 1804–1808 (2000).
6. T. E. Karis, X.-C. Guo, B. Marchon, V. Raman, and Y.-L. Hsiao, Cobalt oxalate formation on thin-film magnetic recording media, *IEEE Trans. Magn.*, 42, 2507–2509 (2006).
7. R.-H. Wang, H. R. Wendt, C. A. Brown, S. Lum, S. McCoy, and T. Karis, Enhanced reliability of hard disk drive by vapor corrosion inhibitor, *IEEE Trans. Magn.*, 42, 2498–2500 (2006).
8. Y. Ma and B. Liu, Lubricant transfer from disk to slider in hard disk drives, *Appl. Phys. Lett.*, 90, 143516 (2007).
9. B. Marchon, X.-C. Guo, T. Karis, H. Deng, Q. Dai, J. Burns, and R. Waltman, Fomblin multidentate lubricants for ultra-low magnetic spacing, *IEEE Trans. Magn.*, 42, 2504–2506 (2006).
10. T. E. Karis, G. W. Tyndall, and M. S. Crowder, Tribology of a solid fluorocarbon film on magnetic recording media, *IEEE Trans. Magn.*, 34, 1747–1749 (1998).
11. T. E. Karis, X.-C. Guo, and J.-Y. Juang, Dynamics in the bridged state of a magnetic recording slider, *Tribology Lett.*, 30, 123–140 (2008).
12. T. E. Karis and X.-C. Guo, Molecular adhesion model for the bridged state of a magnetic recording slider, *IEEE Trans. Magn.*, 43, 2232–2234 (2007).
13. T. E. Karis, X.-C. Guo, E. Marinero, and B. Marchon, Surface chemistry of NiP plated substrates, *IEEE Trans. Magn.*, 41, 3247–3249 (2005).
14. Y. J. Man, S. K. Yu, and B. Liu, Characterization and formation mechanism understanding of asperities to be burnished, *J. Magn. Magn. Mater.*, 303, e101–e105 (2004).
15. R. Sbiaa and S. N. Piramanayagam, Patterned media towards nano-bit magnetic recording: Fabrication and challenges, *Recent Pat. Nanotechnol.*, 1, 29–40 (2007).
16. T. E. Karis, Lubricants for the disk drive industry, in *Lubricant Additives Chemistry and Applications*, 2nd edn., L. R. Rudnick, Ed., CRC Press, Taylor & Francis Group, Boca Raton, FL, pp. 523–581 (2009).
17. J. Ruhe, G. Blackman, V. J. Novotny, T. Clarke, G. B. Street, and S. Kuan, Terminal attachment of perfluorinated polymers to solid surfaces, *J. Appl. Polym. Sci.*, 53, 825–836 (1994).
18. B. Bhushan and C. Kajdas, Mechanism of interaction and degradation of perfluoropolyethers with a DLC coating in thin-film magnetic rigid disks: A critical review, *J. Information Storage Proc. Syst.*, 1, 303–320 (1999).

19. M. Tani, H. Matsumoto, M. Shyoda, T. Kozaki, T. Nakakawaji, and Y. Ogawa, Magnetic Recording Medium, US6605335 (2003).

20. R. J. Waltman, N. Kobayashi, K. Shirai, A. Khurshudov, and H. Deng, The tribological properties of a new cyclotriphosphazene-terminated perfluoropolyether lubricant, *Tribology Lett.*, 16, 151–162 (2004).

21. P. H. Kasai, Degradation of perfluoropoly(ethers) and role of X-1P additives in disk files, *J. Information Storage Proc. Syst.*, 1, 23–31 (1999).

22. N. Tagawa, T. Tateyama, A. Mori, N. Kobayashi, Y. Fujii, and M. Ikegami, Spreading of novel cyclotriphosphazine-terminated PFPE films on carbon surfaces, in *Proceedings of the 2003 Magnetic Storage Symposium, Frontiers of Magnetic Hard Disk Drive Tribology and Technology*, A. A. Polycarpou, M. Suk, and Y.-T. Hsia, Eds., ASME, TRIB-15, 17–20, New York (2003).

23. C. L. Jiaa and Y. Liu, Tribological evaluation and analysis of the head/disk interface with perfluoropolyether and X1-P phosphazene mixed lubricants, *Tribology Lett.*, 7, 1023–8883 (1999).

24. H. Chiba, Y. Oshikubo, K. Watanabe, T. Tokairin, and E. Yamakawa, Tribological characteristics of newly synthesized multi-functional PFPE lubricants, in *Proceedings of the 2005 World Tribology Congress III*, Washington, DC (2005).

25. K. Sonoda, D. Shirakawa, T. Yamamoto, and J. Itoh, The tribological properties of the new structure lubricant at the head-disk interface, *IEEE Trans. Magn.*, 43, 2250–2252 (2007).

26. D. Shirakawa, K. Sonoda, and K. Ohnishi, A study on design and synthesis of new lubricant for near contact recording, *IEEE Trans. Magn.*, 43, 2253–2255 (2007).

27. X.-C. Guo and R. J. Waltman, Mechanism of ultraviolet bonding of perfluoropolyethers revisited, *Langmuir*, 23, 4293–4295 (2007).

28. T. E. Karis, Water adsorption on thin film magnetic recording media, *J. Colloid Interface Sci.*, 225, 196–203 (2000).

29. M. E. Tadros, P. Hu, and A. W. Adamson, Adsorption and contact angle studies. I. Water on smooth carbon, linear polyethylene, and stearic acid-coated copper, *J. Colloid Interface Sci.*, 49, 184–195 (1974).

30. A. W. Adamson, *Physical Chemistry of Surfaces*, John Wiley & Sons, New York, 373 p. (1976).

31. G. T. Barnes, The effects of monolayers on the evaporation of liquids, *Adv. Colloid Interface Sci.*, 25, 89–200 (1986).

32. R. J. Archer and V. K. La Mer, The rate of evaporation of water through fatty acid monolayers, *J. Phys. Chem.*, 59, 200–208 (1955).

33. W. A. Wink, Determining the moisture equilibrium curves of hygroscopic materials, *Ind. Eng. Chem.*, 18, 251–252 (1946).

34. F. E. M. O'Brien, The control of humidity by saturated salt solutions—A compilation of data, *J. Sci. Instrum.*, 25, 73–76 (1948).

35. A. Wexler and S. Hasegawa, Relative humidity-temperature relationship of some saturated salt solutions in the temperature range 0° to 50°C, *J. Res. NBS*, 53, 19–26 (1954).

36. S. Martin, The control of conditioning atmospheres in small sealed chambers, *J. Sci. Instrum.*, 39, 370 (1962).

37. L. Greenspan, Humidity at fixed-points of binary saturated aqueous-solutions, *J. Res. NBS*, 81, 89–96 (1977).

38. J. M. Smith and H. C. Van Ness, *Introduction to Chemical Engineering Thermodynamics*, 3rd edn., McGraw-Hill Book Company, New York, 332 p. (1975).

39. N. Namiki, Y. Otani, H. Emi, and S. Fujii, Particle formation of materials outgassed from silicone sealants by corona-discharge ionizers, *J. Inst. Environmental Sci.*, 39, 26–32 (1996).

40. V. M. Gun'ko, M. V. Borysendko, P. Pissis, A. Spanoudaki, N. Shinyashiki, I. Y. Sulim, T. V. Kulik, and B. B. Palyanytsya, Polydimethylsiloxane at the interfaces of fumed silica and zirconia/fumed silica, *Appl. Surface Sci.*, 253, 7143–7156 (2007).

41. E. Schreck, R. E. Jr. Fontana, and G. P. Singh, Thin film thermocouple sensors for measurement of contact temperatures during slider asperity interaction on magnetic recording disks, *IEEE Trans. Magn.*, 28, 2548–2550 (1992).

42. P. H. Kasai and F. P. Eng, Silicon oxide formation in the disk environment, *J. Information Storage Process. Syst.*, 2, 125–128 (2000).

43. V. Raman, D. Gillis, and R. Wolter, Flyability failures due to organic siloxanes at the head/disk interface, *ASME Trans. J. Tribol.*, 122, 444–449 (2000).

44. R.-W. J. Chia, C. C. Wang, and J.-J. K. Lee, Effect of adatom mobility and substrate finish on film morphology and porosity: Thin chromium film on hard disk, *J. Magn. Magn. Mater.*, 209, 45–49 (2000).

45. M. Smallen, P. B. Mee, A. Ahmad, W. Freitag, and L. Nanis, Observations on electrochemical and environmental corrosion tests for cobalt alloy disc media, *IEEE Trans. Magn.*, 21, 1530–1532 (1985).

46. V. Brusic, M. Russak, R. Schad, G. Grankel, A. Selius, and D. DiMilia, Corrosion of thin film magnetic disk: Galvanic effects of the carbon overcoat, *J. Electrochem. Soc.*, 136, 42–46 (1989).

47. V. Novotny and N. Staud, Correlation between environmental and electrochemical corrosion of thin film nagnetic recording media, *J. Electrochem. Soc.*, 135, 2931–2938 (1988).

48. Q. Dai, B. K. Yen, R. L. White, P. J. Peterson, and B. Marchon, Toward an understanding of overcoat corrosion protection, *IEEE Trans. Magn.*, 39, 2450–2452 (2003).

49. L. Huang, Y. Hung, and S. Chang, Surface and lubricant/overcoat interface properties of the rigid disks after corrosion, *IEEE Trans. Magn.*, 33, 3154–3156 (1997).

50. M.-S. Lin, C. Tsai, Y.-C. Sun, W. Huang, C. M. Wang, and C. Dong, An accelerated test for cobalt migration in thin-film rigid disks, *IEEE Trans. Magn.*, 35, 2703–2705 (1999).

51. Y. Itai, M. Katayama, and Y. Kasamatsu, Method of surface treatment for recording medium, US6511716 (2003).

52. A. J. Prenni, P. J. DeMott, S. M. Kreidenweis, D. E. Sherman, L. M. Russell, and Y. Ming, The effects of low molecular weight dicarboxylic acids on cloud formation, *J. Phys. Chem. A*, 105, 11240–11248 (2001).

53. M. W. Poore, Oxalic acid in $PM_{2.5}$ particulate matter in California, *J. Air Waste Management Assoc.*, 50, 1874–1875 (2000).

54. Y.-L. Cao, D.-Z. Jia, D.-Q. Liu, and X.-Q. Xin, Synthesis and characterization of cobalt oxalate nanorods prepared by one-step solid-state chemical reaction, *Acta Chim. Sin.*, 63, 175–178 (2005).

55. B.-S. Fang, C. G. Olson, and D. W. Lynch, A photoemission study of benzotriazole on clean copper and cuprous oxide, *Surf. Sci.*, 176, 476–490 (1986).

56. A. Johnson and F. Y. Lu, New insights into the application and mechanisms of yellow metal corrosion inhibitors, in *Proceedings of the 8th European Symposium on Corrosion Inhibitors*, pp. 989–998 (1995).

57. T. E. Karis, C. A. Brown, S. Lum, S. McCoy, and R.-H. Wang, Control system to regulate the concentration of vapor in a hard disk drive, US7466514 (2008).

58. K. Kogure, T. Kita, and S. Fukui, Mechanical characteristics of flying head and recording media for 3.2 Gbyte multi-device disk storage, *Rev. Electrical Commun. Laboratories*, 30, 36–45 (1982).

59. K. Miyoshi, Fundamental consideration in adhesion, friction, and wear for ceramic-metal contacts, *Wear*, 141, 35–44 (1990).

60. A. K. Agarwal, C. Y. Shih, M. A. Harper, and C. L. Bauer, Effect of surface coatings on sliding friction and wear of thin-film magnetic recording media, in *Tribology and Mechanics of Magnetic Storage Systems VI*, STLE SP-26, pp. 8–14 (1989).

61. H.-C. Tsai and D. B. Bogy, Critical review: Characterization of diamondlike carbon films and their application as overcoats on thin-film media for magnetic recording, *J. Vac. Sci. Technol. A*, 5, 3287–3312 (1987).

62. R.-H. Wang, V. Raman, P. Baumgart, A. M. Spool, and V. Deline, Tribology of laser textured disks with thin overcoat, *IEEE Trans. Magn.*, 33, 3184–3186 (1997).

63. B. K. Yen, R. L. White, R. J. Waltman, Q. Dai, D. C. Miller, A. M. Kellock, B. Marchon et al., Microstructure and properties of ultrathin amorphous silicon nitride protective coating, *J. Vac. Sci. Technol. A*, 21, 1895–1904 (2003).

64. Q. Dai, B. Marchon, M. A. Scarpulla, R. L. White, and B. K. Yen, Thin film TiSixNy protective layer, US6586070 (2003).

65. B. Fubini, M. Volante, V. Bolis, and E. Giamello, Reactivity towards water of silicon-nitride—Energy of interaction and hydration dehydration mechanism, *J. Mater. Sci.*, 24, 549–556 (1989).

66. C. F. Ye, W. M. Liu, Y. X. Chen, and Z. W. Ou, Tribological behavior of Dy-sialon ceramics sliding against Si3N4 under lubrication of fluorine-containing oils, *Wear*, 253, 579–584 (2002).

67. T. E. Karis, G. W. Tyndall, and R. J. Waltman, Lubricant bonding effects on thin film disk tribology, *Tribol. Trans.*, 44, 249–255 (2001).

68. T. E. Karis, G. W. Tyndall, D. Fenzel-Alexander, and M. S. Crowder, Ellipsometric measurement of solid fluorocarbon film thickness on magnetic recording media, *J. Appl. Phys.*, 81, 5378–5380 (1997).

69. T. E. Karis, G. W. Tyndall, D. Fenzel-Alexander, and M. S. Crowder, Characterization of a solid fluorocarbon film on magnetic recording media, *J. Vac. Sci. Technol. A*, 15, 2382–2387 (1997).

70. R. H. French, Origins and applications of London dispersion forces and hamaker constants in ceramics, *J. Am. Ceram. Soc.*, 83, 2117–2146 (2000).

71. B. W. Ninham and V. A. Parsegian, Van der Waals forces across triple-layer films, *J. Chem. Phys.*, 52, 4578–4587 (1970).

72. L. R. White, R. R. Dagastine, P. M. Jones, and Y.-T. Hsia, Van der Waals force calculation between laminated media, pertinent to the magnetic storage head-disk interface, *J. Appl. Phys.*, 97, 104503 (2005).

73. D. B. Hough and L. R. White, The calculation of Hamaker constants from Lifshitz theory with applications to wetting phenomena, *Adv. Colloid Interface Sci.*, 14, 3–41 (1980).

74. R. R. Dagastine, L. R. White, P. M. Jones, and Y.-T. Hsia, Effect of media overcoat on van der Waals interaction at the head-disk interface, *J. Appl. Phys.*, 97, 126106 (2005).

75. L. D. Landau and E. M. Lifshitz, *Electrodynamics of Continuous Media*, Pergamon Press, New York (1960).

76. W. R. Bowen and F. Jenner, The calculation of dispersion forces for engineering applications, *Adv. Colloid Interface Sci.*, 56, 201–243 (1995).

77. R. R. Dagastine, D. C. Prieve, and L. R. White, Calculations of van der Waals forces in 2-dimensionally anisotropic materials and its application to carbon black, *J. Colloid Interface Sci.*, 249, 78–83 (2002).

78. Wikipedia refractive index. [Online]. http://en.wikipedia.org/wiki/List_of_refractive_indices

79. Wikipedia dielectric constant. [Online]. http://en.wikipedia.org/wiki/Dielectric_constant

80. T. E. Karis and D. Pocker, Surfactants in magnetic recording technology, in *Surfactants in Tribology*, G. Biresaw and K. Mittal, Eds., CRC Press, Taylor & Francis Group, Boca Raton, FL, pp. 59–88 (2008).

81. A. S. Michaels and H. J. Bixler, Flow of gases through polyethylene, *J. Polymer Sci.*, 50, 413–439 (1961).
82. A. M. Scarati and G. Caporiccio, Frictional behaviour and wear resistance of rigid disks lubricated with neutral and functional perfluoropolyethers, *IEEE Trans. Magn.*, 23, 106–108 (1987).
83. E. T. Kuan, D. W. Park, J. Melo, D. Spaulding, J. J. Liu, and K. K. Kim, The corrosion performance of very thin carbon overcoat layers in magnetic media, *IEEE Trans. Magn.*, 40, 3195–3197 (2004).
84. J. Gao, W. D. Luedtke, D. Gourdon, M. Ruths, J. N. Israelachvili, and U. Landman, Frictional forces and Amontons' law: From the molecular to the macroscopic scale, *J. Phys. Chem.*, 108, 3410–3425 (2004).
85. M. Furukawa, J. Xu, Y. Shimizu, and Y. Kato, Mechanism study of scratch-induced demagnetization for perpendicular magnetic disks, *Microsystem Technol.*, 16, 221–226 (2010).
86. S. Y. Chou, P. R. Krauss, and P. J. Renstrom, Nanoimprint lithography, *J. Vac. Sci. Technol. B*, 14, 4129–4133 (1996).
87. T. Bailey, B. J. Choi, M. Colburn, M. Meissle, S. Shaya, J. G. Ekerdt, S. V. Sreenivasan, and C. G. Willson, Step and flash imprint lithography: Template surface treatment and defect analysis, *J. Vac. Sci. Technol. B*, 18, 3572–3577 (2000).
88. D. J. Resnick, D. P. Mancini, S. V. Sreenivasan, and C. G. Willson, Release layers for contact and imprint lithography, *Semiconductor Intl.*, 71–80 (June 2002).
89. T. Omatsu, M. Nishikawa, and K. Moriwaki, Interface binder, resist composition containing the same, laminate for forming magnetic recording medium having layer containing the same, manufacturing method of magnetic recording medium using the same, and magnetic recording medium produced by the manufacturing method, US20090011367 (2009).
90. S. O. Kim, J. J. Solak, M. P. Stoykovich, N. J. Ferrier, J. J. dePablo, and P. F. Nealey, Epitaxial self-assembly of block copolymers on lithographically defined nanopatterned substrates, *Nature*, 424, 411–414 (2003).
91. E. W. Edwards, M. F. Montague, H. H. Solak, C. J. Hawker, and P. F. Nealey, Precise control over molecular dimensions of block-copolymer domains using the interfacial energy of chemically nanopatterned substrates, *Adv. Mater.*, 16, 1315–1319 (2004).
92. E. W. Edwards, M. P. Stoykovich, M. Muller, H. H. Solak, J. J. dePablo, and P. F. Nealey, Mechanism and kinetics of ordering in diblock copolymer thin films on chemically nanopatterned substrates, *J. Polymer Sci. B*, 43, 3444–3459 (2005).
93. K. Ch. Daoulas, M. Muller, M. P. Stoykovich, S.-M. Park, Y. J. Papakonstantopoulos, J. J. dePablo, and P. F. Nealey, Fabrication of complex three-dimensional nanostructures from self-assembling block copolymer materials on two-dimensional chemically patterned templates with mismatched symmetry, *Phys. Rev. Lett.*, 96, 036104 (2006).
94. R. Ruiz, H. Kang, F. A. Detcheverry, E. Dobisz, D. S. Kercher, T. R. Albrecht, J. J. dePablo, and P. F. Nealey, Density multiplication and improved lithography by directed block copolymer assembly, *Science*, 321, 936–939 (2008).

19 Improving Organic Antiwear and Friction Modifier Compounds for Automotive Applications

Frank J. DeBlase

CONTENTS

ABSTRACT

The presence of inorganic compounds or elements, such as sulfur and phosphorous, in automotive lubricants degrades catalytic pollution control devices. Hence, new lubricants must be formulated. During modern lubricant refining processes, natural sulfur-containing antiwear compounds are removed. As a result, fuel economy is diminished and engine wear is increased. Therefore, friction modifiers and antiwear additives are currently being developed. When tested, such additives have improved fuel economy by 2%–3%, while extending engine life. Novel organic antiwear/friction modifiers (OAW-FMs) must not degrade the performance of other additives, especially zinc dialkyldithiophosphate. These concepts are reviewed here. Subsequently, tribological friction and wear data on OAW-FM additives are presented. In addition, new analytical methods such as the Fourier transform infrared microspectroscopic characterization of scar surfaces are discussed. The goal of this chapter is to expand the chemical understanding of these surface active compounds and the manner in which they improve the performance of engine lubricants.

19.1 INTRODUCTION

19.1.1 MODERN ENGINES: FRICTION AND WEAR

Significant challenges exist in developing organic-based lubricant additives that reduce friction and wear in modern automotive engines. In spite of the challenges of an internal combustion engine, a reasonable number of different additives, organic (C, H, N, O), organometallic, and inorganic in nature provide a beneficial enhancement of lubricant performance. The development of friction modifiers and antiwear additives is a result of the systematic analysis of the range of automotive engine stresses. An analysis of an engine design shows that the specific metal–metal contact points in modern engines are subject to a range of pressures and temperatures during normal operation. These stress points can benefit from lubricants that are formulated to bring specific additives to their interfaces. To enhance lubricant performance, friction modifiers reduce friction between sliding contacts and antiwear additives reduce wear at more extreme pressures (EPs) and temperatures.

Figure 19.1 illustrates some of the targeted regions of the engine that benefit from friction modifiers, antiwear additives, or combinations of both. As pointed out, the cam follower (tappet), located at the base of the pushrod (Contact-A), has the highest

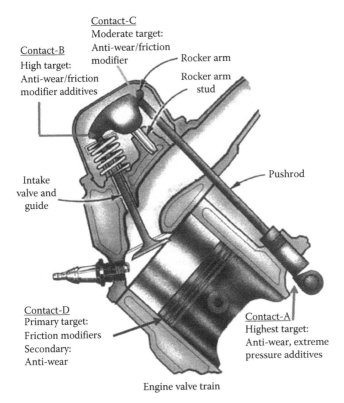

Contact-C
Moderate target:
Anti-wear/friction
modifier

Contact-B
High target:
Anti-wear/friction
modifier additives

Rocker arm

Rocker arm
stud

Intake
valve and
guide

Pushrod

Contact-D
Primary target:
Friction modifiers
Secondary:
Anti-wear

Contact-A
Highest target:
Anti-wear, extreme
pressure additives

Engine valve train

FIGURE 19.1 (See color insert.) Illustration of a valve train typical of a small block Chevrolet V-8 engine. (Courtesy of Joseph N. Valentine.)

pressure by virtue of contact against the rotating cam. It experiences a motion characteristic of a sliding–rotating wear as the cam raises the pushrod. Because of the location of the pivoting fulcrum in the rocker arm, the force applied from the pushrod generates an amplified wear motion at the end closest to the valve. The next wear point (Contact-B), where the rocker arm meets this valve, is then the second highest area of wear. This area and Contact-A both respond well to extreme-pressure additives, and to a lesser extent, to friction modifier additives. The area directly in contact with the pushrod (Contact-C) shows moderately high wear and friction from the pressure exerted and benefits from both antiwear and friction modifier protection. In contrast to the high-pressure regions, the contacts at the piston rings (Contact-D) undergo a more sliding friction with less direct metal–metal force (from expansion of the ring against the cylinder wall) and respond very well to organic friction modifiers. In general, higher-pressure contact points respond less to friction modifiers and require higher concentrations of antiwear additive. Due to the EP conditions, lubrication in these regions of metal-to-metal contact is thinner than the normal friction-modifier boundary lubrication of bilayer molecular dimensions. Thus, these EP regions require inorganic glass-like protective coatings. From the inherent symmetry of engine designs, the illustration and analyses presented for the intake valve hold for the analogous parts of the exhaust side, which operates at even a higher temperature. In no small measure, the stress associated with the friction and wear is magnified by the combustion event, making the challenges even greater. Modern engine oil requires the ability to remove heat and acts as a cooling fluid as well as a lubricant. The contact points discussed above serve as an indication of the complexity and variation of the stresses operating in a normal working engine. It is not an exhaustive list of all engine contact areas. For a detailed discussion of the physics of the many stresses operating within automotive engines, refer to the automotive engineering text written by Taylor and Taylor [1].

19.1.2 ORGANO-METALLIC ANTIWEAR AND FRICTION MODIFIERS (STRUCTURE, MECHANISM THEORIES)

Before presenting the development of new organic antiwear and friction modifier (OAW-FM) additives, the existing organometallic additive chemistry is discussed. These additives serve as a target for the desired level of reduced friction and antiwear performance.

19.1.2.1 Metallic Antiwear

The high-pressure engine contacts discussed are the primary targets for antiwear additives. One of the earliest and effective antiwear additives is zinc dialkyldithiophosphate (ZDDP). Figure 19.2 presents the chemical structure for a typical ZDDP, with 2-ethylhexyl, alkyl groups. ZDDP can have linear and branched alkyl groups, ranging from 1 to 14 carbons in length, mixed or of the same alkyl structure, and can be optimized for typical applications such as automotive crankcase oil, grease, or other machine oils. Some of the typical alkyl groups are 2-butyl, pentyl, hexyl, 1,3-dimethylbutyl, heptyl, octyl, 2-ethylhexyl, 6-methylheptyl,1-methylpropyl, and dodecylphenyl [2].

In modern high-compression engines, when the local pressure within contact areas increases, temperatures will also increase and can reach 200°C [3,4]. At these

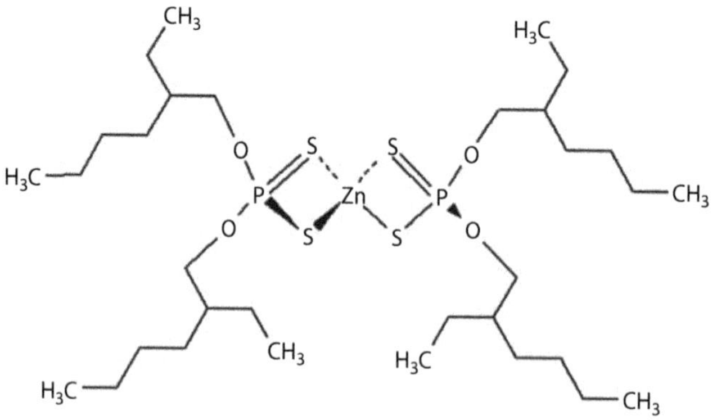

FIGURE 19.2 The chemical structure of a typical zinc dialkylditiophosphate with equivalent 2-ethylhexyl alkyl groups at each phosphate.

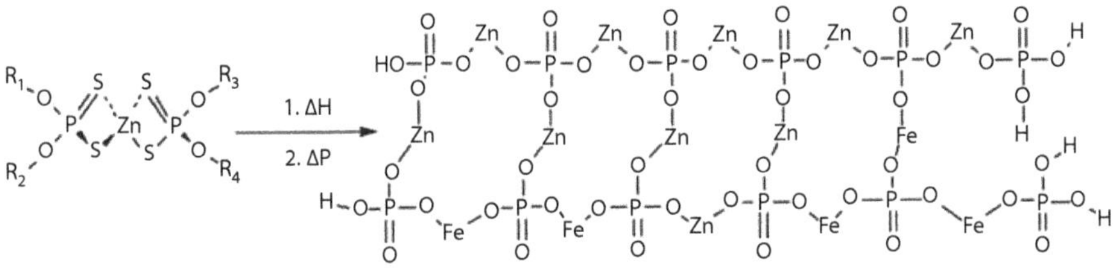

FIGURE 19.3 Protective inorganic glass developed from high-pressure thermal-induced oxidation of ZDDP.

temperatures and under pressure, ZDDP can polymerize to form various types of inorganic glass structures capable of bonding to ferrous metals leaving a durable protective coating. An illustration of this type of transition is presented in Figure 19.3.

19.1.2.2 Metallic Friction Modifiers

In parts of automotive engines with less wear (e.g., sliding contact of pistons) the molecular integrity of the lubricant additive is more likely to remain. Durable organic portions of the friction modifiers can adsorb onto the surface and form various ordered and layered structures. These structures reduce boundary layer friction by allowing metal asperities to slide past each other. Depending on the nature of the friction modifier and its intermolecular London forces, structures range from organic bilayers of alkyl chains with strong van der Waals and dipole–dipole interactions, to chemically transformed soft molybdenum sulfide glasses (similar to the process occurring in ZDDP) capable of deforming under sliding stress.

The chemical variations in organo-molybdenum friction modifiers are specific and designed for different lubricant formulations and applications [5–11]. Traditional molybdenum dithiocarbamate (MoDTC) added to lubricants increases measured vehicle fuel economy by 3%–5%, and structures with increased molybdenum show better friction reduction performance [12]. However, in general, reduction of metals in lubricants is preferred from an emission standpoint to protect pollution control devices. The organo-molybdenum compounds also transform under the boundary lubrication conditions of heat and pressure to generate molecular structures that bind to metal surfaces more efficiently. For example, the molybdenum-based friction modifier (Structure I, Figure 19.4) can break down and form a softer poly(molybdenum sulfide) glass (Structure II, Figure 19.4). It then easily deforms allowing slippage past the weaker disulfide bonds.

The molybdenum friction modifiers shown in Structures I and III (Figures 19.4 and 19.5) involve a friction reduction mechanism different from that of the sulfur-free molybdenum friction modifiers in Structures IV and V (Figure 19.5). These latter two friction modifiers do not contain significant sulfur and cannot form the inorganic polymerized Mo-(S$_2$)-Mo glass structure unless they react with other sulfur-containing additives such as ZDDP, also at the surface. There is evidence that a synergy exists between ZDDP and molybdenum dithiocarbamates and alkylmolybdates [5,6,13]. In addition to friction reduction, some molybdenum friction modifiers are reported to be multifunctional and allow for the reduction in other additives such

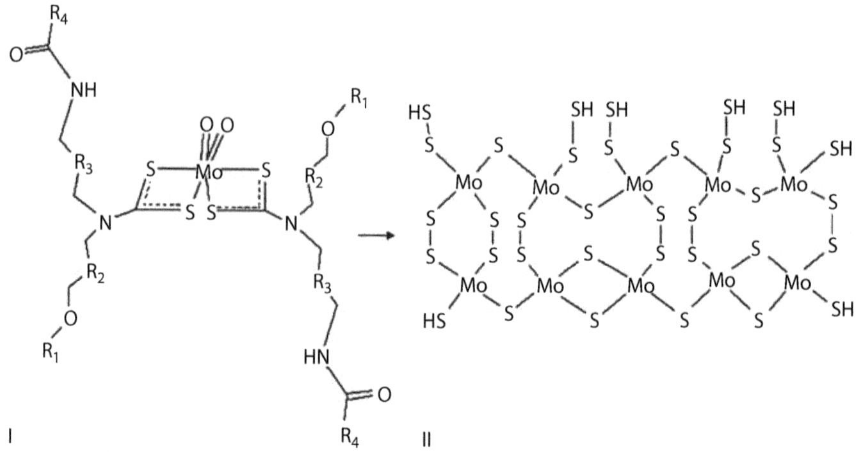

FIGURE 19.4 MoDTC-based friction modifier (I) and it transition to a poly (molybdenum tetrasulfide) residue (II) under high temperature–pressure frictional conditions.

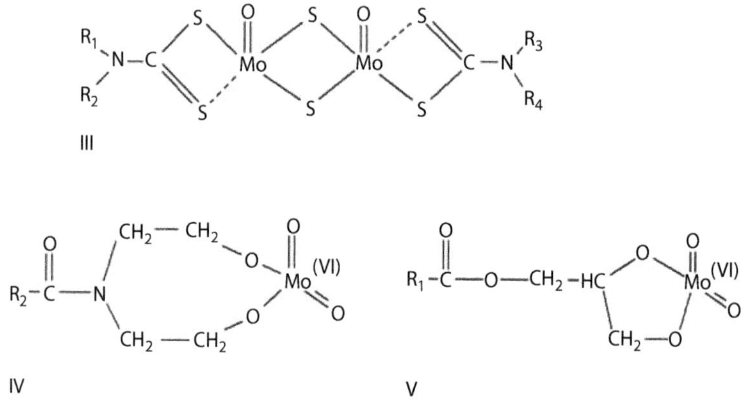

FIGURE 19.5 Three typical molybdenum-based friction-reducing additives with and without sulfur.

as ZDDP antiwear [14–16]. When the ZDDP level is reduced or is unavailable at a surface, then the adsorption of molybdenum oxide directly onto the metal surface (iron and iron oxide) occurs, and the alkyl chains play a larger role in the boundary lubrication. The hydrocarbon chains are more directly involved in boundary lubrication through their chain–chain interaction. This is similar to the class of all-organic-based friction modifiers.

The friction reduction mechanism of these all-organic friction modifiers is a physical chemical process that involves the adsorption of the polar ends of these molecules to the metal surface, while the nonpolar alkyl chains interact by van der Waals intermolecular attractive forces. This is evident from the effectiveness of all-organic friction modifiers with specific alkyl chains such as tall-oil fatty acids, monoglycerides (e.g., glycerol monooleate [GMO]) and fatty alkanolamides. This also holds for the alkyl portion of the molybdenum friction modifiers where their effectiveness is dependent on the carbon chain length of the alkyl groups (R_1 and R_2 of Structures IV and V in Figure 19.5).

Because the all-organic additives are composed of only C, H, O, and N, they can also be used in gasoline to improve fuel economy [17–22]. A portion of the fuel economy additive can survive the combustion event, reach the cylinder wall to improve its boundary lubrication, and finally the lubricant through fuel dilution to fortify its friction modifier content. Fuel economy gains over 2%–3% are achievable when all-organic friction modifiers are added to fuels [23].

19.1.2.3 Impact of Additive Metals on Emission Control Devices

Organometallic ZDDP antiwear and MoDTC friction modifier additives are extremely effective, and are typically used at lower concentrations than all-organic additives. Their performance is a result of the generation of inorganic protective films at metal contact surfaces. However, there are still convincing benefits to reducing the amount of inorganic elements (Mo, Zn, P, S) used in lubricants. Figure 19.6 illustrates the contribution of the portions of the molecular structures of ZDDP and MoDTC to generate chemical species, which in time either poison and inhibit the proper operation of air emission-control devices or generate harmful airborne particulates, particularly in diesel engines.

Phosphorus causes a significant poisoning effect on the Pt in catalytic converters, and sulfur has a negative effect on the three-way catalyst in the after-treatment device. The emission reduction chemistry that occurs in these three-way systems where oxidation catalysts can be composed of Pt, Pd, Rh, and Ce is

1. Nitrogen oxides in the exhaust are catalytically reduced to nitrogen and oxygen:

$$2NO_x \rightarrow xO_2 + N_2$$

2. Carbon monoxide in the exhaust is catalytically oxidized to carbon dioxide:

$$2CO + O_2 \rightarrow 2CO_2$$

3. Unburned hydrocarbons in the exhaust are catalytically oxidized to carbon dioxide and water:

$$C_nH_{(2n+2)} + \left[(3_n + 1)/2\right]O_2 \rightarrow nCO_2 + (n+1)\,H_2O$$

(*Note*: This formula is for straight chain hydrocarbons (no branches, unsaturates, or aromatics)

$$C_{16}H_{34} + 25O_2 \rightarrow 16CO_2 + 18H_2O \text{ (for example)}$$

For lubricants that are used in diesel engines, the ZDDP and MoDTC can affect nitrous oxide traps or selective reduction catalysts (SRCs) used to reduce NO_x

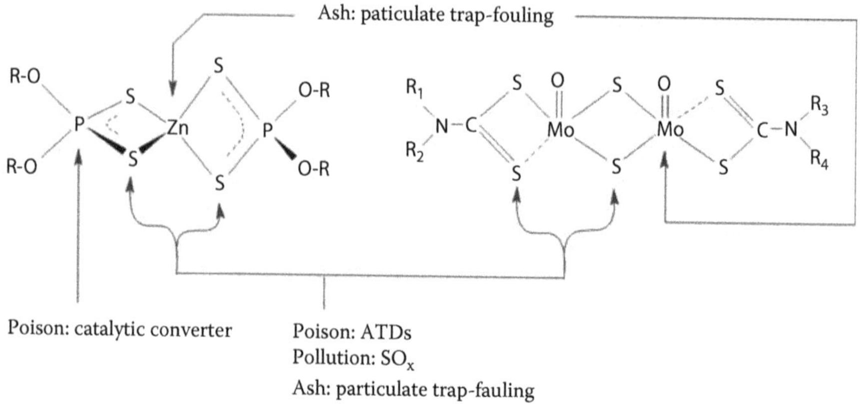

FIGURE 19.6 ZDDP, MoDTC impact on ATDs, generation of pollutants (SO$_x$) and contribution to high levels of particulates fouling traps and increasing harmful emissions.

levels in their after-treatment devices (ATDs). In addition to catalytically converting unwanted emissions to less harmful species, ATDs are combined with exhaust gas filters to remove particulates in diesel engine exhausts. The use of excessive Zn- and Mo-containing lubricant additives leads to fouling of these particulate traps by excessive ash in the diesel exhaust.

Examples of developments in the next generation of ashless, primarily OAW-FM additives follow. The focus will be primarily on recently synthesized organic friction modifiers and antiwear additives with comparisons to ZDDP antiwear and molybdenum friction modifier levels of performance.

19.1.3 CHEMISTRY OF CURRENT ORGANIC ANTIWEAR AND FRICTION MODIFIERS

There are a number of specific requirements to developing all organic based antiwear additives and friction modifiers. The requirements are specifically designed to reduce the overall lubricant levels of S, P, and metals to extend the life of pollution control devices. For this reason, although very effective, ZDDP additive packages must be developed at reduced levels that will maintain emission equipment. While reduction in the levels of Mo, Zn, S, and P will benefit the emission reduction, it is difficult to reach the same level of protection performance with only organic additives. As a result, the additive level needed when using all-organic additives may be two or three times that of the organometallic additives.

19.1.3.1 Organic Antiwear Additive: Alkyl Citrates

Experimental data indicate that at the proper treatment levels (1–2 wt.%), antiwear performance approaching ZDDP is reached for some organic antiwear additives (e.g., alkyl citrates) [24,25]. In some cases, a synergy exists when combined with ZDDP. Figure 19.7 shows a proposed model of the surface organization of alkyl citrates surface films developed for use as either a supplement to ZDDP to reduce the level of zinc or as an additive alone. The alkyl citrates demonstrate a ZDDP

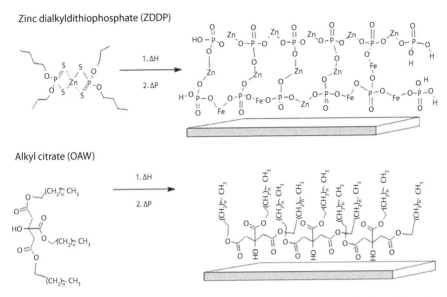

FIGURE 19.7 ZDDP and alkyl citrate antiwear surface protection mechanisms. Note that the polar portions of the alkyl citrate align themselves in a manner to best interact with the metallic surface similar to ZDDP polar metal interactions.

level of performance when used at the proper treatment level. Caution must be used because the volatility at hot oil temperatures increases as the alkyl length decreases. However, antiwear performance increases as the alkyl chain length decreases, so a judicious choice of proper physical chemical balance between volatility and antiwear surface performance must be found.

19.1.3.2 Organic Friction Modifier Additives: Alkyl Esters

As the carbon chain length increases significantly, the application of these surface-active molecules transitions from antiwear to friction modifiers [26–32]. The best friction modifiers include those with alkyl groups large enough to develop mono- and multilayered structures. No doubt, the regularity, organization, and development of molecular layers reduce the friction at the boundary layers.

Friction modifier chemical structures do have an optimum alkyl chain length where either greater or fewer numbers of carbons in the chain result in diminished effectiveness. In 1963, Cameron and Crouch [26] reported that in nature surface-active molecules have a preferred chain length that reduces friction as measured by the maximum scuffing load in a four-ball test. They specifically studied long chain carboxylic acids, amines, and alcohols, in various straight-chain paraffins (hexadecane and tetradecane) and found that the optimum length of a carbon chain was 16 [26].

The friction modifier tribology studies that follow include both organic and inorganic compounds such as alkyl sorbitans [27–29], GMO [30], alkylamine salts [31], alkyl tartrates [24,32], and molybdenum friction modifiers [7–9]. Their structures are given in Figure 19.8. In general, as the alkyl chain length increases, friction reduction dominates antiwear performance.

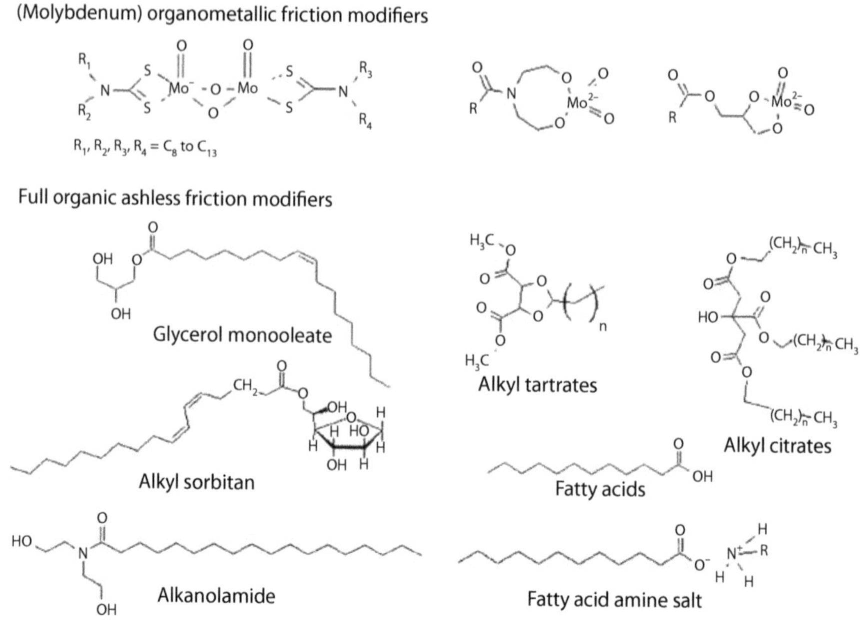

(Molybdenum) organometallic friction modifiers

$R_1, R_2, R_3, R_4 = C_8$ to C_{13}

Full organic ashless friction modifiers

Glycerol monooleate

Alkyl tartrates

Alkyl citrates

Alkyl sorbitan

Fatty acids

Alkanolamide

Fatty acid amine salt

FIGURE 19.8 A comparison of modern organic friction modifier chemistries studied.

19.1.3.3 Friction Modifier Films: Models

The best organic friction modifiers include those with alkyl chains favoring the formation of mono- and multilayered structures. A model of this type of organization is presented (Figure 19.9) using GMO films with two bilayers with an oil soluble region where the nonpolar chains between the monolayers interact. Slippage between monolayers reduces friction between the moving boundaries. The development of multilayered boundary films of friction modifier additives is an accepted explanation for their friction-reducing properties [33]. There is a need, however, to verify at a molecular level, any model developed and to advance organic friction modifier mechanistic theories. New surface analytical tools to accomplish this include techniques such as x-ray absorption near edge spectroscopy (XANES), reflectance–absorbance Fourier transform infrared (FTIR) spectroscopy, sum-wave infrared spectroscopy, and other techniques providing molecular level information such as the composition and organization of both inorganic and organic surface films [34–38].

19.1.3.4 Additive–Additive Compatibility Requirements

One of the key factors affecting the performance of antiwear and friction modifiers that impact required surface film development is their compatibility with other additives such as detergents and dispersants. In fact, some lubricant detergents and dispersants compete with friction modifiers for metal surfaces, and can keep them dissolved in the bulk liquid by forming molybdenum–amine complexes. As a friction modifier, molybdenum–amine complexes are reported to be less effective than MoDTCs [39,40]. Formulators developing modern lubricants must then consider all

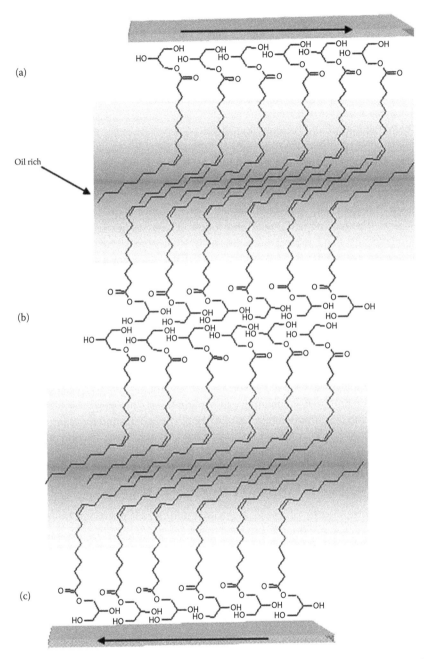

FIGURE 19.9 (See color insert.) A theoretical multilayered structure of GMO, capable of deformation and reforming during a dynamic state, set up between friction surfaces in motion. (a) The polar portion of (GMO) in the first bilayer is bound to the top metal surface, while the nonpolar (oil soluble) chains interact. (b) The polar hydrogen bonding interactions between the ester and hydroxyl functional groups holding the two bilayers together. (c) The bonding of the polar portion of (GMO) to the lower metal surface.

possible additive interactions. There must be a balance between sufficient dispersant to solubilize wear metals (and soot in diesel engines) and insufficient dispersant to reduce the active friction modifier concentration. Especially in diesel engines, soot has a measurable negative effect on some friction modifiers by a wearing action that removes the boundary layer lubrication [40]. In addition to dispersants, compatibility with other existing friction modifier additives (combinations of both molybdenum and organic based) is required and can often show enhanced performance similar to the combinations of alkyl citrates and ZDDP additives. In general, the molybdenum-based friction modifiers respond well at higher temperatures (above 100°C) while organic friction modifiers can show better performance at lower temperatures. As a result, blend combinations can result in a synergistic benefit over a wider temperature range.

Experimental data from some limited representative tribological studies of the OAW-FM (friction reducing) additives follow. These studies focus on the organic friction modifiers and antiwear additives as well as comparisons to a typical molybdenum dithiocarbamate friction modifier and ZDDP antiwear additives.

19.2 EXPERIMENTAL STUDIES OF NEW ANTIWEAR AND FRICTION-REDUCING ADDITIVES

Several bench test methods are used to measure the antiwear and friction-modifier additive performance of formulated lubricants. These correlate to the type of wear and friction occurring in engines, as previously discussed. Wear is measured using the Falex four-ball wear (ASATM D4172) and Cameron–Plint TE77, operating in the wear measurement mode (simulating cam rolling wear). In contrast (simulating sliding friction), the mini traction machine (Stribeck curve data) and the Cameron–Plint TE77, operating under a reduced load, are used to measure the friction coefficients of lubricants with and without friction modifiers.

19.2.1 WEAR STUDIES EXPERIMENTAL BENCH-TEST CONDITIONS

In these studies, the effectiveness of antiwear additives is measured by comparing the average wear scar, which is precisely measured with the aid of a metallurgical microscope. These scars typically range between spherical to somewhat elliptical (for small scars). Measurements are made across two perpendicular directions—the longer and shorter axes of the elliptical scar. Figure 19.10 gives examples of typical wear scars generated with the *two principle tribological bench test methods*—the Falex four-ball wear or (ASTM D 4172) and the Cameron–Plint wear tests (Table 19.1).

19.2.1.1 ASTM D 4172 Falex Four-Ball Wear Test

The Falex four-ball (ASTM D 4172) uses three fixed balls and one rotating during a single isothermal stage performed under a constant load. A load of 392 N is applied to the top ball kept rotating for 1 h at 1200 (rpm) and in contact with the three lower balls immersed in the test oil isothermal at 75°C. To simulate normal crankcase oil aging, 1.0% wt. cumene-hydroperoxide (prowear catalyst) is added to the

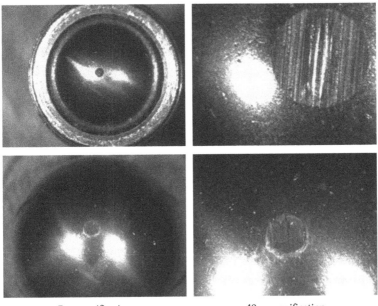

7 × magnification   40 × magnification

FIGURE 19.10 Typical wear experiments with scars generation. Top specimen from the Cameron–Plint wear test. Bottom specimen from the Falex four-ball wear test.

TABLE 19.1
Parameters of Three Stages of Cameron–Plint (Model TE-77 Machine) Operating in Wear Mode: TE-77 Wear Test (Cameron–Plint) Test Conditions[a]

Stages	Load: Newtons (N) (kg)	Temperature	Ramp Time (min)	Hold Time (min)	Frequency (Hz)
1	100 N, 10.2 kg	25°C–50°C	15	15	30
2	100 N, 10.2 kg	50°C–100°C	15	45	30
3	100 N, 10.2 kg	100°C–150°C	15	15	30

[a] Pro-wear cumene catalyst-hydroperoxide added at 1% wt.

test lubricant blended at 60°C for 20 min under agitation. The effect of the prowear catalyst to improve wear discrimination was discussed by Haeeb et al. [41,42] and Rounds [43]. The wear test specimens include three 12.7 mm (0.5 in.) diameter steel balls made of AISI standard E-52100 steel, 64–66 Rockwell-C that are clamped together and fixed in a jig that allows them to be covered with the test lubricant. The fourth top ball (12.7 mm) is pressed within the pocket formed from the lower three fixed balls and is rotated at 1200 (rpm) under a controlled force and temperature.

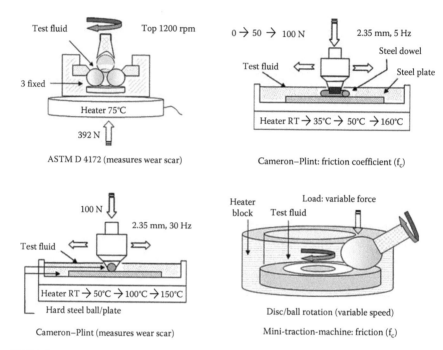

FIGURE 19.11 Four tribology test methods used to study OAW-FMs. (Top left) Falex four-ball wear test. (Bottom left) Cameron–Plint wear configuration. (Top right) Cameron–Plint friction configuration. (Bottom right) PCS instruments mini-traction machine friction measurements (ball on disc).

The average ball scar diameter is calculated by measuring the scar width along two perpendicular axes and averaging the two measurements together. These calculations are repeated for the three fixed balls and the results are further averaged in order to calculate a final overall average wear scar diameter. These measurements are done with a Lecia Stereo Zoom 6, specimen microscope with up to a 4× zoom, and a Mitutoyo digital X–Y translation table. Figure 19.11 illustrates the geometry of this experimental design and gives the specific test condition discussed.

19.2.1.2 Cameron–Plint Wear Test

The Cameron–Plint wear test uses a single ball-on-plate test geometry. Specifically, a 6 mm diameter AISI 52100 steel ball of 800 ± 20 kg/mm^2 hardness and a hardened ground steel plate (Rockwell-C Hardened 60/0.4 µm) is used. Both specimens develop wear scars in three stages under increasing temperature from room temperature to 150°C, first at a 50 N load at 50°C followed by a constant 100 N load when uniformly heated to 160°C in 1 h. As with the case of the four-ball wear test, 1 (wt.%) cumene hydroperoxide is added to simulate normal crankcase oil aging. An illustration of this experimental design is given in Figure 19.11, and the test parameter details, showing each experimental stage are presented in T 1. Measurements of the ball wear scar are made with the aid of Lecia Stereo Zoom 6 specimen microscope with up to 4× zoom, and a Mitutoyo digital X–Y translation

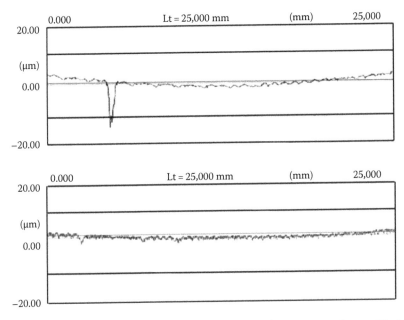

FIGURE 19.12 Wear scar profiles from the Cameron–Plint wear experiment. (Top) Oil without any antiwear additive, 16 μm wear scar depth from scratches on the bottom wear test plate. (Bottom) Same test with oil treated with antiwear additive showing significant reduction in the wear to within the noise of the measurement ±0.5 μm.

table. The plate scar profilometry was done using a Hommelwerke model TKU 300/600 tester. The average ball scar is determined by first aligning the ball under the microscope and measuring the length of the longest ellipsoidal wear scar axis followed by measuring the shorter perpendicular scar axis and averaging the two. The average plate wear scar diameter and the plate depth are measured using the digital profilometer that is rastered across the scar at four equally spaced points along the scar. Figure 19.12 illustrates a typical plate scar profile. The average plate wear scar depth is an additional measure of the degree of wear, although ball scar, in general, has better precision.

19.2.2 FRICTION STUDIES EXPERIMENTAL CONDITIONS

Two tribological tests designed to measure the coefficient of kinetic friction and simulate the sliding motion of piston rings against a cylinder wall (Cameron–Plint friction mode) and the rotating–sliding motion at the rocker arm pushrod area (Ball-on-disc mini traction machine) are described in the following.

19.2.2.1 Cameron–Plint Friction Test

To measure friction by the TE-77 Cameron–Plint method, a dowel under an applied load is rubbed against a steel plate and the coefficient of friction is measured from the force needed to maintain the motion. An illustration of this experimental design is

TABLE 19.2

Cameron–Plint (Model TE-77 Machine) Friction Mode Test Parameters Are Given: TE-77 Friction Test (Cameron–Plint) Test Conditions[a]

Stages	Load: Newtons (N) (kg)	Temperature	Ramp Time (min)	Hold Time (min)	Frequency (Hz)
1	0 (N), 0 kg	25°C–35°C	10	5	0
2	50 (N), 5.1 kg	35°C–50°C	10	5	5
3	100 (N), 10.2 kg	50°C–160°C	60	0	5

In this experiment, a 16 mm long nitrided steel dowel pin (6 mm diameter, RC Hardness 60) is rubbed against a hardened ground steel plate (RC Hardness 60/0.4 µm surface roughness) with a controlled load. The dowel is translated over a fixed amplitude at a set reciprocating frequency in Hz (given below) and a load that increases with increasing temperature in three stages.

[a] No Prowear cumene-hydroperoxide added.

given in Figure 19.11 and the test parameters are detailed in Table 19.2. The sample is first warmed from room temperature to 35°C without load and then kept isothermal at 35°C for an additional 5 min followed by heating to 50°C for 10 min, with 50 N of applied load and a reciprocating force on the dowel operating at 5 Hz for 10 min. In the last stage, the load is increased to 100 N and the sample is steadily heated to 160°C in 1 h while the reciprocating force is maintained at 5 Hz. The load used in these measurements bring the sliding metal surfaces within boundary conditions (where metal surface asperities interact) creating friction conditions, and as temperature is elevated, viscosity goes down from the starting viscosity at 35°C, and the friction in general increases. The usefulness of this test method is to simulate the sliding type friction that occurs in regions such as the piston rings sliding across the cylinder Waals. Engine rolling friction by contrast is simulated by use of the mini-traction machine test.

19.2.2.2 Mini-Traction Machine Test

In this test, a 19 mm hardened stainless steel ball (SAE AISI 52100) is rotated at set speeds and loads against a flat 46 mm diameter polished hardened steel disc (SAE AISI 52100) that is also rotating independently. The speed of both the disc and the rotating ball change the percent of slide to roll ratio (SRR). While maintaining a set SRR, the instrument first operates with the disc running at a higher speed than the ball and is then reversed so the ball turns at a higher speed than the disc. The experimental parameters used include the sample temperature (°C), load (N), and slide-roll-ratio (SRR). The 120°C Stribeck measurement data are based on a 30 N applied load and an SRR of 50. Under these conditions, the mean speed (also called entrainment speed) is slowly reduced from 2000 to 6 mm/s in order to move from hydrodynamic lubrication (bulk lubricant viscosity dependent) to mixed and boundary layer lubrication regimes. An illustration of this mini traction machine experimental

TABLE 19.3

Mini-Traction Test Parameters for Developing Stribeck Curve Data: Mini-Traction Machine Experimental Conditions

Test Phase	Temperature	Load (N)	Speed (mm/min)	Slide-Roll-Ratio	Number of Points
SRR vs. speed	40°C	30	2000	2–70	15
Stribeck curve	40°C	30	6–2000	50	26
SRR vs. speed	80°C	30	2000	2–70	15
Stribeck curve	80°C	30	6–2000	50	26
SRR vs. speed	120°C	30	2000	2–70	15
Stribeck curve	120°C	30	6–2000	50	26

Measurements are made at three temperatures 40°C, 80°C, and 120°C and involve both friction coefficient measurements as a function of increasing the slide-roll-ration and the Stribeck curve (friction vs. increasing entrainment speed). The sequence involved the SRR followed by the Stribeck data repeated at each temperature. The six steps with the experimental conditions are given.

design is given in Figure 19.11, and the test parameter details are shown in Table 19.3. This type of test simulates friction occurring in both the sliding and rolling–sliding region of the engine. For example, it simulates sliding friction at the piston rings (Figure 19.1, Contact-D) and the rolling interfaces of the cam tappet (Figure 19.1, Contact-A), and rocker arm (Figure 19.1, Contact-B and -C).

19.3 NEW ORGANIC ANTIWEAR AND FRICTION REDUCING ADDITIVES: RESULTS AND DISCUSSION

19.3.1 ORGANIC ANTIWEAR ADDITIVE PERFORMANCE

The key attributes of a robust OAW additive are average wear-scar reduction levels in the range of oils formulated with ZDDP, sustained compatibility with overbased calcium sulfonate detergents (i.e., micelles of sulfonate and basic calcium carbonate), and dispersants, such as bis-succinimides. In general, the performance of all other lubricant additives is also preserved. These include ZDDP or other EP additives, antioxidants, molybdenum-based friction modifiers, and additives at lower concentrations such as corrosion inhibitors and silicone antifoam additives.

19.3.1.1 Comparisons to ZDDP

An example of a particularly effective OAW additive is the alkyl citrates, which show many of these attributes. Figure 19.13 is a plot of ASTM D 4172 relative average wear scar differences from data comprising over forty measurements of Group II base oil with and without 1 (wt.%) ZDDP (repeated 14 times), 0.5% ZDDP (repeated eight times), alkyl citrate at 1% without ZDDP, and in combination

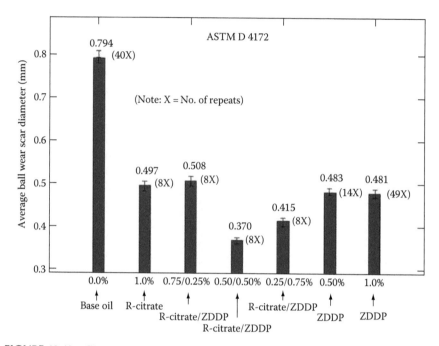

FIGURE 19.13 Characterization of OAW additive performance (ASTM D 4172 Falex four-ball wear) relative to ZDDP. Synergistic effects noted at concentrations of 0.50:0.50 and 0.25:0.75 ratios of alkyl citrate and ZDDP.

with ZDDP at 1:1, 1:4, and 4:1 ratios. In these studies, the total sum of antiwear additive equals 1%. Figure 19.13 indicates when alkyl citrate is used alone as an antiwear additive, 1 (wt.%) is required to reduce the average wear scar from 0.794 to 0.497 mm. When used in a synergistic combination with ZDDP at a 1:1 wt. ratio, it is more effective and the average wear scar level decreases to 0.37. The 1% and 0.5% ZDDP alone are both comparable in performance (0.48 mm) indicating that for this particular oil ZDDP is very effective in interacting and covering the test surface and interface. As levels of polar additives, such as detergents (overbased Ca sulfonates), and alkyl polyamine dispersants, such as poly(isobutenyl succinimides), increase, higher levels of ZDDP are typically needed and a greater difference in performance is observed between 0.5% and 1.0% ZDDP. These typical chemical interactions arise from the transition metal Zn coordinating with the amine portion of dispersants such as tetraethylene tetraamine (TETA). Specifically, the interaction that can take place is illustrated in Figure 19.14 and is discussed in several review articles [44,45].

In contrast, OAW additives such as alkyl citrates do not interact with dispersants/detergents, and if levels of ZDDP are reduced to 0.5%, they can be effective in supplementing their performance. One possible chemical mechanism that can explain the synergy of the combination of alkyl citrate and ZDDP is presented in Figure 19.15. Alkyl citrate provides a chemical means of solvating and isolating the PIB-dispersant from the ZDDP and fills in the gaps on the surface where ZDDP is

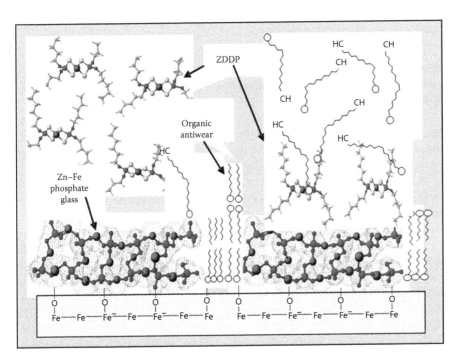

FIGURE 19.14 Coordination interaction of zinc in ZDDP with the nitrogen amine portion contained in a bis-poly(isobutenylsuccinimide) dispersant. This structure helps keep the ZDDP molecule dissolved in the oil and therefore competes with ZDDP adsorption to the meal surface.

FIGURE 19.15 (See color insert.) A combined ZDDP-OAW additive film formation model. While ZDDP inorganic glass protective films form on the metal surface bound to iron oxides, the smaller polar OAW molecules help to fill in the gaps formed in the glass under normal wear action. The resulting film may be reinforced and stronger when both additives are combined.

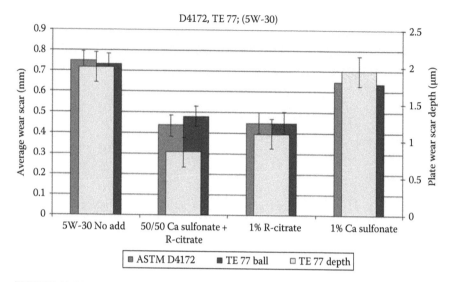

FIGURE 19.16 ASTM D4172 Falex four-ball, and Cameron–Plint TE 77 wear bench test results for an alkyl citrate OAW additive in the presence of a neutral calcium sulfonate lubricant detergent. The data indicate that the two additive systems are compatible. The TE 77 data measure both the ball scar (left axis) and the plate scar depth (right axis) where a profilometer was used to measure the depth.

removed. This type of model would lend itself to experimental measurements by techniques such as XANES in combination with infrared surface reflectance spectroscopy or possibly sum-wave spectroscopy [2,46–48].

Although synergistic with ZDDP, alkyl citrate alone shows a greater decreased performance than ZDDP if its concentration is reduced to 0.5 (wt.%). This is not surprising since an all-OAW additive would not be expected to form the same type of oxidized film as the inorganic glass represented in Figure 19.7.

A better compatibility of trialkyl citrates with detergents as well as with ZDDP is shown in the data presented in Figure 19.16 demonstrating that alkyl citrates in fact do not compete with overbased calcium sulfonates and at a minimum are neutral to the effect of detergents in an oil formulation. The evidence is presented in the four-ball wear ASTM D4172 and Cameron–Plint wear data (*left axis*) as well as in the Cameron–Plint plate wear scar depth (*right axis*). These data indicate, relative to oil formulated without detergent, that 1 (wt.%) detergent has little effect on reducing wear, and a 50:50 (wt. ratio) blend of alkyl citrate and Ca-sulfuonate detergent shows a synergistic benefit that reduces the wear scar. The wear scar is reduced to a level lower than 1% alkyl-citrate alone when measured by the Cameron–Plint wear method and is equally as effective when measured by the ASTM 4172 four-ball wear method. This indicates that the surface cleaning action of detergent chemistry either helps the OAW additive to find bear metal and establish its protective barrier or in combination with detergents is brought to the surface creating a better protective layer. This is a potentially important property since neutral and overbased calcium sulfonate are one of the most vital additives in all oil formulations responsible for neutralizing acids and cleaning varnishes and deposits from oil oxidation.

19.3.2 ADDITIVE FILMS AND SURFACE ANALYTICAL METHODS

Surface sensitive spectroscopies have great potential to unravel the additive chemistry operating at engine friction and wear surfaces. Methods such as microinternal reflectance mid-infrared spectroscopy, grazing angle polarized reflection–absorption spectroscopy, sum-frequency generation (SFG), and sum-frequency spectroscopy (SFS) all advance the understanding of surface chemistry film structure. By combining these vibrational spectroscopic techniques with a complementary approach, such as XANES, it is possible to characterize both the organic and inorganic elements of the tribofilms formed. Specifically, micro-internal reflectance FTIR allows sampling of small areas and small depths of the surface film (to within a fraction of the wavelength) determined by the prism geometry and refractive index of the optical internal reflection element of choice. This occurs by virtue of the evanescent standing wave developed by infrared light reflecting within a prism chosen for its proper refractive index and internal reflectance angle. The standing wave that develops at the surface of the prism–sample interface can interact with the lubricant/additive film and give an internal reflectance absorbance surface spectrum, providing chemical functional group information. External-reflection absorption spectroscopy (RAS) by contrast will penetrate deeper into the films (double-pass transmission through the film). By redundant aperturing (masking the light at the sample and its image), a combination of microscopy and the (RAS) technique can give the chemical composition averaged over the film thickness but limited (based on diffraction) to a very small spatial resolution of the wear film. Micro-external reflection spectroscopy can generate surface maps of the functional groups and chemical changes of films that undergo frictional-wear contact. The region that can be resolved in contrast to sum-wave spectroscopy is limited, however, to the wavelength of the reflected light (25 μm in mid-IR). External-reflection infrared absorption spectroscopy was performed on a Falex four-ball wear scar using a Nicolet-Omni Micro-FTIR system equipped with a narrow band mercury cadmium telluride (MCT) detector, operating in an external reflectance mode using a 250 μm aperture, at a 2 cm^{-1} resolution, and averaging 250 scans per spectrum. The usefulness of this technique in mapping functional group changes over a small area are given in Figures 19.17 through 19.19 where changes in films of GMO friction modifier in oil were observed. These data indicate that changes in the chemical nature can be mapped for specific surface areas showing changes in the carbonyl region typical of ester and oxidation species close to regions of high wear. The spectrum furthest from the contact region appears to show predominantly the hydrocarbon base-oil spectrum. If spatial discrimination is sacrificed for a larger sampling area, mid-infrared light reflected at an optical geometry of grazing angles (high angles of incidence) and parallel polarization yields both the molecular orientation and chemical composition of the surface film at the wear surface. A more recent nonlinear laser spectroscopic technique, sum-wave spectroscopy, has a great potential to probe the chemical composition at film surfaces while maintaining spatial resolutions on the order of angstroms. This technique is particularly sensitive

FIGURE 19.17 (See color insert.) Microscopic image near the end of the wear scar above identifying two points (A and B) and identified with marks where FTIR spectra were collected.

to the interfaces of metal and surface films because the molecules in the bulk are isotropic centrosymmetric, while the interface of a surface is ordered and nonisotropic, and therefore the only region capable of generating a sum-frequency signal. In this technique, the sample interface is probed with two lasers, a fixed wavelength visible laser and a tunable infrared laser. At the additive film surface, the two beams interact and emit photons whose wavelength is the sum of the two incident photons. If a vibrational mode of the additive (AW or FM film) matches the tuned infrared frequency at the interaction site, the emitted photons in resonance with the vibrational modes will exhibit much higher intensity. This then yields the vibrational spectrum at the interface of the film. A surface spectrum will be achievable over an extremely small area and controlled depth if desired [49]. Figure 19.20 illustrates all these useful analytical spectroscopic approaches.

The antiwear model illustrated in Figure 19.15 presents the need for a series of mechanisms to establish a true mixed-additive cooperative boundary-layer protection. Through a complicated series of equilibria, a balance must be maintained between the adsorption and desorption of ZDDP, the OAW, and combinations of the two ZDDP-OAW to the metal surface and the ZDDP-transformed phosphate coated surface [$Fe_x Zn_y (PO_4)_2$]. The kinetics of how these simultaneous adsorption–desorption processes occur can be described by the Langmuir isotherm for the possible species to adsorb onto the metal surface:

$$\frac{d\theta}{dt} = k_a [A](1-\theta) - k_d (\theta) \quad \text{for ZDDP, OAW, ZDDP} - \text{OAW}, Fe_x Zn_y (PO_4)_2 \ldots$$

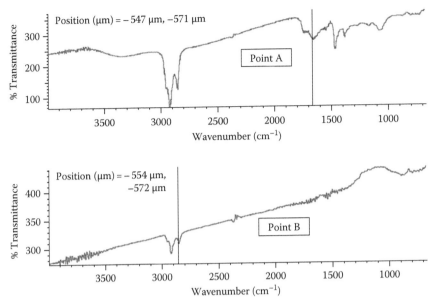

FIGURE 19.18 FTIR spectra of position points (A and B) with features related to ester from GMO and oxidation products in spectrum A and general hydrocarbon oil spectrum of original oil with GMO at 0.5% in point B further away from the wear-scar. Note tilt in the reflectance spectrum due to roughness of the surfaces scattering light at shorter wavelength (smaller wavenumber [cm⁻¹] than at longer wavelength (larger wavenumber cm⁻¹). This is due to a type of debris field caused by the ploughing of metal during the ball wear process, to the outer diameters of the wear scar.

where
θ is the surface coverage fraction
$[A]$ is the concentration of additive in solution
k_a and k_d are the rate constants for adsorption and desorption, respectively
t is the time

Since the rate constants have different temperature coefficients, increasing the temperature can lead to increased, decreased, or unchanged surface coverage [50]. Friction will not develop into wear if the critical minimum surface film fraction is $\theta = 0.5$. Below this value of θ, both friction and wear will increase. This was described for some specific friction modifiers and antiwear lubricant additives including ZDDP [51]. Through a careful testing of structure-performance and compatibility with ZDDP, new OAW additives can be developed. Alkyl citrates are an example of a sulfur- and phosphorus-free additive developed synergistic with ZDDP to maintain adequate θ surface film coverage.

19.3.3 ORGANIC FRICTION MODIFIER ADDITIVE PERFORMANCE

The organic friction modifier compounds (Figure 19.8) are similar to the OAW additive (Figure 19.7) in so far as both have a polar functional group possessing affinity for

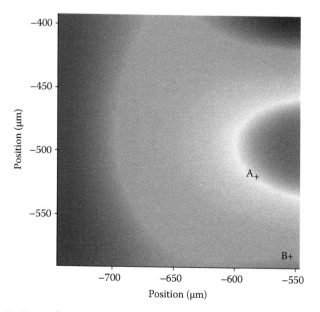

FIGURE 19.19 (See color insert.) Red regions indicate higher absorbance (more C=O concentration), and blue is lower absorbance (less C=O). The map is based on a $40 \times 40\,\mu m$ spot size over the area given.

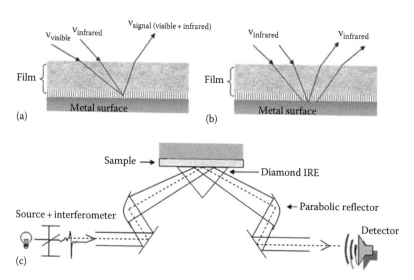

FIGURE 19.20 Optical spectroscopies for tribology film characterization (a) SFS optical geometry, (b) External reflection–absorption FTIR/micro-FTIR spectroscopy, and (c) Mid-FTIR internal reflectance spectroscopy.

metal surfaces, and a nonpolar oil soluble portion. Important differences exist, however, in both the alkyl groups (chain length, branching, and unsaturation) and in the degree of polarity of the polar group. This is important for developing models of multiple monolayers providing slippage and friction reduction at the boundary layer (where the metal asperities of sliding surfaces can come close and are kept from interacting). This kind of ordered structure lubrication is different from hydrodynamic lubrication where surfaces are separate and ride on lubricant liquid alone; friction in that case is the internal friction (viscosity) of the lubricant. The lubricant viscosity is also an important parameter that can affect fuel economy by the work required to overcome this internal viscosity. Its performance at both low and high temperatures are typically improved (reduced at low temperatures and increased at high temperatures) by the use of viscosity improvement lubricant polymer additives (e.g., poly(methacrylates) and ethylene–propylene copolymers). The friction modifier boundary lubrication model for GMO (previously discussed, Figure 19.9), illustrates the multiple monolayer intermolecular organization. The attributes needed to bring about this type of structure with resulting friction reduction near the boundary layer include

1. Self-assembly and re-self-assembly under mechanical operation
2. Planar slippage under stress
3. Regularity of polar and nonpolar groups by chemical design
4. Lubricant penetration within the nonpolar end (oil solubility)
5. Physical chemical adsorption on metal surfaces
6. Prevention of metal–metal contact

In addition to GMO included in Figure 19.8 are the alkyl tartrates, alkyl sorbitans, alkanolamides, and an example of an ionic liquid (titanium salicylate). The reduction of the friction coefficient of formulated engine oils containing these friction modifiers was compared to a molybdenum-based molybdenum dithiocarbamate friction modifer in a number of bench tests. The relative performances of these additives and comparisons to the same oils without any friction modifiers are presented in the Cameron–Plint TE-77 friction data (Figure 19.21) and the mini-traction machine Stribeck curve data (Figure 19.22). These tests were done in fully formulated passenger car motor oil (PCMO) with and without friction modifiers. The reduction in friction is evident for both GMO and alkyl sorbitans, such as tallow sorbitan, and blends of alkyl sorbitans and alkyl tartrates mixtures [27], which show the largest reduction in the friction coefficient (Figure 19.21). These trends are also consistent with the mini-traction Stribeck curve data in Figure 19.22. It is worth noting that as temperatures exceed 120°C (160°C in the Cameron–Plint), the performance of molybdenum friction modifiers decreases when used at low concentrations (350 ppm Mo) in fully formulated 5W-30 with dispersants and detergents. This effect may be due to chemical changes in the structure of the molybdenum dithiocarbamate to create a more soluble Mo-amine complex kept away from the surface when in the presence of amine dispersants. If the concentration of the molybdenum friction modifier approaches 1%, the performance at higher temperatures is

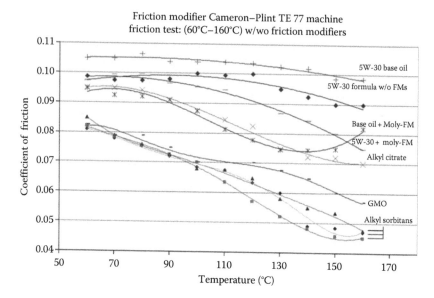

FIGURE 19.21 Cameron–Plint measurements of the effect of 1.0% wt. treatment level of various organic friction modifier additives and 0.035% wt. Molybdenum FM in 5W-30 formulated oil. The base oil alone and fully formulated without a friction modifier is compared to the fully formulated 5W-30 with both molybdenum and several organic friction modifiers. It is most important to reduce friction as the bulk oil viscosity decreases with temperature. This reduction is significant for oils treated with GMO and various alkyl sorbitan formulations or Molybdenum FM at higher treatment levels.

preserved (Figures 19.23 and 19.24). As temperature increases, oil viscosity will decrease causing the boundary layer lubrication to be more important. An increase in the coefficient of friction occurs as temperature increases when there is a lack of boundary layer friction modifier. At treat rates of 1% wt. active concentration, certain combinations of the organic friction modifiers show improvement at these higher temperatures. This would indicate improved molecular diffusion of the surface active molecules to the metal surface and/or some high-temperature chemistry (making the friction modifier more surface-active) is required for performance in both the molybdenum and certain organic types of additives.

19.3.4 OTHER DEVELOPING FRICTION MODIFIER TECHNOLOGIES

Although not completely organic, there are a number of organometalic-based friction modifiers, such as ionic liquids and some nanoparticle systems that are reported to show good performance. The ionic nature of these additives should facilitate adsorption on metal surfaces. A specific example of one ionic liquid (titanium salicylate) is shown in the Cameron–Plint friction and MTM Stribeck data presented in Figures 19.25 and 19.26. The treatment levels for these systems may not be the same as the fully organic-based systems, and this may be critical in stabilizing nanoparticle additive systems. The surface affinity of these ionic species should theoretically

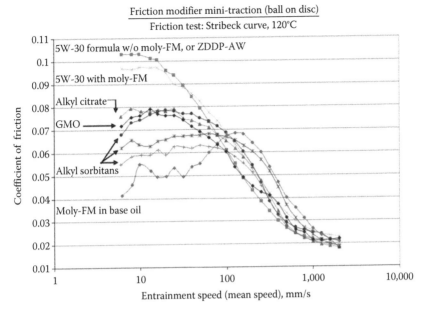

FIGURE 19.22 Comparative data of various organic friction modifiers and molybdenum friction modifier in formulated 5W-30 motor oil. At low entrainment speeds, the measurements are in a boundary layer lubrication regime, while at higher speeds hydrodynamic and mixed lubrication regimes occur. The data indicate that molybdenum FM at 0.035% is most effective when formulated in base oil; however, in a full formula with dispersant, higher treatment levels are needed. The data indicate a Mo-complex with other additives having coordinating elements (such as the nitrogen of dispersants) keeps the friction modifier from reaching the surface. In addition, some additives such as Ca-sulfonate detergents may also compete for the surface.

be higher; however, these types of systems must be stable in the presence of dispersants (metal solvating lubricant additives) that may attract and solvate the metallic portions keeping them from reaching surfaces. A judicious choice of dispersant and effective formulation and treat rate needs to be determined in formulating lubricants with these types of friction modifier compounds.

19.4 SUMMARY AND CONCLUSIONS: PERFORMANCE OF CURRENT ORGANIC ANTIWEAR AND FRICTION MODIFIER ADDITIVES

In summary, we can draw several conclusions regarding new ashless OAW friction modifier additives for automotive applications. The level of ZDDP performance is the target for antiwear additives with an average wear scar below 0.3 μm. Concentrations of 1%–2% of alkyl citrates and some other small polar organic species can approach this target and act synergistically with the ZDDP present. Further, a judicious choice of the organic structure of these additives can result in compatibility with a range of

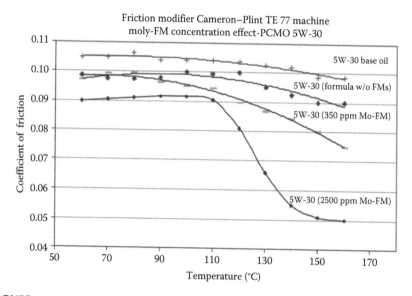

FIGURE 19.23   Above is the Cameron–Plint friction data showing the impact of increasing molybdenum friction modifier concentration from 0.035% to 0.25% wt. in a full-formulated PCMO (5W-30) containing dispersant. Note the greatest drop in the coefficient of friction occurs at an elevated temperature. This supports the idea of changes that occur with the molybdenum friction modifier structure at elevated temperatures.

other lubricant additives present. Specifically, performance and chemical compatibility need to be maintained with neutral and overbased calcium sulfonate detergents (containing reserved calcium carbonate), bis-succinimide dispersants, phenolic and amine based antioxidants, olefin-copolymer viscosity improvers (e.g., ethylene-propylene copolymers), and corrosion inhibitors. No-harm testing for additive compatibility with engine materials (e.g., copper–lead corrosion testing, elastomer swell testing, NACE rust testing, etc.) is required for final lubricant development. In addition, new organic based antiwear and antifriction modifiers need to be compatible with other antiwear and friction modifiers including molybdenum friction modifiers and ZDDPs. This will ensure lubricant–lubricant compatibility between oil changes when different lubricant brands are used.

Friction modifiers have to balance the development of multilayered structures at the metal-to-metal boundary surfaces of high friction when not competing with ZDDP glass film development at extreme-pressure surfaces where wear dominates. The use of alkyl citrates combined with ZDDP in antiwear additive combinations may help fortify this proper balance, resulting in an overall improved performance.

Both the unique alkyl structures and the polar end groups of OAW and friction reduction additives impact the additive's performance. In general, friction modifiers require longer alkyl chains in order to develop van der Waals attractive forces between friction modifier molecules, resulting in organized structures (monolayers, bilayers, etc.). The polar ends of friction modifiers need to interact with the

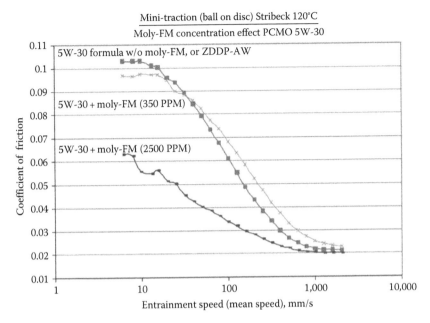

FIGURE 19.24 The Stribeck curve data showing the impact of increasing molybdenum friction modifier concentration in a full-formulated PCMO (5W-30) containing dispersant. Measured at 120°C, these data indicate as the concentration is increased from 350 ppm (0.035% wt.) to 2500 ppm (0.25% wt.) that the performance improves with a reduction in the coefficient of friction. At 0.25% wt., the concentration is adequate for any competing interaction with other additives such as amine dispersants.

metal and metal oxides developed at rubbing surfaces, anchoring the films to the surface, as well as bringing rigidity to multilayer structures. In contrast to these structure-activity requirements of friction modifiers, OAW additives tend to have shorter alkyl chains and need to be small enough to maintain tightly packed molecules at the surface and possible conversion to carbonate films at extreme-pressure surfaces. Future spectroscopic studies may serve to elucidate the chemical transformations that may be occur at high pressures and temperatures by these OAW additives.

In contrast to organic-based friction modifiers, molybdenum-based friction modifiers work with multiple mechanisms including both alkyl chain interactions between molecules followed by transformation into new species at the surfaces ultimately forming molybdenum-based glass films. However, even purely inorganic MoS_2 additives show performance limited to friction reduction and not extreme-pressure antiwear performance as exhibited by ZDDP. These molybdenum sulfur-based films may not be durable enough to consistently separate surfaces under extreme-pressure conditions such as at tappet-pushrod contacts. In addition, there may be competing forces when formulated in the presence of dispersants, such that higher concentrations of molybdenum-based friction modifiers may be required in fully formulated lubricants as compared to base oil performance.

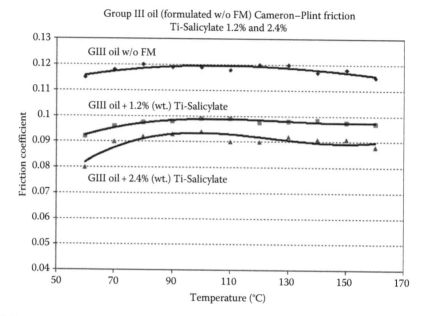

FIGURE 19.25 The reduction in Cameron–Plint friction of a formulated Group III oil containing base-oils, detergents, dispersants, antioxidants, viscosity improvers, pour-point depressants, antifoam, and ZDDP antiwear additives with and without 1.2% and 2.4% of an additive system of ionic liquid: titanium-salicylate and oil.

Finally, organic friction modifiers/antiwear additives will perform well with the following attributes:

- Compatibility with other lubricant additives without forming insoluble salt and haze
- Compatibility with engine materials, i.e., noncorrosive toward metals and elastomers
- Ashless and robust in maintaining engine emission reduction devices
- Sufficiently concentrated (typically higher than neat organometallic additives)
- Synergistic with ZDDP if less effective at extreme-pressure environments
- Formulated considering the dispersant additive interaction role, especially in lubricants with molybdenum dithiocarbamate friction modifiers
- Designed with structures that balance surface activity with lubricant solubility

If all of these considerations are addressed, the basic components of friction developed need to be considered. As described in detail by Suh [52], asperity deformation, adhesion, and metallic plowing can be analyzed and evaluated to assist in developing additive systems capable of interacting/limiting one or more of these components.

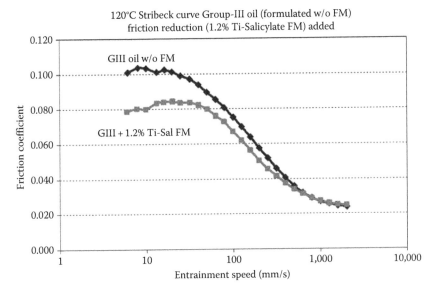

FIGURE 19.26 The Stribeck data showing a reduction in friction of a formulated Group III oil with and without 1.2% of an additive system of ionic liquid titanium-salicylate and oil.

19.5 FUTURE EFFORTS

The future additive developments will no doubt utilize combinations of both microspectroscopic characterization of additive chemical structures at metal–metal frictional interfaces and theoretical computational chemistry to determine both interaction energies and potential structures to explore [53,54]. Some of the latest efforts in developing new additive compounds focus on a number of novel approaches. These include new organometallics with effective ionic structures, utilizing nanostructures, and ionic liquids based on organometallic compounds such as titanium salicylate showing promise [55,56]. For additional information, a number of comprehensive articles are available covering the friction modifier and antiwear areas of ionic liquids, new analytical techniques, and theoretical dynamic models, and many of the ideas and concepts discussed [57–64].

Another area of future interest is friction modifier durability and the loss of performance with oil aging. A glimpse of this effort is presented in experimental data reported in Figure 19.27 that plots the Stribeck boundary-layer friction coefficient as a function of oil aging (lab-scale Nitro-oxidation) and friction modifier type. In these experiments, oils were treated with friction modifiers and subject to lab bench test oxidation followed by characterization of the viscosity and tribology changes. A test developed for simulating aging (Uniroyal Nitro Oxidation Test) treats 550 g of oil with a ferric naphthenate oxidation catalyst. To the test oil (agitated at 500 rpm) is added 8000 ppm of nitric oxide in N_2 at a flow of 100 mL/min combined with dry air at 200 mL/min, and held isothermally at 150°C. The test is continued while the kinematic viscosity at 40°C is monitored and used to halt the test when

Boundary layer friction (Stribeck 120°C)

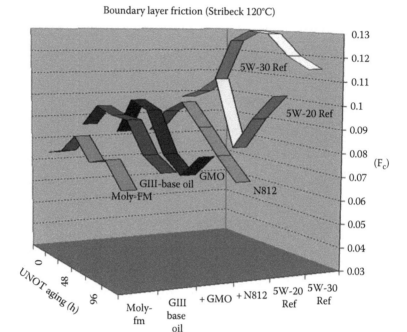

FIGURE 19.27 Friction modifier performance (based on MTM-Stribeck data) measuring the performance durability with lubricant aging during nitro-oxidation. Friction coefficient at boundary lubrication (slowest speed of Stribeck curve) of six samples after every 24 up to 120 h of nitro-oxidation aging. Friction coefficients of Group III base oil, and base oil with molybdenum FM (MoDTC) GMO, alkyl citrate (NL-812). In addition, fully formulated ASTM reference oils 5W-20 and 5W-30 were compared. All samples showed an initial increase in friction followed by a gradual decrease. The GMO treated and ASTM Ref 5W20 increased F_c at 120 h.

the increase in viscosity becomes immeasurable. After an initial period (24–48 h, simulating 2500–5000 vehicle mileage) the viscosity decreases and the coefficient of friction rises. This is then followed by a decrease in the coefficient with further oxidation (96–120 h—simulating high mileage). The latter friction decrease is thought to involve generation of alcohols, aldehydes, and carboxylic acid species from the oxidation of the lubricant to form natural friction modifiers. It is hoped that by monitoring oil changes, additives can be developed to be robust and perform both in fresh unoxidized conditions and after some extended oxidation and viscosity changes. This latter goal is not without its challenges since the lubricant system becomes dynamic with oxidation processes of combustion, generating new chemical species with a range of polarity, from alcohols to corrosive carboxylic acids generating wear metals affecting the solubility of less polar additives, and increasing viscosity to the point of reducing fuel economy.

In general, with the aid of better spectroscopic techniques and molecular modeling, the design of better friction modifiers and antiwear additives is possible. A real understanding of the mechanisms will no doubt depend on molecular level information that will be greatly enhanced by the advances made in molecular spectroscopy and physical chemistry—both experimental and theoretical.

ACKNOWLEDGMENTS

Special thanks to Joseph N. Valentine (Northeast Research and Development Services), for kindly creating the illustration of the small block engine valve-train (Figure 19.1), Bill Wortman (Chemtura Corp.) for his assistance with FT-IR microspectroscopy studies, Dr. Venkat Madabusi (Chemtura Corp.) for assistance in reviewing the current literature, and Dr. Cyril Migdal (Chemtura Corp.) for providing technical discussions of lubricant additive technology in general. In addition, the vital technical support of the tribology-testing laboratory—Brian Fox (Chemtura Corp.), Jon Goodell (Chemtura Corp.), and Mike Maselli (Chemtura Corp.)—is appreciated.

REFERENCES

1. C. Taylor and E. Taylor, *The Internal–Combustion Engine*, 2nd edn., International Textbook Company, Scranton, PA (1961).
2. H. Spikes, The history and mechanisms of ZDDP, *Tribol. Lett.*, 17(3), 469–489 (2004).
3. J. Bell, R. Coy, and H. Spikes, Cryogenic studies of zinc-dialkyldithiophosphate, In: *Proceedings of Japanese International Tribology Conference*, Nagoya, pp. 505–510 (1990).
4. C. Bovington, Fricton, wear and the role of additives in their control, In: *The Chemistry and Technology of Lubricants*, 2nd edn., R.M. Mortier and S.T. Orszulik (eds.), p. 340, Chapman and Hall, London U.K. (1997).
5. M. Muraki and H. Wada, Frictional properties of organomolybdenum compounds in the presence of ZDTPs under sliding conditions, *Tribologist*, 39, 800 (1994).
6. M. Muraki, Y. Yanagi, and K. Sakafuchi, Synergistic effect on frictional characteristics of rolling-sliding conditions due to a combination of molybdenum dialkyldithiocarbamate and zinc dialkyldithiophosphate, In: *Proceedings of International Tribology Conference*, Yokohama, pp. 69–75 (1995).
7. T. Nalesnik and C. Migdal, Oil-soluble molybdenum multifunctional friction modifier additives for lubricant compositions, U.S. Patent 6,103,674 (August 15, 2000, Uniroyal Chemical Company), Middlebury, CT.
8. N. Tanaka, A. Fukushima, Y. Tatsumi, and Y. Saito, Lubricating oil composition, U.S. Patent 5,627,146 (May 6, 1997). Asahi Denka Kogyo K.K., Tokyo, Japan.
9. T. Karol, Organic molybdenum complexes, U.S. Patent 5,137,647 (August 11, 1992) R.T. Vanderbilt Company, Narwalk, CT.
10. T. Karol and S.G. Donnelly, Fuel composition containing organic molybdenum compounds, U.S. Patent 5,628,802 (May 13, 1997), Narwalk, CT.
11. D. Love, R. Schlicht, and J. Biasotti, Lubricant additive, U.S. Patent 4,765,918 (August 23, 1988). Texaco Inc., White Plains, NY.
12. A. Stipanovic and J. Schoonmaker, The impact of organomolybdenum compounds on the frictional characteristics of crankcase engine oils, SAE Paper 932779 (1993).
13. J. Patel, A.J. Stipanovic, and J.P. Schoonmaker, Method of improving the fuel economy characteristics of a lubricant by friction reduction and compositions useful therein, U.S. Patent 5,736,491 (April 7, 1998). Texaco Inc., White Plains, NY.
14. V. Gatto, Molybdenum-containing lubricant additive compositions and processes for making and using same, U.S. Patent 6,645,921 (November 11, 2003). Ethyl Corp., Richmond, VA.

15. V.J. Gatto, Oil soluble molybdenum additives from the reaction of fatty oils and mono-substituted alkylene diamines, U.S. Patent 6,509,303 (January 21, 2003). Ethyl Corp., Richmond, VA.
16. V. Gatto, F.E. Perozzi, and C. Kuo, Lubricants containing molybdenum compounds, phenates, and diaryl amines, U.S. Patent 6,174,842 (January 16, 2001). Ethyl Corp., Richmond, VA.
17. E. Zaweski, Lubricant composition containing mixed fatty acid ester and amide of dieth-anolamine, U.S. Patent 4,439,336 (March 27, 1984). Ethyl Corp., Richmond, VA.
18. R. Schlicht, M. Levin, S. Herbstman, and R.L. Sung, Gasoline composition containing reaction products of fatty acid esters and amines as carburetor detergents, U.S. Patent 4,729,769 (March 8, 1988). Texaco Inc., White Plains, NY.
19. T. DeRosa, B.Kaufman, F. DeBlase, J. Kethcam, M. Rawdon, and M.Cesar, Method of enhancing the low temperature solution properties of a gasoline friction modifier, U.S. Patent 6,524,353 (February 25, 2003). Texaco Development Corporation, White Plains, NY.
20. T. DeRosa, T. Cangelosi, S. Cryvoff, F. DeBlase, T. Hayden, B. Kaufman, M. Rawdon, and Y. Thiel, Friction modifier for poor lubricity fuels, U.S. Patent 6,589,302 (July 8, 2003). Texaco Inc., San Ramon, CA.
21. T. Derosa, B. Kaufman, F. DeBlase, T. Hayden, M. Rawdon, J. Ketcham, Y. Thiel, and M.Cesar, Fuel additive composition for improving delivery of friction modifier, U.S. Patent 6,743,266 (January 1, 2004). Texaco Inc., San Ramon, CA.
22. T. DeRosa, F. DeBlase, B. Kaufman, M. Rawdon, J. Ketcham, and B. Wyman, Fuel composition containing friction modifier, U.S. Patent 6,835,217 (December 28, 2004). Texaco Inc., San Ramon, CA.
23. T. Hayden, C.Ropes, M. Rawdon, and Y. Thiel, The performance of a gasoline friction modifier fuel additive, SAE Paper 2001-01-1961 (2001).
24. C. Migdal and R. Rowland, Lubricant compositions containing hydroxyl carboxylic acid and hydroxy polycarboxylic acid esters, U.S. Patent 7,696,136 (April 13, 2010). Crompton Corp., Middlebury, CT.
25. J. Obiols, L. Pidol, J.-M. Savoie, and I. Rogues de Fursac, Lubrication composition for four stroke engine with low ash content, International Patent Application PCT/FR2008/001668 (February 12, 2008). Total Refining Marketing.
26. A. Cameron and R.Couch, Interaction of hydrocarbon and surface-active agent, Nature 198, 475–476 (1963).
27. F. DeBlase, C.Migdal, G. Mulqueen, and V. Madabusi, Fatty sorbitan ester based friction modifiers, U.S. Patent Application U.S. 12/371872 (February 2009). Chemtura Corporation, Middlebury, CT.
28. M. Dasai, Lubricating oil composition, U.S. Patent 5,064,546 (November 12, 1991). Idemitsu Kosan Company Limited, Tokyo, Japan.
29. G. McLean, Additive concentrate for fuel compositions, U.S. Patent 6,277,158 (August 21, 2001). Exxon Research and Engineering Company, Annandale, NJ.
30. H. Shauck and W. Waddey, Method for improving the fuel economy of internal combustion engines using fuel having hydroxyl-containing ester additive, U.S. Patent 4,617,026 (October 14, 1986). Exxon Research and Engineering Company, Florham Park, NJ.
31. K. Coupland and C.R. Smith, Polymerized fatty acid amine derivatives useful in friction and wear reducing additives, U.S. Patent 4,250,045 (February 10, 1981). Exxon Research and Engineering Company, Florham Park, NJ.
32. J. Kocsis, J. Vilardo, J. Brown, D. Barner, and R. Vickerman, Tartaric acid derivatives as fuel economy improvers and anti-wear agents in crankcase oils and preparations thereof, U.S. Patent 20100081592 (April 1, 2010). The Lubrizol Corporation, Wickliffe, OH.
33. V. Anghel, C. Bovington, and H.A. Spikes, Thick-boundary-film formation by friction modifier additives, Lubr. Sci., 11, 313–335 (2006).

34. M. Suominen, L. Fuller, F. Rodriguez, G. Massoumi, W. Lennard, M. Kasarai, and G. Bancroft, The use of X-ray absorption spectroscopy for monitoring thickness of antiwear film formation from ZDDP, *Tribol. Lett.*, 8, 187–192 (2000).
35. C. Gossirid, J.-M. Martin, K. Vorlot, B. Vacker, T. Mognue, and Y. Yamada, Tribochemical interaction between Zndtp, Modtc and calcium borate, *Tribol. Lett.*, 8(4), 203–212 (2000).
36. Y. Dalia, P. Kalamaras, D. Deckman, and M. Webster, Atomic force microscopy and Raman spectroscopy investigations of additive interactions responsible for anti-wear film formation in a lubricated contact, *Tribol. Trans.*, 49, 108–116 (2006).
37. A. Somaya, J. Pranesh, and B. Aswath, The role of antioxidants and oxidation stability of oils with F-ZDDP and ZDDP, and chemical structure of Tribofilms using XANES, *Tribol. Trans.*, 52, 511–525 (2009).
38. G. Bancroft, Tribological characteristics of ashless dithiocarbamate and its combinations with ZDDP additives in mineral oil, *Tribol. Int.*, 41, 1226–1231 (2008).
39. M. Sasaki, Y. Kishi, T. Hyuga, H. Omata, S. Takeshima, I. Kurihara, and A. Ohashi, Development of high performance heavy-duty diesel engine oil to extend drain intervals: 5W30 fully synthetic oil containing MoDTC, SAE Paper 2000-01-1992 (2000).
40. W. van Dam, P. Kleijwegt, M. Torreman, and G. Parsons. The lubricant contribution to improved fuel economy in heavy duty diesel engines, SAE Paper 2009-01-2856 (2009).
41. J.J. Haeeb, W.W. Rogers, and C.J. May, The role of hydroperoxide in engine wear and the effect of Znddp/dispersant/detergent interactions, SAE Paper 872157 (1987).
42. J.J. Haeeb and W.W. Stover, The role of hydroperoxides in engine wear and the effect of zincdialkyldithiophosphates, *ASLE Trans.*, 30, 419–420 (1987).
43. F. Rounds, Effect of hydroperoxides on wear as measured in four-ball wear test, *Tribol. Trans.*, 36, 297–303 (1993).
44. Z. Zhang, M. Najman, M. Kasrai, G. Bancroft, and E. Yamaguchi, Study of interaction of EP and AW additives with dispersants using XANES, *Tribol. Lett.*, 18, 43–51 (2005).
45. E. Yamaguchi, Z. Zhang, M. Kasrai, and G. Bancroft, Study of the interaction of ZDDP and dispersants using X-ray absorption near edge structure spectroscopy—Part 2: Tribochemical reactions, *Tribol. Lett.*, 15, 385 (2003).
46. G. Periera, A. Lachenwitzer, D. Munoz-Paniagua, M. Kasrai, M.P. Norton, T. Capehart, T.W. Perry, and T.Y. Cheng, Nanoscale chemistry and mechanical properties of tribofilms on Al-Si alloy (A383): Interaction of ZDDP, calcium detergent and molubdenum friction modifiers, *Tribol. Mater. Surf. Interfaces*, 1, 4–14 (2007).
47. A. Nicholls, T. Do, P. Norton, M. Kasrai, and G. Bancroft, Review of the lubrication of metallic surfaces by Zinc dialkyl-dithiophosphates, *Tribol. Int.*, 38, 15–39 (2005).
48. A. Barnes, K. Bartle, R. Vincent, and P. Thibon, A review of zinc dialkyldithiophosphates (ZDDPS) characterization and role in the lubricating oil, *Tribol. Int.*, 34, 389–395 (2001).
49. A. De Simone, Surface specificity of sum frequency spectroscopy, *J. Chem.*, 571, 1 (2005).
50. C. Bovington, Friction, wear and the role of additives in their control, In: *Chemistry and Technology of Lubricants*. R.M. Motier and S.T. Orszulik (eds), p. 320, Blackie Academic and Professional, London (1997).
51. C. Bovington and B. Dacre, The adsorption and reaction of decomposition products of zinc di-isopropyldiophophate on steel, *ASLE Trans.*, 27, 252–258 (1984).
52. N. Suh, *Tribophysics*, Prentice Hall Inc., Englewood Cliffs, NJ (1986).
53. M. Greefield and O. Hiroko, Packing of simulated friction modifier additives under confinement, *Langmuir*, 21, 7568–7578 (2005).

54. J. Davidson, S. Hinchley, S.Harris, A. Parkin, S. Parsons, and P. Tasker, Molecular dynamics simulations to aid the rational design of organic friction modifiers, *Mol. Graph. Model.*, 25, 495–506 (2006).

55. V. Carrick, Titanium compounds and complexes as additives in lubricants, International Patent Application WO 2009/042586 A1, 2009, The Lubrizol Corporation.

56. J. Ana-Eva, B. Maria-dolores, Ionic liquids as lubricants for steel-aluminum contacts at low and elevated temperatures, *Tribol. Lett.*, 26 (2007).

57. J. de Vicente, J.R. Stokes, and H. Spikes, Rolling and sliding friction in compliant, lubricated contact, In: *Proceedings of the Institution of Mechanical Engineers. Part J, Journal of Engineering Tribology*, Vol. 220 J2, pp. 55–1679 (February 2006).

58. K. Komvopoulos and S.A. Pernama, Friction reduction and antiwear capacity of engine oil blends containing zinc dialkyl dithiophosphate and molybdenum–complex additives, *Tribol. Trans.*, 49, 151–165 (2000).

59. H. Spikes, Origins of the friction and wear properties of antiwear additives, *Lubr. Sci.*, 18, 223–230 (2006).

60. J.M. Martin, C. Grossiord, T. Le Mogne, S. Bec, and A. Tonck, The two-layer structure of Zndtp tribofilms Part I: AES, XPS and XANES analyses, *Tribol. Int.*, 34, 523–530 (2001).

61. A. Morina and A. Neville, Understanding the composition and low friction tribofilm formation/removal in boundary lubrication, *Tribol. Int.*, 40, 1696–1704 (2007).

62. J.Q. Hu, X.Y. Wei, J.B. Yao, L. Han, and Z.M. Zong, Evaluation of molybdate ester as a synergist for arylamine antioxidant lubricants, *Tribol. Int.*, 39, 1469–1473 (2006).

63. A. Neville, A. Morina, T. Haque, and M. Voong, Compatibility between tribological surfaces and lubricant additives—How friction and wear reduction can be controlled by surface/lube synergies, *Tribol. Int.*, 40, 1680–1695 (2007).

64. J. Qu, P.J. Blau, S. Dai, H. Luo, H.M. Meyer III, and J.J. Truhan, Tribological characteristics of aluminum alloys sliding against steel lubricated by ammonium and imidazolium ionic liquids, *Wear*, 267, 1226–1231 (2009).

20 Antimicrobial–Surface Activity Relationship of Novel Di-Schiff Base Cationic Gemini Amphiphiles Bearing Homogeneous Hydrophobes

Nabel A. Negm

CONTENTS

ABSTRACT

In this review, the antimicrobial activities of surface-active agents were reviewed, as well as their mode of action on different microorganisms. A novel series of cationic gemini amphiphiles containing di-Schiff base species were synthesized and their chemical structures were determined using different analytical tools. Their surface properties were determined using surface tension measurements. The adsorption and micellization thermodynamic parameters were calculated using Gibb's equations at 25°C. The surface parameters were also determined including critical micelle concentration, effectiveness, efficiency, maximum surface excess, minimum surface area, interfacial tension, and emulsification power. The synthesized cationic gemini amphiphiles were evaluated as bactericides for Gram-negative and Gram-positive bacteria and also against sulfate-reducing bacteria (SRB). The results of cytotoxicity investigations of the synthesized compounds against targeted bacterial strains were promising and completely dependent on the surface activity of these compounds.

20.1 INTRODUCTION

In recent years, new classes of amphiphilic molecules have emerged and have attracted the attention of various industrial and academic research groups. One of these classes is gemini or dimeric surfactants, which have two hydrophobic groups per molecule, separated by a covalently bonded spacer. The name Gemini was coined in 1991 by Menger, who has reviewed the chemistry of these remarkable compounds [1]. The name surfactant typically means the amphiphiles.

These surfactants have also been referred to as bipolar or bisquaternary ammonium in the case of cationic surfactants. The first reports on dimeric surfactants concerned bisquaternary ammonium halide surfactants. Their biological activity in aqueous solutions was studied, and micellar solutions of these surfactants were used to catalyze chemical reactions [2].

Most studies were reported on the surface tension of the aqueous solutions of dimeric surfactants for critical micelle concentration determination and assessment of their capacity to reduce the surface tension of water [3,4]. These surfactants appear to be better in certain properties than corresponding and more conventional monomeric surfactants that are made up of one hydrophilic and one hydrophobic group, which may include one or more alkyl chain. Gemini surfactants have much higher surface activity than the corresponding monomeric surfactants at the same molar concentrations.

In gemini surfactants, the hydrocarbon tail may have various structures, such as long chain, short chain, aromatic, branched aromatic, heterocyclic, or sugar nucleus. Based on the structure of the polar or hydrophilic part, gemini surfactants can be classified as cationic, anionic, nonionic, and zwitterionic.

Depending on the environmental factors which control the application where these surfactants will be used, the type of gemini surfactants plays a vital role. Cationic surfactants are used for a number of stabilization problems and have found many technical applications. Because of the predominant negatively charged nature of natural colloids and surfaces including microorganisms, cationic surfactants form strong adsorption layers and increase the hydrophobicity of the surfaces of these materials or microorganisms. On the other hand, anionic gemini surfactants have found many applications in the industrial fields including emulsification and foaming processes.

20.2 STRUCTURE

The structure of gemini surfactants depends on the nature of the hydrophilic groups.

20.2.1 CATIONIC GEMINI SURFACTANTS

Figure 20.1 represents a cationic gemini surfactant molecule. The structure of a cationic gemini surfactant is denoted by the two symbols (m and n); m is referred to the hydrophobic chain length or the tail length, while n represents the spacer length between the two positively head groups.

Several investigators [5–8] have made considerable effort in the field of cationic gemini surfactants synthesis. The structures were varied from ordinary diquaternary ammonium surfactants containing different alkyl chains as hydrophobic chains with aliphatic spacer groups also with different lengths.

Some new features were added to cationic gemini surfactants employing new synthetic routes and by introducing different spacer groups containing different functional groups. The first such attempt involved introducing nonionic spacers consisting of poly (ethylene glycol) [9] with different molecular weights varying from 9 to 45 PEO units as shown in Figure 20.2.

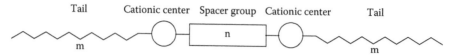

FIGURE 20.1 Representative structure of the cationic gemini surfactant molecules. m is the hydrophobic chain length, n is the spacer length between the two positively head groups.

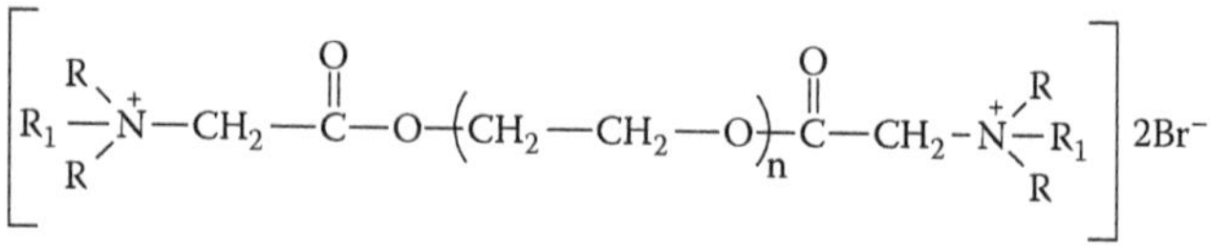

$$\left[R_1 - \overset{\displaystyle R}{\underset{\displaystyle R}{N}} - CH_2 - \overset{O}{\overset{\|}{C}} - O - (CH_2 - CH_2 - O)_n - \overset{O}{\overset{\|}{C}} - CH_2 - \overset{\displaystyle R}{\underset{\displaystyle R}{N}} - R_1 \right] 2Br^-$$

FIGURE 20.2 Gemini surfactants containing nonionic spacer group. n = 10, 15, 25 and R=CH$_2$CH$_2$OH R$_1$=C$_{11}$H$_{23}$COOCH$_2$CH$_2$ (lauric), C$_{15}$H$_{31}$COOCH$_2$CH$_2$ (palmitic) and C$_{17}$H$_{35}$COOCH$_2$CH$_2$ (stearic).

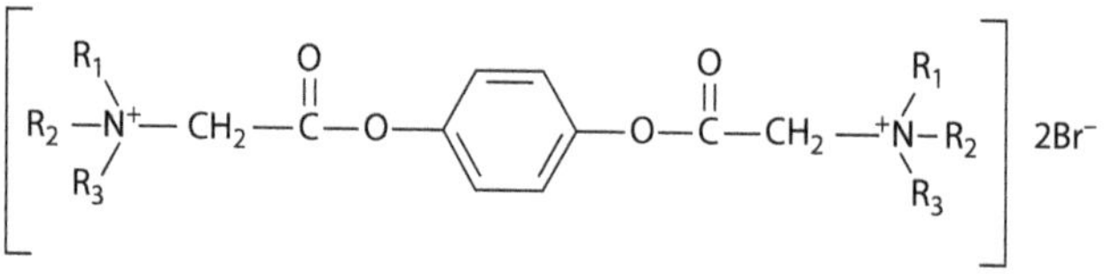

$$\left[R_2 - \overset{\displaystyle R_1}{\underset{\displaystyle R_3}{N^+}} - CH_2 - \overset{O}{\overset{\|}{C}} - O - \bigcirc - O - \overset{O}{\overset{\|}{C}} - CH_2 - \overset{\displaystyle R_1}{\underset{\displaystyle R_3}{N^+}} - R_2 \right] 2Br^-$$

FIGURE 20.3 Gemini surfactants containing aromatic spacer group. R$_1$=C$_{11-17}$-COO-CH$_2$-; R$_2$=CH$_3$; R$_3$=CH$_2$CH$_2$OH.

The second attempt to improve the applicability of the cationic gemini surfactants was by introducing some strongly adsorbing moieties in their chemical structures which are the aromatic polycyclic nucleus [10]. The resulting amphiphilic molecules showed extremely strong adsorption tendency at the interfaces and, consequently, high efficiency in corrosion inhibition of metallic surfaces against different corrosive environments. The chemical structures of these compounds were represented as shown in Figure 20.3.

Another attempt made in the field of synthesis of cationic gemini surfactants was the introduction of sugar nucleus as spacer groups in order to increase their compatibility with the biological membranes and human tissues [11]. The synthesis protocol for these surfactants depended on the synthesis of the dihaloester sugar derivatives (glucose and fructose) according to coupling of the halo-ester with the sugar nucleus as shown in Figure 20.4.

Quaternization was then performed between these derivatives and triethanol-amine alkanoate in appropriate solvents. The structure of the resulting product is shown in Figure 20.5. The structure was confirmed using a combination of various analytical data including elemental, FTIR, ^{1}H-NMR, and mass spectroscopy fragmentation patterns.

The synthesized novel gemini cationic surfactants were highly compatible with the cellular membranes of the microorganisms and showed high biological activities including antibacterial and antifungal actions at extremely low dose (5 ppm).

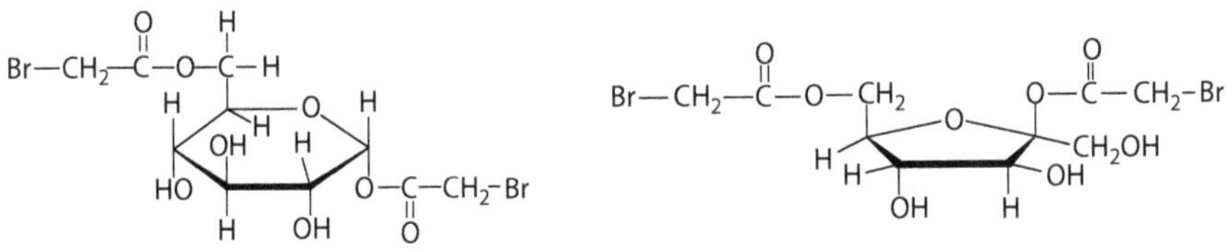

FIGURE 20.4 Different sugar-based dibromoesters.

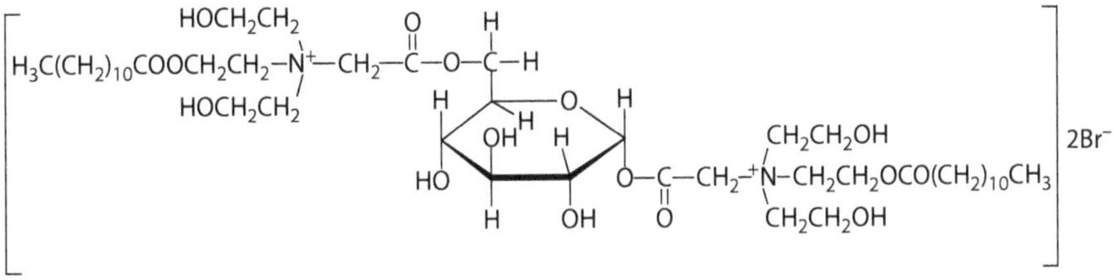

FIGURE 20.5 Sugar-based cationic gemini surfactants (dodecyl glucose derivative).

The most compatible cationic gemini surfactants with the living membranes of the different microorganisms were those which contain amino acids in their chemical skeletons including the essential amino acids. These attempts produced a series of environmentally friendly amphiphiles which easily degraded biologically by the action of the environmental bacteria and their biodegradation products were non-toxic both to the plants as well as the soil [12]. The chemical structures of these compounds are given in Figure 20.6.

Perez et al. synthesized cationic gemini amphiphiles derived from arginine [13]. These gemini quaternaries showed antibacterial activity and acute toxicity against *Daphnia crustaceous* organisms. The synthesis scheme and products of these arginine-based cationic gemini amphiphiles is shown in Figure 20.7.

20.2.1.1 Surface Activity of Cationic Gemini Surfactants

Kim et al. synthesized a series of cationic gemini surfactants with two alkyl chains and two ammonium groups and investigated their surface active properties [14]. These surfactants showed excellent surface active properties similar to anionic types. These cationic gemini surfactants were synthesized by condensation of epichloro-hydrin and two different tertiary amines as shown in Figure 20.8. The measured surface properties of these cationic gemini surfactants is summarized in Table 20.1.

The conclusions from the surface activity study of these compounds (Table 20.1), which can be generalized to most of the gemini cationic surfactants, are summarized as follows:

1. These compounds show good water solubility and their Kraft points lay below 0°C, except those with longer hydrophobic chain lengths.
2. The critical micelle concentrations of these compounds in molar concept are smaller by about one to two orders of magnitude than those of

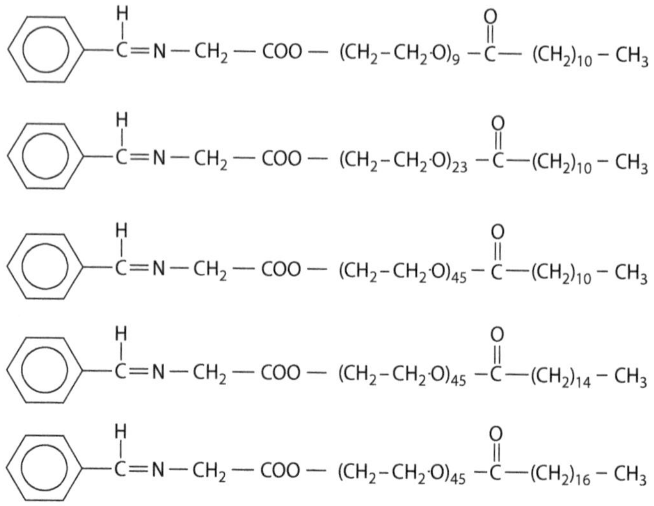

FIGURE 20.6 Chemical structure of environmentally friendly amino acid nonionic surfactants.

conventional monoquaternary ammonium salts. As recognized in cationic gemini surfactants, due to the presence of the two alkyl chains, it is likely that diquaternary ammonium salts exhibit large intermolecular hydrophobic interaction which makes it easier to form aggregates in water than the monoquaternary ammonium salts [15].

3. Compounds with decyl and dodecyl chains show maximum ability to lower the surface tension of their solutions. Their effectiveness values (γ_{cmc}) increased slightly with increasing the alkyl chain length in their homologous series (the effectiveness is the difference between the values of the surface tension of twice-distilled water and surfactant solution at the critical micelle concentration).

4. The efficiency values (Pc_{20} which is the concentration of the surfactant in the solution capable to depress the surface tension by 20 mN/m relative to twice-distilled water, 71.8 mN/m) are higher by about one order of magnitude than those of monoquaternary ammonium compounds with the same alkyl chain length. The Pc_{20} values are a useful parameter to measure the efficiency of adsorption of surfactant at the air-water interface. The values of Pc_{20} of the diquaternary ammonium compounds in several studies indicate their efficient adsorption at the interface than the monoquaternary ammonium compounds [16].

5. The foamability and foam stability of the diquaternary ammonium compounds with decyl tails is low at the critical micelle concentration (CMC) value. While the foamability of the conventional dodecyl trimethyl ammonium chloride is almost zero, but that of dodecyl derivative of the

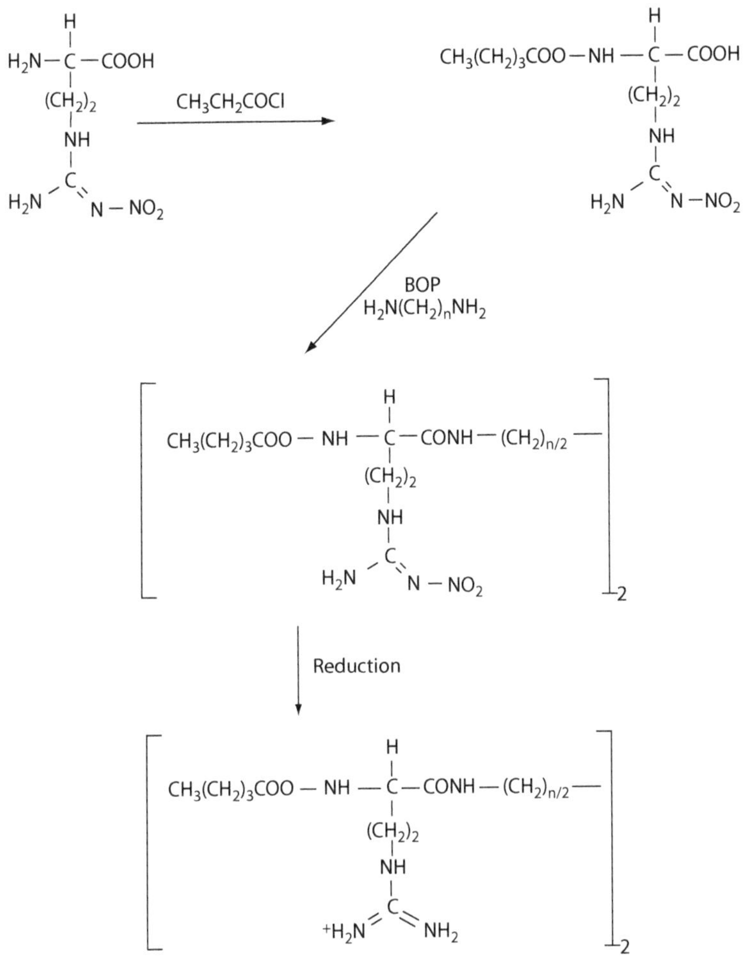

FIGURE 20.7 Cationic gemini amphiphiles derived from arginine.

diquaternary ammonium compounds is much higher than even that of the sodium dodecyl sulfate.

6. Conventional alkyl trimethyl ammonium chloride cationic surfactants show very little foam below their CMC values. On the contrary, dodecyl and tetradecyl gemini cationic surfactants show very high foamability when the spacer groups are short (two or three carbon atoms).

The aggregation behavior of the gemini cationic surfactants attracted considerable attention from scientists, especially in the field of enhanced oil recovery and drug delivery systems. Different methods were used to determine the aggregates including surface tension measurements, light scattering, NMR, and electron spectroscopy.

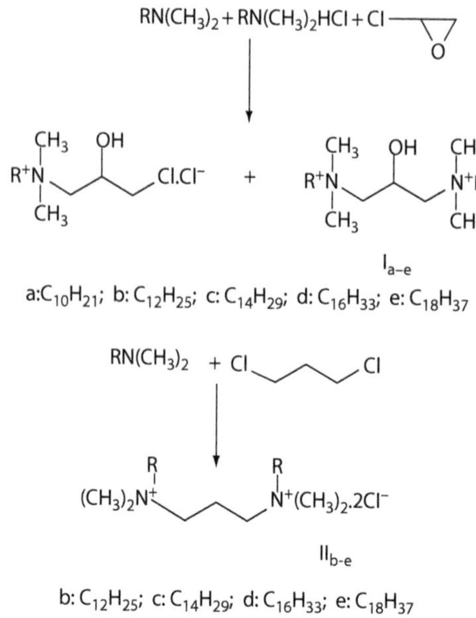

$$RN(CH_3)_2 + RN(CH_3)_2HCl + Cl\overset{O}{\triangle}$$

$$a:C_{10}H_{21};\ b:C_{12}H_{25};\ c:C_{14}H_{29};\ d:C_{16}H_{33};\ e:C_{18}H_{37}$$

$$RN(CH_3)_2 + Cl\diagdown\diagup Cl$$

$$(CH_3)_2N^+\diagdown\diagup N^+(CH_3)_2.2Cl^-$$

$$II_{b-e}$$

$$b:C_{12}H_{25};\ c:C_{14}H_{29};\ d:C_{16}H_{33};\ e:C_{18}H_{37}$$

FIGURE 20.8 Condensation of epichlorohydrin and two different tertiary amines.

TABLE 20.1
Surface Properties of Cationic Gemini Surfactants

Compound	R	Kraft Point, °C	CMC, mol/L	π_{cmc}, mN/m	Pc_{20}	Foam, mL 0 min	Foam, mL 30 min
I_a	$C_{10}H_{21}$	<0	0.0032	36.5	2.9	280	0
I_b	$C_{12}H_{25}$	<0	0.00078	37	3.2	280	270
I_c	$C_{14}H_{29}$	<0	0.00014	39	4.4	270	270
I_d	$C_{16}H_{33}$	<0	0.000019	42.2	5.3	100	10
I_e	$C_{18}H_{37}$	<0	0.000011	44	5.6	0	0
II_b	$C_{12}H_{25}$	<0	0.00098	39.2	3.3	270	270
II_c	$C_{14}H_{29}$	<0	0.00011	41.8	4.5	270	260
II_d	$C_{16}H_{33}$	<0	0.000015	42	5.4	60	10
II_e	$C_{18}H_{37}$	<0	0.000012	43.5	5.7	0	0
$C_{12}H_{25}N^+(CH_3)_3Cl^-$		<0	0.012	39	2.6	20	0
$C_{14}H_{29}N^+(CH_3)_3Cl^-$		<0	0.0045	41	2.7	–	–
$C_{16}H_{33}N^+(CH_3)_3Cl^-$		<0	0.0013	42.5	2.9	–	–

R, alkyl chains; CMC, critical micelle concentration (mol/L); π_{cmc}, the difference between surface tension of surfactant solution at CMC and bidistilled water (mN/m); Pc_{20}, log concentration of surfactant solution which depresses the surface tension by 20 units; 0 min, 30 min: foam height after 0–30 min of shaking the surfactant solution.

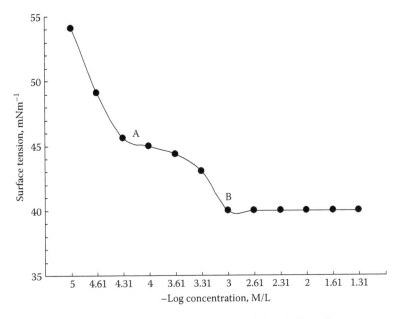

FIGURE 20.9 Typical surface tension profile of the gemini cationic surfactants.

The surface tension profile of the gemini cationic surfactants (Figure 20.9) showed two characteristic points (A and B); the concentration seems to be constant for each homologous series of cationic gemini at concentration (A). This was referred to a large aggregation type which is much larger than the individual spherical micelles which formed at concentration (B). This large aggregative form varies from multilayer, single wall cylinder and double wall cylinder to the double wall tubular structure [17–20]. Conductivity measurements confirmed the formation of these aggregates in conjugation with the electron microscopic data.

20.2.1.2 Solubilization

Solubilization is an important phenomenon required in tertiary oil recovery and detergency. Several investigators showed that cationic gemini surfactants were better solubilizing agents than the conventional cationic surfactants. They referred the higher capacity for solubilization to the geometrical structures and aggregates in their solutions at their critical micelle concentration values and even at concentrations higher than their CMC values. The tendency of the gemini cationic aggregates for oil solubilization is significantly higher than the conventional mono-quaternary surfactants. The gemini cationic surfactants showed higher tendency toward solubilization of toluene than n-hexane due to the strong ion-dipole interaction with toluene [21].

The influence of structure on the solubilization efficiency of some synthesized nonconventional symmetrical gemini cationic surfactants containing nonionic spacer groups with different ethylene oxide units of two different materials (polar and nonpolar solubilizates) was described briefly [9].

The solubilization curves of the nonpolar and polar solubilizates using these gemini cationics are represented in Figures 20.10 and 20.11, using the turbidity and time to follow up the solubilizing systems (NTU is "nephelometry turbidimitry unit," which measures the transparency of the solution).

The author [9] gave a brief description of the solubilization process in light of the solubilization curves (Figures 20.10 and 20.11) as follows:

1. In the case of nonpolar solubilizate (paraffin oil), the first stage turbidity is located at very low NTU value, which indicates low adsorption of nonpolar molecules at the micellar surface due to its polar nature (polyethylene

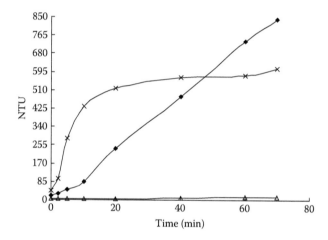

FIGURE 20.10 Solubilization behaviour of nonpolar solubilizate (paraffin oil) using the gemini cationics contain different nonionic species (n=polyethylene glycol units), where ◊: n=10, x: n=15, and Δ: n=25.

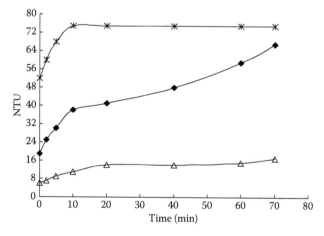

FIGURE 20.11 Solubilization behaviour of polar solubilizate (octanol) using the gemini cationics contain different nonionic species (n=polyethylene glycol units), where ◊: n=10, x: n=15, and Δ: n=25.

TABLE 20.2

Slopes of Solubilization Curves of Studied Gemini Cationics in Figures 20.10 and 20.11 at 25°C

Surfactant	Nonpolar Solubilizate (Paraffin Oil)			Nonpolar Solubilizate (Octanol)		
	n=10	n=10	n=10	n=10	n=15	n=25
m=12	86	86	86	1.2	1.2	0.4
m=16	10	10	10	0.6	0.8	0.8
m=18	16	16	16	3.2	0.8	1.2

"m" is the hydrophobic chain length, and "n" is the number of ethylene glycol units in the spacer group between the two positively head groups.

oxide layer). Increasing the time of experiment, i.e., at the second stage, penetration occurred into the micellar cores. It was found that increasing the HLB values of the synthesized surfactants increases the solubilization extent of their micelles. Generally, the smaller hydrophobic chains showed higher efficiency in solubilizing nonpolar substrates and vice versa.

2. However, in case of polar solubilizate (octanol), the initial stage solubilization is located at high NTU values indicating high adsorption of octanol molecules to the polar surface of the micelles. At the second stage of solubilization, the turbidity of the system is gradually increased by smaller rate. This is due to the hydrogen bonds formed between surfactant and octanol molecules. These hydrogen bonds decrease the octanol penetration into the micellar core. Hence, the turbidity of the system stays at relatively low values. The kinetics of the solubilization process for both polar and nonpolar solubilizates using the studied gemini cationic amphiphiles as solubilizing agents is shown in Table 20.2.

20.2.2 ANIONIC GEMINI SURFACTANTS

The anionic gemini amphiphiles are characterized by the presence of two negatively charged head groups separated by a spacer group. These head groups are mostly phosphate, sulfonate, sulfate, or phosphonate groups.

The following examples represent the different types of the anionic gemini amphiphile structures which have been synthesized in the last 20 years:

1. Phosphate type anionic gemini surfactants (a and b)

2. Phosphonate type anionic gemini surfactants (c and d)

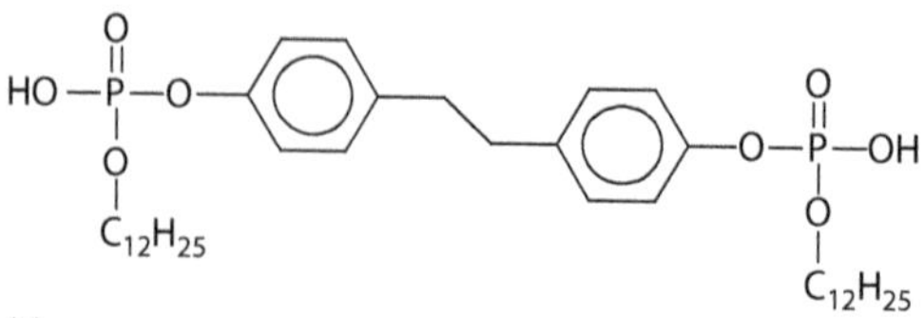

(c)

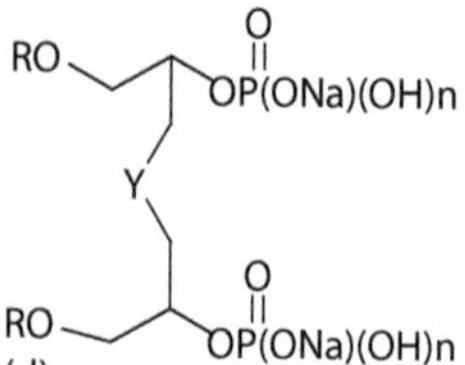

(d)

3. Sulphonate type anionic gemini surfactants (e, f, and g)

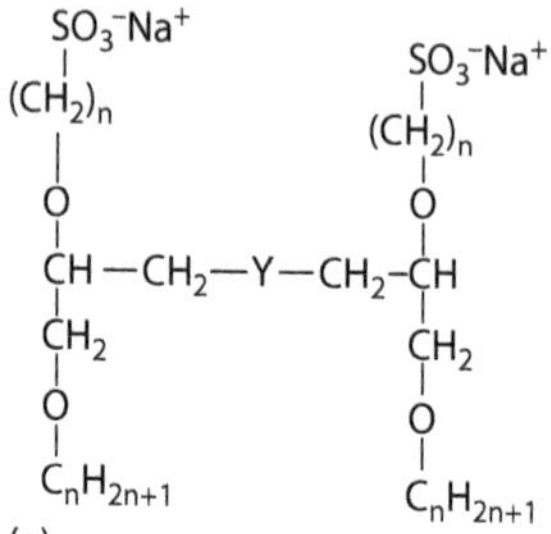

(e)

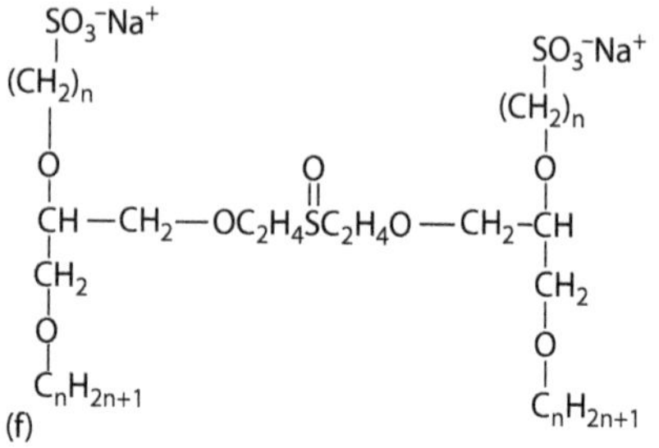

(f)

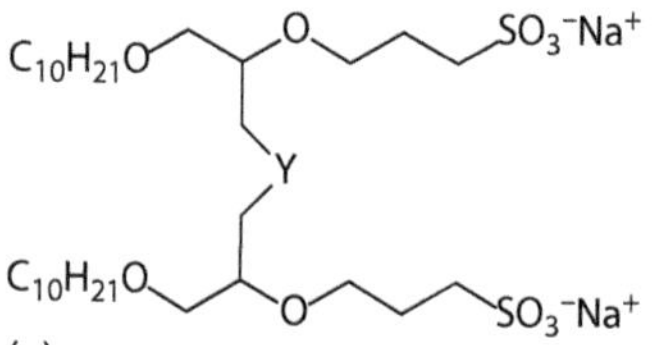

(g)

4. Sulfate type anionic gemini surfactants (h and i)

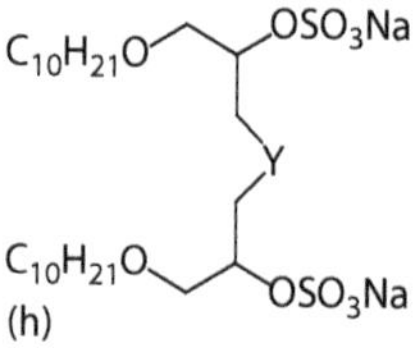

(h)

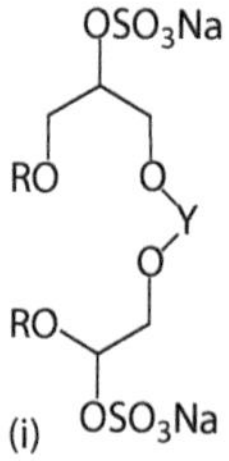

(i)

5. Carboxylate type anionic gemini surfactants (j and k)

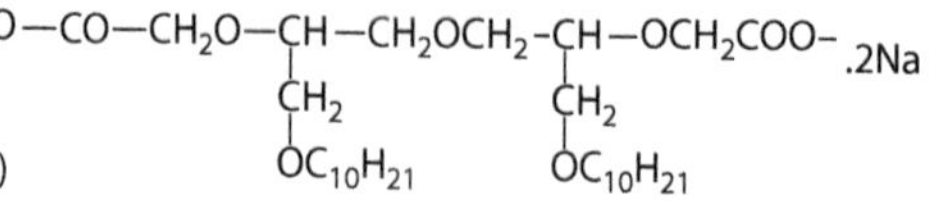

(j)

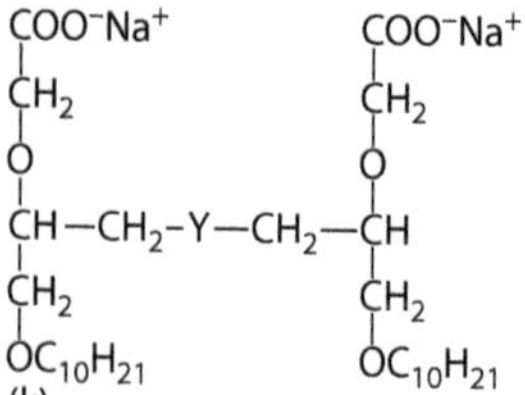

(k)

20.3 BIOLOGICAL ACTIVITY OF GEMINI AMPHIPHILES

20.3.1 ANTIMICROBIAL PROPERTIES OF GEMINI CATIONIC SURFACTANTS

Quaternary ammonium compounds have a broad spectrum of antimicrobial activity, against both Gram-positive and Gram-negative bacteria, yeast, mold, viruses, and protozoans. Because of the variety of quaternary ammonium compounds and test condition variations, such as microbial density, presence of organic matter, temperature, exposure time, and so on, it is difficult to make general comments regarding efficacy. Mixtures of different homologous of the quaternary ammonium compounds probably have enhanced antimicrobial profile and are often more potent than the pure quaternary itself.

20.3.1.1 Antibacterial Properties

It was recognized that Gram-positive bacteria differ from Gram-negative bacteria in susceptibility to the action of quaternary ammonium compounds [22]. In recent studies, quaternary ammonium compounds exhibited marked bactericidal activity on Gram-positive bacteria and less pronounced activities on the Gram-negative bacteria. The three bacterial genera, *Staphylococcus aureus*, *Lactobacillus*, and *Streptococcus*, were killed within 10 min at 167 ppm by most of the cationics tested. This shows not only the differences in activity between Gram-positive and Gram-negative bacterial strains but also among genus and species of Gram-negative bacteria. Other studies dealt with special types of pathogenic bacterial species as the work of Wang et al. [23], which studied the effect of various alkyl benzyl dimethyl ammonium chloride on *Camphlobacter jejuni* and also on *Legionella pneumophila*, the organism which was associated with the Legionnaires' disease [24]. The study showed this organism targeted by different types of quaternary ammonium compounds, and 125 ppm dose produced a good inhibitory effect for 1 min. While some studies showed that alkyl dimethyl dichlorobenzyl ammonium chloride is efficient bacteriostatic at 1 ppm and the dose of 62 ppm is sufficient as bactericide for the same microorganism [25]. The antimicrobial properties of low doses of quaternary ammonium compounds were evaluated against *Listeria innocua* and *Listeria monocytogenes* in solution and on dry surfaces. When the organism was in suspension, 40 ppm of dimethyl benzyl ammonium chloride was found to be effective but was not sufficient to kill the organisms dried on hard surfaces [26].

20.3.1.2 Antifungal Properties

Quaternary ammonium compounds also possess fungicidal activity. It was found that a mixture of n-alkyl (60% C_{14}, 30% C_{16}, 5% C_{12}, and 5% C_8)-dimethyl benzyl ammonium chloride and n-alkyl (68% C_{12} and 32% C_{14})-dimethyl benzyl ethyl benzyl ammonium chloride were effective at 488 ppm against the fungus *Trichophyton mentagrophytes* [27]. A mixture of benzalkonium chloride (commercial mixture of benzyl trialkyl ammonium chloride) and cetyl dimethyl ethyl ammonium chloride was effective against *Candida albicans*, *Trichophyton rubru*m, *Epidermophyton floccosum*, and *Fusarium* sp. but was not effective against other fungi including *Mucor* sp., *Aspergillus niger*, and *Microsporum canis* [28].

Some synthetic nonconventional antifungal quaternary ammonium compounds
derived from triethanolamine monoalkanoate were tested against *Candida albicans*
and *Aspergillus flav.* at 1, 2.5, and 5 ppm. The results were dependent on the surface
activity of these biocides and the most efficient derivative was the compound which
contained hexadecyl side chain [29].

In a separate study, the fungistatic and algistatic activity of about 164 quaternary ammonium compound was determined [30]; some of these results are listed in
(Table 20.3).

20.3.1.3   Antiviral Properties

Quaternary ammonium compounds are virucidal for lipophilic viruses such as the
myxo virus, paramyxo virus, adeno virus, herpes virus, and pox virus strains [31].
Laboratory evaluations of various disinfectants have been conducted against animal
viruses [32]. Among the disinfectants tests were two quaternaries of alkyl dimethyl
benzyl ammonium chlorides with different alkyl chains, containing 67% C_{12}, 25%
C_{14}, 7% C_{16}, 1% C_{18} and the other 60% C_{14}, 30% C_{16}, 5% C_{12}, 5% C_{18} were tested.
As might be expected, their virucidal activity was similar, in that both inactivated
pseudorabies virus and transmissible gastroenteritis virus (TGEV). However, both
failed to inactivate porcine parvovirus. Pseudo rabies virus and TGEV are lipophilic
viruses that have a lipid envelope and are sensitive to most disinfectants. On the
other hand, hydrophilic viruses such as porcine parvo are nonenveloped and are
resistant to some disinfectants [32].

The efficacy of quaternary ammonium compounds against hepatitis-A virus
dried on surfaces has been compared to that of other commonly used disinfectants
[33]. Of the five quaternary ammonium compounds tested, only one was effective
against hepatitis A virus. With the acidic quaternaries, pH of 0.42 (hydrochloric acid
was present at a level of 23%), the efficacy of this quaternary ammonium compound
formulation apparently was due to its high acidity and not to the action of the quaternary ammine itself. The efficacy of n-alkyl dimethyl benzyl ammonium chloride
(50% C_{14}, 40% C_{12}, and 10% C_{16}) against bacteriophage also has been studied [34].
The investigators applied the quaternary ammonium compounds as an aerosol to
surfaces inoculated with streptococcus phages 144F and 18–16. Whereas 500 ppm
quaternary dose provided some inactivation, little or no inactivation was observed
at the 1000–2000 ppm doses. In an attempt to improve activity, the chelating agent
ethylene diamine tetraacetic acid disodium salt (EDTA-2Na) was incorporated at
200 ppm into the 2000 ppm solution of the quaternary ammonium salts and the combination applied as an aerosol. However, the inactivation of bacteriophage was no
greater than that of quaternaries without EDTA-2Na [35].

20.3.1.4   Other Biocidal Activities

The lethality of quaternary ammonium compounds on the parasite *Plasmodium falciparum* has been demonstrated in vitro [36]. Because quaternary ammonium compounds have demonstrated algistatic activity [30] and have only little deleterious
effect on materials of construction, they are often used to treat water in cooling
towers. Quaternary ammonium compounds are also registered with the U.S. EPA as
swimming pool algicides at doses shown in Table 20.4.

TABLE 20.3
Inhibiting Concentrations (ppm) of Some Quaternary Ammonium Compounds for Some Bacteria and Fungi

Microorganisms	Benzethonium Chloride[a]	Benzalkonium Chloride[b]	Quaternary Ammonium Compounds			
			Dodecyl Trimethyl Ammonium Chloride	Dodecyl Benzyl Dimethyl Ammonium Chloride	Cocobenzyl Dimethyl Ammonium Chloride	Dodecyl Dimethyl Ammonium Chloride
Bacteria						
E. coli	1000	200	500	750	225	225
P. fluorescens	300	300	500	750	225	750
Bacillus subtilis	3	3	5	2	2	0.7
Staphylococcus aureus	3	4	5	2	2	7
Fungi						
Aspergillus niger	300	60	500	75	20	75
Chaetomium globosum	30	10	50	7	7	7
Myrothecium verrucaria	300	40	500	150	20	20
Trichoderma viridae	200	80	500	75	20	20

[a] Commercial mixture of benzyl trialkyl ammonium chloride.
[b] Commercial mixture of benzyl trialkyl ammonium chloride.

TABLE 20.4
Recommended Maximum Algistatic Concentrations (ppm) of Some Selected Quaternary Ammonium Compounds[a]

Quaternary Ammonium Compound	Algae			
	Anabaena cylindrica	*Chlorella vulgaris*	*Oscillatoria tenuis*	*Stigeoclonium* sp.
Benzthionium chloride	1.0	3.0	1.0	1.0
Benzalkonium chloride	1.0	1.0	0.6	0.7
Dodecyl trimethyl ammonium chloride	5.0	50	0.5	5.0
Dodecyl benzyl dimethyl ammonium chloride	0.2	0.2	0.2	0.2
Dodecyl dimethyl ammonium chloride	0.2	2.0	0.7	0.7

[a] Recommended by U.S. EPA.

The cationic gemini amphiphiles showed antimicrobial activity with no odor or toxicity, when used in the recommended doses. They are quite stable and do not contain any phenol, iodine, active chlorine, mercury, or other heavy metals. But, their activities are still limited to inhibiting the growth of various microorganisms such as bacteria, fungi, viruses, and yeasts. However, they do not fulfill the requirements of a universal antimicrobial agent. In some cases, they do not provide complete protection against the different types of microorganisms due to the acquired immunity of some microorganisms as a result of long-term uses.

20.3.2 ANTIMICROBIAL RESISTANCE

The development of resistance to antibacterial agents is often associated with therapeutic pharmaceuticals, but antimicrobial compounds used for preservation and disinfection have also been implicated. Among these compounds are iodine, phenolics, anionics, and quaternary ammonium compounds.

Resistance to the antimicrobial properties of quaternary ammonium compounds has been widely reported, beginning in the early 1950s. Among the early reports of demonstrated resistance was that of *Serratia marcescens* to benzalkonium chloride. *S. marcescens* resistance was increased by repeated culture media containing benzalkonium chloride. In a recent report by Nishikawa et al. [37], the microorganisms isolated from the soil exhibited resistance towards benzalkonium chloride. They identified one microorganism, *Enterobacter cloacae*, which grew in the presence of 10% quaternary ammonium compound. Evidence for the bacterial metabolism of the benzalkonium chloride was not found. This resistance was lost when these microorganisms were cultured in the absence of benzalkonium chloride.

There are several explanations for the organisms surviving exposure to chemicals designed to be lethal to the organisms.

Russel et al. [38–40] proposed two possible mechanisms for such resistance which they describe as: intrinsic and acquired resistance. Intrinsic resistance is related to the structural and chemical composition of the outer layers of the cells, which may provide an effective barrier to the entry of antibacterial agents. Acquired resistance results from genetic changes in the bacterial cell and arises either by mutation or by the acquisition of genetic material from another cell. Russell and Day [41] suggested that enterobacteria resistance is associated with the outer cell layers, which limit penetration of the chemicals. Other investigators [42] have agreed with this hypothesis and found that the phospholipid content and fatty and neutral lipid content of the bacterial cell wall of certain Gram-negative bacteria increased the resistance to benzalkonium chloride.

Similar mechanisms were proposed for resistance to quaternary ammonium compounds due to modification of the lipid composition of the outer membrane for *Pseudomonas aeruginosa* [43].

Attachment to surfaces also appears to be an important factor in microorganisms surviving exposure to quaternary ammonium compounds. Dhaliwal et al. [44] evaluated the efficacy of two different quaternary ammonium-based disinfectants at three concentrations against biofilms of *E. coli*, *Listeria monocytogenes*, *Staphylococcus aureus*, and *Salmonella enteritidis* on six substrates. The study revealed that the type of the quaternary ammonium compound, type of the microorganism and also the type of the surface have distinguishable effect on the antimicrobial activities of these compounds.

The pH also plays a vital role in the efficacy of the quaternary ammonium compounds as antimicrobial agents for the different microorganisms. Acidic quaternary ammonium compounds were found to be more effective than the neutral pH quaternaries on *L. monocytogenes* biofilm [45]. Few studies were published which reported self-resistance of some common microorganisms.

20.4 FEATURES

Hence the need for other types of antimicrobial agents became apparent because of different types of bacterial genera which have immunity for several types of the mono or gemini quaternary ammonium antimicrobial agents.

The design of the new antimicrobial agents in this work is based on the use of a highly efficient antimicrobial class of biocides, which is, the schiff bases antimicrobial agents. Schiff bases are characterized by the presence of the functional group (-CH=N-) which has a high ability for adsorption at interfaces including the cellular membrane of the microorganisms. Also, introducing cationic sites in their chemical skeleton increases their solubility in an aqueous phase. In addition, the presence of the two hydrophobic chains increases their surface activity as amphiphiles.

The idea of using schiff base derivatives in their amphipathic form had not been tried as antimicrobial agents before. The available literature on the amphiphilic schiff bases has been concerned only with the effect of the structure on their surface activities.

Schiff bases were used first early in the last century as biocide for different microorganisms. But, some limitations were found in their use due to their lack of solubility in aqueous phases which restricts their application as commercial biocides.

Several attempts were made to increase their water solubility through complexation with transition metal nitrates or metal halides.

Our previous studies were concerned with the synthesis of schiff base amphiphiles bearing different functionalities in order to increase their solubility and improve their surface activity for use in several applications. The results of these studies showed two features:

1. Schiff base amphiphiles containing highly conjugated systems showed moderate solubility in aqueous media which could be improved by varying the pH of the medium.
2. Their effectiveness in the different fields, mainly corrosion inhibition and as antimicrobial agents was increased due to their solubility.

But, the problem was still there due to immunity of some microorganisms, especially the Gram-negative bacteria due to their unique cellular membrane.

The approach was directed towards increasing the biocidal activity of these amphiphiles through complexation of these schiff base amphiphiles with different transition metals. The results were promising in several cases, but in some cases, the results were very weak relative to the parent cationic schiff bases. This was due to the mode of complexation and the type of metal ions participation in the crystalline structure of the complexes produced. Also, the presence of transition metal ions in the chemical structure of these biocides may restrict their wide application due to mineral accumulation in the environment.

In accordance with our previous studies [12,46–49], a novel di-Schiff base cationic gemini amphiphiles bearing homogeneous hydrophobes were synthesized and characterized using elemental analyses, FTIR, and ^{1}H-NMR spectroscopic analysis.

20.5 SYNTHESIS

Ethylene diamine, 4-N,N-diethyl aminobenzaldehyde, chloroacetic acid, octyl, decyl, hexadecyl, and octadecyl alcohols were analytical grade obtained from Sigma Aldrich and were used as supplied.

20.5.1 Synthesis of Fatty Chloroacetic Esters

Equimolar amounts of chloroacetic acid and fatty alcohols, namely, octyl-, decyl-, hexadecyl-, and octadecyl alcohols were esterified in sufficient amount of xylene and 0.1% benzene sulfonic acid as a catalyst. The reaction was completed by removal of the calculated amount of water of the reaction. Then, the reaction product (Figure 20.12), was extracted using diethyl ether in a separating funnel and dried in a vacuum oven at temperature of 60°C [50].

$$ClCH_2COOCH_2(CH_2)_n CH_3$$

$$n = 7, 9, 15 \text{ and } 17$$

FIGURE 20.12 Alkyloxycarbonylmethyl chloride esters.

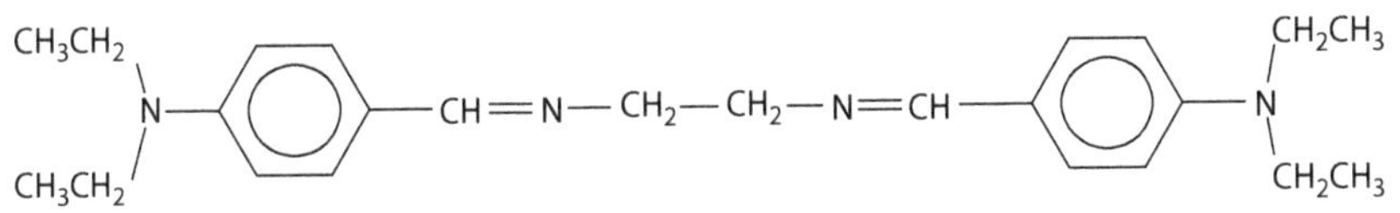

FIGURE 20.13   The synthesized di-Schiff base.

20.5.2   SYNTHESIS OF DI-SCHIFF BASE

Ethylene diamine (0.1 mol) and 4-N,N-diethyl aminobenzaldehyde (0.2 mol) were refluxed in absolute ethanol for 8 h. The reaction mixture was left overnight to precipitate the yellow crystalline product di-Schiff base (Figure 20.13) and then filtered. The produced di-Schiff bases were washed twice with ethanol/diethyl ether mixture (50/50 vol.) and dried in a vacuum oven at 50°C [12,48].

20.5.3   SYNTHESIS OF DI-SCHIFF BASE CATIONIC GEMINI AMPHIPHILES

Fatty alcohol chloroacetate (0.2 mol) and the synthesized di-Schiff base were refluxed in absolute ethanol for 12 h, then left overnight to precipitate the produced cationic gemini amphiphiles and filtered. Recrystallization of the products was done in ethyl alcohol to produce the desired cationic gemini amphiphiles [51–54]. The synthesized compounds were denoted as SB8DQ, SB10DQ, SB12DQ, SB16DQ, and SB18DQ and their chemical structures are given in Figure 20.14.

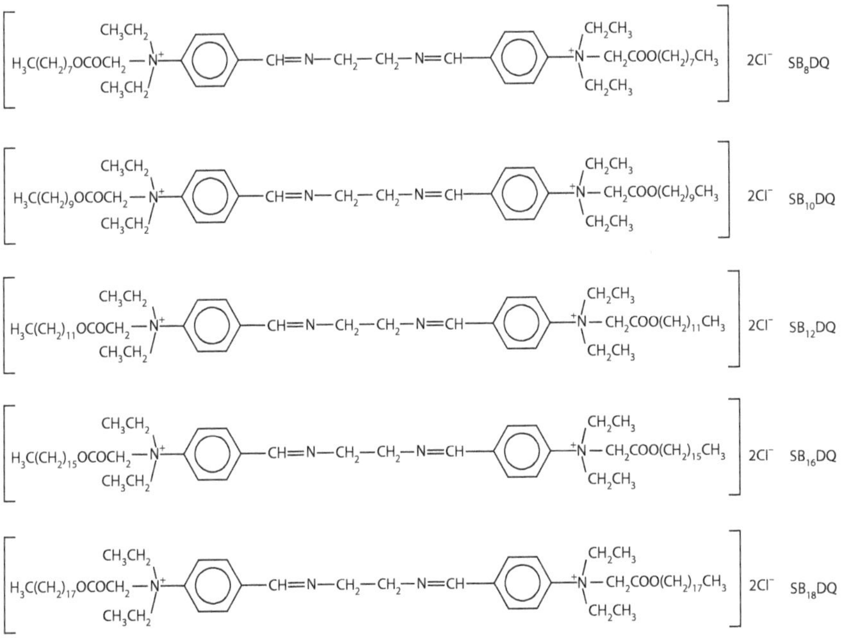

FIGURE 20.14   The synthesized cationic gemini di-Schiff base amphiphiles.

20.6   MEASUREMENTS

20.6.1   SURFACE AND INTERFACIAL TENSION

Surface tension measurements were conducted on freshly prepared inhibitors solutions in a concentration range of 0.1–0.0001 mol/L at 25°C using a Du-Nuoy Tensiometer-Kruss-K6. Interfacial tension measurements were conducted on solutions of gemini di-Schiff base cationic surfactants (10 mL) in presence of 10 mL of paraffin oil at 25°C [55].

20.6.2   ANTIMICROBIAL STUDIES

The synthesized gemini di-Schiff base cationic surfactants were screened for their antimicrobial activity against bacteria and fungi using agar well diffusion method [56,57].

20.6.3   GROWING OF MICROORGANISMS

The bacterial strains were cultured in a nutrient medium, while the fungi strains were cultured in malt medium. For bacteria, the broth medium was incubated for 24 h. As for fungi, the broth medium was incubated for approximately 48 h, with subsequent filtering of the culture through a thin layer of sterile sintered glass G2 before the solution was used for inoculation.

20.6.4   MEASUREMENTS OF RESISTANCE AND SUSCEPTIBILITY

For preparation of discs and inoculation, 1.0 mL of inocula were added to 50 mL of agar medium (40°C) and mixed. The agar was poured into 120 mm petri dishes and allowed to cool to room temperature. Wells (6 mm in diameter) were cut in the agar plates using proper sterile tubes. The wells were filled to the surface of agar with 0.1 mL of the synthesized gemini di-Schiff base cationic surfactants (5 mg/mL dimethyl formamide). The plates were left on a leveled surface, incubated for 24 h at 30°C for bacteria and 48 h for fungi, and then the diameter of the inhibition zones measured. The inhibition zone is the area where the microorganisms cannot grow due to the influence of the biocide. The inhibition zone formed by these compounds against the bacterial strain determined the antibacterial and antifungal activities of the synthetic compounds. The mean value obtained for three replicates determined the zone of inhibition of each sample [58].

20.6.5   MICROORGANISMS

The antimicrobial activity of the synthesized surfactants was tested against *Staphylococcus aureus* (NCTC-7447), *E. coli* (NCTC-1041), *Bacillus subtillus* (NCIB-3610), and *Pseudomonas aeruginosa* (NCIB-9016) and *Desulfomonas pigra* as a common strain of SRB. The fungicidal activity was tested against *Aspergillus niger* (Ferm-Bam C-21) and *Candida albicans*.

20.7 RESULTS AND DISCUSSION

20.7.1 SYNTHESIS

The chemical structures of the synthesized gemini di-Schiff base cationic surfactants were confirmed by elemental analyses, FTIR spectroscopic analyses and ^{1}H-NMR spectroscopy. The elemental analysis data showed the purity of the synthesized compounds within 98.73%. The FTIR spectra showed the following absorption bands: 2950, 2840, 1738, 1640, 1280, and 860 cm^{-1} corresponding to stretching of CH_3, CH_2, C=O, C=N, and C–N, and phenyl groups, respectively. While, the ^{1}H-NMR spectra showed signals of the characteristic protons of the functional groups and their integration equal to the number of the protons in the chemical structures. The spectral analyses data confirmed the chemical structures of the synthesized gemini di-Schiff base cationic amphiphiles. The molecular weight and elemental analysis data of the gemini di-Schiff base cationic amphiphiles is provided in Table 20.5.

20.7.2 SURFACE ACTIVITY

The variation of the surface tension against concentration of the synthesized di-Schiff base cationic gemini amphiphiles at 25°C were performed at concentration higher than 0.00009 mol/L, so that the two inflection points of the characteristic gemini cationic curves did not appear but only one inflection point.

The surface tension profile showed two characteristic regions. The intercept determines the CMC. The low surface tension values of the synthesized amphiphiles can be attributed to the similarity of the hydrophobic chains in their chemical structures [10]. In addition, the presence of the benzene rings in the molecule imparts these amphiphiles higher hydrophobic character leading to high depression in the surface tension of their solutions. The highest depression in the surface tension values was found in dodecyl derivative while the lowest was observed in the case of hexadecyl derivative. The octadecyl derivative showed weak surface activity, which was attributed to its low solubility in water. The critical micelle concentration values of the synthesized gemini di-Schiff base cationic amphiphiles varied in the range of 2.45 and 6.46 mmol/L at 25°C.

Also, it is clear that increasing the hydrophobic chain length decreases the surface tension considerably (see Figure 20.15). This was true for the derivatives containing shorter hydrophobic chains than hexadecyl chain. But, in case of the hexadecyl derivative, the surface tension increased extraordinarily (Figure 20.15), which was attributed to two reasons. The first is the decrease in the solubility of the compound due to the increase in the carbon content. The second is the coiling of the carbon chain which decreases its actual size to that of the octyl chain.

In the relation between the hydrophobic chain length and the critical micelle concentration values of the synthesized gemini di-Schiff base cationic amphiphiles (Figure 20.16), it is clear that the depression in the CMC of these amphiphiles was accompanied by the increase of the hydrophobic chain length.

The lowest CMC value was observed for dodecyl derivative (SB12DQ) at 2.45 mmol/L. The hexadecyl derivative (SB16DQ) showed the highest CMC value

TABLE 20.5

Elemental Analyses of Synthesized Di-Schiff Base Cationic Gemini Amphiphiles

Compound	M.wt.	Carbon, %		Hydrogen, %		Nitrogen, %		Chlorine, %	
		Calculated	Found	Calculated	Found	Calculated	Found	Calculated	Found
SB8DQ	791.57	66.73	65.89	9.16	9.05	7.08	6.99	8.95	8.84
SB10DQ	848.06	67.98	67.13	9.51	9.39	6.61	6.53	8.36	8.26
SB12DQ	904.16	69.07	68.21	9.81	9.69	6.20	6.12	7.84	7.74
SB16DQ	1016.37	70.90	70.01	10.31	10.19	5.51	5.44	6.98	6.89
SB18DQ	1072.48	71.67	70.77	10.53	10.40	5.23	5.16	6.61	6.53

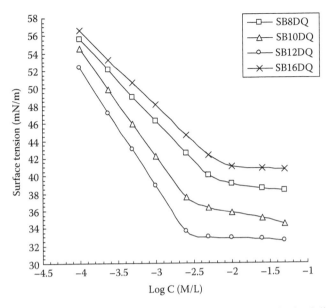

FIGURE 20.15 Surface tension versus log concentration of the synthesized di-Schiff base cationic gemini amphiphiles at 25°C.

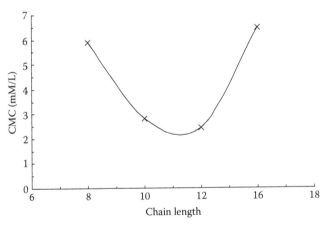

FIGURE 20.16 Effect of the hydrophobic chain length on the critical micelle concentration of the synthesized di-Schiff base gemini amphiphiles.

at 6.46 mmol/L at 25°C. The effectiveness of the synthesized cationic diquaternary amphiphiles was also in same order as the CMC. This was in agreement with the cationic amphiphiles behaviors. A lower depression in the surface tension at the critical micelle concentration was observed for the dodecyl derivative which shows their high surface activity.

The decyl- and dodecyl derivatives (SB10DQ, SB12DQ) have higher surface concentration (Γ_{max}) and lower area occupied by their molecules at the interface (A_{min}).

Meanwhile, the octyl- and hexadecyl derivatives (SB8DQ, SB16DQ) have lower sur-face concentration (Γ_{max}) and higher area occupied by their molecules at the interface (A_{min}). This behavior is strongly related to the hydrophobicity of the surfactant mol-ecules which was clearly described in our earlier study [59].

The thermodynamic parameters of adsorption and micellization processes were also calculated [60–62]. The adsorption and micellization free energies were always found in negative indicating that the two processes had occurred spontaneously. Also, it was observed that the adsorption energies were more negative than the micelliza-tion energies indicating the tendency of these cationics toward the first process than the second. The adsorption energy for both decyl and dodecyl (SB10DQ, SB12DQ) derivatives are much lower (larger negative numbers) than those of octyl and hexa-decyl derivatives (SB8DQ, SB16DQ). This makes them more suitable for adsorp-tion applications including corrosion inhibition, prevention of microbial growth, and phase transfer catalysis. While octyl and hexadecyl derivatives (SB8DQ, SB16DQ) were more suitable for micellar applications including solubilization and biologically drugs delivery systems.

20.7.3 Overview of Bacterial Cell Membranes

20.7.3.1 Components of Bacterial Cell Wall

Lipids are one class of large biological molecules that form the cellular membranes. They are grouped together because they share one important chemical property: they have little or no affinity for water. The hydrophobic behavior of lipids is based on their molecular structure. Although they may have some polar bonds associated with oxygen, lipids consist mostly of hydrocarbons. Smaller than true (polymeric) macromolecules, lipids vary in both form and function, and include such materials as waxes and certain pigments. In this chapter, we will focus on three classes of lipids: fats, steroids, and phospholipids.

20.7.3.1.1 Fats (Triacylglycerols) and Phospholipids

Phospholipids are major components of the cell membrane. They are similar to fats but have only two fatty acids rather than three. The third hydroxyl group of glycerol is joined to a phosphate group, which is negative in electrical charge. Additional small molecules, usually charged or polar, can be linked to the phosphate group to form a variety of phospholipids. Phospholipids are described as being amphipathic, having both a hydrophobic and a hydrophilic region. Their tails, which consist of hydrocarbons, are hydrophobic and are excluded from water. Their heads, however, which consist of the phosphate group and its attachments, are hydrophilic and have an affinity for water.

20.7.3.1.2 Steroids

Steroids are characterized by a carbon skeleton consisting of four fused rings. One steroid, cholesterol, is a common component of animal cell membranes and func-tions to help stabilize the membrane (Figures 20.17 and 20.18). It is also the precur-sor from which other steroids are synthesized. Thus, cholesterol is a crucial molecule in animals although high levels of it in the blood may contribute to atherosclerosis.

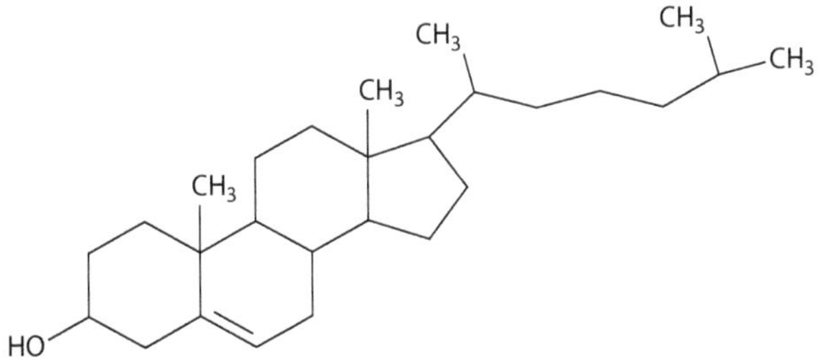

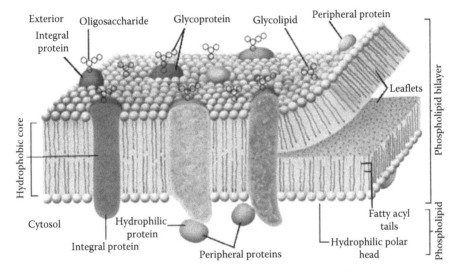

FIGURE 20.17   Chemical structure of steroids.

FIGURE 20.18 (See color insert.)   Distribution of steroids in the cell membrane.

20.7.3.1.3   Teichoic Acids

Wall teichoic acids (Figure 20.19) are found only in certain Gram-positive bacteria (such as *Staphylococci*, *Streptococci*, *Lactobacilli*, and *Bacillus* spp.). So far, they have not been found in Gram-negative organisms. Teichoic acids are polyol phosphate polymers, with either ribitol or glycerol linked by phosphodiester bonds. Substituents on the polyol chains can include D-alanine (ester linked), N-acetylglucosamine, N-acetylgalactosamine, and glucose. These substituents are characteristic for a particular bacterial species and can act as a specific antigenic determinant. Teichoic acids are covalently linked to the peptidoglycan. These highly negatively charged polymers of the bacterial wall can serve a cation-sequestering mechanism.

20.7.3.1.4   Peptidoglycan Structure and Function

Unique features of almost all prokaryotic cells are cell wall peptidoglycan and the specific enzymes involved in its biosynthesis. These enzymes are target sites for inhibition of peptidoglycan synthesis by specific antibiotics. The primary chemical

CH₂O
|
HCOAla
|
HCOH
|
HO   HCOR   HO — P = O
| |
HO — P — OCH₂
‖
O
(a)

CH₂O
|
HCOAla
|
HCOH
|
HCOR   HO — P = O
|
OH₂C
⌋ₙ
Ribitol teichoic acid

CH₂OH
|
HCOAla
|
HCOH
|
HCOR
|
OH₂C

O
‖
P   CH₂O
| |
HO   ROCH
OH₂C

O
‖
P   CH₂O
| |
HO   ROCH
OH₂C
⌋ₙ

(b)   Glycerol teichoic acid

FIGURE 20.19 Teichoic acid derivatives in the cell membrane.

structures of peptidoglycans of both Gram-positive and Gram-negative bacteria have
been established (Figure 20.20). It consist of a glycan backbone of repeating groups of
beta-1,4-linked disaccharides of beta-1,4-N-acetyl muramyl-N-acetyl glucosamine.
Tetrapeptides of L-alanine-D-isoglutamic acid-L-lysine (or diaminopimelic acid)-n-
alanine are linked through the carboxyl group by amide linkage of muramic acid residues of the glycan chains; the delta-alanine residues are directly cross-linked to the
epsilon-amino group of lysine or diaminopimelic acid on a neighboring tetrapeptide,

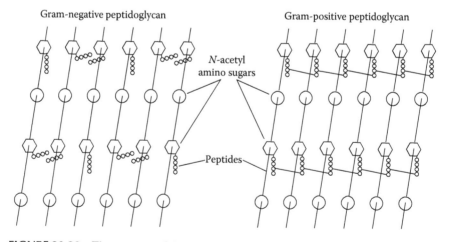

FIGURE 20.20 The structure of the peptidoglycan in different bacterial genera.

or they are linked by a peptide bridge. In *S. aureus* peptidoglycan, a glycine pentapeptide bridge links the two adjacent peptide structures. The extent of direct or peptide-bridge cross-linking varies from one peptidoglycan to another. The *Staphylococcal* peptidoglycan is highly cross-linked, whereas that of *E. coli* is much less, and has a more open peptidoglycan mesh. The diamino acid providing the epsilon-amino group for cross-linking is lysine or diaminopimelic acid, the latter being uniformly present in Gram-negative peptidoglycans. A peptidoglycan with a chemical structure substantially different from that of all eubacteria has been discovered in certain archaebacteria. Instead of muramic acid, this peptidoglycan contains talosaminuronic acid and lacks the beta-amino acids found in the eubacterial peptidoglycans. Interestingly, organisms containing this wall polymer are insensitive to penicillin, an inhibitor of the transpeptidases involved in peptidoglycan biosynthesis in eubacteria.

20.7.3.1.5　Lipopolysaccharides

One of the major components of the outer membrane of Gram-negative bacteria is lipopolysaccharide (endotoxin), a complex molecule consisting of a lipid anchor, a polysaccharide core, and chains of carbohydrates. Sugars in the polysaccharide chains confer serologic specificity.

20.7.3.2　Types of Bacterial Cell Membranes

20.7.3.2.1　Gram-Positive Cells

Most Gram-positive bacteria have a relatively thick (about 20–80 nm), continuous cell wall (often called the sacculus), which is composed largely of peptidoglycan (also known as mucopeptide or mucein). In thick cell walls, other cell wall polymers (such as the teichoic acids, polysaccharides, and peptidoglycolipids) are covalently attached to the peptidoglycan.

20.7.3.2.2　Gram-Negative Cells

The cell walls of Gram-negative bacteria are more chemically complex, thinner, and less compact. Peptidoglycan makes up only 5%–20% of the cell wall, and is not the outermost layer. The peptidoglycan layer in Gram-negative bacteria is thin (about 5–10 nm thick); in *E. coli*, the peptidoglycan is probably only a monolayer thick. Outside the peptidoglycan layer in the Gram-negative envelope is an outer membrane structure (about 7.5–10 nm thick). In most Gram-negative bacteria, this membrane structure is anchored noncovalently to lipoprotein molecules (Braun's lipoprotein), which, in turn, are covalently linked to the peptidoglycan. The lipopolysaccharides of the Gram-negative cell envelope form part of the outer leaflet of the outer membrane structure. The function of the cellular membrane is mainly to allow the diffusion of the material necessary for the biological reactions and for excretion of the wastes produced from the reactions. The controlling of the two processes is performed through selective permeability. Control of the selectivity of the polar species to enter or get out from the cell is carried out by the charged amino acids (teichoic acid). Control of the selectivity of nonpolar materials is carried out by the phospholipids and peptidoglycans. When the selective permeability of the cellular membrane is disturbed for any reason, the biological reactions and activities in the cell are diminished which leads to the death of the microorganism. The role of the biocides against bacteria and fungi is to disturb or destroy the selective permeability of these membranes in order to kill these microorganisms. This typically happens in the presence of cationic surfactants [63].

20.7.4 CYTOTOXICITY OF CATIONIC SURFACTANTS

The germicidal activity of quaternary ammonium compounds is directly related to the chemical properties of the cationic surfactants. These properties include a reduction in surface tension and formation of ionic aggregates, which results in changes in conductivity and solubility [64]. Since quaternary ammonium compounds are positively charged, the cationics are attracted to negatively charged materials such as bacterial proteins. Proteins comprise a large portion of the bacterial cell, both structural components of the cell and bacterial enzymes [65]. Disorganization and denaturation of essential proteins and release of nitrogen and phosphorus containing cellular constituents [66] were among the first explanation for the mode of antibacterial activity of the quaternary ammonium compounds. Electrophoresis has been used to study the interaction of various chemicals with bacteria. Dyar and Ordal [67] investigated the effect of cetyl pyridinium chloride on 10 different Gram-positive and Gram-negative bacteria. They found that as the concentration of the quaternary ammonium salt gradually increased, the mobility of the cells toward the cathode decreased and eventually the charge was reversed and these cells moved toward the anode. McQuillen [68] confirmed the findings of Dyar and Ordal using the Gram-negative bacteria *E. coli*. McQuillen found that the Gram-positive bacteria *Staphylococcus aureus* and *Streptococcus faecalis* reacted differently. McQuillen explained that the difference was due to release of molecules from the cell due to changes in cell permeability resulting from interaction with the quaternary ammonium compounds.

Hugo [65] presented a good explanation for the mechanism of antimicrobial action of the quaternary ammonium compounds, discussing five possible modes of microbial inhibiting action:

1. Direct effects on proteins including denaturation and disruption
2. Effect on metabolic reactions
3. Effect on cell permeability and membrane damage
4. Stimulatory effect on the glycolysis reactions
5. Effect on the enzyme system, maintaining a dynamic cytoplasmic membrane

The cytotoxicity of the gemini cationic surfactants has been reviewed in several works and their mode of action on bacteria and fungi was described. These works dealt with the action of these cationics from different points of view.

20.7.5 CYTOTOXICITY OF STUDIED DI-SCHIFF BASE CATIONIC GEMINI SURFACTANTS

The synthesized cationic di-Schiff base cationic diquaternary surfactants showed high antibacterial and antifungal efficacy against the tested microorganisms. The charts in Figure 20.21 represent the antibacterial activities of the tested four

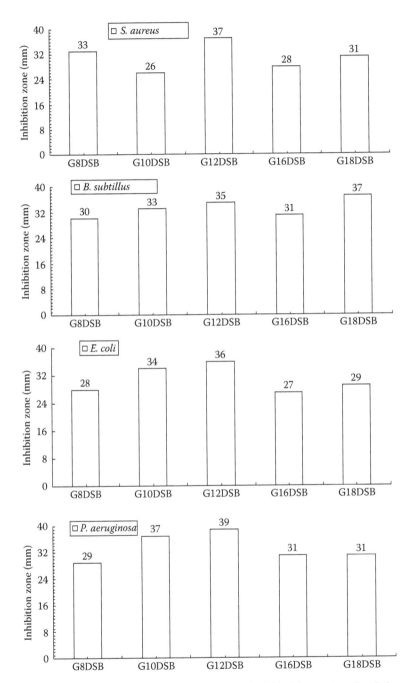

FIGURE 20.21 Bactericidal activity of the synthesized biocides against *Staphylococcus aureus*, *Bacillus subtillus*, *E. coli*, and *Pseudomonas aeruginosa*.

TABLE 20.6
Antibacterial Activity of Synthesized Di-Schiff Base Cationic Gemini Amphiphiles at 5 mg/mL Expressed in Diameter of the Inhibition Zone

Compound	SB8DQ	SB10DQ	SB12DQ	SB16DQ	SB18DQ
E. coli (NCTC-1041)	28	34	36	27	28
Bacillus subtillus (NCIB-3610)	30	33	35	31	37
Staphylococcus aureus (NCTC-7447)	33	26	37	28	28
Pseudomonas aeruginosa (NCIB-9016)	29	37	39	31	31
Desulfomonas pigra	18	23	25	19	17

gemini cationic di-Schiff bases amphiphiles against the following bacterial strains: *Staphylococcus aureus*, *Bacillus subtillus*, *E. coli*, and *Pseudomonas aeruginosa*).

For *Staphylococcus aureus*, *Pseudomonas aeruginosa* and *E. coli* the dodecyl derivative (SB12DQ) showed the maximum inhibition zone diameter of 37, 39, and 36 mm, respectively. On the other hand, the octadecyl derivative (SB18DQ) showed the maximum inhibition efficacy against *Bacillus subtillus* with an inhibition zone diameter of 37 mm (Table 20.6).

The high antibacterial efficacy of the synthesized cationic gemini surfactants can be attributed to several factors including structural factors and interfacial factors:

1. The structural factors include the aliphatic hydrocarbon side chains, the benzene ring nucleus, azomethine groups, and positively charged head groups (N^+).
2. The interfacial factors include the low effectiveness (γ_{cmc}), low efficiency (Pc_{20}), high surface concentration, and low surface area.

The presence of the aliphatic side chains (hydrophobic chains) increases the adsorption of these inhibitors at the cellular membranes of the microorganisms due to the similarity between these chains and the alkyl chains in bacterial lipid layers. Such adsorption is facilitated by the presence of the positively charged head groups which are electrostatically attached to the negatively charged amino acids (teichoic acid) in these membranes. This electrostatic interaction neutralizes the cellular protein at the cellular membranes and causes a disturbance in the distribution of charges on these membranes.

Furthermore, the high interfacial activity of these inhibitors eases the interaction between the environmental components and the cellular membrane by depressing their selective permeability. Hence, the biological reaction and the metabolic pathways are disturbed in the microorganisms which lead to their death. The most compatible hydrocarbon chain length with bacterial lipids is the dodecyl chain length which exhibits the highest efficacy against the studied bacterial strains.

From the interfacial activity point of view, the most efficient interfacial compound is the dodecyl derivative (SB12DQ). This can be referred to the higher effectiveness (π_{cmc}) values and also higher maximum surface excess (Table 20.7). The higher

TABLE 20.7

Surface and Adsorption Thermodynamics Properties of Synthesized Di-Schiff Base Cationic Gemini Amphiphiles at 25°C

Compound	CMC, mmol/L	π_{cmc}, mN/m	Pc_{20}, mmol/L	Γ_{max}, mol/K. 10^{10}	A_{min}, nm^2	ΔG_{mic}, kJ/mol	ΔG_{ads}, kJ/mol
SB8DQ	5.89	32.10	0.309	0.810	2.076	−7.17	−7.66
SB10DQ	2.82	34.25	0.186	1.050	1.582	−8.19	−11.69
SB12DQ	2.45	38.20	0.115	1.180	1.408	−8.25	−11.46
SB16DQ	6.46	29.70	0.457	0.733	2.266	−7.04	−7.60

π_{cmc}, the difference between surface tension of surfactant solution at CMC and bidistilled water Γ_{max}, the maximum surface excess which is the concentration of surfactant molecules at the interface at CMC A_{min}, average area occupied by each surfactant molecules at the interface ΔG_{mic} and ΔG_{ads}, standard free energies of micellization and adsorption.

effectiveness values indicate tendency of adsorption of the targeted compounds at different interfaces including bacterial membranes (Figure 20.22).

Also, the higher surface excess (Γ_{max} in Table 20.7), revealed that these compounds were adsorbed in a high concentration at the interfaces. Hence, they will be highly efficient against the bacterial microorganisms.

On the other hand, (SB18DQ) showed different behaviors and unexpected higher antibacterial activity towards *Bacillus subtillus* and *E. coli* which may be due to the difference in the biochemical structure of the cellular membranes of these two strains.

The values of the inhibition zones of the synthesized di-Schiff base against two potent fungal strains *Aspergillus niger* and *Candida albicans* ranged between 27 and 38 mm. This indicates a high cytotoxic efficacy of these cationic gemini surfactants against the studied fungi (Table 20.8).

The values of inhibition zone diameter showed that the cytotoxic efficacy of these compounds was strongly related to their interfacial properties especially the effectiveness and the maximum surface excess (Γ_{max} in Table 20.6).

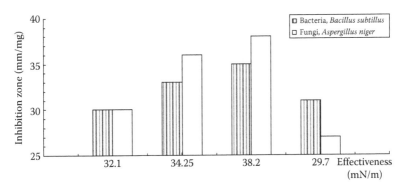

FIGURE 20.22 Effect of effectiveness (π_{cmc}) of the synthesized di-Schiff base cationic gemini surfactants on their antimicrobial activity.

TABLE 20.8
Antifungal Activity of Synthesized
Di-Schiff Base Cationic Gemini
Surfactants at 5mg/mL Expressed as
Inhibition Zone Diameter (mm)

Compound	Aspergillus niger, (Ferm-Bam C-21)	Candida albicans
SB8DQ	30	27
SB10DQ	36	34
SB12DQ	38	37
SB16DQ	27	29
SB18DQ	29	28

20.7.6 Sulfate-Reducing Bacteria

Sulfate-reducing bacteria comprise several groups of bacteria that use sulfate as an oxidizing agent, reducing it to sulfide. Most sulfate-reducing bacteria can also use other oxidized sulfur compounds such as sulfite and thiosulfate, or elemental sulfur. This type of metabolism is called dissimilatory since sulfur is not incorporated or assimilated into any organic compounds. Sulfate-reducing bacteria have been considered as a possible way to deal with acid mine waters that are produced by other bacteria.

The sulfate-reducing bacteria have been treated as phenotypic group for identification purposes. They are found in several different phylogenetic lines. Three lines are included among the *Proteobacteria*, all in the delta subgroup: *Desulfobacterales*, *Desulfovibrionales*, and *Syntrophobacterales*. A fourth group including *thermophiles* is given its own phylum, the *Thermodesulfo* bacteria. The remaining sulfate reducers are included with other bacteria among the *Nitrospirae* and the Gram-positive *Peptococcaceae* for instance *Thermodesulfovibrio* and *Desulfotomaculum*, respectively. There is also a single genus of *Archaea* capable of sulfate reduction, *Archaeoglobus*. Some bacteria such as *Proteus*, *Campylobacter*, *Pseudomonas,* and *Salmonella* have the ability to reduce sulfur but can also use oxygen and other terminal electron acceptors. Others, such as *Desulfuromonas*, use only sulfur. Some bacteria can use both elemental sulfur and sulfate as a source of energy.

Generally SRB require a complete absence of oxygen and a highly reduced environment to function efficiently. However, they circulate (probably in a resting state) in aerated waters, including those treated with chlorine and other oxidizers, until they find an "ideal" environment supporting their metabolism and multiplication.

Many SRB strains also contain hydrogenase enzymes, which allow them to consume hydrogen. Most common strains of SRB grow best at temperatures from 25°C to 35°C. A few thermophilic strains are capable of functioning efficiently at more than 60°C.

Tests for the presence of SRB have traditionally involved growing the organisms on laboratory media, quite unlike the natural environment in which they were found. These laboratory media will only grow certain strains of SRB, and even then some samples require a long time before the organisms will adapt to the new growth conditions. As a result, misleading information has been obtained regarding the presence or absence of SRB in field samples.

SRB have been implicated in the corrosion of cast iron and steel, ferrite stainless steels, 300 series stainless steels (also very highly alloyed stainless steels), copper nickel alloys, and high nickel molybdenum alloys. They are almost always present at corrosion sites because they are in soils, surface water streams and different deposits in general.

Where *Thiobacillus* bacteria are associated with corrosion, they are almost always accompanied by SRB. Thus, both types of organisms are able to draw energy from a synergistic sulfur cycle. The fact that two such different organisms, one a strict anaerobe that prefers neutral pH, and the other an aerobe that produces and thrives in an acid environment, can coexist, demonstrates that individual organisms are able to form their own microenvironment within an otherwise hostile larger world.

Desulfomonas pigra strain used in this evaluation is a famous genus of SRB in the oil fields and causes a lot of damage for the metallic constructions. Preventing their growth and their fatal action required some special antibacterial agents which also showed minor side effects in case of environmental considerations. Regarding the antibacterial activities of the synthesized cationic gemini biocides, their efficacy against SRB was found to be considerably high. The values of inhibition zones of the synthesized di-Schiff base gemini cationics (Table 20.9) showed high efficiency of these compounds as SRB biocides relative to the most traditional SRB biocides used in the oilfields. The resistivity of the SRB bacterial strains and the rigidity of their cellular membranes did not form a barrier against the high efficiency of these compounds.

TABLE 20.9
Antibacterial Activity of Synthesized Di-Schiff Base Cationic Gemini Amphiphiles at 5 mg/mL against SRB Bacterial Type (*Desulfomonas pigra*) in Terms of Inhibition Zone Diameter (mm)

Compound	*Desulfomonas pigra*
SB8DQ	18
SB10DQ	23
SB12DQ	25
SB16DQ	19
SB18DQ	17

The cytotoxic efficacy of the synthesized di-Schiff base cationic gemini surfactants displayed excellent inhibition tendency toward different bacterial genera and fungi. The inhibition tendency can be ascribed to the unique chemical structure of these compounds due to the presence of the azomethine groups and the diquaternary groups [69]. The presence of these groups increases the adsorption of the molecules at the microorganism's cellular membrane and the presence of the chloride ions (counter ions) helps penetrate these molecules easily through these membranes. Hence the normal biological reactions in the cells are strongly interfered with, and a quick death of these microorganisms occurs.

20.8 CONCLUSIONS

Antimicrobial agents play a number of critical roles in tribology, including prevention of microbial attack on lubricant ingredients and preserving the performance of the lubricant formulation. They prevent the corrosion of metals in machinery and oil field structures due to attack by sulfate-reducing bacteria. In addition, their application is very wide in the protection of plants, animals, and human from the attack of lethal microorganisms.

This chapter explained the surface activity and antimicrobial efficacy of different types of surface-active agents. Also, it focused on the gemini type of anionic and cationic surface active agents. In addition, the chapter described the synthesis, surface activity, and antimicrobial activity of a novel type of diquaternary surfactants containing two Schiff base groups. The synthesized compounds showed high biocidal activity against different bacterial strains (Gram-positive and Gram-negative) and fungi. Also, the biocidal activity of these compounds was considerable against sulfate-reducing bacteria.

ACKNOWLEDGMENTS

I would like to thank Prof. R. Miller (Max Planck Institute of Colloids and Interfaces, Potsdam-Golm, Germany) and Prof. Amal A. Hafiz (Egyptian Petroleum Research Institute, Petrochemicals Department) for their assistance during the preparation of this chapter.

REFERENCES

1. F.M. Menger and J.S. Keiper, Gemini-tenside, *Angew. Chem.* 112, 1980 (2000).
2. C.A. Bunton, L. Robinson, J. Schaak and M.F. Stam, Catalysis of nucleophilic substitutions by micelles of dicationic detergents, *J. Org. Chem.* 36, 2346–2351 (1971).
3. F. Devinsky and I. Lacko, Formation of some bis-quaternary ammonium salts of glycerine derivative, *Tenside Surf. Deterg.* 27, 334–339 (1990).
4. F. Devinsky, F.B. Bitterova, and I. Lacko, Surface activity and micelle formation of some new bisquaternary ammonium salts, *J. Colloid Interface Sci.* 114, 314–319 (1986).
5. R. Zana, Micellization of nonionic surfactant dimers and of the corresponding surfactant monomers in aqueous solution, *Adv. Colloid Interface Sci.* 97, 205–212 (2002).
6. R. Zana, Dimeric (Gemini) surfactants: Effect of the spacer group on the association behavior in aqueous solution, *J. Colloid Interface Sci.* 248, 203–220 (2002).

7. A.D. Miller, The problem with cationic liposome/micelle-based non-viral vector systems for gene therapy, *Angew Chem.* 110, 1862–1868 (1998).
8. A.D. Miller, Cationic liposomes for gene therapy, *Angew. Chem. Int. Ed.* 37, 1768–1773 (1998).
9. N.A. Negm, Solubilization, surface active and thermodynamic parameters of Gemini amphiphiles bearing nonionic hydrophilic spacer, *J. Surf. Deterg.* 10, 71–78 (2007).
10. N.A. Negm and A.S. Mohamed, Surface and thermodynamic properties of diquaternary bola-form amphiphiles containing aromatic spacer, *J. Surf. Deterg.* 7(1), 23–31 (2004).
11. N.A. Negm and A.S. Mohamed, Synthesis, characterization and biological activity of sugar-based gemini cationic amphiphiles, *J. Surf. Deterg.* 11(3), 215–222 (2008).
12. N.A. Negm and M.F. Zaki, Synthesis and characterization of some amino acid derived Schiff-bases bearing nonionic species as corrosion inhibitors for carbon steel in 2N HCl, *J. Dispers. Sci. Technol.* 30, 5–11 (2009).
13. L. Perez, J.L. Ribosa, A. Manresa, C. Solans, and M.R. Infante, Synthesis, aggregation, and biological properties of a new class of gemini cationic amphiphilic compounds from arginine, bis(args), *Langmuir* 12, 5296–5302 (1996).
14. T.S. Kim, T. Hirao, and I. Ikeda, Preparation of bis-quaternary ammonium salts from epicholorohydrin, *J. Am. Oil Chem. Soc.* 73, 67–73 (1996).
15. Y.P. Zhu, K. Ishihara, A. Masuyama, Y. Nakatsuji, and M. Okahara, Preparation and properties of double-chain bis(quaternary ammonium) compounds, *Yukagaku* 42, 161–168 (1993).
16. M.J. Rosen, *Surface and Interfacial Phenomena*, 2nd edn. Wiely Press, New York (1989).
17. O. Rist, A. Rike, L. Ljonesm, and H.J. Carlsen, Synthesis of novel diammonium Gemini surfactants, *Molecules* 6, 979–986 (2001).
18. M. Tanaka, T. Ishida, T. Araki, A. Masuyama, Y. Nakatsuji, M. Okahara, and S. Terabe, Double-chain surfactant as a new and useful micelle-forming reagent for micellar electrokinetic chromatography, *J. Chromatogr.* 648(2), 469–475 (1993).
19. D. Ono, S. Yamamura, M. Nakamura, and T. Takeda, Preparation and properties of bis-(sodium sulfate) types of cleavable surfactants derived from diethyl tartrate, *J. Ole. Sci.* 54(1), 51–62 (2005).
20. R. Zana and Y. Talmon, Dependence of aggregate morphology on structure of dimeric surfactants, *Nature* 362, 228–236 (1993).
21. T. Dam, J.B. Engbert, J. Karthauser, S. Karaborw, and N.M. Van, Synthesis, surface properties and oil solubilisation capacity of cationic gemini surfactants, *Colloid. Surf. A.* 118, 41–49 (1996).
22. Z. Baker, R.W. Harrison, and B.F. Miller, Action of synthetic detergents on the metabolism of bacteria, *J. Exp. Med.* 73, 249–256 (1941).
23. W.L. Wang, B.W. Powers, N.M. Luechtefeld, and M.J. Blaster, Effects of disinfectants on *Campylobacter jejuni*, *Appl. Environm. Microbiol.* 45, 1202–1209 (1983).
24. W.L. Wang, M.J. Blaster, J. Cravens, and M.A. Johnson, The microorganism: Growth, survival, and resistance of the Legionnaires disease bacterium, *Ann. Inter. Med.* 90, 614–621 (1979).
25. J.J. Miller, W.E. Brown, and V.J. Krieger, Laboratory methods for testing bacteriostatic and bacteriocidal effects of water-treatment biocides on *Legionella pneumophila*, *Dev. Ind. Microbiol.* 22, 763–768 (1981).
26. M. Best, M.E. Kennedy, and F. Coates, Efficacy of a variety of disinfectants against *Listeria* spp., *Appl. Environ. Microbiol.* 56, 377–383 (1990).
27. D.F. Greene and A.N. Petrocci, Formulating quaternary cleaner disinfectants to meet EPA requirements, *Soap Cosm. Chem. Spec.* 56, 33–41 (1980).
28. B. Terleckyj and D.A. Axler, Quantitative neutralization assay of fungicidal activity of disinfectants, *Antimicrob. Agent Chemother.* 31, 794–802 (1987).

29. N.A. Negm and I.A. Aiad, Synthesis and characterization of multifunctional surfactants in oil-field protection applications, *J. Surf. Deterg.* 10, 87–93 (2007).

30. H.J. Hueck, D.M.M. Adema, and J.R. Wiegman, Biological effects of surfactants: Part 6—Effects of anionic, non-ionic and amphoteric surfactants on a green alga (Chlamydomonas), *Appl. Microbiol.* 14, 308–314 (1966).

31. G.C. Lavelle, Evaluation of an antimicrobial soap formula for virucidal efficacy in vitro against human immunodeficiency virus in a blood–virus mixture, *Chem. Times Trends* 10, 45–52 (1987).

32. T.T. Brown, Laboratory evaluation of selected disinfectant as virucidal agents against porcine parvovirus, and transmissible gastroenteritis virus, *J. Vet. Res.* 42, 1033–1041 (1981).

33. N.J. Mbithi, V.S. Springthorpe, and S.A. Sattar, Comparative in vivo efficiencies of hand-washing agents against hepatitis A virus (HM-175) and poliovirus type 1 (Sabin), *Appl. Environ. Microbiol.* 56, 3601–3909 (1990).

34. E.L. Sing, P.R. Elliker, and W.E. Sandine, Comparative destruction of airborne lactic bacteriophage, *J. Milk Food Technol.* 27, 101–108 (1964).

35. S. Watkins, H.A. Hays, and P.R. Elliker, Virucidal activity of hypochlorides, quaternary ammonium compounds and other iodophors against bacteriphage of *Streptococcus cremoris*, *J. Milk Food Technol.* 20, 84–93 (1957).

36. M.L. Ancelin, H.J. Vial, and J.R. Philippot, Quaternary ammonium compounds efficiently inhibit *Plasmodium falciparum* growth in vitro by impairment of choline transport, *Biochem. Pharmacol.* 34, 4068–4077 (1985).

37. K. Nishikawa, S. Oi, and T. Yamamoto, A bacterium resistant to benzalkonium chloride, *Agri. Biol. Chem.* 43, 2473–2481 (1979).

38. A.D. Russell, Mechanisms of bacterial resistance to biocides, *Int. Biodeterio. Biodegrad.* 36(3–4), 247–252 (1995).

39. A.D. Russell, U. Tattawasart, J.Y. Maillard, and J.R. Furr, Possible link between bacterial resistance and use of antibiotics and biocides, *Antimicro. Agents Chemother.* 42(8), 2151 (1998).

40. G. Donnell and A.D. Russell, Antiseptics and disinfectants: Activity, action and resistance, *Clini. Microbiol. Rev.* 12(1), 147–168 (1999).

41. A.D. Russell and M.J. Day, Antibiotic and biocide resistance in bacteria, *Microbios* 85(342), 45–52 (1996).

42. Y. Sakagami, H. Yokoyama, H. Nishimura, Y. Ose, and T. Tashima, Mechanism of resistance to benzalkonium chloride by *Pseudomonas aeruginosa*, *Appl. Environ. Microbiol* 55, 2036–2043 (1989).

43. M.V. Jones, T.M. Herd, and H.J. Christie, Resistance of *Pseudomonas aeruginosa* to amphoteric and quaternary ammonium biocides, *Microbios.* 58, 49–54 (1989).

44. D.S. Dhaliwal, J.L. Cordier, and L.J. Cox, Impedimetric evaluation of the efficiency of disinfectants against biofilms, *Lett. Appl. Microbiol.* 15, 217–224 (1992).

45. E.P. Krysinski, L.J. Brown, and T.J. Marchisello, Effect of cleaners and sanitizers on listeria monocytogenes attached to product contact surfaces, *J. Food Prot.* 55, 246–257 (1992).

46. N.A. Negm, A.F. El-Farargy, M.F. Zaki, S.A. Mahmoud, and N.R. Abdel Rahman, Cationic Schiff base amphiphiles: 1. Synthesis, characterization and surface activities of cationic surfactants bearing Schiff base groups and their Mn(II), Cu(II) and Co(II) complexes, *Egypt. J. Petrol.* 17, 15–25 (2009).

47. N.A. Negm and M.F. Zaki, Corrosion inhibition efficiency of nonionic Schiff base amphiphiles of p-aminobenzoic acid for aluminum in 4N-HCl, *Colloid. Surf. A: Physicochem. Eng. Aspec.* 322, 97–102 (2008).

48. N.A. Negm and M.F. Zaki, Structural and biological behaviors of some nonionic Schiff-base amphiphiles and their Cu(II) and Fe(III) metal complexes, *Colloid. Surf. B: Biointerface* 64, 179–184 (2008).

49. I.A. Aiad and N.A. Negm, Some corrosion inhibitors based on Schiff base surfactants for mild steel equipments, *J. Disper. Sci. Technol.* 30, 1142–1148 (2009).
50. E.M. Azzam, N.A. Negm, and E.A. Gad, Surface and solubilization activities of 1-amino-2-alkyloxy–naphthalene-4-sodium sulfonates, *Adsorp. Sci. Technol.* 22(8), 663–669 (2004).
51. K.C. Emregul, A.A. Akay, and O. Atakol, The corrosion inhibition of steel with Schiff base compounds in 2M HCl, *Mater. Chem. Phys.* 93, 325–331 (2005).
52. M.G. Hosseini, S.F. Mertens, M. Ghorbani, and M.R. Arshadi, Asymmetrical Schiff bases as inhibitors of mild steel corrosion in sulphuric acid media, *Mater. Chem. Phys.* 78, 800–806 (2003).
53. K.C. Emregul and O. Atakol, Corrosion inhibition of iron in 1 M HCl solution with Schiff base compounds and derivatives, *Mater. Chem. Phys.* 92, 373–379 (2004).
54. D. Ono, S. Yamamura, M. Nakamura, and T. Takeda, Preparation and properties of bis(sodium sulfate) types of cleavable surfactants derived from diethyl tartrate, *J. Ole. Sci.* 54(1), 51–58 (2005).
55. N.A. Negm, A.A. Hafiz, and M.Y. El-Awady, Influence of structure on the cationic polytriethanol ammonium bromide derivatives. I. Synthesis, surface and thermodynamic properties, *Egypt. J. Chem.* 47(4), 369–378 (2004).
56. R.E. Cooper, *Analytical Microbiology*, F.W. Kavanageh (ed.), I and II Academic Press, New York and London (1972).
57. National Committee for Clinical Laboratory Standards; (1997). Methods for dilution antimicrobial susceptibility tests for bacteria that grow aerobically. Approved standard M7-A4. National Committee for Clinical Laboratory Standards, Wayne.
58. A.A. Hafiz, N.A. Negm, and M.Y. Elawady, Influence of structure of the cationic polytriethanol ammonium bromide derivatives. III. Biological activity, *Egypt. J. Chem.* 48(2), 245–253 (2005).
59. N.A. Negm, Surface activities and electrical properties of long chain diquaternary bolaform amphiphiles, *Egypt. J. Chem.* 45, 483–496 (2002).
60. R. Oda, I. Hucb, and J. Sauveur, Gemini surfactants, the effect of hydrophobic chain length and dissymmetry, *Chem. Commun.* 21, 2105–2112 (1997).
61. R.C. Bazito and O.A. El-Seoud, Sugar-based cationic surfactants: Synthesis and aggregation of methyl 2-acylamido-6-trimethylammonio-2,6-dideoxy-d-glucopyranoside chlorides, *J. Surf. Deterg.* 4(4), 395–403 (2001).
62. M.J. Rosen, *Surface and Interfacial Phenomena*, 2nd edn. Wiely, New York, 151 (1991).
63. J. Parekh, P. Inamdhar, R. Nair, S. Baluja, and S. Chanda, Synthesis and antibacterial activity of some Schiff bases derived from 4-aminobenzioc acid, *J. Serb. Chem. Soc.* 70, 1155–1163 (2005).
64. A.N. Petrocci, *Disinfection, Sterilization and Preservation* (S.S. Block), Lea and Febiger, Philadelphia, PA (1983).
65. W.B. Hugo, *Surface Active Agents in Microbiology*, SCI Monograph 19, Society of the Chemical Industry, London (1965).
66. R.D. Hotchkiss, The nature of the bactericidal action of surface-active agents, *Ann. N.Y. Acad. Sci.* 46, 479–485 (1946).
67. M.T. Dyar and E.J. Ordal, Electrokinetic studies on bacterial surfaces: I. Effects of surface-active agents on electrophoretic mobilities of bacteria, *J. Bacteriol.* 51, 149–156 (1946).
68. K. McQuillen, Some aspects of structure and function in cytoplasm, *Biochem. Biophys. Acta* 5, 643 (1950).
69. R. Nair, A. Shah, S. Baluja, and S. Chanda, Synthesis and antibacterial activity of some Schiff base complexes, *J. Serb. Chem. Soc.* 71(7), 733–746 (2006).

Index

Milton Keynes UK
Ingram Content Group UK Ltd.
UKHW051013071024
449327UK00012B/220

9 780367 382896